THE MOTION, EVOLUTION OF ORBITS, AND ORIGIN OF COMETS

INTERNATIONAL ASTRONOMICAL UNION
UNION ASTRONOMIQUE INTERNATIONALE

SYMPOSIUM No. 45

HELD IN LENINGRAD, U.S.S.R., AUGUST 4–11, 1970

THE MOTION, EVOLUTION OF ORBITS, AND ORIGIN OF COMETS

EDITED BY

G. A. CHEBOTAREV AND E. I. KAZIMIRCHAK-POLONSKAYA

Institute for Theoretical Astronomy, Leningrad, U.S.S.R.

AND

B. G. MARSDEN

Smithsonian Astrophysical Observatory, Cambridge, Mass., U.S.A.

D. REIDEL PUBLISHING COMPANY

DORDRECHT-HOLLAND

1972

Published on behalf of
the International Astronomical Union
by
D. Reidel Publishing Company, Dordrecht, Holland

Library of Congress Catalog Card Number 73–179895

ISBN-13: 978-94-010-2875-2   e-ISBN-13: 978-94-010-2873-8
DOI: 10.1007/978-94-010-2873-8

To the memory of

MIKHAIL FEDOROVICH SUBBOTIN

(1893–1966)

and

SAMUIL GDAL'EVICH MAKOVER

(1908–1970)

PREFACE

The many papers by Soviet authors have been translated into English by A. P. Kirillov, N. A. Nikiforova, E. A. Voronov, and others. Some of the papers were translated by the authors themselves. The discussion records have been prepared at the Institute for Theoretical Astronomy by V. K. Abalakin, N. A. Belyaev, A. P. Kirillov, V. A. Shor, E. A. Voronov, N. S. Yakhontova, and others. The three papers published in French have been carefully checked by B. Milet. The final editing has been done at the Smithsonian Astrophysical Observatory, and we thank J. H. Clark, P. D. Gregory, J. E. Kervick, and G. Warren for retyping much of the material.

Our special thanks are due to the D. Reidel Publishing Company for the excellent care they have taken in printing these proceedings of IAU Symposium No. 45.

G. A. CHEBOTAREV
E. I. KAZIMIRCHAK-POLONSKAYA
B. G. MARSDEN

INTRODUCTION

The idea to organize a Symposium on 'The Motion, Evolution of Orbits, and Origin of Comets' dates back to the IAU thirteenth General Assembly, held in 1967 in Prague. Owing to the impossibility of completing during the General Assembly the discussion on the problem of orbital evolution of comets Professor G. A. Chebotarev, then the newly elected President of IAU Commission 20, initiated the organization of the international symposium in Leningrad where the full scope of cometary problems might be considered from the viewpoint of celestial mechanics. This idea was warmly welcomed by the participants of Commission 20. Since this was to be the first international symposium on this subject, it was decided that it should encompass the following objectives.

(1) It would be essential to specify those problems of cometary dynamics presenting the greatest difficulties and to acquaint ourselves with the methods for their solution employed by astronomers of various schools and working in many different countries.

First of all, there is the problem of constructing analytical and numerical theories of cometary motion. The main obstacles in numerical theories are connected with the differential correction of orbits, the accumulation of errors in numerical integration, consideration of nongravitational effects, and the need for improvement in the accuracy of observations and in the ephemeris service. For proper consideration of the nongravitational effects it is necessary to know the causes responsible for them. This renders it desirable to have, on the one hand, an intimate acquaintance with cometary observations, and on the other hand, knowledge of the physical structure of cometary nuclei, the phenomena that take place in cometary atmospheres, the laws governing the mass loss from comets, and so on.

(2) The second objective was to discuss matters that may be solved by the coordinated effort of astronomers of different specialities. Among such questions are the distribution of the work necessary for preparing a new catalogue of cometary orbits and a new cometography, the establishment of strong links between theoreticians and observers, and the exchange of all essential data on the subject, in order to ensure the active development both of the observation and ephemeris service and of theoretical studies on cometary motion.

(3) There are also numerous problems of cometary astronomy that interact with those of other minor bodies of the solar system. For this reason we wanted to discuss the achievements and difficulties in the adjacent fields of dynamics of asteroids and meteor streams. This could lead to a more profound clarification of the peculiarities of their motion and the role played by the major planets in the orbital evolution of all the minor bodies of the solar system on a time-scale of centuries, millennia, and even longer.

(4) Another objective was the detailed examination of controversial or little-investigated problems, such as the study of all the steps in the evolution of cometary

orbits, research on the relationship between the long-period and short-period comets, analysis of the capture and eruption theories, discussion of the diffusion theory of nearly parabolic and long-period comets, consideration of stellar perturbations on the orbits of comets belonging to the so-called Oort cloud, and the investigation of the stability, structure, dimensions, and origin of the cloud. And finally, we wished to discuss, from the viewpoint of both celestial mechanics and astrophysics, the origin of comets and other minor bodies of the solar system.

Consideration of all these problems would indisputably be of value in the solution of the main problem of the cosmogony of the solar system as a whole.

Accordingly, IAU Symposium No. 45 was held in Leningrad during 4–11 August, 1970.

We gratefully acknowledge the help received from Dr L. Perek, General Secretary, and Dr C. de Jager, Assistant General Secretary, concerning the approval of the Symposium by the IAU Executive Committee. We are also much obliged to Dr de Jager for providing and allocating travel grants for many scientists, thus enabling them to participate in the Symposium.

The Presidium of the Academy of Sciences of the U.S.S.R., and particularly E. R. Mustel, Chairman of the Astronomical Council, were also most helpful in organizing and financing the Symposium.

The international Organizing Committee was appointed at an early stage in the preparation of the Symposium. It consisted of G. A. Chebotarev, Chairman (U.S.S.R.), M. Bielicki (Poland), M. P. Candy (Australia), V. V. Fedynskij (U.S.S.R.), E. I. Kazimirchak-Polonskaya (U.S.S.R.), L. Kresák (Czechoslovakia), J. Kovalevsky (France), B. G. Marsden (U.S.A.), E. Roemer (U.S.A.), and F. L. Whipple (U.S.A.).

B. G. Marsden displayed exceptional activity in the organization of the Symposium and performed the most valuable work of finally editing the Proceedings of Symposium No. 45 for publication.

The local Organizing Committee was: G. A. Chebotarev (Chairman), V. K. Abalakin, N. A. Belyaev, N. A. Bokhan, O. V. Dobrovol'skij, G. N. Duboshin, V. V. Fedynskij, E. I. Kazimirchak-Polonskaya, V. N. Lebedinets, †S. G. Makover, K. A. Shtejns, and N. S. Yakhontova.

We also appreciate the generous assistance of the following personnel of the Institute for Theoretical Astronomy, Leningrad: Yu. V. Batrakov, V. A. Ivakin, Ts. G. Khajmovich, S. V. Men'shchikova, T. B. Sabanina, and I. Ts. Zvyagin.

The programme for the Symposium was extensive and many-sided. The Symposium was divided into 14 sessions, each of which was concerned with a particular set of problems, and the sequence ensured the logical development of the entire range of problems. The duties of chairmen during the sessions were performed by: V. K. Abalakin (U.S.S.R.), M. Bielicki (Poland), G. A. Chebotarev (U.S.S.R.), A. Z. Dolginov (U.S.S.R.), E. Everhart (U.S.A.), V. V. Fedynskij (U.S.S.R.), E. I. Kazimirchak-Polonskaya (U.S.S.R.), W. J. Klepczynski (U.S.A.), B. Yu. Levin (U.S.S.R.), B. G. Marsden (U.S.A.), E. Rabe (U.S.A.), K. A. Shtejns (U.S.S.R.), P. Stumpff (Germany), S. K. Vsekhsvyatskij (U.S.S.R.), and F. L. Whipple (U.S.A.).

A total of 162 scientists participated in the Symposium, representatives of Argentina (1), Belgium (1), Czechoslovakia (3), France (3), Germany (1), The Netherlands (1), Norway (1), Poland (5), Sweden (1), U.S.A. (10), and U.S.S.R. (135).

More than 80 papers and communications were presented. All those presented in English or French were followed by abstracts in Russian, and those presented in Russian were supplemented by similar abstracts in English. Discussion in both English and Russian followed each of the reports.

Most of the reports and discussions were interpreted by A. P. Kirillov and E. A. Voronov, who are specialists in translating scientific and technical literature; some of the interpreting was done by V. K. Abalakin. A. P. Kirillov also took part in organizing the Symposium.

A series of excursions, organized in association with INTOURIST, enabled participants to visit the Main Astronomical Observatory of the U.S.S.R. Academy of Sciences at Pulkovo, Leningrad and its suburbs, and the art treasures of the Hermitage and Russian Museum. At Pulkovo the participants attended the ceremonial unveiling of the memorial to M. F. Subbotin, former Director of the Institute for Theoretical Astronomy, an outstanding scientist and author of classical works on celestial mechanics. The participants also visited the tombs of the distinguished Leningrad astronomers G. N. Neujmin, a former Director of the Pulkovo Observatory, and S. G. Makover and D. K. Kulikov, of the Institute for Theoretical Astronomy.

The Symposium was marked by its exceptionally warm atmosphere; it strengthened the friendly relations between the participants and permitted us to define the paths for our mutual cooperation on cometary problems in the future.

Leningrad   E. I. KAZIMIRCHAK-POLONSKAYA
October 1970

TABLE OF CONTENTS

B. *Numerical Methods*

C. *Determination of Orbits*

PART III / MOTIONS OF THE SHORT-PERIOD COMETS

A. *Planetary Perturbations and Nongravitational Effects*

B. *Determination of Planetary Masses*

PART IV / PHYSICAL PROCESSES IN COMETS

B. *Possibility of Common Origin*

CONCLUDING DISCUSSION   519

LIST OF PARTICIPANTS

V. K. Abalakin, Institute for Theoretical Astronomy, Leningrad, U.S.S.R.
E. L. Akim, Institute for Applied Mathematics, Moscow, U.S.S.R.
N. V. Alekseeva, Kazan, U.S.S.R.
H. Alfvén, Royal Institute of Technology, Stockholm, Sweden
V. M. Amelin, Institute for Theoretical Astronomy, Leningrad, U.S.S.R.
D. A. Andrienko, Astronomical Observatory, Kiev University, Kiev, U.S.S.R.
I. V. Andrsh, Pedagogical Institute, Kirovograd, U.S.S.R.
R. L. Aptekar', A. F. Ioffe Institute of Physics and Technology, Leningrad, U.S.S.R.
I. S. Astapovich, Kiev University, Kiev, U.S.S.R.
P. B. Babadzhanov, Institute of Astrophysics, Dushanbe, U.S.S.R.
A. S. Baranov, Institute for Theoretical Astronomy, Leningrad, U.S.S.R.
Yu. V. Batrakov, Institute for Theoretical Astronomy, Leningrad, U.S.S.R.
I. R. Bejtrishvili, Astrophysical Observatory, Abastumani, U.S.S.R.
L. M. Belous, Northwestern Extramural Polytechnic Institute, Leningrad, U.S.S.R.
N. A. Belyaev, Institute for Theoretical Astronomy, Leningrad, U.S.S.R.
M. Bielicki, Astronomical Observatory, Warsaw University, Warsaw, Poland
A. F. Bogorodskij, Astronomical Observatory, Kiev University, Kiev, U.S.S.R.
N. A. Bokhan, Institute for Theoretical Astronomy, Leningrad, U.S.S.R.
V. A. Bordovitsin, Polytechnic Institute, Tomsk, U.S.S.R.
T. V. Bordovitsyna, Tomsk University, Tomsk, U.S.S.R.
M. P. Boris, Sternberg Astronomical Institute, Moscow, U.S.S.R.
M. M. Bredov, A. F. Ioffe Institute of Physics and Technology, Leningrad, U.S.S.R.
N. M. Bronnikova, Pulkovo Observatory, Leningrad, U.S.S.R.
L. E. Bykova, Tomsk University, Tomsk, U.S.S.R.
G. A. Chebotarev, Institute for Theoretical Astronomy, Leningrad, U.S.S.R.
V. M. Chepurova, Sternberg Astronomical Institute, Moscow, U.S.S.R.
N. S. Chernykh, Crimean Astrophysical Observatory, Crimea, U.S.S.R.
K. I. Churyumov, Astronomical Observatory, Kiev University, Kiev, U.S.S.R.
M. S. Davis, Triangle Universities Computation Center, Research Triangle Park, N.C., U.S.A.
H. Debehogne, Observatoire Royal de Belgique, Bruxelles, Belgique
J. Delcourt, Centre National d'Etudes des Télécommunications, Issy-les-Moulineaux, France
A. A. Demenko, Astronomical Observatory, Kiev University, Kiev, U.S.S.R.
A. Deprit, Boeing Scientific Research Laboratories, Seattle, Wash., U.S.A.
A. D. Derbeneva, Institute of Astrophysics, Dushanbe, U.S.S.R.
L. K. Dirikis, Latvian State University, Riga, U.S.S.R.
M. A. Dirikis, Astronomical Observatory, Latvian State University, Riga, U.S.S.R.
A. Z. Dolginov, A. F. Ioffe Institute of Physics and Technology, Leningrad, U.S.S.R.

S. V. Drozdov, Pedagogical Institute, Novgorod, U.S.S.R.

G. N. Duboshin, Sternberg Astronomical Institute, Moscow, U.S.S.R.

F. K. Edmondson, Goethe Link Observatory, Indiana University, Bloomington, Ind., U.S.A.

R. P. Eremenko, Institute for Theoretical Astronomy, Leningrad, U.S.S.R.

Yu. V. Evdokimov, Kazan University, Kazan, U.S.S.R.

E. Everhart, Department of Physics, University of Denver, Denver, Colo., U.S.A.

V. V. Fedynskij, Astronomical Council, U.S.S.R. Academy of Sciences, Moscow, U.S.S.R.

A. M. Fominov, Institute for Theoretical Astronomy, Leningrad, U.S.S.R.

I. V. Galibina, Institute for Theoretical Astronomy, Leningrad, U.S.S.R.

S. Gąska, University of Toruń Observatory, Toruń, Poland

G. Guigay, Observatoire de Marseille, Marseille, France[†]

O. Havnes, Astronomical Institute, Utrecht, The Netherlands

M. N. Itskov, Institute for Theoretical Astronomy, Leningrad, U.S.S.R.

V. A. Ivakin, Institute for Theoretical Astronomy, Leningrad, U.S.S.R.

M. Ya. Joseleva, Institute of Mines, Leningrad, U.S.S.R.

E. A. Kajmakov, A. F. Ioffe Institute of Physics and Technology, Leningrad, U.S.S.R.

N. J. Kamarnitskaya, Polytechnic Institute, Odessa, U.S.S.R.

L. A. Katasev, Institute for Experimental Meteorology, Obninsk, U.S.S.R.

A. A. Kaverin, Pedagogical Institute, Irkutsk, U.S.S.R.

E. I. Kazimirchak-Polonskaya, Institute for Theoretical Astronomy, Leningrad, U.S.S.R.

F. B. Khanina, Institute for Theoretical Astronomy, Leningrad, U.S.S.R.

K. V. Kholshevnikov, Leningrad University, Leningrad, U.S.S.R.

W. J. Klepczynski, U.S. Naval Observatory, Washington, D.C., U.S.A.

V. N. Klevetskij, Riga, U.S.S.R.

N. G. Kochina, Institute for Theoretical Astronomy, Leningrad, U.S.S.R.

E. D. Kondrat'eva, Kazan University, Kazan, U.S.S.R.

V. P. Konopleva, Main Astronomical Observatory, Ukrainian Academy of Sciences, Kiev, U.S.S.R.

Yu. I. Koptev, A. F. Ioffe Institute of Physics and Technology, Leningrad, U.S.S.R.

I. M. Korovin, Institute for Cosmic Research, Moscow, U.S.S.R.

V. N. Korpusov, Institute for Experimental Meteorology, Obninsk, U.S.S.R.

E. N. Kramer, Astronomical Observatory, Odessa State University, Odessa, U.S.S.R.

L. Kresák, Astronomical Institute, Slovak Academy of Sciences, Bratislava, Czechoslovakia

N. V. Kulikova, Institute for Experimental Meteorology, Obninsk, U.S.S.R.

I. P. Kuz'min, Institute for Cosmic Research, Moscow, U.S.S.R.

I. N. Latyshev, Astronomical Observatory, Leningrad University, Leningrad, U.S.S.R.

L. K. Lautsenieks, Latvian State University, Riga, U.S.S.R.

A. K. Lavrukhina, Geochemical Institute, Moscow, U.S.S.R.

I. A. Lebedeva, Institute for Theoretical Astronomy, Leningrad, U.S.S.R.

[†] It is with regret that we record the death of Dr Guigay on 19 January 1971.

E. N. Lemekhova, Institute for Theoretical Astronomy, Leningrad, U.S.S.R.
B. Yu. Levin, O. Schmidt Institute of Physics of the Earth, Moscow, U.S.S.R.
M. L. Lidov, Institute for Applied Mathematics, Moscow, U.S.S.R.
S. I. Luchich, Polytechnic Institute, Chelyabinsk, U.S.S.R.
N. G. Magnaradze, Tbilisi University, Tbilisi, U.S.S.R.
V. A. Magone, Riga, U.S.S.R.
M. Z. Markovich, Polytechnic Institute, Kalinin, U.S.S.R.
B. G. Marsden, Smithsonian Astrophysical Observatory, Cambridge, Mass., U.S.A.
V. I. Mayorova, Institute for Theoretical Astronomy, Leningrad, U.S.S.R.
E. P. Mazets, A. F. Ioffe Institute of Physics and Technology, Leningrad, U.S.S.R.
S. V. Men'shchikova, Institute for Theoretical Astronomy, Leningrad, U.S.S.R.
M. A. Merzlyakova, Pedagogical Institute, Kherson, U.S.S.R.
N. M. Merzlyakova, Planetarium, Kherson, U.S.S.R.
B. Milet, Observatoire de Nice, Nice, France
V. A. Minin, U.S.S.R. Academy of Sciences, Moscow, U.S.S.R.
D. O. Mokhnach, Leningrad, U.S.S.R.
A. Mrkos, Kleť Observatory, České Budějovice, Czechoslovakia
E. E. Mukin, Moscow University, Moscow, U.S.S.R.
V. F. Myachin, Institute for Theoretical Astronomy, Leningrad, U.S.S.R.
P. E. Nacozy, Department of Aerospace Engineering and Engineering Mechanics, Austin, Tex., U.S.A.
G. K. Nazarchuk, Kiev Planetarium, Kiev, U.S.S.R.
N. V. Nevskaya, Institute for Theoretical Astronomy, Leningrad, U.S.S.R.
E. M. Nezhinskij, Institute for Theoretical Astronomy, Leningrad, U.S.S.R.
L. E. Nikolskaya, Leningrad, U.S.S.R.
T. K. Nikolskaya, Institute for Theoretical Astronomy, Leningrad, U.S.S.R.
L. E. Nikonova, Kazan University, Kazan, U.S.S.R.
V. I. Orel'skaya, Institute for Theoretical Astronomy, Leningrad, U.S.S.R.
A. A. Orlov, Sternberg Astronomical Institute, Moscow, U.S.S.R.
F. Kh. Perlin, Leningrad, U.S.S.R.
G. G. Petrov, A. F. Ioffe Institute of Physics and Technology, Leningrad, U.S.S.R.
E. M. Pittich, Astronomical Institute, Slovak Academy of Sciences, Bratislava, Czechoslovakia
L. Yu. Pius, Institute for Theoretical Astronomy, Leningrad, U.S.S.R.
E. N. Polyakhova, Leningrad University, Leningrad, U.S.S.R.
E. Rabe, University of Cincinnati Observatory, Cincinnati, Ohio, U.S.A.
L. N. Radlova, Moscow, U.S.S.R.
H. K. Raudsaar, Astronomical Observatory, Tartu, U.S.S.R.
V. G. Rijves, Tartu University, Tartu, U.S.S.R.
T. B. Sabanina, Institute for Theoretical Astronomy, Leningrad, U.S.S.R.
A. L. Shakh-Budagov, A. F. Ioffe Institute of Physics and Technology, Leningrad, U.S.S.R.
Sh. G. Sharaf, Institute for Theoretical Astronomy, Leningrad, U.S.S.R.
V. I. Sharkov, A. F. Ioffe Institute of Physics and Technology, Leningrad, U.S.S.R.
L. M. Sherbaum, Astronomical Observatory, Kiev University, Kiev, U.S.S.R.
T. K. Shinkarik, Ivanovo, U.S.S.R.

M. Ya. Shmakova, Institute for Theoretical Astronomy, Leningrad, U.S.S.R.[†]
T. A. Sholokhova, Institute for Theoretical Astronomy, Leningrad, U.S.S.R.
V. A. Shor, Institute for Theoretical Astronomy, Leningrad, U.S.S.R.
K. A. Shtejns, Astronomical Observatory, Latvian State University, Riga, U.S.S.R.
L. M. Shul'man, Main Astronomical Observatory, Ukrainian Academy of Sciences, Kiev, U.S.S.R.
A. N. Simonenko, Committee on Meteorites, U.S.S.R. Academy of Sciences, Moscow, U.S.S.R.
G. Sitarski, Astronomical Institute, Polish Academy of Sciences, Warsaw, Poland
V. I. Skripnichenko, Institute for Theoretical Astronomy, Leningrad, U.S.S.R.
P. V. Slavenas, Vilnius University, Vilnius, U.S.S.R.
A. S. Sochilina, Institute for Theoretical Astronomy, Leningrad, U.S.S.R.
N. A. Solovaya, Sternberg Astronomical Institute, Moscow, U.S.S.R.
L. I. Sorokina, Institute for Theoretical Astronomy, Leningrad, U.S.S.R.
A. K. Sosnova, Institute for Experimental Meteorology, Obninsk, U.S.S.R.
V. A. Strel'tsov, Kaliningrad, U.S.S.R.
P. Stumpff, Max-Planck-Institut für Radioastronomie, Bonn, West Germany
M. L. Sveshnikov, Institute for Theoretical Astronomy, Leningrad, U.S.S.R.
V. M. Tabachnik, Pedagogical Institute, Odessa, U.S.S.R.[†]
I. P. Tarasashvili, Polytechnic Institute, Tbilisi, U.S.S.R.
V. V. Terent'ev, Leningrad University, Leningrad, U.S.S.R.
V. P. Tomanov, Pedagogical Institute, Gor'ki, U.S.S.R.
J. Trulsen, The Auroral Observatory, University of Tromsø, Tromsø, Norway
L. L. Tsarakova, Institute for Theoretical Astronomy, Leningrad, U.S.S.R.
V. S. Ural'skaya, Sternberg Astronomical Institute, Moscow, U.S.S.R.
A. V. Vasil'eva, Institute for Theoretical Astronomy, Leningrad, U.S.S.R.
E. A. Vorob'ev, Pedagogical Institute, Ul'yanovsk, U.S.S.R.
E. N. Vorob'eva, Pedagogical Institute, Ul'yanovsk, U.S.S.R.
E. I. Vsekhsvyatskaya, Astronomical Observatory, Kiev University, Kiev, U.S.S.R.
S. K. Vsekhsvyatskij, Astronomical Observatory, Kiev University, Kiev, U.S.S.R.
G. G. Vysotskaya, Leningrad University, Leningrad, U.S.S.R.
F. L. Whipple, Smithsonian Astrophysical Observatory, Cambridge, Mass., U.S.A.
J. M. Witkowski, University Observatory, Poznań, Poland
N. S. Yakhontova, Institute for Theoretical Astronomy, Leningrad, U.S.S.R.
G. T. Yanovitskaya, Main Astronomical Observatory, Ukrainian Academy of Sciences, Kiev, U.S.S.R.
M. S. Yarov-Yarovoi, Sternberg Astronomical Institute, Moscow, U.S.S.R.
D. K. Yeomans, Astronomy Program, University of Maryland, College Park, Md., U.S.A.
P. E. Zadunaisky, University of La Plata and Instituto T. Di Tella, Buenos Aires, Argentina
D. V. Zagrebin, Leningrad, U.S.S.R.
I. E. Zal'kalne, Astronomical Observatory, Latvian State University, Riga, U.S.S.R.
A. F. Zausaev, Institute of Astrophysics, Dushanbe, U.S.S.R.

† It is with regret that we record the death of Mrs Shmakova on 31 August 1971 and of Dr Tabachnik on 6 October 1971.

I. D. Zhongolovich, Institute for Theoretical Astronomy, Leningrad, U.S.S.R.
S. S. Zhuravlev, A. F. Ioffe Institute of Physics and Technology, Leningrad, U.S.S.R.
K. Ziolkowski, Computing Centre, Polish Academy of Sciences, Warsaw, Poland
I. D. Zosimovich, Astronomical Observatory, Kiev University, Kiev, U.S.S.R.
M. S. Zverev, Pulkovo Observatory, Leningrad, U.S.S.R.

Papers or abstracts were also submitted by the following astronomers, who were unable to attend the Symposium:

J. L. Brady, Lawrence Radiation Laboratory, University of California, Livermore, Calif., U.S.A.
M. P. Candy, Perth Observatory, Bickley, Western Australia
O. V. Dobrovol'skij, Institute of Astrophysics, Dushanbe, U.S.S.R.
V. G. Fesenkov, Committee on Meteorites, U.S.S.R. Academy of Sciences, Moscow, U.S.S.R.[†]
G. E. O. Giacaglia, Department of Aerospace Engineering and Engineering Mechanics, University of Texas, Austin, Tex., U.S.A.
P. Herget, University of Cincinnati Observatory, Cincinnati, Ohio, U.S.A.
V. N. Lebedinets, Institute for Experimental Meteorology, Obninsk, U.S.S.R.
V. S. Safronov, O. Schmidt Institute of Physics of the Earth, Moscow, U.S.S.R.
Z. Sekanina, Smithsonian Astrophysical Observatory, Cambridge, Mass., U.S.A.
A. T. Sinclair, Royal Greenwich Observatory, Herstmonceux Castle, Hailsham, Sussex, England

[†] It is with regret that we record the death of Dr Fesenkov on 12 March 1972.

1. EVOLUTION OF COMETARY ORBITS ON A COSMOGONIC TIME SCALE

G. A. CHEBOTAREV

Institute for Theoretical Astronomy, Leningrad, U.S.S.R.

Abstract. A review of modern concepts of the evolution of cometary orbits for the time span of some million years is presented. These concepts are based on the hypothesis of the existence of the Oort cloud, the theory of diffusion developed by Shtejns, and the work by Kazimirchak-Polonskaya concerning the influence of the major planets in transforming cometary orbits. All these ideas are currently being developed with the aid of rigorous mathematical methods by cometary researchers in various parts of the world.

1. Celestial Mechanics and the Origin of Comets

Celestial mechanics is the branch of astronomy that is devoted to the investigation of the motions of the bodies of the solar system. There exists, however, one problem in celestial mechanics for the solution of which we have to go outside the solar system. This is the problem of the 'evolution of cometary orbits', it being the problem of the 'origin of comets' from a cosmogonical point of view.

The principal difficulty encountered in the solution of the problem is the relatively short 'orbital' life of the short-period comets, and this necessitates the solution of another problem, that of finding the source that replenishes these objects.

Three possible sources exist: (1) eruption processes occurring on the surfaces of the bodies of the solar system; (2) the outer region of the solar system, inaccessible to direct observation; and (3) interstellar space. We are therefore in a position to formulate three essentially different cosmogonical hypotheses on the origin of comets, and each of these points of view has its own ardent supporters.

The principles taken as the basis for cometary investigations at the Institute for Theoretical Astronomy are: the hypothesis on the existence of the Oort cloud, the problem of stellar perturbations, the 'diffusion theory' and the 'capture theory'.

All four stages may be investigated by the rigorous mathematical methods of celestial mechanics and, what is perhaps even more important, the major points of the theory can be compared with observations.

2. The Oort Cloud

The existence of the Oort cloud was assumed by many astronomers long before Oort discussed the matter in 1950. The Oort cloud is a natural consequence of several of the cosmogonic hypotheses on the origin of the solar system. The theory proposed by O. Schmidt provides a simple explanation for the formation of cometary nuclei at the periphery of the protoplanetary cloud (Levin, 1960, 1963). Oort's assumption that the cloud of cometary nuclei originated together with the asteroids as the result of the explosion of a major planet existing between Mars and Jupiter therefore

Chebotarev et al. (eds.), The Motion, Evolution of Orbits, and Origin of Comets, 1–5. All Rights Reserved.
Copyright © 1972 by the IAU.

becomes superfluous. This somewhat artificial supposition was indispensable when cometary nuclei were thought to be stony bodies. It became unnecessary after Whipple and Dubyago had formulated the icy model for a cometary nucleus. According to Oort the cometary cloud must contain about 10^{11} cometary nuclei moving in orbits of various inclinations and eccentricities. The semimajor axes of the orbits are confined within a distance of 30 000 to 100 000 AU (Oort, 1963), the maximum density being at about 50 000 AU from the Sun.

An estimate of the total mass of the cometary cloud is naturally rather unreliable. For cometary nuclei of mean mass 10^{15} g, the total mass is about 10^{26} g, or about 1/60 the mass of the Earth. At its formation, however, the mass of the cometary cloud could have been two orders of magnitude greater.

3. Stellar Perturbations

The Oort cloud is dynamically unstable, and individual stars often pass through it. As was shown by Chebotarev (1966), the motion of a comet (with eccentricity $e = 0.6$) is only possible at distances of less than 80 000 AU from the Sun. The boundaries of the cometary cloud are therefore confined within the limits 60 000 to 100 000 AU. This result is in a good agreement with Oort's data, obtained by entirely different methods. Sekanina (1968a) extended the classical concept of the solar sphere of action (Chebotarev, 1963, 1964) for an n-body problem, where $n = 43$ (the Sun and 42 stars situated in its vicinity). The mean radius of the sphere turned out to be 1.5 parsec, but local variations in the distribution of the stars caused a compression to 0.6 parsec (120 000 AU) in the direction of α Centauri. For this reason the outer regions of Oort's cometary cloud cannot be stable.

Finally, the problem of the boundaries of Oort's cloud has been studied by Antonov and Latyshev (1971). For the galactic potential they adopted the value obtained from observations ($F = M/r^{1.4}$) rather than the Newtonian potential $F = M/r^2$ used by Chebotarev.

In the investigation of the stability of the Oort cloud the irregular perturbations from individual stars must also be taken into consideration. Vsekhsvyatskij (1954) was the first to make a sufficiently accurate estimate of the number of stars systematically passing through the Oort cloud; see Table I. This table shows that during the lifetime of the solar system some 3000 stars at a distance of 50 000 AU

TABLE I

Frequency of stellar passages through
the Oort cloud

Distance (AU)	Time (yr)
30 000	5 000 000
50 000	1 700 000
100 000	500 000
150 000	230 000
200 000	110 000

and up to 20 000 stars at 150 000 AU might have passed through the cometary cloud. Vsekhsvyatskij concludes that the perturbations by these stars might destroy the cloud in a few million years. But Nezhinskij (1971) has shown that the half-life of the Oort cloud as a result of the cumulative effect of stellar perturbations is 1.1×10^9 yr. This is about one-fifth the lifetime of the solar system. This confirms that when the solar system was formed the mass of the cometary cloud must have been several times greater than the mass of the Earth.

Stellar perturbations on the motions of comets can be considered in two ways:

(1) with the star fixed relative to the Sun and moving comet (Fesenkov, 1951; Shtejns, 1955; Makover, 1964); or

(2) with the comet fixed relative to the Sun and moving star (Shtejns and Sture, 1962).

Since the velocity of the comet ($v = 0.2$ km s^{-1}) at the periphery of the solar system is two orders less than the velocity of the passing star ($V = 20$ km s^{-1}), we might assume that the second method is the more realistic. Sekanina (1968b), however, has shown that both dynamical arrangements yield statistically identical results, namely, small changes in semimajor axis and large changes in perihelion distance. Only very close stellar encounters (within 1000 AU) are capable of appreciably changing the character of the cometary orbit within the Oort cloud. As a result of those large perturbations some of the cometary nuclei are forced into orbits with perihelion distances less than the radius of Jupiter's orbit, and the comets may thus become observable from the Earth.

4. The Theory of Diffusion

When a comet from the Oort cloud passes into the region of the major planets the perturbations by the stars become less important than those by the planets. As a result of the planetary perturbations the semimajor axis of the orbit may decrease, or the comet may be ejected outside the solar system into interstellar space. The accumulation of small planetary perturbations is random and requires the application of nonclassical methods of celestial mechanics. Actually, only the semimajor axes are involved in this diffusion (Shtejns, 1964, 1965). Shtejns has formulated three laws of diffusion:

(1) cometary orbits with small semimajor axes also have small inclinations (Oort was the first to state this law);

(2) Orbits with large perihelion distances have on the average small eccentricities;

(3) The number of 'new' comets, i.e., those approaching the Sun for the first time, increases with decreasing perihelion distance. For instance, at perihelion distance $q = 1$ AU about 30% of the comets are 'new', while at $q = 4.5$ AU only 3 to 5% are 'new'.

The laws of diffusion have been checked by Shtejns and Kronkalne (1964) on 20 000 fictitious comets.

5. The Theory of Capture

The theory of capture in its classical form was advanced by Laplace in 1796. He thought that comets entered the solar system from interstellar space and that some of them could be captured by the planets if a sufficiently close approach occurred, i.e., the parabolic orbit would be transformed into an ellipse. This theory gained wide popularity in the last quarter of the nineteenth century, following the classical work by Tisserand, Schulhof, H. A. Newton, Callandreau, and others. They demonstrated that such a 'capture' could take place for a comet passing through Jupiter's sphere of action. However, the low probability of this was found to be in serious disagreement with the observations, and the idea was untenable for other reasons too, such as the absence of short-period comets with retrograde orbits.

Only recently has the evolution of orbits of comets that entered the inner planetary system been fully understood (Kazimirchak-Polonskaya, 1967a, 1967b, 1968; Belyaev, 1967a, 1967b). As a rule, a comet approaching the Sun has already passed through the solar system a number of times before, and in accordance with the theory of diffusion it cannot have a parabolic orbit with an arbitrary inclination to the ecliptic plane. By studying the motions of a number of comets over the interval 1660–2060 Kazimirchak-Polonskaya has obtained for the first time a real picture of the evolution of cometary orbits within the orbit of Pluto. This appears to be a remarkable confirmation of the modern concept of the capture theory.

6. Comparison with Observations

The main difficulty in conclusively proving the capture theory is that no comets have been observed with perihelion distances greater than 4 to 5 AU. Table II lists those with the largest perihelion distances.

TABLE II

The comets with the largest perihelion distances

Comet	q (AU)	Comet	q (AU)
1957 IV	5.5	1925 VI	4.2
1948 III	4.7	1942 VIII	4.1
1954 V	4.5	1956 I	4.1
1957 VI	4.4	1729	4.1
1959 X	4.3	1936 I	4.0

A comet becomes accessible to observation only after it approaches inside the orbit of Jupiter. When comparing theory with observations some caution is necessary because it is impossible to detect, not only the comets within the Oort cloud, but also the great majority of comets inside Pluto's orbit.

It is an observational fact, however, that the aphelia of the vast majority of comets lie at distances exceeding 20 000 AU. It may be inferred that a cometary cloud exists at the periphery of the solar system. The density of the cloud is uncertain. It is probable that the cometary cloud is supplemented by so-called 'interstellar' comets that might have originated in the vicinity of other stars and were ejected afterwards into interstellar space (Sekanina, 1968b).

It is important to stress here that we do not know anything about the existence of major planets outside Pluto's orbit (Chebotarev, 1972), and any such planets would play an important role in the evolution of cometary orbits.

7. Conclusions

Modern celestial mechanics allows us to outline certain aspects of the theory of the origin of comets. The Oort cloud, situated at the periphery of the solar system, provides a constant supply of observable comets. The evolution of cometary orbits is determined by perturbations from the stars and planets. Because of the perturbations many comets are ejected outside the solar system while others remain forever the 'prisoners' of the major planets.

The problem of the genetic relationship between comets and minor planets involves the physical structure of cometary nuclei. We are of opinion that comets and minor planets are objects of absolutely different types. The minor planets are the debris of parental protoplanets that originated between the orbits of Jupiter and Mars at the formation of the solar system. There are difficulties, however, in the classification of some objects (e.g., Hidalgo).

References

Antonov, V. A. and Latyshev, I. N.: 1971, *Astron. Zh.* **48**, 854.
Belyaev, N. A.: 1967a, *Byull. Inst. Teor. Astron.* **10**, 696.
Belyaev, N. A.: 1967b, *Astron. Zh.* **44**, 461.
Chebotarev, G. A.: 1963, *Astron. Zh.* **40**, 812.
Chebotarev, G. A: 1964, *Astron. Zh.* **41**, 983.
Chebotarev, G. A: 1966, *Astron. Zh.* **43**, 435.
Chebotarev, G. A.: 1972, *Byull. Inst. Teor. Astron.* **13** (in press).
Fesenkov, V. G.: 1951, *Astron. Zh.* **28**, 98.
Kazimirchak-Polonskaya, E. I.: 1967a, *Trudy Inst. Teor. Astron.* **12**, 63.
Kazimirchak-Polonskaya, E. I.: 1967b, *Astron. Zh.* **44**, 439.
Kazimirchak-Polonskaya, E. I.: 1968, *Astronomie* **82**, 217, 323, 432.
Levin, B. Yu.: 1960, *Priroda Moskva* No. 9.
Levin, B. Yu.: 1963, *Vopr. Kosmogonii*, No. 9.
Makover, S. G.: 1964, *Byull. Inst. Teor. Astron.* **9**, 525.
Nezhinskij, E. M.: 1971, *Byull. Inst. Teor. Astron.* **13**, 31.
Oort, J. H.: 1963, in *The Moon, Meteorites and Comets*, Vol. IV of the series: *The Solar System* (ed. by B. M. Middlehurst and G. P. Kuiper), University of Chicago Press, Chicago and London, p. 665.
Sekanina, Z.: 1968a, *Bull. Astron. Inst. Czech.* **19**, 223.
Sekanina, Z.: 1968b, *Bull. Astron. Inst. Czech.* **19**, 291.
Shtejns, K. A.: 1955, *Astron. Zh.* **32**, 282.
Shtejns, K. A.: 1964, *Uch. Zap. Latr. Gos. Univ.* **68**, 39.
Shtejns, K. A.: 1965, *Zemlya i Vselennaya* No. 5.
Shtejns, K. A. and Kronkalne, S.: 1964, *Acta Astron.* **14**, 311.
Shtejns, K. A. and Sture, S. Ya.: 1962, *Astron. Zh.* **39**, 506.
Vsekhsvyatskij, S. K.: 1954, *Astron. Zh.* **31**, 537.

PART I

OBSERVATIONS AND EPHEMERIDES

2. COMETARY OBSERVATIONS AND VARIATIONS IN COMETARY BRIGHTNESS

S. K. VSEKHSVYATSKIJ

Astronomical Observatory, Kiev University, Kiev, U.S.S.R.

Abstract. The cometary observation service is surveyed. Absolute magnitudes and other physical characteristics are given for the comets that appeared during 1965–1970. The general laws of cometary brightness, in particular the secular decrease in brightness, are discussed. Remarks are made about the standardization of cometary photometry.

The establishment of the cometary observation service, undertaken in the International Quiet Sun Year (IQSY) and afterwards, led to a considerable increase both in the number of observations and in the number of observatories participating in these observations. The expansion of cometary studies was due, first of all, to the ephemerides that were calculated, determinations of orbits and the dissemination of information being accomplished mainly by the International Astronomical Union's Central Bureau for Astronomical Telegrams (B. G. Marsden), as well as by the Comet and Computing Sections of the British Astronomical Association (S. W. Milbourn, C. Dinwoodie), Japanese and Czechoslovak astronomers, the Comet Centre in Kiev and the Institute for Theoretical Astronomy in Leningrad.

As for the determination of positions, it is necessary to mention the outstanding contribution of the American astronomers, who during recent years have developed an efficient system for observation of both new and periodic, and especially the faint, comets, using for this purpose large telescopes and other modern devices (G. Van Biesbroeck, E. Roemer, Smithsonian satellite-tracking network).

The international comet service acknowledges also the numerous series of observations made by the astronomers from Tokyo and elsewhere in Japan (K. Tomita, T. Seki and others), the outstanding successes of the Japanese discoverers of new comets, the observations at the Skalnaté Pleso and Kleť Observatories (A. Mrkos, M. Antal), in Nice (B. Milet), at the Crimean Astrophysical Observatory (N. S. Chernykh), and in the Southern Hemisphere, with observations in Córdoba (Z. M. Pereyra), Santiago (C. Torres), Bickley (M. P. Candy), and elsewhere.

In the U.S.S.R. observations both for determining positions and for studying the physical nature of comets were made also at the observatories in Alma-Ata, Abastumani, Byurakan, Dushanbe, Kiev (Kiev University and the Academy of Sciences Observatories), and more recently in Pulkovo, Moscow, Odessa and at a number of other observatories.

Astronomers in Kiev have made use of the large telescopes at the observatories in the southern and mountainous regions of the U.S.S.R. As a result, in 1969 over 200 exposures were made for investigating cometary structure and brightness and for determining positions. In the course of this programme the short-period comet 1969 IV Churyumov-Gerasimenko was discovered.

Chebotarev et al. (eds.), The Motion, Evolution of Orbits, and Origin of Comets, 9–15. All Rights Reserved.
Copyright © 1972 by the IAU.

At the Institute for Theoretical Astronomy ephemerides were calculated for periodic comets Faye and Ashbrook-Jackson (these comets being recovered in 1969 and 1970), as well as for P/Encke and P/Giacobini-Zinner. Systematic calculation of the ephemerides of new comets was organized at ITA in 1965, numerous orbits determined from three and more observations being published in the Kiev *Comet Circulars*. In these *Circulars* are published positions reduced with the help of the BESM-4 computer at ITA from the measurements obtained at the Crimean Astrophysical Observatory and at a few other observatories in the U.S.S.R.

Since 1963–1964 the Comet Centre in Kiev has published 105 *Comet Circulars* that contain, in addition to positions and ephemerides, information about cometary brightness and the various physical processes in comets. They also give short accounts of the results of cometary investigations in the U.S.S.R. In Kiev the absolute magnitudes and other physical features of comets were determined (Vsekhsvyatskij, 1958, 1966a, 1967). Table I shows the values of H_{10} (absolute magnitude), D_1 (reduced

TABLE I

Absolute magnitudes (H_{10}), reduced head diameters (D_1) and
maximum tail length (S) for comets 1965–1970

	Comet	H_{10}	D_1	S (AU)
1965 I	P/Tsuchinshan 1	14.3	0.2	—
1965 II	P/Tsuchinshan 2	12.1	0.3	0.000 1
1965 III	P/Wolf-Harrington	12.1	0.5	0.000 5
1965 IV	P/Tempel-Tuttle	13.6	—	—
1965 V	P/Reinmuth 1	12.1	—	—
1965 VI	P/Klemola	13.4	0.2	—
1965 VII	P/de Vico-Swift	14.4	0.3	0.000 5
1965 VIII	Ikeya-Seki	6.2	4.2	1.3
1965 IX	Alcock	9.1	0.7	—
1966 I	P/Giacobini-Zinner	11.9	—	—
1966 II	Barbon	5.5	5.9	0.33
1966 III	P/Van Biesbroeck	10.1	0.7	—
1966 IV	Ikeya-Everhart	7.5	7	0.003
1966 V	Kilston	4.5	6	0.095
1966 VI	P/Neujmin 1	11.0	0.4	
1967 I	P/Grigg-Skjellerup	14.8	—	—
1967 II	Rudnicki	9.7–10.7	2.1	0.095
1967 III	Wild	10.3	2.8	0.001
1967 IV	Seki	10.5	2.0	0.001
1967 V	P/Tuttle	10.0	2.2	—
1967 VI	P/Arend	13.3	—	—
1967 VII	Mitchell-Jones-Gerber	5.1	13.4	0.11
1967 VIII	P/Borrelly	12.5	—	—
1967 IX	P/Finlay	12.5	—	—
1967 X	P/Tempel 2	10.4	2	0.000 5
1967 XI	P/Reinmuth 2	11.7	0.8	—
1967 XII	P/Wolf	12.6	—	—
1967 XIII	P/Encke	11.4	1.3	—
1967 XIV	P/Wirtanen	14.3	0.1	—

TABLE I (continued)

	Comet	H_{10}	D_1	S (AU)
1968 I	Ikeya-Seki	4.4	9.'5	0.045
1968 II	P/Schwassmann-Wachmann 2	10.1	1.0	—
1968 III	Wild	7.5	1	0.004
1968 IV	Tago-Honda-Yamamoto	10.0–11.0	2.8	0.002
1968 V	Whitaker-Thomas	10.3	1.8	—
1968 VI	Honda	5.5	6–9.9	0.041
1968 VII	Bally-Clayton	7.3–8.0	4.0	0.005
1968 VIII	P/Perrine-Mrkos	15.6	0.9	—
1968 IX	Honda	6.9	2.8	0.002
1969 I	Thomas	5.8	2.7–3.8	0.012
1969 II	P/Gunn	8.9	—	—
1969 III	P/Harrington-Abell	14.3	—	—
1969 IV	P/Churyumov-Gerasimenko	10.4	1.3	0.004
1969 V	P/Honda-Mrkos-Pajdušáková	11.6	1.0	—
1969 VI	P/Faye	10.8	—	—
1969 VII	Fujikawa	8.3	5.0	0.032
1969 VIII	P/Comas Solá	11.6	2	—
1969 IX	Tago-Sato-Kosaka	6.0	7	0.220
1970a	Daido-Fujikawa	8.5	2	0.074
1969i	Bennett	4.0	5	0.193
1969b	Kohoutek	7.0	2.4	0.01
1970f	White-Ortiz-Bolelli	6.0	—	0.86
1970m	Suzuki-Sato-Seki	8.7	4	0.002
1970g	Abe	4.5	5	0.15

diameter of the head) and S (maximum tail length) for the comets appearing in 1965–1970 (Vsekhsvyatskij and Il'ichishina, 1970). These cometographies also record various observations illustrating physical activity in comets.

It has been found that variations in the integral brightness and observations of the plasma tails of comets may give important data about solar activity and conditions in interplanetary space. Study of the data from P/Schwassmann-Wachmann 1 by the method of superposition of epochs has shown that this comet's flares in brightness have a recurrence period of 25–30 days, there being in many cases a direct correspondence between the increases in brightness and geomagnetic disturbances.

During the previous 50 years at least 54 comets showed these flares and other variations in brightness that demonstrate the influence of corpuscular streams in interplanetary space (Vsekhsvyatskij, 1966b). The results indicate that during certain periods the radial streams in the Sun's outer corona reach out even beyond Jupiter's orbit. A great number of cometary brightness curves have been studied in Kiev. The outcome of this investigation is that the variation in brightness may serve to record the dynamical characteristics of the corpuscular field in interplanetary space, not only in the vicinity of the ecliptic plane, but at great distances from it as well.

Despite the considerable frequency of flares and fluctuations in brightness it has been found that the integral brightness (i.e., the total brightness of the head) of the overwhelming majority of comets depends on the mean law $y = 10$ with $y = 2.5\,n$ conforming to the empirical law $I = I_0^{-n} \Delta^{-2}$. Both direct analysis of photometric curves and study of the distribution of the absolute magnitudes H_{10} lead to $y = 10$ as the most probable value. The question of the changes in the brightness of cometary nuclei is not so clear. Great variations in the luminosity of nuclei (central condensations) may depend on a comet's icy structure and chemical composition (or cometary 'snow') and probably to a great extent on the conditions in interplanetary space (i.e., on the density and velocity of particles in corpuscular streams and the solar wind, the strength of magnetic fields in the streams and interplanetary space – a consequence of the solar flare activity – and solar and galactic cosmic radiation). The latter is proven by the clear-cut dependence of flares in the brightness of comets upon the level of solar activity.

The number of individual comets with well determined absolute magnitudes now amounts to 603. Taking into consideration only the first appearances of short-period comets we obtain the distribution of the absolute magnitudes H_{10} for these 603 comets shown in Figure 1. The range of H_{10} is from -2 or -3 to $+16.4$. The mode of the distribution corresponds to the value $H_{10} = 6.5$. Beyond the region $H_{10} = 5$ or 6 one should observe a systematic decrease in the number of intrinsically faint objects as the result of observational selection. By extrapolating the premaximum curve of the distribution (well represented by the expression $n = 3.24 \times 1.84^{H_{10}}$) we find that the true number of comets brighter than absolute magnitude 17 that

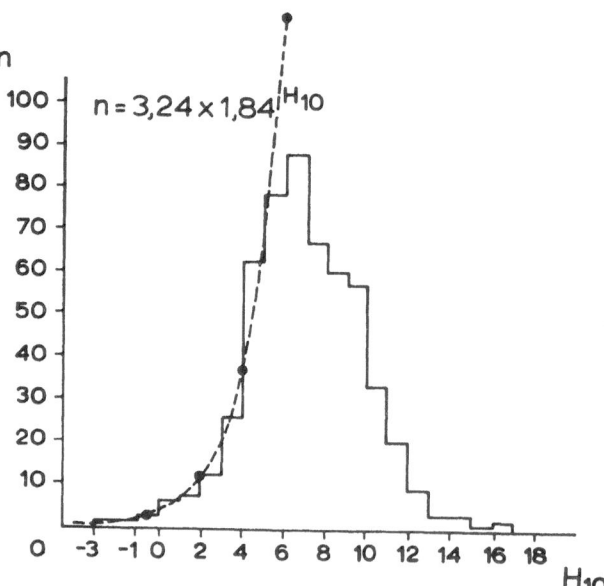

Fig. 1.   Distribution of absolute magnitudes of 603 comets (until 1970). Only the first apparitions of periodic comets are included.

passed through perihelion during 1700–1969 with perihelion distance $q < 2$ to 3 AU amounts to 10^5. Thus, the great majority of intrinsically faint comets is unobserved, on account of the apparent faintness of comets of larger perihelion distances, disappearance in twilight, and other conditions of visibility. From this we suppose that the real number of cometary objects should depend on the law $n \sim d^{-k}$ (where d is the comet nucleus diameter, and $k \sim 3$ to 4 from asteroid and meteorite data). The distribution of H_{10} with q does not show systematic deviations from the mean, and this is an argument which confirms the mean law $y = 10$.

Figure 2 shows the distribution of H_{10} for the first and latest appearances of 102 periodic comets with $P < 200$ yr (with 1969 data included). Comets observed at but

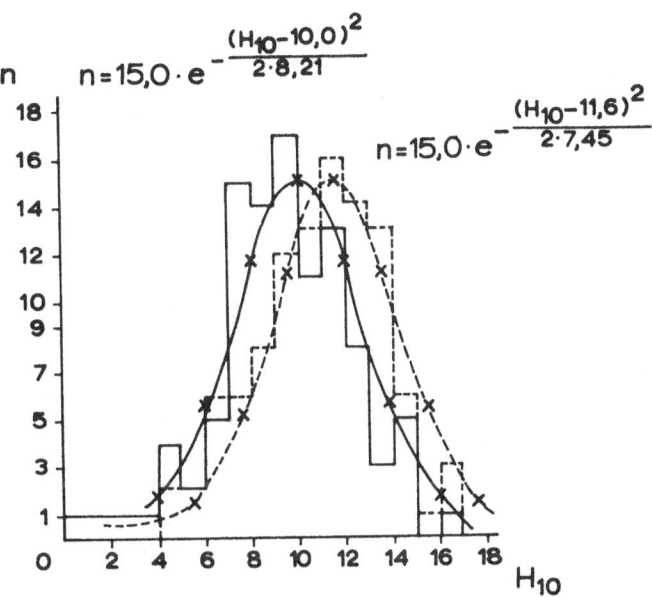

Fig. 2.   Comparison of absolute magnitude distribution curves for the first (solid lines) and latest (broken lines) appearances of periodic comets.

one perihelion passage are included in both plots. There is some similarity between the distribution of H_{10} for short-period comets at their first appearances and the curve for all known comets (i.e., for parabolic and long-period comets). The curve for the latest appearances differs greatly in character and becomes rather asymmetric. The mean epoch for the first appearances is 1877.6, and that for the latest or single appearances is 1927.8. The equations for normal distributions that represent the curves are indicated in the figure.

Thus, on the average, during the 50 years the mode has changed from 10.0 to 11.6, indicating a decrease of brightness of 1.6 magnitudes for some mean period of revolution $P = 10.9$ yr. This corresponds to a diminution of brightness (i.e., disintegration of comets) of 0.3 to 0.4 magnitude per revolution. The decrease in brightness of individual comets fluctuates from less than 0.1 to 1.0 magnitude per revolution.

When making statistical investigations of cometary characteristics one should keep in mind, not only the conditions of visibility (i.e., observational selection), but also the recent improvement in observational techniques, efficient sky patrols and the possibility now of observing very faint comets.

In order to make reliable determinations of integral and nuclear magnitudes it is necessary to standardize observations carefully. Even nowadays observers often estimate the brightness only approximately and do not indicate whether they are referring to the head, central condensation or nucleus. Because of this divergences, amounting to 2, 3, or even 4 magnitudes, arise in the estimates by different observers.

Of no less significance can be real effects of comet flare activity, which can appear both in the integral brightness and in the brightness of the central condensation. By uncritical use of the observational material some investigators have acquired a sceptical attitude to the study of cometary activity and the systematic decrease in cometary brightness. It is necessary to set up standard procedures for the calibration of photographic observations of the brightness of comets and systematically to conduct photoelectric and image-tube photometric observations also.

The results must be free from all biases that investigators (who in most cases are not those who make the observations) may have about the systematic diminution of brightness of periodic comets. As we have already stated, the last effect is sometimes masked by real and large fluctuations in brightness.

The question of the systematic decrease in brightness of periodic comets is of very great importance. The data gathered over historic times prove the most important proposition of cometary astronomy: that comets disintegrate rapidly, and that in centuries, if not in decades, short-period comets become inaccessible to observation, not only because of changes in their orbits, but more from the exhaustion of their supplies of cometary ices. It is necessary to study the changes in the orbits of short-period comets in order to find out the secular variations of their physical properties. In this sense historical studies – analysis of observations of the past – are to be recommended.

On the other hand, establishment of a cometary observation service today requires definite standardization, not only of the methods of determination of positions, but also of the ways of determining the physical parameters of comets. The first attempts of such a type were made in connection with the IQSY, when comet observations were included in the international programme. This work seems to be continuing.

To determine the integral brightness of comets, photoelectric devices on powerful telescopes, as well as photographic systems (astrographs, Schmidt and Maksutov telescopes), should be used to survey standard areas of stars or regions containing galaxies or planetary nebulae. When publishing the observations mention should be made of the region of the spectrum to which the estimates refer. The procedure used by Roemer in Flagstaff and more recently in Tucson should be followed when measuring the brightness of cometary nuclei with large telescopes.

For comets that are brighter than magnitude 11 or 12 it is necessary to continue making visual (extrafocal) brightness estimates using both wide-field instruments (comet-seekers, binoculars) and refractors of aperture 150–250 mm, so that comparison can be made with turn-of-the-century observations in the so-called Holet-schek system.

Despite the considerable success of the international cometary service, particularly in recent years, we cannot consider the present system of cometary observations and discoveries to be perfect. Because of the insufficient number of precise positions, especially of faint comets, we are not able to determine orbits with accuracy sufficient to clarify the true role and nature of nongravitational forces and to study the peculiarities of individual comets as indicators of conditions in interplanetary space.

It is clear that a number of measures should be undertaken to increase activity in searching for new comets. Approaches should be made to individual observatories and societies of amateur astronomers in order to develop visual comet hunting in different countries (particularly in the Southern Hemisphere) and to inspire success such as that recently in Japan and during the late 1940's and 1950's in Czechoslovakia. On the other hand it is necessary to call to the attention of all comet observers the desirability of accompanying their publication of the positions of a comet with a short description of the comet's characteristics (the integral brightness, brightness of central condensation, head diameter, presence and length of tail, its structure, etc.). At present, important information about cometary phenomena often remains lost in the observers' notebooks.

References

Vsekhsvyatskij, S. K.: 1958, *Fizicheskie Kharakteristiki Komet*, Moscow. (English translation: 1964, *Physical Characteristics of Comets*, Jerusalem.)
Vsekhsvyatskij, S. K.: 1966a, *Komety 1954–1960*, Moscow.
Vsekhsvyatskij, S. K.: 1966b, *Mem. Soc. Roy. Sci. Liège Ser. 5* **12**, 57.
Vsekhsvyatskij, S. K.: 1967, *Komety 1961–1965*, Moscow.
Vsekhsvyatskij, S. K. and Il'ichishina, N. I.: 1970, *Komety 1965–1969* (in press).

3. COMETARY BRIGHTNESS VARIATIONS AND CONDITIONS IN INTERPLANETARY SPACE

D. A. ANDRIENKO, A. A. DEMENKO, I. M. DEMENKO,
and I. D. ZOSIMOVICH

Department of Astronomy, Kiev University, Kiev, U.S.S.R.

Abstract. The study of cometary brightness variations can be a good method for determining conditions in interplanetary space. In this work we compare the light curves of comets with curves showing variations in the geomagnetic field at times when the comets are near the ecliptic plane. We have used photometric observations made of 29 comets between 1881 and 1937. It is shown that an increase in the brightness of a comet is associated with an increase in the geomagnetic activity index and consequently with the influence of the solar corpuscular streams and solar wind.

It is well known that short-term variations take place in the integral brightness of comets. These variations are also known to be associated with solar activity. The investigation of cometary brightness variations has been described in detail by Vsekhsvyatskij (1958, 1966) and Dobrovol'skij (1966), but the processes causing cometary outbursts and brightness variations, and the role of the solar wind and corpuscular radiation, as well as photon solar radiation, are still open for discussion.

This study is an attempt to evaluate the influence of the corpuscular radiation of the Sun on the outbursts observed in the integral brightness of comets. It is known that the disturbances in the magnetic field of the Earth are generally connected with the corpuscular activity of the Sun. The presence or absence of a correlation between the Earth's magnetic activity index and cometary outbursts will also give an unambiguous answer as to the influence of solar corpuscular radiation upon variations in cometary brightness. To solve the problem we made use of the series of photometric observations of comets listed by Bobrovnikoff (1942), who gave photometric data for a number of comets observed between 1858 and 1937.

The comparison of cometary activity with the geomagnetic activity index was carried out with allowance made for the geometric position of the comet with reference to the ecliptic plane. For each comet the times of nodal passage, the ecliptic latitudes and longitudes during the period of observation and a number of other geometric parameters were computed. The cometary light curves were compared with the geomagnetic activity index, due regard being given to the different ecliptic longitudes of the comets and the Earth. Thus, the moment of each observed cometary outburst was correlated with the geomagnetic field index corresponding to the date for which the same area of the solar surface was visible both from the Earth and the comet. The daily shift in ecliptic longitude was calculated as described by Demenko (1971).

The analysis shows that there is a high correlation between the two processes, especially when the comets were observed near the ecliptic plane. This can easily be seen from the curves shown in Figure 1. The continuous line shows the inter-

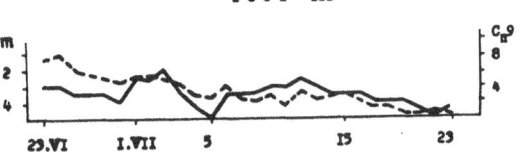

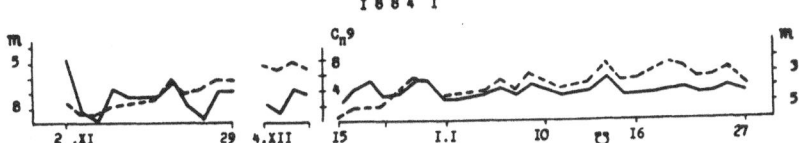

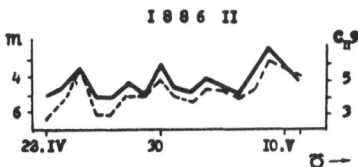

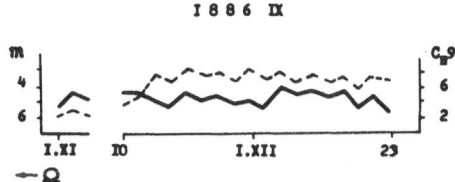

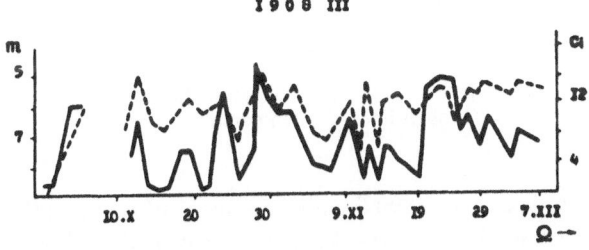

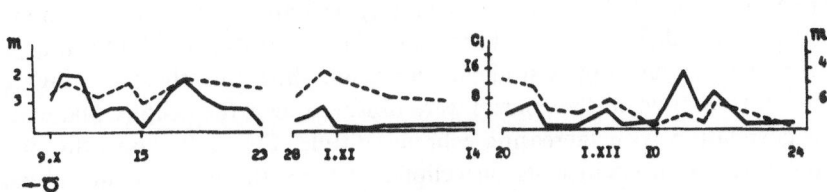

Fig. 1a–f

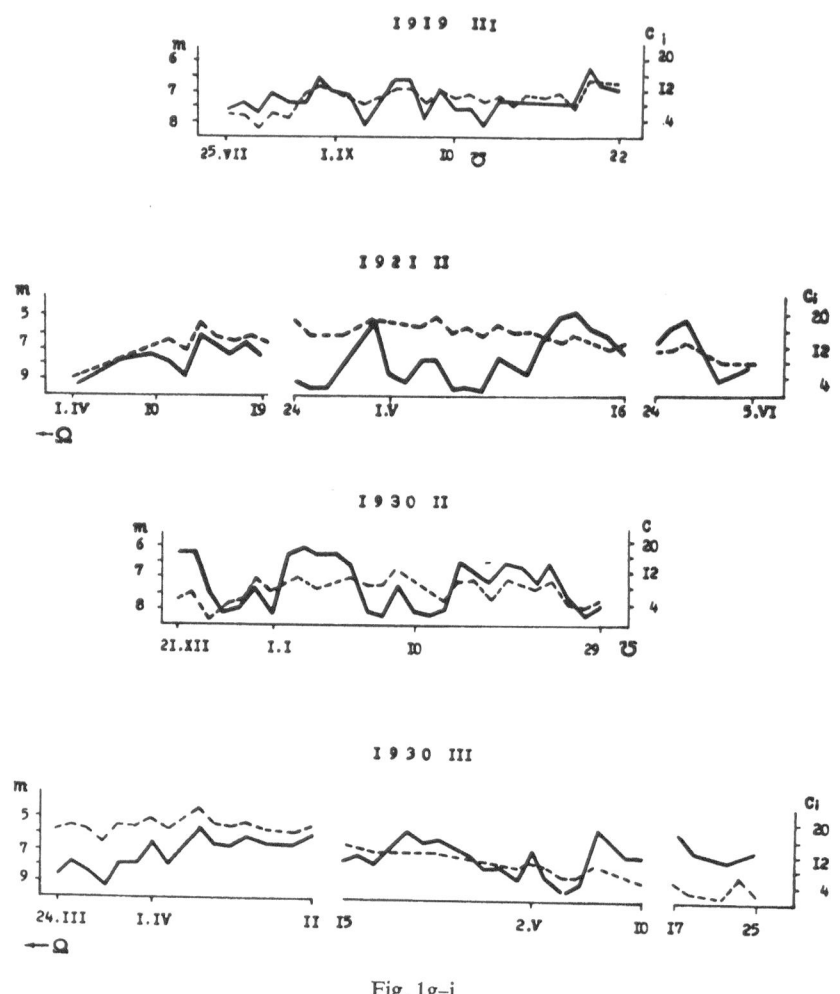

Fig. 1g–j

Fig. 1a–j. Correlation of geomagnetic disturbances (solid lines) with brightness variations (broken lines) for ten comets.

national geomagnetic index, since 1890 from Chapman and Bartels (1940) and before then from Zosimovich and Andrienko (1971); the broken line shows the cometary magnitudes, based on Bobrovnikoff's raw data.

The correlation coefficients r and the probability 0.68 and 0.90 confidence intervals for these coefficients, defined as by Shchigolev (1969) and Zajdel' (1968), are given in Table I for the various comets studied. The results show that there is a high correlation ($r = 0.8$ to 0.9) when the comets were observed near the ecliptic plane and a decrease in correlation with increasing ecliptic latitude. Figure 2 shows the dependence of the correlation coefficients on ecliptic latitude, the points being obtained by averaging the coefficients in 5° latitude steps.

TABLE I

Correlation of cometary brightness variations with geomagnetic activity

Comet	Inclination of orbit to ecliptic	Investigated latitude range	Correlation coefficient	Confidence interval σ_r for r probability 0.68	probability 0.90
1881 III	63°	− 5° to +36°	0.56	± 0.13	0.10 < σ_r < 0.18
1884 I	74	+80 +49	0.10	± 0.25	0.20 0.35
		+45 +21	0.22	± 0.13	0.10 0.18
		+17 −20	0.63	± 0.15	0.12 0.21
1886 II	84	+70 +28	0.83	± 0.08	0.06 0.11
1886 IX	102	+ 6 +20	0.82	± 0.15	0.12 0.21
		+29 +37	0.81	± 0.15	0.12 0.21
		+43 +76	0.30	± 0.28	0.22 0.39
1893 II	160	+23 +29	0.32	± 0.22	0.16 0.28
1899 I	146	−33 −34	0.35	± 0.29	0.23 0.41
		+33 +28	0.35	± 0.29	0.23 0.41
1900 II	62	− 1 +23	0.61	± 0.15	0.12 0.21
1902 III	156	+11 +17	0.44	± 0.15	0.12 0.21
		+17 +24	0.44	± 0.21	0.16 0.28
1903 IV	85	+ 2 + 9	0.13		
		+11 +17	0.19	± 0.15	0.12 0.21
		+20 +32	0.40	± 0.25	0.20 0.35
		+34 +70	0.10	± 0.13	0.08 0.14
1906 VII	56	+15 +42	0.20	± 0.13	0.15 0.27
1907 IV	9	− 1 − 2	0.69	± 0.22	0.16 0.28
		− 3 − 6	0.61	± 0.14	0.12 0.21
		− 6 − 9	0.65	± 0.13	0.10 0.18
1908 III	140	+13 +10	0.55	± 0.13	0.10 0.18
		+10 + 7	0.93	± 0.04	0.03 0.05
1911 II	148	+28 +30	0.20	± 0.13	0.10 0.18
		+29 +17	0.63	± 0.20	0.16 0.28
1911 V	34	+34 +27	0.32	± 0.25	0.20 0.35
		+27 −18	0.79	± 0.09	0.07 0.12
		−22 −34	0.32	± 0.25	0.20 0.35
1912 II	80	+16 +49	0.44	± 0.20	0.16 0.28
		+50 +79	0.34	± 0.19	0.15 0.27
		+78 +58	0.14	± 0.25	0.20 0.35
1913 IV	143	+30 +25	0.28	± 0.17	0.14 0.24
1913 VI	41	+ 5 +17	0.10		
1914 II	24	+10 +15	0.53	± 0.14	0.10 0.18
1914 V	68	+14 +29	−0.16	± 0.19	0.15 0.27
		+29 +44	−0.46	± 0.12	0.10 0.18
1915 II	55	+20 +11	0.22	± 0.13	0.10 0.18
		+11 + 5	0.88	± 0.07	0.06 0.10
1917 II	149	+21 + 5	0.66	± 0.14	0.10 0.18
1917 III	26	+10 + 5	0.04		
		+ 5 + 1	0.35	± 0.25	0.20 0.35
1919 III	19	+ 9 +17	0.56	± 0.13	0.10 0.18
1921 II	132	+14 +36	0.66	± 0.13	0.10 0.18
		+36 +47	0.30	± 0.20	0.16 0.28
1930 II	124	+55 + 4	0.68	± 0.10	0.08 0.14
1930 III	67	+30 +65	0.86	± 0.06	0.04 0.06
		+59 +33	0.46	± 0.18	0.14 0.25
		+27 +23	0.34	± 0.25	0.20 0.35
1936 II	79	+74 +40	−0.20	± 0.13	0.10 0.18
		+40 +11	−0.20	± 0.13	0.10 0.18
		+ 9 − 9	0.51	± 0.18	0.14 0.25
1937 II	26	+20 +25	0.38	± 0.29	0.23 0.41
		+25 +20	−0.78	± 0.15	0.12 0.21
1937 V	146	+27 +33	−0.80	± 0.15	0.12 0.21
		+33 +22	0.31	± 0.14	0.10 0.18
		+18 +15	0.47	± 0.27	0.22 0.38

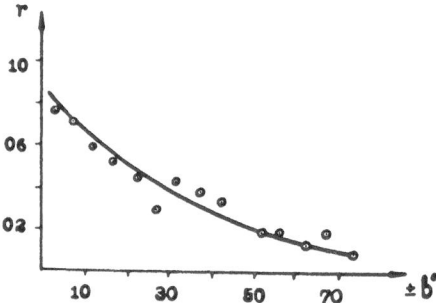

Fig. 2.   Dependence of the correlation coefficients r on ecliptic latitude $(\pm b)$.

The correlation of cometary flares with the geomagnetic activity index varies in different parts of the 11-year solar activity cycle. Figure 3 shows the dependence of

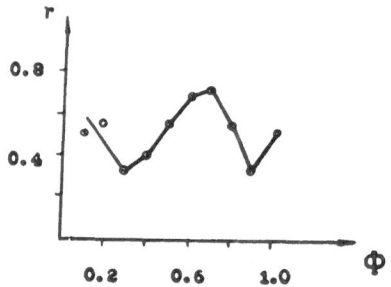

Fig. 3.   Variation of the correlation coefficients r during the solar cycle.

the correlation coefficients on the phase of the solar cycle at times when the comets were close to the ecliptic plane. It is clear that the greatest value of r occurs when solar activity is decreasing, and the smallest value falls near solar maximum. This is in good agreement with the hypothesis that the corpuscular streams are most stable when solar activity is decreasing (Zosimovich, 1965). From analysis of the relation between cometary outbursts and disturbances in the geomagnetic field one can arrive at the following conclusions:

(1) Outbursts and variations in integral cometary brightness are closely connected with the solar corpuscular radiation. Cometary activity and changes in the magnetic field should be compared with due regard to the geometric position of the comet. The best results are achieved when a comet is observed at small ecliptic latitudes, because the comet and the Earth are influenced by the same corpuscular structures of the solar corona.

(2) From the correlation coefficients obtained, one can evaluate the angular dimensions of the solar corpuscular streams. From Figure 2 and Table I it follows that the smallest value of r for which the correlation may be considered substantial is 0.35.

This brings us to the conclusion that the half-angle of the cone for the latitude dispersion of the corpuscular streams may approach 40°.

(3) It follows from Figure 3 that study of the interaction between the interplanetary plasma and cometary material gives evidence for the stability of the solar corpuscular streams. Despite the fact that the difference in the times of corpuscular interaction between the Earth and a comet may approach two weeks, the correlation coefficients remain rather high. It gives independent proof of the result obtained earlier (Zosimovich, 1965), where the problem of the stability of the streams was investigated by means of geomagnetic data.

(4) Thus, the investigation of cometary brightness variations seems to be very important in connection with the problem of determination both of the geometrical characteristics of the solar corpuscular radiation and of the physical parameters of the solar wind and corpuscular streams at points distant from the Sun and at heliographic latitudes too high for direct measurements to be possible.

(5) The correlation of geomagnetic disturbances with cometary outbursts is best at small heliocentric distances (see Figure 4). The correlation practically vanishes at

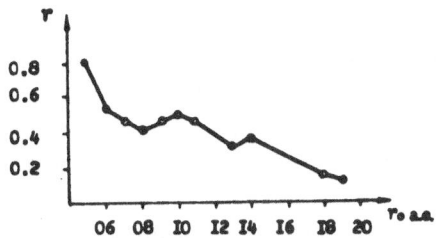

Fig. 4.   Dependence of the correlation coefficients r on heliocentric distance ($r_\odot$).

distances exceeding 1.3 AU. This may be explained by the dissociation of corpuscular streams at greater distances from the Sun, where the structure of corpuscular streams seems to change.

References

Bobrovnikoff, N. T.: 1942, *Contr. Perkins Obs.* No. 16.
Chapman, S. and Bartels, J.: 1940, *Geomagnetism*, Oxford, Vol. 2.
Demenko, I. M.: 1971, *Probl. Kosmich. Fiz.* No. 6.
Dobrovol'skij, O. V.: 1966, *Komety*, Nauka, Moscow.
Shchigolev, B. M.: 1969, *Matematicheskaya Obrabotka Nablyudenij*, Moscow.
Vsekhsvyatskij, S. K.: 1958, *Fizicheskie Kharakteristiki Komet*, Moscow.
Vsekhsvyatskij, S. K.: 1966, *Problemy Kometnoj Fotometrii*, *Inf. Byull.* No. 10, 3.
Zajdel', A. N.: 1968, *Elementarnye Otsenki Oshibok Izmerenij*, Leningrad.
Zosimovich, I. D.: 1965, *Geofiz. Astron.* No. 8.
Zosimovich, I. D. and Andrienko, D. A: 1971, *Probl. Kosmich. Fiz.* No. 6.

4. OBSERVATIONS OF COMETS AT THE CRIMEAN ASTROPHYSICAL OBSERVATORY

N. S. CHERNYKH

Crimean Astrophysical Observatory, Crimea, U.S.S.R.

Abstract. The cometary observing programme at the Crimean Astrophysical Observatory is described.

Regular observation of bright comets with the 40-cm astrograph of the Crimean Astrophysical Observatory was initiated by the author in the autumn of 1965. The successful development of this work was facilitated by the following two factors:

(1) The ephemeris service for new comets was organized in autumn 1965 at the department of minor planets and comets of the Institute for Theoretical Astronomy, providing

(a) information to observers about newly discovered comets;

(b) urgent reductions of measurements of cometary positions from photographic plates;

(c) calculation of cometary ephemerides and dissemination of these to observers.

(2) Two years earlier systematic large-scale observations of minor planets were initiated with the 40-cm double astrograph. By the time work began on comets all the necessary tools for handling the photographic plates were available: star atlases, catalogues, a measuring engine, and considerable experience that could also be applied to cometary observations.

The main aim of the observations is measuring positions and obtaining direct photographs for the study of cometary structure. The establishment of constant links with the Kiev Comet Centre has had a particular stimulating effect.

The Zeiss double astrograph, with focal length 160 cm and relative aperture 1/4, was used for the purpose. The scale of 129″/mm readily permits an accuracy of measurement of 0″.3 to 0″.5. The limiting magnitude of the instrument for stars is 18, and for diffuse objects like comets it is 16 or 17.

By means of a guide telescope (aperture 15 cm, magnification 225×) we can directly track a comet brighter than magnitude 5 or 6. When taking plates of faint comets, the instrument is guided by the Metcalf method. A micrometer with a scale division of 2″ and offset motions enables the instrument to have an additional motion relative to the guide star of up to 30″ per minute.

The first stage in the treatment of observations is the selection of reference stars and the measurement of the plates. This is done at the Observatory. Both the observer and the members of the ITA Crimean team engaged in observations of minor planets with the same instrument take part in this work. The coordinates of the reference stars are obtained from the Yale catalogues, the Smithsonian Astrophysical Observatory catalogue and sometimes from the AGK$_2$. Four reference stars are selected for each position of a comet. The plates are measured with an SIP-5 instrument, which has glass scales and spiral micrometers. The accuracy of reading the scales is up to 0.2 μ. The measurements are recorded on forms and sent to ITA, where the next stage takes place – the computation of spherical coordinates using a BESM-4 computer. The computations are made using a programme for the six plate-constant method with the simultaneous determination of dependences.

The interval between photographing a comet and the production of an exact position depends almost entirely on the time needed for the data to reach Leningrad by mail. In some special cases we telephone the data so it can be used for computing an orbit the same day. Urgent information about our observations is printed regularly on the *Comet Circulars*, published in Kiev by the cometary studies group.

From September 1965 to May 1970, 23 comets were observed with the double astrograph and a total of 350 positions obtained for them; see Table I. The constant use of the instrument for intensive observations of minor planets does not permit us to pay proper attention to the short-period comets. This is why the list of observed comets contains more new comets than returning periodic comets.

TABLE I

Cometary observations at the Crimean Astrophysical Observatory

Comet		Interval of observations	Number of positions
1965 VIII	Ikeya-Seki	1965 Sept. 26–Dec. 3	24
1965 IX	Alcock	1965 Sept. 28–Oct. 2	7
1966 III	P/Van Biesbroeck	1966 June 15–July 20	4
1966 V	Kilston	1966 Aug. 13–Nov. 6	40
1966 II	Barbon	1966 Aug. 21–Oct. 21	21
1966 IV	Ikeya-Everhart	1966 Sept. 13–Oct. 13	17
1967 II	Rudnicki	1966 Oct. 19–Dec. 30	25
1967 III	Wild	1967 March 5–11	5
1967 XIII	P/Encke	1967 Aug. 2–16	17
1967 XI	P/Reinmuth 2	1967 Sept. 30–Oct. 6	3
1968 II	P/Schwassmann-Wachmann 2	1968 Feb. 1–21	3
1968 I	Ikeya-Seki	1968 Feb. 4–May 27	49
1968 IV	Tago-Honda-Yamamoto	1968 May 6–27	9
1968 V	Whitaker-Thomas	1968 June 28–July 26	5
1968 VI	Honda	1968 July 26–Sept. 15	16
1968 VII	Bally-Clayton	1968 Aug. 27–Oct. 23	17
1969b	Kohoutek	1969 Aug. 4–1970 Apr. 28	36
1969 VII	Fujikawa	1969 Aug. 18–Sept. 14	10
1969 IV	P/Churyumov-Gerasimenko	1969 Nov. 8–15	4
1969 VI	P/Faye	1969 Nov. 14	2
1969 IX	Tago-Sato-Kosaka	1970 Jan 30–March 3	8
1969 VIII	P/Comas Solá	1970 Mar. 13	2
1969i	Bennett	1970 Apr. 1–28	24

Series of photographs of comets Ikeya-Seki 1965 VIII, Ikeya-Seki 1968 I, Honda 1968 VI, Tago-Sato-Kosaka 1969 IX and Bennett 1969i were obtained in order to investigate the detailed structure of these comets.

Acknowledgments

The author would like to express his sincere thanks to the members of the department of minor planets and comets of ITA engaged in the ephemeris service and to the Crimean team of the Institute for their constant and effective assistance.

5. L'OBSERVATION DES COMÈTES À L'ASTROGRAPHE DE L'OBSERVATOIRE DE NICE

B. MILET

Observatoire de Nice, Nice, France

Abstract. The programme of cometary observations at the Nice Observatory is described. Observations are made with the double astrograph by the Trépied-Metcalf method, offset motions of up to half a degree per hour being possible. Five reference stars are measured, and the reductions are made by the least-squares dependence method.

Les problèmes essentiels qui se posent à un observateur de comètes désirant obtenir des positions précises se rapportent d'une part à la technique employée lors de la prise du cliché et d'autre part, à la précision de la mesure, surtout lorsqu'il s'agit d'objets trop brillants pour donner une image quasi-stellaire.

1. Observation

Jusqu'en avril 1970, et chaque fois que cela était rendu nécessaire par le mouvement propre de l'objet, nous avons utilisé la méthode Trépied-Metcalf par déplacement pré-calculé des fils du micromètre. Avec l'astrographe de Nice, il est encore possible d'effectuer des déplacements en x et en y jusqu'à une vitesse de l'ordre du degré par jour; toutefois, même dans ces conditions, le guidage se fait par sauts de 1 division du tambour micrométrique, ce qui correspond à 13 μ et l'image d'une étoile sera en 'marche d'escalier'.

Aussi, le problème ayant été posé aux responsables des équipes électronique et technique de l'Observatoire, il a été possible de réaliser un système d'entraînement automatique, non de l'instrument, mais de l'un des chassis de l'astrographe. Orientable dans le sens du mouvement, par repérage en degrés, sur le bâti de l'instrument, le porte-plaque est entraîné par un moteur réglable en vitesse, sur des glissières réalisées avec le maximum de précision. A l'aide d'un générateur TBF (très basse fréquence) et d'un moteur pas-à-pas le déplacement micron par micron peut se faire à l'aide d'impulsions jusqu'à la fréquence de 300 par seconde, c'est-à-dire que l'on peut suivre un objet se déplaçant jusqu'à un demi-degré à l'heure.

En outre, plus particulièrement lors des observations des comètes intenses, les clichés sont pris en plusieurs couleurs. Nous utilisons des plaques Kodak, avec filtre, dans les bandes suivantes: IIaO + filtre GG13 avec un maximum vers 3800 Å, 103aG + filtre VJ3 vers 5500 Å, et 103aE + filtre VR1 vers 6500 Å. Ainsi, il est possible de mettre en évidence la variation de l'indice de couleur de la comète en fonction de sa distance au Soleil.

2. Mesure des clichés

On utilise une machine Bouty, à règles graduées, permettant théoriquement la mesure au 0,1 μ. Dans le cas des comètes de magnitude inférieure à 12, il est assez

Chebotarev et al. (eds.), The Motion, Evolution of Orbits, and Origin of Comets, 25–26. All Rights Reserved.
Copyright © 1972 by the IAU.

facile de centrer l'image et d'obtenir une précision de mesure du même ordre que celle que l'on a pour les étoiles, c'est-à-dire $\pm 2\,\mu$. Mais s'il s'agit d'objets dont la trace sur les clichés est de l'ordre du centimètre, il faut alors trouver le moyen d'obtenir une position relativement précise: on adapte sur la machine à mesurer un système optique donnant un faisceau de lumière parallèle intense et la position du noyau, ou de la zone de concentration maximale, est alors aussi bien définie que celle des étoiles, souvent mieux, car dans le cas d'une pose longue ou d'un mouvement rapide de la comète, les images stellaires sont très allongées. Dans ces conditions ces dernières sont mesurées en x et en y à leurs extrémités et la valeur moyenne est adoptée pour le calcul.

3. Réduction

C'est généralement par rapport à cinq étoiles que la position est calculée. On choisit au mieux ces étoiles de référence, de façon qu'elles forment un pentagone dont le centre de gravité soit voisin de l'objet. La réduction, sur l'ordinateur IBM de l'Observatoire de Nice, est faite par les moindres carrés d'une part en se rapportant à ce centre de gravité par écarts de x, y à la moyenne des coordonnées, réduisant ainsi les équations à un système à deux inconnues. On calcule les constantes dont l'une est l'écart au coefficient d'échelle dont on tient compte à priori, et l'autre est fonction de l'orientation du cliché. L'utilisation des formules de Banachiewicz permet de déterminer l'erreur quadratique moyenne sur chaque constante. D'autre part, la position de la comète est fournie par la méthode des dépendances rapportée aux cinq étoiles. Les deux résultats doivent être identiques, qu'ils soient calculés par les cracoviens ou les dépendances.

Généralement deux poses permettent, pour une même plaque, d'obtenir deux positions. Les mêmes étoiles, avec les mêmes valeurs α et δ, étant utilisées pour les deux réductions et les résidus sur chaque étoile étant calculés, il est possible de distinguer les erreurs dues aux positions, ou mouvements propres, qui alors se retrouvent avec des valeurs analogues pour les deux poses, et les erreurs dues aux mesures qui, indépendantes, donnent des valeurs différentes des résidus.

Discussion

N. M. Bronnikova: What is the limiting magnitude of your instrument?

B. Milet: About 17 or 18, for objects of stellar appearance.

F. L. Whipple: Have you measured cometary positions from photographs in two colours to determine whether they depend upon the colour, i.e., upon the specific gases that may surround the nucleus in the inner coma?

B. Milet: I have found systematic differences, although this may perhaps be explained by the fact that we are sometimes obtaining the position of the nucleus and at other times the mean position of the whole cometary head.

6. PHYSICAL OBSERVATIONS OF THE SHORT-PERIOD COMET 1969 IV

K. I. CHURYUMOV and S. I. GERASIMENKO

Department of Astronomy, Kiev University, Kiev, U.S.S.R.

Abstract. The new short-period comet Churyumov-Gerasimenko, discovered by the authors on plates taken by the Kiev University cometary expedition to Alma-Ata in September 1969, was systematically photographed with fast telescopes at Byurakan and Alma-Ata until March 1970. Measurements were made of the photographic magnitude of the photometric nucleus, as well as of the photographic and photovisual integral magnitudes. The variations in nuclear magnitude were found to be well correlated with changes in the total sunspot area. The integral photometric parameters are $H_y = 11.91 \pm 0^m.54$ and $n = 4.0 \pm 0.8$ (in the photographic spectral region). Deviations of the tail axis from the prolonged radius vector were considerable. A spectrogram shows the continuum and emission of CN, C_2 and C_3 in the head, the continuum and a single emission (perhaps CO^+) in the tail.

During August and September 1969 the authors took part in the third Kiev University expedition to the Alma-Ata Astrophysical Institute (Churyumov and Gerasimenko, 1970). The purpose of the expedition was to carry out visual and photographic searches for new comets in the morning and evening Everhart (1967) zones and also to make photographic observations of the well-known short-period comets Faye (1969 VI), Comas Solá (1969 VIII), Honda-Mrkos-Pajdušáková (1969 V) and the two new comets Kohoutek (1969b) and Fujikawa (1969 VIII). The observations were made with a 50-cm $f/2.4$ Maksutov telescope and a 17-cm $f/1$ Schmidt camera. Altogether we took about 100 plates suitable for integral photometry and the determination of exact positions of the above-mentioned comets.

Still in Alma-Ata, we noted on September 20 a cometary object of magnitude 13 on a plate taken September 11 for P/Comas Solá (Figure 1). Back in Kiev on October 22 we found that the object was 1°.8 from the position of P/Comas Solá given in the ephemeris. Then we saw P/Comas Solá close to its ephemeris position, suggesting that the object we had noted was a new comet. Examination of other plates for P/Comas Solá – two on September 9 and two on September 21 – immediately revealed the new object, and although it was near the edge, it still had its cometary appearance and showed motion among the stars (Figure 2). Vsekhsvyatskij cabled news of our discovery to the IAU Central Bureau for Astronomical Telegrams. The new comet was given the preliminary designation 1969h, and later the final designation 1969 IV.

On the basis of the first exact positions, reduced by Shmakova (Leningrad) from our measurements, Marsden (U.S.A.) calculated six ephemerides from two parabolic and four elliptical orbits. On October 31 the new comet was photographed by Scovil (U.S.A.). This observation, and then observations by Seki (Japan), Roemer (U.S.A.), and Milet (France), closely confirmed one of the elliptical orbits. Later elements of the elliptical orbit of comet 1969 IV are as follows (Marsden, 1969):

Chebotarev et al. (eds.), The Motion, Evolution of Orbits, and Origin of Comets, 27–34. All Rights Reserved.
Copyright © 1972 by the IAU.

$$T = 1969 \text{ Sept. } 11.029 \text{ ET}$$

$$\left.\begin{array}{l} \omega = 11°192 \\ \Omega = 50.353 \\ i = 7.145 \end{array}\right\} \ 1950.0$$

$$q = 1.28483 \text{ AU}$$

$$a = 3.50094 \text{ AU}$$

$$e = 0.63301$$

$$n° = 0.150462$$

$$P = 6.55 \text{ yr}$$

Thus the new comet proved to be the 73rd member of the Jupiter family of short-period comets. The fact that the comet had a close encounter with Jupiter in 1959 (Marsden, 1969) is very interesting. Vsekhsvyatskij (1969) points out that powerful cataclysms take place in the Jupiter system and that this probably led to the production of the new comet one and a half revolutions before its discovery.

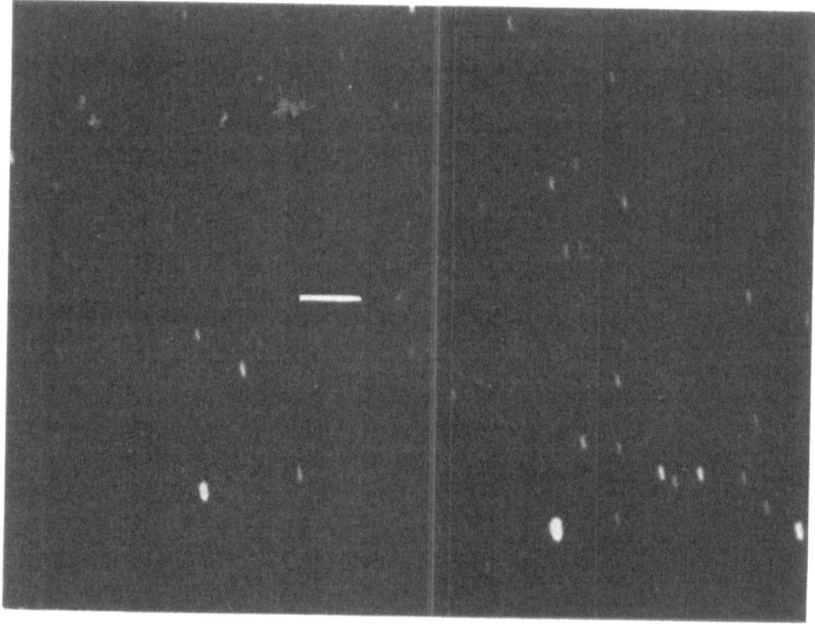

Fig. 1.   The plate on which comet 1969 IV was found (1969 Sept. 11.92010 UT).

From the beginning of November 1969 scientists from the Astronomy Department of Kiev University (Vsekhsvyatskij, Gerasimenko, Afanas'ev) systematically observed comet 1969 IV at southern Soviet observatories (Alma-Ata, Byurakan). The purpose of these observations was to obtain a continuous photometric series (which is very important in the case of periodic comets) for determining integral magnitudes of the comet, to study the comet's physical structure and to obtain accurate positions.

During this period visibility conditions made it possible to photograph the comet at low zenith distances (less than 25°), which is essential for the photography of faint, extended objects. Most of the material suitable for photometric work was obtained with the 20-cm and 50-cm Schmidt cameras at the Byurakan Astrophysical Observatory (Vsekhsvyatskij, Afanas'ev) and the 50-cm Maksutov telescope at the Alma-Ata

Astrophysical Institute (Afanas'ev, Gerasimenko, Churyumov). During September 1969–March 1970 more than 50 plates were taken on 34 nights. About 45 plates were

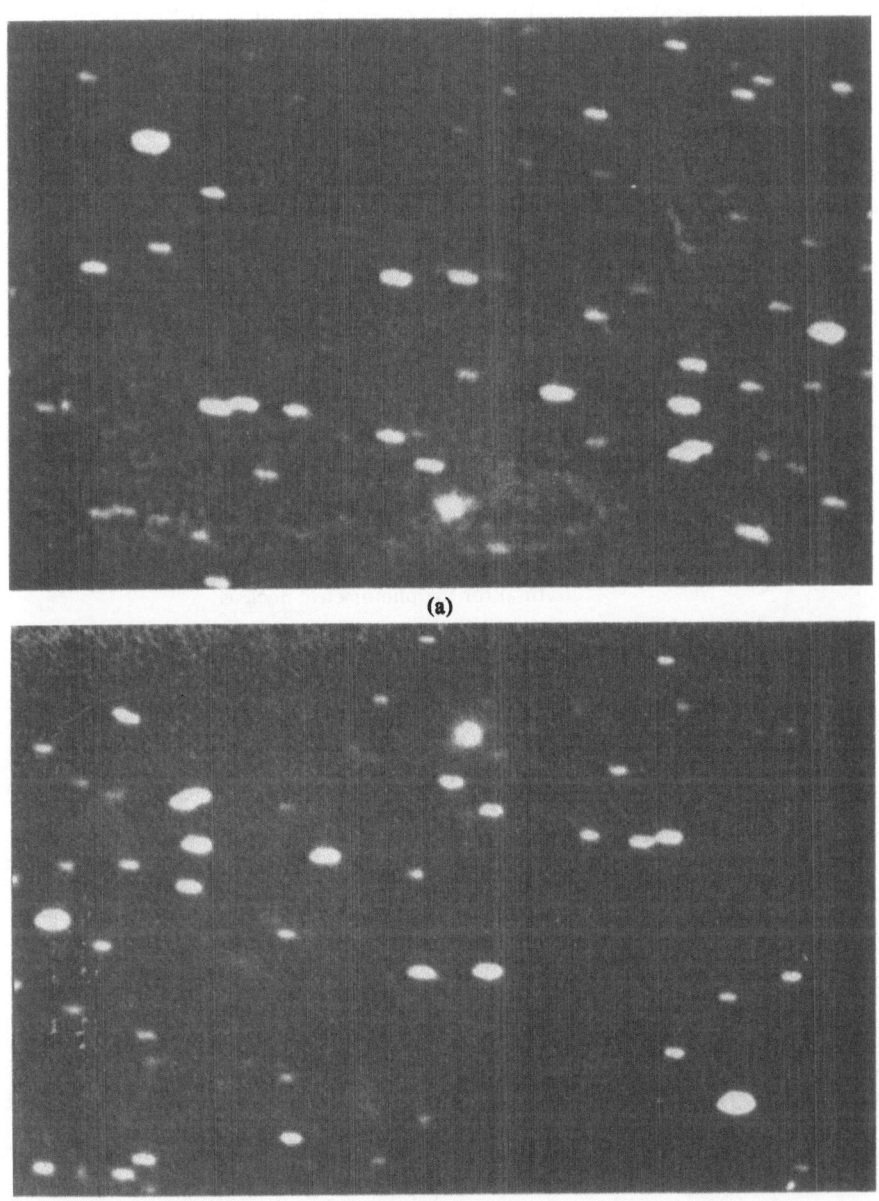

(a)

(b)

Fig. 2.   Comet 1969 IV (a) 1969 Sept. 21.92752; (b) 1969 Sept. 21.94795.

selected for determining nuclear magnitudes of the comet. Integral photographic magnitudes were obtained from nine plates taken at times of good atmospheric transparency and giving reliable extrafocal standards.

On most of the plates the comet had a strongly pronounced nucleus 0.06 mm (50-cm Schmidt) and 0.10 mm (50-cm Maksutov telescope) in diameter. For our microphotometric measurements we used a circular diaphragm of diameter 0.12 mm (corresponding to 1 to 2×10^4 km at the comet). For standards we obtained extrafocal exposures of star clusters (Coma Berenices, the Pleiades, NGC 2632, NGC 1647, NGC 1628, NGC 2264) and the North Polar Sequence.

From Kodak O-aO and ORWO ZU-2 plates (without filters) we reduced the B-magnitudes of the standard clusters to the international photographic system in the following way (Martynov, 1966):

$$m_{pg} = -0.18 + 1.09\,(B-V) + V.$$

In Table I, we give m_{pg} for the photometric nucleus of comet 1969 IV over the entire period of the observations and also the magnitudes h_{10} reduced to unit distances from the Earth and Sun. (On some nights when several plates were taken the average values of m_{pg} are listed.)

TABLE I

Apparent photographic magnitude (m_{pg}) and absolute magnitude (h_{10}) for the photometric nucleus

1969		m_{pg}	h_{10}	1969/70		m_{pg}	h_{10}
Sept.	10	14.67	12.83	Dec.	11	14.98	12.27
	12	15.05	13.23		14	13.95	11.21
	22	14.13	12.31		15	16.02	13.25
					16	16.09	13.31
Nov.	16	14.38	11.98		19	14.55	11.74
	17	15.32	12.92		20	14.34	12.52
	18	15.85	13.45				
	19	14.54	12.11	Feb.	6	15.60	12.08
	20	14.45	12.01		7	15.27	11.72
	22	14.23	11.76		8	15.96	12.37
	23	16.33	13.84		9	15.72	12.06
					13	16.23	12.46
Dec.	6	14.97	12.33				
	7	15.79	13.14	Mar.	2	16.17	12.08
	8	15.75	13.10		3	16.64	12.51
	9	15.43	12.77		4	16.44	12.29
	10	15.22	12.53		6	16.30	12.09

In Figure 3 we plot the absolute magnitude h_{10} of the photometric nucleus of the comet and also the sunspot area S against the time. The solar data are taken from *Solnechnye Dannye* and *Solar Phenomena*. We took into consideration the difference in heliographic longitude between the Earth and the comet. Since $i = 7°$, the difference in heliographic latitude is negligible.

As can be seen from Figure 3, the variations in nuclear magnitude correlate well with the sunspot area changes, except in February. The amplitude of the variations in the brightness of the photometric nucleus was about 2 magnitudes, indicating great activity in the comet. The strongest outburst took place during December 8–14, and the increase in sunspot area at the time was not substantial – although it is possible that there was a considerable output of energetic particles. Besides, it could also be that internal causes of comet activity, such as reactions between radicals produced by the radiation (Donn and Urey, 1957) or explosions of the frozen NH radical (Rice and Freamo, 1951), play a part.

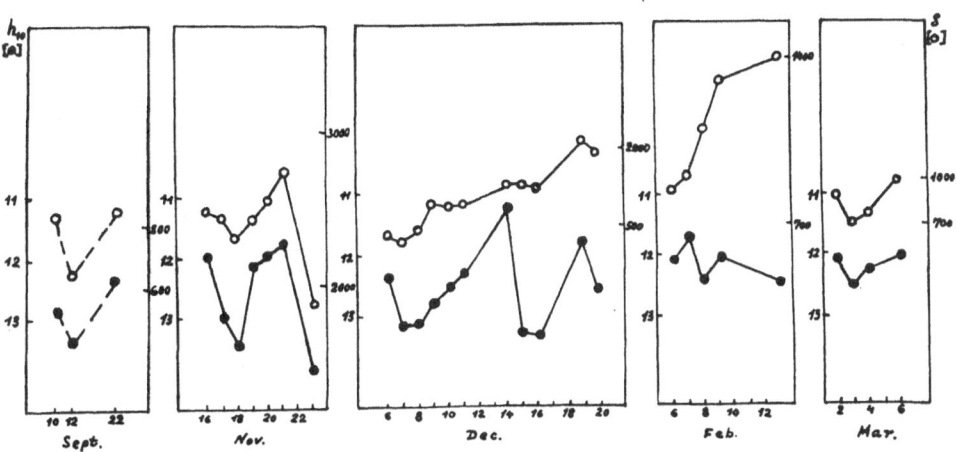

Fig. 3. Variations of the absolute photometric nuclear magnitude (h_{10}) and the sunspot area (S) with time.

The sensitive response of the comet to variations in the dynamical situation in interplanetary space resulted in the 'synchronous' reaction of the comet's brightness to fluctuations in the solar activity index (especially in November). In accordance with the idea of a quasi-stationary system of corpuscular streams generated in the solar corona (Vsekhsvyatskij *et al.*, 1965) it can be assumed that during the outbursts (e.g., Nov. 18–22, Dec. 8–14 and 16–19, Feb. 6–7 and 8–9, Mar. 3–6) the comet travelled through coronal rays (corpuscular streams) and experienced the increased density and energy of the particles. The action of these particles accelerated the sublimation of cometary ices and the processes of gas dissociation and ionization in the comet's atmosphere. The enhancement of different forms of solar activity (sunspot areas, flares, flocculi) leads to an increase in the total number of the fast particles in the solar corona, and these replenish the coronal rays.

As Vsekhsvyatskij has pointed out, the curves for the photometric nuclear brightness of comet 1969 IV show a tendency for a 27-day recurrence period in the comet outbursts: Sept. 22–Nov. 16 (55 days), Nov. 16–Dec. 14 (29 days), Dec. 14–Feb. 7 (54 days), Feb. 7–Mar. 5 (27 days). We also note a 27-day recurrence period in the minima: Nov. 18, Dec. 15, Feb. 8.

The absence of a difference in time between the nuclear magnitude variations and the changes in the solar activity index can be explained by assuming that the output of particles causing the nuclear outburst does not have to take place at the moment when the active region on the Sun is being generated. To cause an outburst when the comet is 1.5 AU from the Sun, particles travelling with velocities of about 1000 km s^{-1} must leave the deepest layers of the photosphere two days before the formation of the active region.

From nine plates Afanas'ev has determined the absolute integral magnitude and the photometric parameters H_y and n. The results of the measurements of m_{pg} and S_a are given in Table II. S_a is the area of the cometary image in square minutes of arc. H_y and n were found by least squares to be:

$$H_y = 11.91 \pm 0.54, \quad n = 4.0 \pm 0.8.$$

The large residuals are an indication of the high activity in the comet. Table III lists some photovisual (V) magnitudes, obtained from plates taken with the Maksutov telescope at Alma-Ata.

TABLE II

Apparent photographic integral magnitude (m_{pg}) and total area of the cometary image (S_a) in square minutes of arc

1969/70		m_{pg}	S_a
Dec.	7.06	13.95	13.1
	11.08	13.33	13.1
	15.04	13.88	13.6
	19.08	14.46	14.2
Feb.	5.92	15.81	10.7
	6.86	15.82	10.7
	7.94	16.08	10.7
	8.91	15.57	10.7
	12.92	16.07	10.7

Over the entire period of our observations the comet had a narrow, straight tail, probably of type I. Its length ranged from 1′ to 11′. On November 16 Burnasheva (Crimean Astrophysical Observatory) noted that there was a fan-like tail. Some characteristics related to the structure of the comet are listed in Table IV, d being the diameter of the coma, P the position angle of the tail, and s the length of the tail. The tail axis deviated from the prolonged radius vector by up to $\pm 20°$, which is evidence

TABLE III

Photovisual V magnitudes

1969		V
Nov.	16.90	12.72
	16.93	12.55
	16.95	12.74
	17.91	12.84
	17.94	12.69
Dec.	6.98	12.84
	7.04	13.30
	7.99	13.35
	8.11	13.23

TABLE IV

Coma diameter (d), and position angle (P)
and length (s) of the tail

1969/70		d	P	s
Nov.	16.90	1′52	298°24	2′4
	16.93	1.43	300.90	2.4
	16.95	0.99	297.05	6.3
	17.91	1.72	295.82	10.6
	17.93	1.28	297.24	3.5
	19.08	1.11	292.20	5.5
Dec.	6.09	0.55	306.54	4.5
	6.98	1.15	288.70	4.1
	7.99	1.46	301.01	9.4
	8.11	1.02	300.37	11.2
	10.05	0.60	301.02	1.7
	11.08	0.42	295.50	7.4
	16.08	1.48	303.89	3.4
	19.08	0.65	305.37	3.8
Feb.	6.86	1.87	294.92	6.0
	7.94	2.03	297.40	6.3
	8.91	2.18	300.66	7.2
	12.92	1.48	279.04	5.1
Mar.	1.83	0.24	293.62	4.2

for strong interaction between the comet tail plasma and electromagnetic interplanetary fields (Hoffmeister, 1943).

On November 16 a spectrogram of the comet was obtained by Lipovetskij with the 100-cm Schmidt camera (1°5 prism) at Byurakan. Since the dispersion was very low (1800 Å/mm at $H\gamma$), only the continuous spectrum of the head and tail was noted. The emissions of CN, C_2 and apparently C_3 are visible on a contact print (using the photographic Sabatier effect). Only a single emission becomes visible in the tail (perhaps CO^+). We plan to study the spectrum in more detail.

The substantial variations in the nuclear and integral head magnitudes of comet

1969 IV, the structural changes in the head, and the presence of the long tail are evidence for high activity and therefore the comet's comparative youth.

Acknowledgment

The authors wish to thank S. K. Vsekhsvyatskij for helpful comments.

References

Churyumov, K. I. and Gerasimenko, S. I.: 1970, *Priroda, Moskva*, No. 4.
Donn, B. and Urey, H. C.: 1957, *Mem. Soc. Roy. Sci. Liège* **18**, 124.
Everhart, E.: 1967, *Astron. J.* **72**, 716.
Hoffmeister, C.: 1943, *Z. Astrophys.* **22**, 265.
Marsden, B. G.: 1969, *IAU Circ.* Nos. 2181, 2187.
Martynov, D. Ya.: 1966, *Kurs Prakticheskoj Astrofiziki*, Moscow.
Rice, F. O. and Freamo, M.: 1951, *J. Am. Chem. Soc.* **73**, 5529.
Vsekhsvyatskij, S. K.: 1969, in *Problemy Sovremennoj Kosmogonii*, Moscow.
Vsekhsvyakskij, S. K., Nikolskÿ, G. M., Ivanchuk, V. I., Nesmyanovich, A. T., Ponomarev, E. A., Rubo, G. A. and Cherednichenko, V. I. 1965, *Solnechnaya Korona i Korpuskulyarnoe Izluchenie v Mezhplane tom Prostranstve*, Kiev.

7. ON ESTABLISHING AN INTERNATIONAL SERVICE FOR COMETARY OBSERVATIONS AND EPHEMERIDES

M. P. CANDY

Perth Observatory, Bickley, Western Australia

I agree entirely that there is a need for such a service. As the proposal has come from the Institute for Theoretical Astronomy, is it intended that the service should be organized from there? Consideration should be given to having subsidiary centres with identical information available on request – as for the double-star service.

The service should:

(1) Publish all ephemerides: there is not room for this on the *IAU Circulars*.

(2) Publish all positions: the *IAU Circulars* are not suitable for extended series of observations.

(3) Remind individual observers when a comet is accessible and in need of observation.

(4) Encourage observers to communicate positions promptly.

(5) Actively collect all ephemerides and observations from the current literature, in case they have not been communicated.

(6) Provide orbit investigators with positions on request, saving them the trouble of searching the literature.

The service could regularly publish references to all papers with relevance to comets (including physical and spectroscopic research) with brief abstracts, if possible.

Comet positions obtained at the Perth Observatory are all sent to the IAU Bureau, prior to publication in our *Communications*. We can easily send a copy to another centre at the same time. We would, in any case, always supply positions to investigators who asked for them.

8. GENERAL REMARKS ON ORBIT AND EPHEMERIS COMPUTATION

B. G. MARSDEN

Smithsonian Astrophysical Observatory, Cambridge, Mass., U.S.A.

Abstract. Present and future requirements concerning ephemeris prediction for comets (particularly the short-period comets) are discussed, with reference to both positions and magnitudes.

The recoveries of periodic comets at their various returns to perihelion are usually made with long-focus reflecting telescopes. Time on such instruments is at a premium, however, and since the usable fields are so small, it is very helpful if the positions predicted for the comets can be accurate to 5′ or better (Roemer, 1968). Of course, there are special problems with comets that have previously made only one appearance, or that have not been observed for several decades, or that experience large, irregular nongravitational anomalies in their motions – a particularly troublesome situation because it cannot be anticipated – but in general to obtain an accuracy of 5′ is not really asking for very much. It is perhaps not widely appreciated that, except in 1855, P/Encke was always recovered within 3′ of Encke's predicted positions. The same accuracy was achieved with Möller's prediction for the fourth apparition of P/Faye and with Leveau's prediction for the third apparition of P/d'Arrest, the latter in spite of that comet's having been missed at its previous return and being strongly perturbed by Jupiter. All this was accomplished a century ago and more! There are of course several examples in more recent times where similar accuracies have been achieved using logarithms or desk calculators.

But there are a good many other instances where observers have had to waste time over predictions that are considerably in error as the result of careless computations, even when high-speed automatic machines have been utilized. It is certainly desirable to try to allow for the nongravitational effects in some way. However, to obtain predictions accurate to 5′ it is certainly not essential to have a sophisticated computer program to determine these effects. If a comet has a revolution period of less than 20 yr and has been observed at its two most recent returns, it is merely necessary to verify that the perturbations – those by Jupiter and Saturn, at any rate – have been calculated correctly, both between the two recent apparitions and afterwards. One need not try to obtain a least-squares fit to the observations: just check that the residuals are reasonably small. The next predicted perihelion time will then in general be good to 0.1 day, for there are few such comets where the nongravitational effects over one revolution amount to more than this; and unless the comet comes close to the Earth, the predicted ephemeris should easily be within 5′ of the truth. It may even be possible to refine the prediction by applying to the perihelion time the nongravitational correction ΔP (Marsden, 1972, Table I).

If a comet has been missed at some of its recent returns, either since its latest apparition, or between its two latest apparitions, or both, one can still often carry out this procedure with success, applying to the perihelion time instead the correction

$$\Delta T = \tfrac{1}{2} n(n + N)\Delta P,$$

where n is the number of revolutions since the last apparition and N the number of revolutions between the last two apparitions. Of course, this procedure becomes less reliable when n or N is large or if there have been close approaches to Jupiter.

In any case, the orbits of the short-period comets are continually requiring attention, now as much as ever before. Because of the nongravitational effects, we shall probably never arrive at the situation we have with most of the minor planets, where ephemerides can be predicted half a century ahead and the objects easily recovered then. With new short-period comets still being discovered at the rate of seven per decade, with a larger proportion being successfully recovered at subsequent returns, and with several of the comets given up long ago as lost also being recovered, the amount of necessary computational work is rapidly increasing. There is a need for greater cooperation among those in situations where they are able to provide satisfactory predictions for periodic comets. Duplication of effort has in the past been stressed as important. Perhaps some duplication is still desirable; but if the predictions are satisfactory, it should no longer be necessary, and efforts can be put toward assuring that reliable computations are made on more comets. Fortunately, we do nowadays have the high-speed computers, and the coordinates of the perturbing planets can be easily read from magnetic tape or directly generated by the computers. Some progress has been made with putting cometary observations into machine-readable form: we have at the Smithsonian Astrophysical Observatory nearly 6000 positions of short-period comets (as well as 4000 positions of long-period comets) on punched cards; for 38 of the 62 comets of more than one appearance we have on cards observations made at all the apparitions since discovery, and there are only 14 cases where we do *not* have observations at every apparition since 1925. All the observations have been precessed to equinox 1950.0, the file is kept up to date and is in fact complete for all comets that have appeared since the beginning of 1965 – except that many observations known to be erroneous have been removed.

As for the long-period, single-apparition comets the principal computational effort is in quickly obtaining orbits to yield reliable ephemerides. This phase of the work is nowadays conducted at least as satisfactorily as before. Since new comets are usually quite bright most of the astrometric observations are made with small to moderate wide-field instruments, and it is encouraging to note that there has of late been a considerable increase in the number of such observations made from the Southern Hemisphere. If a long-period comet remains under observation for at least four or five months, it is customary for a 'definitive' orbit to be calculated, in which all available observations are discussed and planetary perturbations taken into account. An important reason for doing this has been to verify whether comets originate in the solar system. The calculation of the 'original', and also the 'future', barycentric orbits for past comets is now simplified by means of the tabulation by Everhart and Raghavan (1970). But this work is not so important as previously, and because of nongravitational effects one certainly cannot regard the few original orbits found to be hyperbolic to several times their mean errors as evidence for interstellar origin: a strong reason for believing that we can indeed detect nongravitational effects in the motions of comets 1957 III and 1960 II (Marsden, 1972) is that allowance for them quite definitely makes the original orbits elliptical, and this was not the case when nongravitational terms were omitted. In future, the emphasis on the orbits of long-period comets should perhaps be to ascertain whether nongravitational effects are more generally detectable.

One aspect of ephemeris calculation that continues to be unsatisfactory is magnitude prediction. For faint comets, particularly short-period comets whose recoveries are expected, ephemerides often give magnitudes that are far too bright. One must bear in mind again the fact that these comets are most likely to be observed photographically with long-focus reflectors, and further, that the observers would like to limit the exposure times so that they obtain cometary images that are suitable for astrometric measurement; ideally, the magnitude should be that of the nucleus. On the other hand, magnitudes so calculated are very misleading for comets that come bright enough to be seen in small telescopes; such magnitudes are discouraging, not only for amateurs, but also for professionals who wish to make physical observations of the comets.

Since there are observable at any given time a lot of very faint comets and not more than two or three bright comets it seems preferable to try basically to predict the 'nuclear' magnitudes. In a few cases an observer really comes close to observing a true nucleus, and one can use a magnitude formula that varies according to the inverse square of heliocentric distance (perhaps with a phase effect); more usually, however, what the observer reports is more consistent with an inverse fourth-power of variation with heliocentric distance, and occasionally even an inverse sixth-power law may be required. For bright comets one should try to predict the 'total' magnitude as well; indeed, for newly discovered comets nuclear estimates are rarely available. The total magnitude will almost certainly have a stronger dependence on heliocentric distance, and considering the nature of the comet one might reasonably choose an inverse fourth-, sixth- or eighth-power law. If both total and nuclear magnitudes are being predicted, there seems little point in listing sets of numbers differing only by a constant; one should use two different laws, and an individual observer can then select the one that suits him best, adjusting the constant as he desires. Whatever is done in practice, however, it should be clearly indicated whether 'total' or 'nuclear' magnitude is meant, and a convenient notation would be one based on the telegraphic code, using m_1 for total magnitudes and m_2 for nuclear magnitudes.

References

Everhart, E. and Raghavan, N.: 1970, *Astron. J.* **75**, 258.
Marsden, B. G.: 1972, this Symposium, p. 135.
Roemer, E.: 1968, *Trans. IAU* **13B**, 115.

PART II

GENERAL METHODS OF ORBIT THEORY

A. ANALYTICAL METHODS

9. A SERIES-SOLUTION METHOD FOR COMETARY ORBITS

P. E. NACOZY

*Dept. of Aerospace Engineering and Engineering Mechanics,
University of Texas, Austin, Tex., U.S.A.*

Abstract. A series-solution method for highly-eccentric perturbed orbits using a modified form of Hansen's method of partial anomalies is presented. Series in Chebyshev polynomials in the eccentric anomaly of a comet and the mean anomaly at an epoch of a planet provide a theory valid to first order with respect to the masses. The first-order solution becomes a reference solution about which higher-order perturbations are obtained by the method of successive approximations. The first-order solutions are valid approximations for long durations of time, whereas the higher orders are valid only over the interval of time that is selected for the Chebyshev expansions. The method is somewhat similar to Encke's method of special perturbations except that for each successive interval of time perturbations about a first-order solution are calculated instead of perturbations about a conic solution.

1. Introduction

The application of conventional methods of general perturbations to the highly eccentric orbits of comets and certain asteroids presents several difficulties. Two of the most severe difficulties are the slowness of convergence of the series-solutions, if they converge at all, and the inability of low-order solutions to be valid over long intervals of time.

The convergence of the developments may be strengthened by expansions in the eccentric anomaly of the comet and by segmentation of the reference orbit of the comet following the ideas of Hansen (1856) and Gyldén (1870). But, for accuracy over longer intervals of time, higher-order solutions must be obtained.

The present paper discusses a method of general perturbations that utilizes expansions in Chebyshev polynomials in the eccentric anomaly and segmentation of the orbit to strengthen convergence. The method incorporates the Picard method of successive approximations to attain a high-order solution.

The use of Chebyshev polynomials and Picard's method for the solution of nonlinear differential equations in initial and boundary value problems has been well studied (see, for example, Clenshaw and Norton, 1963; Fox and Parker, 1968). The technique has been applied to the planetary system, with notable success, by Carpenter (1966) and Broucke (1969).

This paper presents a method that yields series-solutions of cometary orbits in double Chebyshev series and utilizes a reference solution comprising both the zeroth-order (conic motion) and the first-order solution (with respect to the disturbing masses). This reference solution approximates more closely the true solution than does a conic reference solution. The method allows the perturbing planets to have their actual nonelliptical motion. The method is semianalytical in that the series have numerical coefficients.

Chebotarev et al. (eds.), The Motion, Evolution of Orbits, and Origin of Comets, 43–51. All Rights Reserved.
Copyright © 1972 by the IAU.

2. Expansion in Chebyshev Polynomials

The method of general perturbations presented here may be applied to any formulation of the differential equations of motion of a comet. For this discussion, we will define the perturbations by second-order differential equations in rectangular coordinates. For most other equations and coordinates, all of the following will apply with minor modifications. Let

$$\mathbf{r} = \mathbf{r}_0 + \delta\mathbf{r},$$

where $\mathbf{r}$ is the actual position vector to the comet, $\mathbf{r}_0$ the position vector to the reference conic, and $\delta\mathbf{r}$ the perturbation of the comet. Let $\mathbf{r}'_j$ be the position vector to the jth perturbing planet and allow p perturbing planets. The heliocentric perturbation equations may be simply written as

$$\frac{d^2}{dt^2}\,\delta\mathbf{r} = -\mu\left(\frac{\mathbf{r}}{|\mathbf{r}|^3} - \frac{\mathbf{r}_0}{|\mathbf{r}_0|^3}\right) + \mu\sum_{j=1}^{p} m'_j\left(\frac{\mathbf{r}'_j - \mathbf{r}}{|\mathbf{r}'_j - \mathbf{r}|^3} - \frac{\mathbf{r}'_j}{|\mathbf{r}'_j|^3}\right) \tag{1}$$

$$= \boldsymbol{\varphi}(\mathbf{r}_0, \mathbf{r}; \mathbf{r}'_1, \mathbf{r}'_2, \ldots, \mathbf{r}'_p).$$

In Equation (1), μ is the gravitational constant and m'_j the mass of the jth perturbing planet, and $\boldsymbol{\varphi}$ is a vector function. Let us confine the following discussion to the presence of only one perturbing planet. The generalization to p perturbing planets is straightforward.

Let us transform the independent variable of Equation (1) to the eccentric anomaly u of the reference conic using the relation

$$dt = (r_0/n_0 a_0)\,du, \tag{2}$$

where $r_0 = |\mathbf{r}_0|$, and n_0 and a_0 are the mean motion and semimajor axis of the reference conic, respectively. The function $\boldsymbol{\varphi}$ of Equation (1) is also transformed through Kepler's equation to a function of the eccentric anomaly u of the comet, as will be shown later. Let us proceed formally to integrate Equation (1) by the Picard method of successive approximations. Let the subscript n denote the nth iterate. Let $\{\mathbf{r}\}_{n+1}$ denote the position vector determined by the nth iterate. Define

$$\{\boldsymbol{\varphi}\}_n \equiv \boldsymbol{\varphi}(\mathbf{r}_0, \{\mathbf{r}\}_n; \mathbf{r}'), \tag{3a}$$

and

$$\boldsymbol{\varphi}_n \equiv r_0 \times \{\boldsymbol{\varphi}\}_n = \boldsymbol{\varphi}_n(\mathbf{r}_0, \{\mathbf{r}\}_n; \mathbf{r}'). \tag{3b}$$

Then, using Equations (1), (2), (3a), and (3b), the position vector $\{\mathbf{r}\}_{n+1}$ is given by

$$\{\delta\mathbf{r}\}_n = \frac{1}{n_0^2 a_0^2}\int^u r_0\,du^* \int^{u^*} \boldsymbol{\varphi}_n\,du^* + \mathbf{a}_n t + \mathbf{b}_n, \tag{3c}$$

$$\{\mathbf{r}\}_{n+1} = \mathbf{r}_0 + \{\delta\mathbf{r}\}_n; \qquad n = 0, 1, 2, \cdots. \tag{3d}$$

If the function $\boldsymbol{\varphi}$ is continuous and satisfies the Lipschitz condition in a domain of u, or, from Equation (1), if neither orbital intersections nor collisions occur between comet and planet in an interval of time, then the process defined by Equations (3) is convergent to the true solution of Equation (1) in this interval (see Ince, 1956, pp. 62–92). In Equation (3c), the symbols $\mathbf{a}_n$ and $\mathbf{b}_n$ are vector constants of integration determined during the nth iterate. They will also converge to the true constants of integration satisfying the desired set of boundary conditions.

If we restrict our discussion for the moment to the zeroth iterate of the process (3), we have, since $\{\mathbf{r}\}_0 = \mathbf{r}_0$,

$$\boldsymbol{\varphi}_0 = \boldsymbol{\varphi}_0(\mathbf{r}_0; \mathbf{r}'). \tag{4}$$

Equations (3) and (4) define a quantity $\{\mathbf{r}\}_1$, which is the first-order solution with respect to the masses. With the assumption, which will be removed later, that the perturbing planet moves on a conic, the functions $\mathbf{r}_0$ and $\mathbf{r}'$ are periodic in the eccentric anomaly of the comet u, and the mean anomaly of the planet l', respectively, both in 2π. Hence, the function $\boldsymbol{\varphi}_0$, defined by Equation (4), may be represented by a Fourier series with arguments u and l'. The series may be written as

$$\boldsymbol{\varphi}_0 = \sum_{i=0}^{\infty} \left[\sum_{j=0}^{\infty} \boldsymbol{\alpha}_{ij} \, \frac{\sin}{\cos} (ju) \right] \frac{\sin}{\cos} (il'). \tag{5}$$

The notation $\dfrac{\sin}{\cos}$ denotes a sine and cosine Fourier series. The coefficients, here and in the following series, are vector quantities. Let

$$l' = \frac{n'}{n_0} (u - e_0 \sin u - c) + c', \tag{6}$$

where n' is the mean motion of the planet, and c and c' define the mean anomalies of the comet and planet, respectively, at an epoch. If we choose the epoch at a perihelion passage of the comet, then, $c = 0°$ and c' is equal to the mean anomaly of the planet at this epoch.

Substituting Equation (6) into the series of Equation (5), it can be shown with the use of Bessel functions that Equation (5) may be transformed into

$$\boldsymbol{\varphi}_0 = \sum_{i=0}^{\infty} \left[\sum_{j=-\infty}^{\infty} \boldsymbol{\beta}_{ij} \, \frac{\sin}{\cos} \left(i\frac{n'}{n_0} + j \right) u \right] \frac{\sin}{\cos} ic'. \tag{7}$$

Equations (3c) and (7) yield the first-order perturbation $\{\delta\mathbf{r}\}_0$. The position vector $\{\mathbf{r}\}_1$ is given by Equation (3d), accurate to first order. Series with the eccentric anomaly of the perturbed body as one argument, as in the series of Equation (7), have often been used in methods of general perturbations to obtain first-order perturbations (see Hansen, 1857; Herget, 1948).

If n'/n_0 is irrational, $\boldsymbol{\varphi}_0$, defined by Equation (7), is not periodic in u. If n'/n_0 is rational $(=p/q)$, $\boldsymbol{\varphi}_0$ is periodic in u with period $2\pi q$. If n'/n_0 is irrational or if q is large, the coefficients $\boldsymbol{\beta}_{ij}$ or Equation (7) cannot be determined in an efficient manner by harmonic analysis, valid for all u. Rather, one must determine first the $\boldsymbol{\alpha}_{ij}$ of Equation (5) and then determine the $\boldsymbol{\beta}_{ij}$ by use of the Bessel functions.

Another difficulty is encountered when we proceed to determine the higher orders by the process of Equations (3). The first-order position vector $\{r\}_1$, again assuming no fundamental commensurabilities, is not periodic in u, due to the appearance of secular perturbations. Hence, the function φ_1 will not be periodic in u, and the co-efficients of the Fourier expansion of φ_1 in the arguments u and l' cannot be determined efficiently by harmonic analysis, valid for all u. This difficulty renders the Picard method of Equation (3) not applicable without some modification.

Both of the above difficulties may be removed by requiring validity of the solution only over a finite interval of the independent variable. With this relaxation, direct harmonic analysis of all orders of the solution is efficient and the method of Picard iteration may be applied. Also, if the perturbing planet cannot be approximated by conic motion, the same may be said of the variable c' in the expansion of φ_1.

Any finite interval may be chosen. Hansen, in his memoir on general perturbations for cometary orbits, considering only first-order perturbations, chooses intervals of the independent variable less than one revolution of the comet (Hansen, 1856; Nacozy, 1969). Carpenter (1966), for high-order perturbations of the outer planets, chooses an interval of 200 yr. The criterion for the choice of the size of the interval depends mainly on the desired rapidity of convergence of the series.

In the present discussion, we will allow any size interval to be chosen in both variables u and c', either less than, equal to, or greater than 2π radians. The intervals in u and c' will be denoted by $[u_1, u_2]$ and $[c_1', c_2']$, respectively. It is assumed that φ_n is not naturally periodic in either $[u_1, u_2]$ or $[c_1', c_2']$. The Fourier series expansions of φ_n in the intervals introduces periods $[u_1, u_2]$ and $[c_1', c_2']$ in u and c'. Hence, the expansion is not a valid representation for φ_n outside the chosen intervals $[u_1, u_2]$ and $[c_1', c_2']$. Also, the induced periodicities require the Fourier expansions to represent discontinuities at the points u_1, u_2, c_1' and c_2'. This is undesirable since a discontinuous function has a slowly convergent Fourier series representation. Also, points of discontinuity are inaccurately represented by a truncated Fourier series. An accurate representation of the end points of the solution is extremely important in the continuation of the solution to subsequent (or previous) intervals.

To eliminate the discontinuities, Hansen introduced transformations to new independent variables, which he called the partial anomalies. Identically motivated, we will introduce similar transformations. To remove the discontinuities of the function φ at the end points of the interval $[u_1, u_2]$ we introduce the transformation

$$\frac{2(u - u_1)}{u_2 - u_1} = 1 + \cos \bar{x}, \tag{8}$$

$$-\pi \leq \bar{x} \leq 0, \qquad u_1 \leq u \leq u_2.$$

The function φ_n, $n = 0, 1, 2, \ldots$, no matter what the interval in u, becomes a continuous and periodic function of $\bar{x}$. Also, due to the transformation (8), φ_n becomes an even function of $\bar{x}$. The Fourier expansion of φ_n in $\bar{x}$, valid only in the interval $[u_1, u_2]$, is

$$\varphi_n = \sum_{i=0}^{\infty} \left(\sum_{j=0}^{\infty} \gamma_{ij} \cos j\bar{x} \right) \begin{matrix} \sin \\ \cos \end{matrix} \, ic'.$$

But $\cos j\bar{x} = T_j(x)$, where $x = \cos \bar{x}$ and $T_j(x)$ is the Chebyshev polynomial of the jth degree in x. The transformation to x is given by

$$u = \tfrac{1}{2}(u_2 - u_1)(1 + x) + u_1, \tag{9a}$$

$$u_1 \leq u \leq u_2, \qquad -1 \leq x \leq 1.$$

The variable c' may be transformed similarly. We introduce the transformation

$$\frac{2(c' - c_1')}{c_2' - c_1'} = 1 + \cos \bar{y},$$

$$-\pi \leq \bar{y} \leq 0, \qquad c_1' \leq c' \leq c_2'.$$

The expansion of the function φ_n, valid in the intervals $[u_1, u_2]$ and $[c_1', c_2']$, becomes

$$\varphi_n = \sum_{i=0}^{\infty} \sum_{j=0}^{\infty} \delta_{ij} T_j(x) \cos i\bar{y}.$$

But $\cos i\bar{y} = T_i(y)$, where $y = \cos \bar{y}$ and $T_i(y)$ is the Chebyshev polynomial of the ith degree in y. The transformation to y is given by

$$c' = \tfrac{1}{2}(c_2' - c_1')(1 + y) + c_1' \tag{9b}$$

$$c_1' \leq c' \leq c_2', \qquad -1 \leq y \leq +1.$$

The expansion of φ_n becomes

$$\varphi_n = \sum_{i=0}^{\infty} \sum_{j=0}^{\infty} \delta_{ij} T_j(x) T_i(y), \tag{10}$$

$$-1 \leq x \leq 1, \qquad -1 \leq y \leq 1.$$

The interpretation of the variable y in Equation (10) is as follows. The variable y is related to the variable c' through Equation (9b) and c' is defined by Equation (6). In Equation (6), we have chosen c' to be the value of l' at the instant when $u = 0$. That is, c' is the mean anomaly of the planet at the time of perihelion passage of the comet. With this value for c', Equation (6) gives the mean anomaly of the planet l' as the comet traverses its reference orbit. Both comet and planet move with mean motions n_0 and n', respectively. With the interval $u_2 - u_1 = 2\pi$, introducing Equation (9a) into Equation (6) yields

$$l' = (n'/n_0)[\pi(1 + x) + e_0 \sin \pi x] + c'. \tag{11}$$

Equation (11) gives l' only in the interval $c' \leq l' \leq c' + 2\pi(n'/n_0)$, for the interval $u_2 - u_1 = 2\pi$ and $-1 \leq x \leq 1$. Hence, c' is the value of l' at the beginning of the chosen interval. For previous or subsequent traversals through the interval, c' must be given the value of l' at the instant when the comet begins traversing the interval. The corresponding value of y is given by Equation (9b). During each interval c' (or y) remains constant but changes from one interval to the next. If the eccentric anomaly of the comet (and hence the mean anomaly) moves through 2π radians between successive entries of the interval, then from one traversal to the next,

$$c' = c' + 2\pi(n'/n_0).$$

Since the variables x and y are linearly related to the variables u and c' by Equations (9), the series (10), truncated to a finite number of terms, is merely a representation of the function φ_n by a polynomial in u and c', valid in the intervals $u_2 - u_1$ and $c_2' - c_1'$.

3. The First-Order Solution

Consider the intervals $0 \le u \le 2\pi$ and $0 \le c' \le 2\pi$. Let $u = 0$ correspond to the epoch and let c' correspond to the value l' at epoch. We are choosing the epoch to be a perihelion passage of the comet and the interval in u to be a revolution of the comet from perihelion to perihelion on the reference conic. The mean anomaly of the perturbing planet l' is a function of both u and c', or x and y through Equations (9). The function φ_0 may be expanded, by double harmonic analysis, into a truncated, double Chebyshev series in the form of Equation (10) with numerical coefficients. We then proceed with the zeroth iterate of the process of Equations (3). We have

$$\varphi_0 = \sum_{j=0}^{m_1} \sum_{i=0}^{m_2} \delta_{ij} T_i(x) T_j(y), \qquad du = \pi \, dx,$$

$$\frac{d}{dt}\{\delta\mathbf{r}\}_0 = \frac{1}{n_0 a_0} \int^u \varphi_0 \, du = \sum_{j=0}^{m_1} \left(T_j(y) \sum_{i=1}^{m_2} \varepsilon_{ij} T_i(x) \right) + \mathbf{a}_0, \qquad (12)$$

where

$$\varepsilon_{ij} = \frac{\pi}{n_0 a_0} \times \frac{1}{2i}(\delta_{i-1,j} - \delta_{i+1,j}),$$

and $\mathbf{a}_0$ is a constant of integration. The radius vector of the reference conic is

$$r_0 = |\mathbf{r}_0| = a_0(1 - e_0 \cos u)$$

and may be expanded in a Chebyshev series in x. Since the expansion for r_0 converges faster than the expansion for φ_0, we have $m_3 < m_2$, and

$$r_0 = \sum_{\ell=0}^{m_3} d_\ell T_\ell(x).$$

From Equation (2), the integral of Equation (12) becomes

$$\{\delta\mathbf{r}\}_0 = \frac{\pi}{n_0 a_0} \sum_{j=0}^{m_1} T_j(y) \int^u \left(\sum_{i=1}^{m_2} \varepsilon_{ij} T_i(x) \right)$$

$$\times \left(\sum_{\ell=0}^{m_3} d_\ell T_\ell(x) \right) dx + \frac{1}{n_0} \mathbf{a}_0(\pi(x+1) + e_0 \sin \pi x) + \mathbf{b}_0. \qquad (13)$$

The coefficient of the constant of integration $\mathbf{a}_0$ in Equation (13) is the time, t, in terms of the variable x. The integrand of Equation (13), upon multiplication, becomes a series of Chebyshev polynomials of degree $m_2 + m_3 = m_4$. Let the symbol $\boldsymbol{\theta}_{ij}$ denote the coefficients of the product series. The coefficients $\boldsymbol{\theta}_{ij}$ may be expressed in terms of the coefficients $\boldsymbol{\varepsilon}_{ij}$ and d_ℓ by straightforward relations. The integral of Equation (13) then becomes

$$\{\delta \mathbf{r}\}_0 = \sum_{j=0}^{m_1} \sum_{i=1}^{m_4} \lambda_{ij} T_i(x) T_j(y) + \frac{1}{n_0} \mathbf{a}_0(\pi(x+1) + e_0 \sin \pi x) + \mathbf{b}_0, \qquad (14)$$

where

$$\lambda_{ij} = \frac{\pi}{n_0 a_0} \times \frac{1}{2i}(\boldsymbol{\theta}_{i-1,j} - \boldsymbol{\theta}_{i+1,j}).$$

Equation (14) yields the first-order solution for the perturbation $\{\delta \mathbf{r}\}_0$. At the epoch, $u=0$ ($x=-1$), c' (or y) is equal to the value of l' when $u=0$. Denote this value by y^0. The constants of integration, $\mathbf{a}_0$ and $\mathbf{b}_0$, are determined so that

$$\{\delta \mathbf{r}\}_0 = 0, \qquad \frac{\mathrm{d}}{\mathrm{d}t}\{\delta \mathbf{r}\}_0 = 0, \qquad (15a)$$

at $u=0$ ($x=-1$). Proceeding to the subsequent (or previous) revolution, the variable c' and the constants of integration of the first-order solution of Equation (14) must be reevaluated. For the second evaluation, c' (or y) is equal to the value of l' at the instant of time of the subsequent (or previous) perihelion passage of the comet on the reference conic, or at $u=2\pi$ (or $u=-2\pi$). That is, $c'=c'+2\pi(n'/n_0)$ at $u=2\pi$. Denote this new value of c' by y^1. The constants $\mathbf{a}_0$ and $\mathbf{b}_0$ are determined so that

$$\{\delta \mathbf{r}\}_0 \bigg|_{\substack{x=-1 \\ y=y^1}} = \{\delta \mathbf{r}\}_0 \bigg|_{\substack{x=+1 \\ y=y^0}} \qquad (15b)$$

and

$$\frac{\mathrm{d}}{\mathrm{d}t}\{\delta \mathbf{r}\}_0 \bigg|_{\substack{x=-1 \\ y=y^1}} = \frac{\mathrm{d}}{\mathrm{d}t}\{\delta \mathbf{r}\}_0 \bigg|_{\substack{x=+1 \\ y=y^0}}. \qquad (15c)$$

And for all previous or subsequent revolutions the determination of y, $\mathbf{a}_0$, and $\mathbf{b}_0$ proceeds in the same manner. We now possess a reference solution, valid to first order for all time.

4. The Higher-Order Solution

We now proceed to obtain a higher-order solution by the Picard method of successive approximations defined by Equations (3). For each revolution or interval in u, the successive approximations begin with the first-order solution as the reference solution, not with the zeroth-order solution or reference conic. Using the first-order solution given by Equation (14), with the constants of integration determined by Equations (15) for the revolution for which we desire the higher-order solution, the

Picard iteration yields the higher-orders and more accurate constants of integration by continual application of Equations (15). Since the accuracy of the constants depends on the accuracy of the solution of the previous or subsequent revolution we must begin at the epoch and proceed backward or forward in time.

The value of c' (or y) during the Picard iteration remains fixed for each revolution or interval in u since it is defined here as the mean anomaly of the planet at perihelion passage of the comet *on the reference ellipse.*

For iterates past the first, the motion of the perturbing planets may be allowed to deviate from the elliptic motion assumed in the first-order solution. Since the special values for the harmonic analysis of φ_n, $n > 0$, depend on *numerical values* of r_j' for each iteration, values of r_j' not derived from elliptic motion may be used for $n > 0$. This relaxation is admitted due to the fact that validity of the higher-order solutions exists only for a definite interval of time. Also, for the higher iterates, the variable y may be given its actual numerical value appropriate to the interval under consideration. This allows the expansions to collapse from double series in x and y to single series in x for the higher orders.

For the first-order solution to be valid for all intervals, the double series and conic motion for the planets must be retained. It is, of course, preferable to use mean conics for both the comet and the planets.

It may be noted that for the higher-order iterates, the interval may include more than one interval of the first-order solution. Including more intervals would necessitate more iterates to the desired convergence, but the added efficiency of the calculations may prove advantageous. Also, it is known that in the Picard iteration using Chebyshev polynomials one may begin in the first iterates with a fewer number of terms in the series and increase the number of terms in the series for the higher iterates as convergence is approached (see, for example, Norton, 1964). Hence, the first-order solution may contain far fewer terms than the higher-order solutions. Another comment, noted by Broucke (1969), is that the solution may be modified or corrected in only one or two additional iterations without having to do the complete numerical integration over again.

If the comet encounters actual close approaches or if the converged solution begins to deviate substantially from the conic and first-order reference solution, one must then perform a rectification of the reference solution. The conic and first-order solution must be recalculated. The process is analogous to rectification in Encke's method of special perturbations. But rectification will not be required as often as it would be for Encke's method since the present method incorporates a more accurate reference solution. Also, the perturbations about this reference solution will generally be much smaller than perturbations about conic motion in the same interval of time.

Acknowledgments

I wish to thank P. Herget of the University of Cincinnati, and V. Szebehely, G. Giacaglia, and T. Feagin of the University of Texas at Austin, for several suggestions and for many helpful discussions.

The financial support of the National Science Foundation, Grant No. GP-17369, and the Office of Naval Research, Grant No. N00014-67-A-0126-0007, is gratefully acknowledged.

References

Broucke, R.: 1969, *Celes. Mech.* **1**, 110.
Carpenter, L.: 1966, NASA TN D-3168.
Clenshaw, D. W. and Norton, H. J.: 1963, *Comput. J.* **6**, 88.
Fox, L. and Parker, I. B.: 1968, *Chebyshev Polynomials in Numerical Analysis*, Oxford University Press, London.
Gyldén, H.: 1870, *Bull. Acad. Imp. Sci. St. Petersb.* **14**, 195.
Hansen, P.: 1856, *Suppl. Compt. Rend. Acad. Sci. Paris* **1**, 121.
Hansen, P.: 1857, *Abh. Königl. Sächs. Ges. Wiss. Math.-Phys. Kl.* **3**, 41.
Herget, P.: 1948, *The Computation of Orbits*, Cincinnati.
Ince, E. L.: 1956, *Ordinary Differential Equations*, Dover Publ., New York.
Nacozy, P.: 1969, *Astron. J.* **74**, 544.
Norton, H. J.: 1964, *Comput. J.* **7**, 76.

Discussion

G. N. Duboshin: Did you perform any comparison with numerical integration?

P. E. Nacozy: I have not compared the method with solutions obtained by numerical integration. But it appears that the method will provide solutions with more uniform and definitive error bounds due to the properties of Chebyshev polynomial series. It also appears that some advantages may result by using a reference solution that includes the first-order perturbations, as well as the conic motion.

10. ON THE APPLICATION OF HANSEN'S METHOD OF PARTIAL ANOMALIES TO THE CALCULATION OF PERTURBATIONS IN COMETARY MOTIONS

V. I. SKRIPNICHENKO

Institute for Theoretical Astronomy, Leningrad, U.S.S.R.

Abstract. The main points of the method of partial anomalies are described. The method makes it possible to obtain analytical solutions for comets moving in orbits with high eccentricities. The method may be applied, in particular, to systems in which the orbits of the disturbed and disturbing bodies have points equidistant from the Sun. The method is applied to the calculation of the general perturbations by Saturn on P/Tuttle.

In order to increase the convergence of the series employed to express the general perturbations of orbits with high eccentricities Hansen (1856) suggested a method that caused the variations of the functions expanded to decrease. This involved division of the orbit of the disturbed body into segments, and in each of the segments the classical variables (the true, eccentric or mean anomalies) were substituted by new ones – partial anomalies – such that as they varied from 0 to 2π any coordinate of the disturbed body would vary only within the range corresponding to the boundaries of the segment.

Each segment is thus characterized by its own rapidly convergent series for the coordinates of the disturbed body developed in a single argument – the partial anomaly related to the segment. The coordinates of the disturbing body are expressed in two-argument series; one of the arguments, a partial anomaly, varies from segment to segment; while the other argument C', common to all the segments, has the value of the mean anomaly of the disturbing body at the time of passage of the disturbed body through its perihelion. The magnitude of the variation of this argument from revolution to revolution is given by

$$\Delta C' = 2\pi(n'/n),$$

where n and n' are the mean motions of the disturbed and disturbing bodies, respectively.

Two-argument series for the calculation of general perturbations may be obtained on the basis of the above series. If the orbits do not intersect in the segment of the disturbed body orbit, the convergence with respect to the partial anomaly of all the series used for the calculation of perturbations may be arbitrarily accelerated by reducing the segment in size, i.e., by dividing the orbit of the disturbed body into more segments. However, when a segment contains points close to the intersection of the orbits, the convergence of the expansions for the negative powers of the mutual distance with respect to the second argument remains slow even with an arbitrary reduction in the segment size. This requires operating with two-argument series containing many terms with respect to one of the variables, and it prevented Hansen from completing his calculations of the perturbations by the Earth on P/Encke.

Those who followed Hansen – Gyldén, Asten, Backlund, and Wittram – tried to by-pass this obstacle and to increase the convergence of the series by using as the second argument, not C', but some other variable introduced by means of elliptic functions. However, the increased convergence with respect to the second argument brought about some difficulty in the determination of the constants of integration. Owing to the tremendous amount of computation and great difficulties involved in the use of these complicated series for the calculation of perturbations – should such series have been obtained – the calculations, started by Asten, Backlund, and Wittram, of the general perturbations on P/Encke by Jupiter were never completed. Subsequent attempts at using the method were also abandoned.

The first to make full use of the method was Nacozy (1969), who determined the Earth's perturbations on P/Encke by means of a computer. At about the same time the method of partial anomalies had been fully mastered by the author, and half of the necessary calculations had been made on the BESM-4 computer (Skripnichenko, 1970). The difference in approach lies in the fact that Nacozy determined the perturbations by means of harmonic analysis, applying it to particular values of the perturbations as soon as they had been calculated, whereas the author carried out all the calculations analytically by manipulation of Fourier series with numerical coefficients. The programme for applying the method of partial anomalies on the BESM-4 was developed, and using the programme the series were calculated for computing the perturbations by Saturn on the motion of P/Tuttle.

The procedure for performing the calculation was as follows. The initial sets of elements, corresponding to the moment 1885 Oct. 6.5, were as shown in Table I.

TABLE I

Initial elements

P/Tuttle		Saturn	
$M =$	1°73396	$M' =$	359°49935
$\omega =$	206.77678	$\omega' =$	339.89543
$\Omega =$	270.54314	$\Omega =$	113.41486
$i =$	54.33026	$i' =$	2.48610
$a =$	5.7422705 AU	$a' =$	9.5692934 AU
$e =$	0.8215463	$e' =$	0.0568010

Saturn's elements for this moment were taken from the collection of major planet elements available in the department of minor planets and comets at the Institute for Theoretical Astronomy. The elements of P/Tuttle for the same moment were obtained by numerical integration, using the computer programme by Belyaev and including the disturbing effects of eight planets: the initial set of the comet's elements, referred to the time of passage through perihelion in September 1885, was taken from Porter (1961).

The comet's orbit was divided into six segments according to the set of elements for the moment 1885 Oct. 6.5. Two points of division were selected at the orbit's perihelion and aphelion, respectively. Two other points, corresponding to true anomalies of 160° and 170°, separate the portion of the ellipse where the distance between the orbits is the shortest. The remaining two points are symmetric to these with respect to the major axis of the ellipse.

On the basis of the same set of elements, the values of the Fourier coefficients for the coordinates of the comet and Saturn, as well as those for the expansion of the square of their mutual distance, were found for each segment, by means of the usual formulae. To save computation time and to simplify application of the method, double harmonic analysis was employed for transition from the latter expansion to series for the negative powers of the mutual distance. This was done to avoid numerical calculations in one argument and analytical calculations in the other (Hansen, 1856), which is a possible alternative. Further, accurate values of negative powers were derived from approximate values by the method described by Brouwer and Clemence (1961). The expansion for the inverse cube of mutual distance in the 160°–170° segment contains 10 harmonics in the partial anomaly and 60 harmonics in the second argument, and it includes all the coefficients larger than 0".001.

These expansions were used to derive the right-hand sides of the differential equations for Hansen's five elements. We have integrated these equations and found integration constants which, when the method of partial anomalies is used, are single-argument series in the second argument C'. Subsequently, first-order perturbations were obtained in the three Hansen coordinates: δz in the mean anomaly, ν in the logarithm of the radius vector, and δs in the sine of the heliocentric latitude. Particular checks, which are quite easy to find in this case (Asten, 1872), confirm the correctness of the calculations. The series for the perturbations are contained on magnetic tape but are not illustrated here because they are too cumbersome.

The next and final stage will include a thorough comparison of these perturbations with numerical integration. This will enable us to ascertain the time interval over which the series can be used and found reasonably accurate.

References

Asten, E.: 1872, *Mem. Acad. Imp. Sci. St. Petersb.* **18**, No. 10.
Brouwer, D. and Clemence, G. M.: 1961, *Methods of Celestial Mechanics*, Academic Press, New York and London.
Hansen, P. A.: 1856, *Suppl. Compt. Rend. Acad. Sci. Paris* **1**, 121.
Nacozy, P. E: 1969, *Astron. J.* **74**, 544.
Porter, J. G.: 1961, *Mem. Br. Astron. Assoc.* **39**, No. 3.
Skripnichenko, V. I.: 1970, *Byull. Inst. Teor. Astron.* **12**, 16.

11. ORBITAL CHARACTERISTICS OF COMETS
PASSING THROUGH THE 1:1 COMMENSURABILITY
WITH JUPITER

E. RABE

University of Cincinnati Observatory, Cincinnati, Ohio, U.S.A.

Abstract. When, in consequence of a close approach, a comet of the Jupiter group changes its osculating semimajor axis from $a > 1$ to $a < 1$ ($a' = 1$ for Jupiter's orbit), or vice versa, then the normal case is that of an abrupt change from one side of Jupiter's orbit to the other one. Under special conditions, however, temporary capture into satellite or 'Trojan' status is possible. P/Slaughter-Burnham, the first known comet in temporary 1:1 resonance with Jupiter, sheds some light on the requirements for Trojan captures. In consideration of the recent finding that the Trojan 'cloud' around L_4 contains probably as many as 700 Trojans brighter than magnitude 20.9, it is suggested that at least some comets of the Jupiter group may have originated among these accumulations around L_4 and L_5.

The available evidence indicates that most comets of the Jupiter group are able, through close approaches to Jupiter, to change from outside into inside orbits, and vice versa, in the sense of changing their osculating semimajor axis a from $a > 1$ to $a < 1$, or from $a < 1$ to $a > 1$, respectively. For convenience, Jupiter's mean solar distance a' is chosen here as the unit of length. The actual cases on record in the literature suggest that during such passages through Jupiter's orbit the resulting changes in a tend to be quite large, regardless of the other orbital changes involved. This probably explains the fact that until recently no comets had been observed in temporary captures either as Jupiter satellites (for one or more complete revolutions) or 'Trojans'. Such captures require a to approach the narrow range $0.95 \leq a \leq 1.05$ (approximately) in a rather special or almost asymptotic manner, or from rather selective initial conditions, as indicated for example by the results of many numerical integrations for asteroids experiencing temporary captures as Jupiter satellites (Hunter, 1967). As to capture into temporary 'Trojan' status, Thüring's (1959) numerical integrations of certain evidently unstable librations about the equilateral points of the restricted three-body problem provided a first demonstration of the possibility of such particular transfers through the 1:1 commensurability of the heliocentric orbital periods involved, or through $a = a' = 1$.

While the basic long- and short-period families of libration orbits, in the rotating reference frame of the restricted problem, are stable against very small superposed oscillations, certain limits of stability exist for the more complicated (and in reality more frequent) librations which incorporate substantial long- *and* short-period components. Further complications are introduced by the ellipticity of Jupiter's orbit and by additional planetary perturbations. For unstable librations involving sufficiently large short-period oscillations superposed on any one of the well-known reference orbits of long period, it has been shown (Rabe, 1966) that such motions would indeed exhibit the spiraling features noted by Thüring (1959). The same result is easily established for sufficiently large long-period fluctuations superposed on a reference libration of short period. As the total libration amplitude increases due to the spiraling effects, Jupiter is approached ever more closely, and all numerical integrations of such unstable librations support indeed the conclusion that the temporary Trojan status begins and ends with a particularly close approach to the primary of mass μ. In terms of heliocentric, osculating elements, it has been known for a long time that in the long-period librations the semimajor axis a oscillates about a mean value close to $a' = 1$, and that, in the Sun–Jupiter case, the amplitude Δa of this long-period oscillation must be smaller than about 0.052, for any stable orbit around L_4 or L_5. In the light of the results for the motion of P/Slaughter-Burnham, to be discussed next, it appears that the requirement $0.95 \leq a \leq 1.05$ determines the limits, not only for the existence of stable solutions of long period, but also those for the possibility of a temporary or unstable Trojan status for motions incorporating a substantial short-period component.

Marsden kindly provided me with the osculating values of the elements a, e, and i, at 10 000-day intervals, from his original numerical integration of the motion of P/Slaughter-Burnham over more than 1400 years (Marsden, 1970) and also agreed to continue the backward integration for the further exploration of the orbital development of this first known comet in an extended 1:1 resonance with Jupiter. It was found that the librational stage begins around the year -450, after a very close approach to Jupiter, to about 0.02 Jupiter units. The end of the librational motion will come near the year 2075, with an approach to about 0.06 Jupiter units. Now it should be stressed that the early results far away from the basic epoch in 1959 are very uncertain, due to the limited accuracy of the elements as determined from all the 1958/59 observations and the semiapproximate recovery observations on two consecutive nights in 1969. Nevertheless, the results are of definite mathematical interest, thanks to the high computational accuracy employed, because they give us the true dynamical history of a small body for which these initial elements would be rigorously correct. Any subsequent improvement of the elements of the real comet Slaughter-Burnham may, of course, lead to a modification of the early librational features depicted in Figures 1 and 2, and to a changed total duration of the librational stage, which presently is found to cover about 2500 years. Sitarski (1968) has also remarked on the 1:1 resonance with Jupiter and the fact that the two objects are currently on opposite sides of the Sun.

Figure 1, provided by Marsden as an extended version of his original diagram (Marsden, 1970, Figure 8), shows the long-period component of the libration in longitude, in the form of the difference $\lambda - \lambda_J$ of the mean longitudes of the comet and of Jupiter, respectively. Since through the librational stage the comet's orbital eccentricity e increases from about 0.43 to roughly 0.50, accompanied by a simultaneous decrease of the inclination i from $16°$ to $8°$, very substantial short-period oscillations have to be superposed on the $\lambda - \lambda_J$ of Figure 1, in order to obtain the rather complicated true longitude libration. Actually, the short-period libration component dominates in size during the initial centuries of the long lasting association with L_5, which begins near the year 0, but which around the year 1400 changes

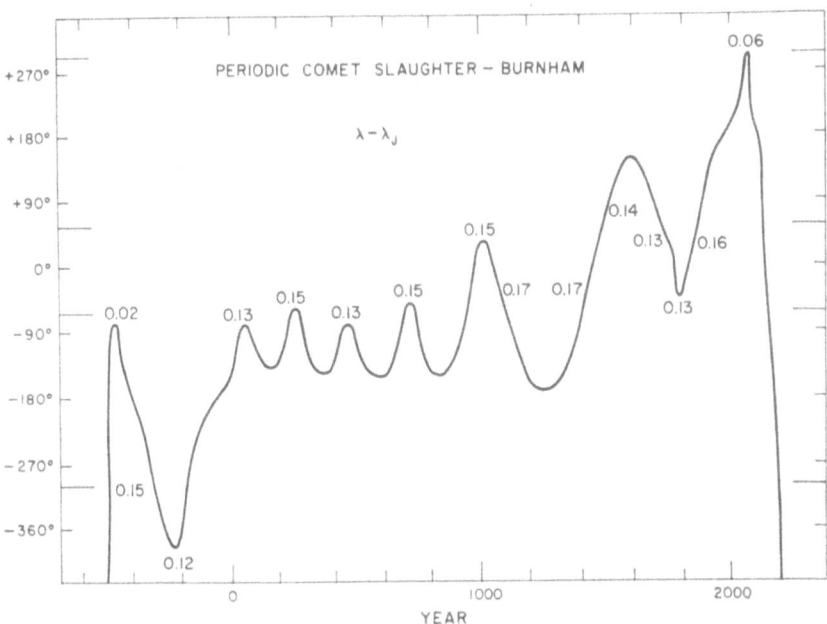

Fig. 1. The libration argument $\lambda - \lambda_J$ for P/Slaughter-Burnham. The long horizontal lines directed inward from the sides represent the Lagrangian points L_4 and L_5, at $\lambda - \lambda_J = \pm 60°$ and $\pm 300°$.

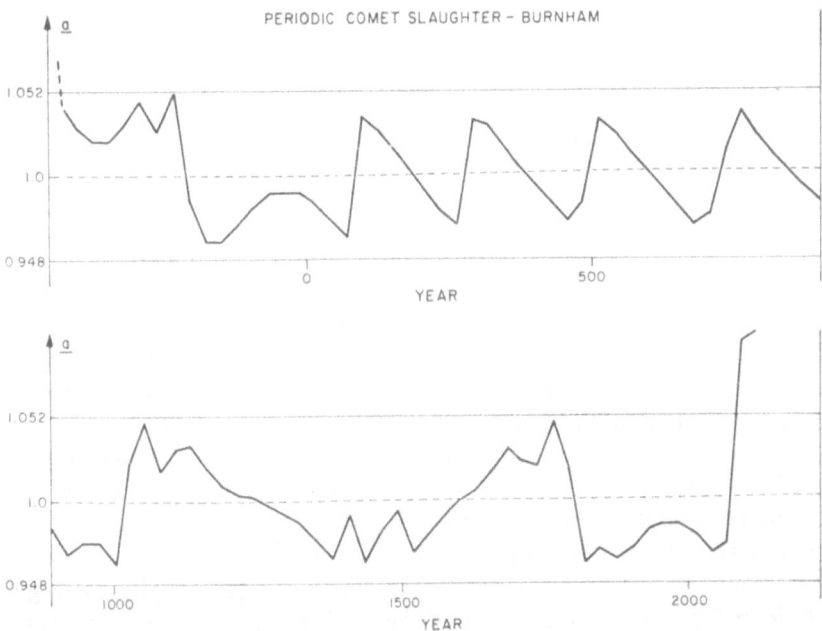

Fig. 2. The osculating semimajor axis a of P/Slaughter-Burnham. The dashed horizontal line represents Jupiter's mean solar distance $a' = 1.0$.

over into a more short-lived association with L_4. The spiraling effect is clearly exhibited in the $\lambda - \lambda_J$ of Figure 1, with a minimum amplitude appearing not far from the year 100. The recurring approaches to Jupiter have been indicated by the approximate minimum distances involved. It is seen that the two terminating approaches are much closer than the other 13 between them.

The 10 000-day values of a have been plotted in Figure 2. However, they have been connected simply by straight lines, which accounts for the lack of smoothness. The most striking feature is the strict compliance with the requirement $0.95 \leq a \leq 1.05$ over the whole librational stage of the motion, and the abrupt termination of this stage when these limits are penetrated. It appears that the long-period component is the one of principal significance for the maintenance and duration of the librational status. Some light seems to be shed on the requirements for capture into such a libration by those 'convex' arcs of the a-curve in Figure 2 which are curved away from the horizontal line for $a = 1.0$. These arcs appear directly after capture and before escape, but also just before the comet's entry into an extended period of close association with L_5. The dashed line $a = 1.0$ is being approached but not reached by the central part of each of these arcs. The times of closest approach to $a = 1.0$ in the various arcs of this sort coincide with those times when the $\lambda - \lambda_J$ of Figure 1 passes through the opposition point $\pm 180°$, at least approximately. It seems thus that temporary libration capture becomes possible when a approaches the range $0.95 \leq a \leq 1.05$ in such a manner, that a minimum (or maximum) of a is reached inside this range, but in addition to this also at a time when the mean longitude of the comet or asteroid differs from that of Jupiter by approximately $180°$. This evidently is a rather restrictive condition and may account for the rareness of such captures. Some less pronounced convex arcs are indicated in Figure 2 around the times when $\lambda - \lambda_J = 0$, or when the mean longitude of the comet passes Jupiter during one of the few 'transfers' (as from L_5 to L_4) which can be recognized in Figure 1.

The case of P/Slaughter-Burnham shows that under the right conditions even comets in rather eccentric orbits may enter the Trojan-type 1:1 resonance for a limited period of time, but that the probability of such an occurrence appears to be small. On the other hand, over very long periods of time, most of the known comets of the Jupiter group may occasionally experience the appropriate perturbations in the all-important element a to make such a Trojan capture possible. The Jacobi 'constants' C of practically all comets of the Jupiter group lie in the range between 2.5 and 3.0, so that their temporary Trojan status, involving a more or less substantial short-period component, is quite conceivable. For P/Slaughter-Burnham, the e of order 0.5 is reflected in $C = 2.72$. On the other hand, a capture into satellite status requires a C-value just slightly larger than 3, or relatively small values of e and i immediately before and after capture, so that such an event is rather unlikely for most of the Jupiter comets. Some asteroids are better candidates for such an occurrence.

Quite recently it has been estimated that about 700 Trojan planets brighter than magnitude 20.9 are associated with the Lagrangian point L_4 preceding Jupiter (van Houten *et al.*, 1970). Thus it seems that the Trojans are much more numerous than originally thought, or that relatively dense 'clouds' of these asteroids exist in the neighborhoods of the equilateral points. Obviously, then, a great number of small bodies has come into existence in these regions during the earlier stages of the solar system. The Trojans which remained there in rather stable orbits are characterized by small eccentricities, up to about $e = 0.15$. If additional Trojans with somewhat larger e-values or/and larger libration amplitudes Δa also existed there long ago, but eventually escaped from libration due to various perturbing effects, they would unavoidably have tended to transfer into motions very similar to those of the present comets of the Jupiter group. The C-values of such escapers would indeed tend to fall into just that range, between 3.0 and about 2.5, which we find associated with the Jupiter comets. Furthermore, for the 38 comets recently listed by Marsden (1967) the values of the relevant elements are such that, *if* passage through $a = 1$ is assumed to happen at some time without a change in the (generally much smaller) inclination i, many eccentricities would at such time attain greatly reduced values e_1, in compliance with the Tisserand criterion. A small fraction of these e_1-values even falls into the range $0.00 \leq e \leq 0.15$ of the known Trojan planets. Consequently it may be suggested that at least some of the comets of the Jupiter group may have *originated* in these fairly dense Trojan clouds and may indeed never have departed very far from the vicinity of Jupiter's orbit. Such an origin would help to explain the complete absence of retrograde orbits among the comets of the Jupiter group. Finally, the anomalous distribution of the longitudes of perihelion noted by Marsden (1967) for the comets of the Jupiter group is very similar to the one noted by Thüring (1951) for the Trojan planets. In both groups, the perihelia prefer the semicircle centered approximately near Jupiter's perihelion longitude π' and containing also the longitudes $\pi' \pm 60°$.

It should be noted that van Houten *et al.* (1970) use the notation L_5, instead of L_4, for the Lagrangian point preceding Jupiter.

References

Hunter, R. B.: 1967, *Monthly Notices Roy. Astron. Soc.* **136**, 267.
Marsden, B. G.: 1967, *Science* **155**, 1207.
Marsden, B. G.: 1970, *Astron. J.* **75**, 206.
Rabe, E.: 1966, *Mitt. Astron. Ges.* No. 21, 125.
Sitarski, G.: 1968, *Acta Astron.* **18**, 419.
Thüring, B.: 1951, *Astron. Nachr.* **279**, 217.
Thüring, B.: 1959, *Astron. Nachr.* **285**, 71.
van Houten, C. J., van Houten-Groenveld, I., and Gehrels, T.: 1970, *Astron. J.* **75**, 659.

Discussion

H. Alfvén: If you integrate the orbit of this comet two or three thousand years backward, there arises the question as to how long the comet's actual lifetime is. I understand from Vsekhsvyatskij's catalogue that he estimates that cometary lifetimes are of the order of decades or centuries, and this seems to be in agreement with other results. Has it any realistic meaning to integrate for such long time intervals?

E. Rabe: The closer a comet comes to the Sun, the more volatile material it loses, but one can perhaps expect something to be left over. How long an individual comet lives is an open question. But it is a physical problem completely separate from the mathematical one.

B. G. Marsden: It may well be that our integration of the orbit of the real P/Slaughter-Burnham is unreliable as long ago as the year −450, but comparison with another integration from rather different starting conditions suggests that its general features are certainly correct back to about 1000 years ago. The main difficulties with this kind of calculation stem from close approaches to Jupiter and nongravitational effects. But close approaches of P/Slaughter-Burnham to Jupiter have not occurred for at least 1000 years, and with a perihelion distance as large as 2.5 AU the nongravitational effects are certainly very small, and so, presumably, is the comet's rate of decay.

S. K. Vsekhsvyatskij: The main point is that the secular decreases in brightness show that comets are physically unstable and hence rather young objects. We do not know those comets of Jupiter's family that might have existed in the central region of the solar system more than a hundred or few hundred years ago. Besides, it is impossible to explain the appearance of short-period comets and the peculiarities of their motions by the capture hypothesis, if we take into account the low probability of transformation of a long-period orbit into a short-period one. As to the idea that these comets could be formed within Jupiter's libration region, this is refuted by the whole of our knowledge of the conditions in interplanetary space. The solar photon and corpuscular radiation should result in the sublimation of cometary ices and the disintegration of comets at distances up to 11 or 12 AU. In the vicinity of Jupiter cometary ices exist only on the surfaces of Jupiter's satellites.

L. Kresák: P/Van Biesbroeck moves in an orbit closely resembling that of P/Slaughter-Burnham, not only by an approximate 1:1 resonance with Jupiter and a similar eccentricity, but also by a very similar value of the libration argument, putting it at present opposite Jupiter. I wonder if anybody has tried to obtain long-term perturbations for this comet; possibly the results may be analogous to those obtained for P/Slaughter-Burnham.

B. G. Marsden: We have found that P/Van Biesbroeck was not librating in the past, perturbations by Jupiter having caused significant changes in the orbit not too long ago. I do not know about the future motion.

G. A. Chebotarev: What is the reason for the absence of retrograde motions among the comets of the Jupiter family?

E. Rabe: If some of these comets were originally members of 'Trojan' clouds, then they would be in direct orbits only. All the bodies associated with the libration points are in direct orbits because they have a mean angular distance of 60° from Jupiter and move in the same direction as Jupiter.

12. THE MOTIONS OF BODIES CLOSE TO COMMENSURABILITIES WITH JUPITER

A. T. SINCLAIR

Royal Greenwich Observatory, Herstmonceux Castle, Hailsham, Sussex, England

Abstract. The theory of the libration motion of a body close to a commensurability in mean motions with Jupiter is briefly described, and the importance of such motions in that they enable the body to avoid close approaches to Jupiter is discussed. By means of numerical integrations of the equations of motion some of the work that has been done on librations in orbits of low eccentricities is extended to higher eccentricities, so that the results can be applied to cometary orbits. By making some approximations the results are expressed in a very simplified form which makes it possible to compare the sizes of libration regions. This is done for six interior commensurabilities, namely the 3:1, 2:1, 5:3, 3:2, 7:5 and 4:3 commensurabilities.

Chebotarev et al. (eds.), The Motion, Evolution of Orbits, and Origin of Comets, 61. All Rights Reserved.
Copyright © 1972 by the IAU.

13. ON THE MOTION OF SHORT-PERIOD COMETS IN THE NEIGHBOURHOOD OF JUPITER

V. M. CHEPUROVA

Sternberg Astronomical Institute, Moscow, U.S.S.R.

Abstract. An intermediate orbit is developed for studying the hyperbolic jovicentric motion of a comet in the vicinity of Jupiter, where the effect of Jupiter's oblateness may become significant. A comparison is made with some of the results obtained by other investigators on the motions of specific periodic comets.

An intermediate orbit has been constructed (Chepurova, 1970) on the basis of the solution, advanced by Aksenov *et al.* (1961, 1963), of the generalized problem of two fixed centres. If the force function

$$U = \frac{fm}{2}\left(\frac{1}{r_1} + \frac{1}{r_2}\right)$$

is written in the form of the series

$$U = \frac{fm}{r}\left[1 + \sum_{k=1}^{\infty} \frac{\gamma_{2k}}{r^{2k}} P_{2k}\left(\frac{z}{r}\right)\right],\tag{1}$$

where f is the gravitational constant, m the mass of the central body, and $r=(x^2+y^2+z^2)^{1/2}$ the radius vector of a moving, massless particle; $r_1=[x^2+y^2+(z-ci)^2]^{1/2}$ and $r_2=[x^2+y^2+(z+ci)^2]^{1/2}$, where $i=\sqrt{-1}$ and c is the constant of the generalized two fixed centres problem; $P_{2k}(z/r)$ are Legendre polynomials, and $\gamma_{2k}=(-1)^k(c^{2k}/2)$. Specifically,

$$\gamma_2 = -J_2 R_0^2,$$

where J_2 is the second harmonic of the potential of an axially symmetrical body, and R_0 is the equatorial radius of the body. As is well known, the potential of the axially symmetrical body has the form

$$V = \frac{fm}{r}\left[1 + \sum_{k=2}^{\infty} J_k \left(\frac{R_0}{r}\right)^k P_k\left(\frac{z}{r}\right)\right],\tag{2}$$

where the zonal harmonics J_k describe the gravitational field of the body (in the case of the planets of the solar system J_2 is much larger than all the other terms of the series). Then U approximates V sufficiently well. Comparison of calculations by the formulae of the intermediate orbit with the results of numerical integration of the differential equations of the motion

$$\ddot{x} = \frac{\partial U}{\partial x}, \qquad \ddot{y} = \frac{\partial U}{\partial y}, \qquad \ddot{z} = \frac{\partial U}{\partial z}\tag{3}$$

has shown that the intermediate orbit, which is the solution of Equations (3) and therefore involves the second harmonic of the potential of the central body, approximates the motion of the massless point in the gravitational field of the oblate body with axial symmetry much better than a Keplerian hyperbola.

The rectangular coordinates of a point moving along the intermediate orbit are expressed by the formulae

$$x = (\zeta^2 + c^2)^{1/2}(-\sin \varphi \sin \Omega + \cos \varphi \cos \Omega \cos i)$$
$$y = (\zeta^2 + c^2)^{1/2}(\sin \varphi \cos \Omega + \cos \varphi \sin \Omega \cos i) \qquad (4)$$
$$z = \zeta s \cos \varphi,$$

where

$$\zeta = a(1 - e\bar{e}) \frac{1 + \kappa \cos \psi}{1 + \bar{e} \cos \psi}$$
$$\varphi = \psi(1 + \nu) + \bar{\omega} + \omega(\psi)$$
$$\Omega = \Omega_0 + \zeta(\psi),$$

and the variable ψ is connected with the time t by the equations

$$M = \bar{M}_0 + \bar{n}\Delta t = e^* \tan E - \log \tan \left(\frac{E}{2} + \frac{\pi}{4}\right) + f_1(\psi)$$

$$(5)$$

$$E = 2 \arctan n' \tan \frac{\psi}{2}.$$

Here a, e, s, $\bar{\omega}$, Ω_0, $\bar{M}_0$ and $\bar{n}$ are constants that become the Keplerian elements (or quantities connected with them) when $c=0$ and so may be called the elements of the intermediate orbit; $\omega(\psi)$, $\zeta(\psi)$ and $f_1(\psi)$ are trigonometric series in ψ with coefficients that depend on the quantities $\bar{e}$, κ, ν, e^* and n', as well as on the elements a, e and s of the intermediate orbit.

It is known that short-period comets of the Jupiter family may occasionally approach Jupiter so closely that the planet's oblateness, which usually influences them very little, begins to affect their motions. When investigating such very close approaches to a major planet it is usual to apply a planetocentric method for the calculation of the perturbations. After familiarizing herself with the work of Kazimirchak-Polonskaya (1961) the author had the idea of taking the first approximation to the motion in such cases to be the above-mentioned intermediate orbit, instead of a Keplerian hyperbola. This is the first attempt to construct this kind of analytical theory for the investigation of cometary motion, and thus for the time being we have not taken into account any other perturbations.

Cometary coordinates obtained from the formulae of the intermediate orbit have been compared with those given earlier by other investigators, notably Dubyago's (1950) investigation of P/Brooks 2 and Kamieński's (1949, 1956) investigation of P/Wolf over several close approaches to Jupiter. Some of the results are given in the Tables I, II and III, the coordinates being referred to Jupiter's equator and measured in AU. The tables show that the results are satisfactory enough, but in the future it will be necessary to take into account the perturbations by the Sun and the other planets.

TABLE I

Coordinates of P/Brooks 2 at $t = 1886$
November 3.0

	Dubyago	Intermediate orbit
x	− 0.081157	− 0.082498
y	+ 0.277657	+ 0.277604
z	+ 0.088909	+ 0.095393

The comet approached Jupiter on 1886
July 20.7657 to a distance of 0.0009636
AU.

TABLE II

Coordinates of P/Wolf at $t = 1757$
February 16.0

	Kamieński	Intermediate orbit
x	− 0.1403911	− 0.1434908
y	+ 0.1052629	+ 0.1047257
z	+ 0.0932434	+ 0.0921376

The comet approached Jupiter on 1757
January 0.2 to a distance of 0.0758 AU.

TABLE III

Coordinates of P/Wolf at $t = 1875$ June
24.0

	Kamieński	Intermediate orbit
x	+ 0.0643464	+ 0.0643479
y	+ 0.0119418	+ 0.0119423
z	+ 0.1156719	+ 0.1156721

The comet approached Jupiter on 1875
June 8.7 to a distance of 0.1180 AU.

References

Aksenov, E. P., Grebenikov, E. A., and Demin, V. G.: 1961, *Iskusstv. Sputniki Zemli* No. 8, 64.
Aksenov, E. P., Grebenikov, E. A., and Demin, V. G.: 1963, *Astron. Zh.* **40**, 363.
Chepurova, V. M.: 1970, *Byull. Inst. Teor. Astron.* **12**, 216.
Dubyago, A. D.: 1950, *Trudy Astron. Obs. Kazan* No. 31.
Kamieński, M.: 1949, *Bull. Acad. Polon. Sci. Lettres Ser. A* 61.
Kamieński, M.: 1956, *Acta Astron.* **6**, 153.
Kazimirchak-Polonskaya, E. I.: 1961, *Trudy Inst. Teor. Astron.* **7**.

Discussion

D. V. Zagrebin: Why have you done the expansion in Legendre polynomials, which are divergent – as was shown by Lyapunov?

V. M. Chepurova: I have done it in the traditional way. I am going to try other expansions in my future work.

G. N. Duboshin: The expansion of the potential gravitational field in Legendre polynomials is convergent, as was shown in the classical theory of gravitation, and it was quite appropriate for Chepurova to apply this expansion – just as it has been applied in all the papers on the motions of artificial satellites. Nevertheless it is possible that expansion in Lamé functions will be more rapidly convergent. But we lack now the appropriate mathematical operators to apply Lamé expansions directly.

14. SECULAR PERTURBATIONS ON PERIODIC COMETS

G. E. O. GIACAGLIA

Dept. of Aerospace Engineering and Engineering Mechanics,
*University of Texas, Austin, Tex., U.S.A.**

Abstract. In this work we propose an approach to the calculation of secular perturbations on a comet, valid for any eccentricity (less than unity) and any inclination (not zero). The effect of several planets is discussed, but we suppose that there are no significant close approaches to any planet.

1. Disturbing Function

The force function due to a planet – considered to be moving in a circular orbit around the Sun – corresponds in the secular sense to the force exerted upon the comet by a loaded ring. It is therefore necessary to obtain the potential of such a ring, giving an expression valid for distances less than or greater than the radius of the ring. The classical development in spherical harmonics has therefore to be abandoned.

According to Figure 1, let a' be the radius of the ring, δ the latitude of the comet P, α its longitude, r its distance from the Sun (center of the ring), Δ the distance between P and P' (a point on the ring with longitude α') and ψ their mutual elongation.

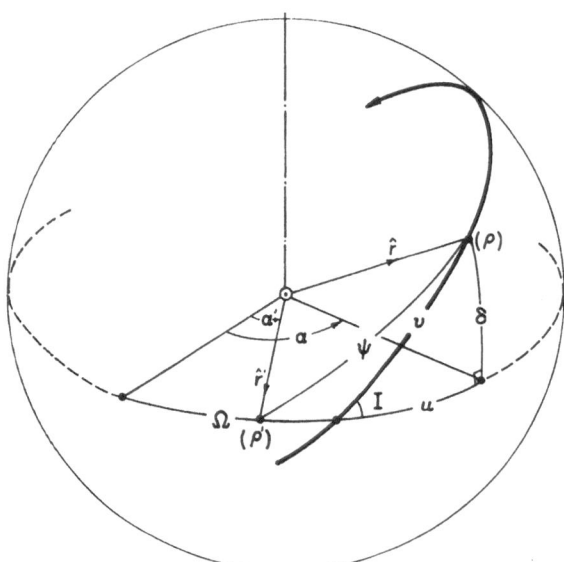

Fig. 1.   Geometry of the problem.

* On leave from the University of São Paulo and University of Campinas, São Paulo, Brazil.

It follows that

$$\cos \psi = \cos \delta \cos (\alpha' - \alpha)$$

$$\Delta^2 = r^2 + a'^2 - 2a'r \cos \psi$$

and, using osculating elements,

$$r = a(1 - e \cos E),$$

where E is the eccentric anomaly of the comet, a the semimajor axis and e the eccentricity. The maximum value of Δ is, at any instant,

$$A = a' + a(1 + e), \tag{1}$$

so that we write

$$\Delta^2 = A^2\{1 - [B + C \cos (\alpha' - \alpha)]/A^2\}, \tag{2}$$

where

$$B = \rho - r^2$$
$$\rho = a(1 + e)[a(1 + e) + 2a'] \tag{3}$$
$$C = 2a'r \cos \delta.$$

Evidently, unless $\Delta=0$, the expansion by the binomial formula of Δ^{-1} will converge for all values of r, so that one can write the series

$$\Delta^{-1} = A^{-1} \sum_{k=0}^{\infty} \binom{-\frac{1}{2}}{k}(-1)^k A^{-2k}[B + C \cos (\alpha' - \alpha)]^k$$

or

$$\Delta^{-1} = A^{-1} \sum_{k=0}^{\infty} \sum_{j=0}^{k} \binom{-\frac{1}{2}}{k}\binom{k}{j}(-1)^k A^{-2k} B^{k-j} C^j \cos^j (\alpha' - \alpha). \tag{4}$$

The force function of the loaded ring (mass m') is given by

$$U = \int_0^{2\pi} G\Delta^{-1}\lambda'a' \, d(\alpha' - \alpha), \tag{5}$$

where $\lambda' = m'/2\pi a'$ and G is the gravitational constant.

Since now

$$\int_0^{2\pi} \cos^{2j+1} x \, dx = 0$$

$$\int_0^{2\pi} \cos^{2j} x \, dx = 2\pi \frac{(2j - 1)!!}{2^j j!},$$

it follows that

$$U = \frac{Gm'}{A} \sum_{k=0}^{\infty} \sum_{j=0}^{[k/2]} (-1)^k \binom{-\frac{1}{2}}{k} \binom{k}{2j} \frac{(2j-1)!!}{2^j j!} A^{-2k} B^{k-2j} C^{2j}, \tag{6}$$

where $[k/2]$ is the integer part of $k/2$.

Now let $v = \omega + f$ be the argument of the latitude (ω the argument of perihelion, f the true anomaly) and Ω the longitude of the ascending node. We have the relations

$$\cos \delta \cos (\alpha - \Omega) = \cos v$$
$$\cos \delta \sin (\alpha - \Omega) = \sin v \cos I,$$

where I is the inclination of the osculating orbit. Therefore

$$C^{2j} = 4^j a'^{2j} r^{2j} (\cos^2 I + \sin^2 I \cos^2 v)^j,$$

and also

$$B^{k-2j} = (\rho - r^2)^{k-2j} = \sum_{p=0}^{k-2j} \binom{k-2j}{p} (-1)^p \rho^{k-2j-p} r^{2p},$$

so that one finds

$$U = \frac{Gm'}{A} \sum_{k=0}^{\infty} \sum_{j=0}^{[k/2]} \sum_{p=0}^{k-2j} \sum_{s=0}^{j} K_{kjps}^* A^{-2k} \rho^{k-2j-p} a'^{2j} r^{2(p+j)}$$

$$\times \cos^{2(j-s)} I \sin^{2s} I \cos^{2s} v, \tag{7}$$

where

$$K_{kjps}^* = \frac{(-1)^p (2k)! (2j+1)}{4^k k! p! (k-2j-p)! j! s! (j-s)!}. \tag{8}$$

We change the index notation by setting

$$p + j = q, \qquad K_{kjps}^* = K_{kj,q-j,s}^* = K_{kjqs} \tag{9}$$

so that

$$U = \frac{Gm'}{A} \sum_{k=0}^{\infty} \sum_{j=0}^{[k/2]} \sum_{q=j}^{k-j} \sum_{s=0}^{j} K_{kjqs} A^{-2k} \rho^{k-j-q} a'^{2j}$$

$$\times \cos^{2(j-s)} I \sin^{2s} I r^{2q} \cos^{2s} v. \tag{10}$$

The long period and secular perturbations on the comet are generated by the average U over the mean anomaly of the comet, that is

$$U_s = \frac{1}{2\pi} \int_0^{2\pi} U \, dl,$$

assuming, of course, no commensurability exists between the mean motion of the comet and that of the planet. It has long been known, for example, that the ratio of the periods of P/Encke and Jupiter is very close to 5/18 (Whipple, 1940). Although of very high degree, this commensurability may produce a large coefficient in the perturbations because of the excessive eccentricity of the comet (about 0.85). On the other hand, no attempt has been made to determine the average of U analytically because, in the classical sense, it would require an expansion in powers of e, making any results invalid. Brouwer (1947) did obtain a good approximation to the secular variations of P/Encke by evaluating a numerical average of the disturbing function over the period of the system (five revolutions of Jupiter), and he obtained good agreement with Whipple's empirical results. This indicates that, at least in this case, Jupiter perturbations would suffice to give a good first approximation for the orbit.

The force function U we have obtained is valid everywhere except on the ring it-self – which would imply the possibility of collision. The excluded cases are therefore those which correspond to the condition $\Delta = 0$ or

$$r = a', \qquad \psi = 0$$

or

$$r = a', \qquad \delta = 0, \qquad \alpha = \alpha'.$$

For the ring potential this reduces to

$$r = a', \qquad \delta = 0.$$

The first of these gives the equation

$$1 - e \cos E = a'/a,$$

which can be satisfied in the cases when

$$|1 - a'/a| \le e.$$

2. Average Disturbing Function

Next, in order to avoid any expansion in powers of the eccentricity, the function U_s will be obtained using the eccentric anomaly as independent variable.

Let

$$Q_{2q,2s} = \frac{1}{2\pi} \int_0^{2\pi} r^{2q} \cos^{2s} (f + \omega) \, dl, \tag{11}$$

where

$$q = 0, 1, 2, \ldots, k - 2j$$
$$s = 0, 1, 2, \ldots, j$$
$$j = 0, 1, \ldots, [k/2]$$
$$k = 0, 1, 2, \ldots$$

Considering the relation

$$dl = \frac{r}{a} \, dE,$$

$$Q_{2q,2s} = \frac{a^{2q}}{2\pi} \int_0^{2\pi} \left(\frac{r}{a}\right)^{2q+1} \cos^{2s} (f + \omega) \, dE. \tag{12}$$

If we let $\varepsilon = \exp(iE)$, we find that

$$\left(\frac{r}{a}\right)^{2q+1} \cos^{2s} v = \frac{1}{2^{2s}} \sum_{j=0}^{2s} \sum_{k=0}^{2(s-j)} \sum_{\gamma=k}^{k+2q+1} \sum_{\alpha=0}^{k} \sum_{\beta=0}^{\gamma-k} L_{j,k,\gamma,\alpha,\beta}^{2q,2s}$$

$$\times\, e^{2(s-k-j)+\gamma}(1-\eta)^{k-\alpha}(1+\eta)^{\alpha}$$

$$\times \exp\left[-2i(s-j)\omega\right]\varepsilon^{\gamma-2(\alpha+\beta)}, \tag{13}$$

where $\eta = \sqrt{1-e^2}$ and

$$L_{j,k,\gamma,\alpha,\beta}^{2q,2s} = (-2)^{-\gamma}\binom{2s}{j}\binom{2s-2j}{k}\binom{2q+1}{\gamma-k}\binom{k}{\alpha}\binom{\gamma-k}{\beta}. \tag{14}$$

Contributions to the integral $Q_{2q,2s}$ come from even values of γ, so that the corresponding terms are

$$\left[\left(\frac{r}{a}\right)^{2q+1}\cos^{2s} v\right]_{\text{even}} = \frac{1}{2^{2s}} \sum_{j=0}^{2s} \sum_{k=0}^{2(s-j)} \sum_{\gamma=[(k+1)/2]}^{[(k+1+2q)/2]} \sum_{\alpha=0}^{h} \sum_{\beta=0}^{2\gamma-k} L_{j,k,2\gamma,\alpha,\beta}^{2q,2s}$$

$$\times\, e^{2(s-k-j+\gamma)}(1-\eta)^{k-\alpha}(1+\eta)^{\alpha}$$

$$\times \exp\left[-2i(s-j)\omega\right]e^{2(\gamma-\alpha-\beta)},$$

and therefore, setting $\gamma = \alpha + \beta$,

$$2^{2s}Q_{2q,2s} = a^{2q} \sum_{j=0}^{2s} \sum_{k=0}^{2(s-j)} \sum_{\gamma=[(k+1)/2]}^{[(k+1+2q)/2]} \sum_{\alpha=0}^{k} L_{j,k,2\gamma,\alpha,\gamma-\alpha}^{2q,2s}$$

$$\times\, e^{2(s-k-j+\gamma)}(1-\eta)^{k-\alpha}(1+\eta)^{\alpha} \exp\left[-2i(s-j)\omega\right]$$

$$= 2^{2s}a^{2q} \sum_{\beta=0}^{s} M_{2q,2s}^{\beta} \cos 2\beta\omega, \tag{15}$$

where

$$2^{2s}M_{2q,2s}^{0} = a^{2q} \sum_{\gamma=0}^{q} L_{s,0,2\gamma,0,\gamma}^{2q,2s}\, e^{2\gamma}$$

and, for $\beta \neq 0$,

$$M_{2q,2s}^{\beta} = \frac{1}{2^{2s-1}} \sum_{k=0}^{2\beta} \sum_{\gamma=[(k+1)/2]}^{[(k+1+2q)/2]} \sum_{\alpha=0}^{k} L_{s-\beta,k,2\gamma,\alpha,\gamma-\alpha}^{2q,2s}$$

$$\times\, e^{2(\beta-k+\gamma)}(1-\eta)^{k-\alpha}(1+\eta)^{\alpha}. \tag{16}$$

We finally arrive at the average disturbing function U_s, given by

$$U_s = \frac{Gm'}{A} \sum_{k=0}^{\infty} \sum_{j=0}^{[k/2]} \sum_{q=j}^{k-j} \sum_{s=0}^{j} \sum_{\beta=0}^{s} N_{kjqs}^{\beta} A^{-2k}$$

$$\times\, \rho^{k-j-q}a'^{2j}a^{2q} \cos^{2(j-s)} I \sin^{2s} I \cos 2\beta\omega, \tag{17}$$

where

$$N_{kj,2q,2s}^{\beta} = K_{kjqs}M_{2q,2s}^{\beta}, \tag{18}$$

the factors on the right-hand side being defined in Equations (8), (9) and (16).
We also remember that

$$A = a' + a(1 + e), \qquad \rho = (A - a')(A + a') = A^2 - a'^2.$$

3. An Example

As an example we develop U_s up to $k=4$. Rearranging terms properly, we find

$$U_s = \frac{\mu'}{A} \left\{ \left(1 + \frac{1}{2}\frac{\rho}{A^2} + \frac{3}{8}\frac{\rho^2}{A^4} + \frac{5}{16}\frac{\rho^3}{A^6} + \frac{35}{128}\frac{\rho^4}{A^8} \right) \right.$$

$$- \frac{1}{2}\frac{a^2}{A^2} \left(1 + \frac{3}{2}\frac{\rho}{A^2} + \frac{15}{8}\frac{\rho^2}{A^4} + \frac{35}{16}\frac{\rho^3}{A^6} \right) \left(1 + \frac{3}{2}e^2 \right)$$

$$+ \frac{3}{8}\frac{a^4}{A^4} \left(1 + \frac{5}{2}\frac{\rho}{A^2} + \frac{35}{8}\frac{\rho^2}{A^4} \right) \left(1 + 5e^2 + \frac{5}{16}e^4 \right)$$

$$- \frac{5}{16}\frac{a^6}{A^6} \left(1 + \frac{7}{2}\frac{\rho}{A^2} \right) \left(1 + \frac{21}{2}e^2 + \frac{105}{8}e^4 + \frac{35}{16}e^6 \right)$$

$$+ \frac{35}{128}\frac{a^8}{A^8} \left(1 - 18e^2 + \frac{189}{4}e^4 + \frac{105}{4}e^6 + \frac{315}{128}e^8 \right)$$

$$+ \frac{3}{4}\frac{a^2 a'^2}{A^4} \left(1 + \frac{5}{2}\frac{\rho}{A^2} + \frac{35}{8}\frac{\rho^2}{A^4} \right) \left[\left(1 + \frac{3}{2}e^2 \right) \right.$$

$$\left. - \frac{1}{2} \left(1 - e + \frac{5}{2}e^2 \right) \sin^2 I \right]$$

$$- \frac{15}{8}\frac{a^4 a'^2}{A^6} \left(1 + \frac{7}{2}\frac{\rho}{A^2} \right) \left[\left(1 + 5e^2 + \frac{5}{16}e^4 \right) \right.$$

$$\left. + \left(-\frac{1}{2} + \frac{1}{2}e - 3e^2 + \frac{5}{8}e^4 \right) \sin^2 I \right]$$

$$+ \frac{105}{32}\frac{a^6 a'^2}{A^8} \left[\left(1 + \frac{21}{2}e^2 + \frac{105}{8}e^4 + \frac{35}{16}e^6 \right) \right.$$

$$\left. - \frac{1}{2} \left(1 + \frac{21}{2}e^2 + \frac{105}{8}e^4 + \frac{35}{16}e^6 \right) \sin^2 I \right]$$

$$+ \frac{105}{64}\frac{a^4 a'^4}{A^8} \left[\left(1 + 5e^2 + \frac{5}{16}e^4 \right) - \left(\frac{3}{4} + \frac{81}{16}e^2 + \frac{1}{2}e^4 \right) \sin^2 I \right.$$

$$\left. \left. - \frac{1}{2} \left(1 - \frac{1}{4}e^2 - \frac{3}{4}e^4 \right) \sin^4 I \right] \right\}$$

$$+ \frac{\mu'}{A} \left\{ \frac{3}{4}\frac{a^2 a'^2}{A^4} \left(1 + \frac{5}{2}\frac{\rho}{A^2} + \frac{35}{8}\frac{\rho^2}{A^4} \right) \frac{e}{2} \left(1 + \frac{3}{2}e \right) \sin^2 I \right.$$

$$- \frac{15}{8}\frac{a^4 a'^4}{A^6} \left(1 + \frac{7}{2}\frac{\rho}{A^2} \right) \frac{e}{2} \left(1 + \frac{17}{4}e + \frac{21}{8}e^3 \right) \sin^2 I$$

$$- \frac{105}{32} \frac{a^6 a'^2}{A^8} \left(1 + 6e^2 + \frac{45}{8} e^4 + \frac{25}{32} e^6\right) \sin^2 I$$

$$+ \frac{105}{64} \frac{a^4 a'^4}{A^8} \left[\frac{1}{2}\left(1 + 5e^2 + \frac{5}{16} e^4\right) + \left(-\frac{3}{2} + 3e^2 + \frac{53}{32} e^4\right) \sin^2 I\right.$$

$$\left. + \left(1 - \frac{1}{4} e^2 - \frac{3}{4} e^4\right) \sin^4 I\right]\right\} \cos 2\omega$$

$$+ \frac{\mu'}{A} \left\{\frac{105}{64} \frac{a^4 a'^4}{A^8} \left[\frac{1}{2}\left(1 - \frac{23}{8} e^2 + \frac{15}{8} e^4\right)\right.\right.$$

$$\left.\left. - \frac{1}{2}\left(1 - \frac{1}{4} e^2 - \frac{3}{4} e^4\right) \sin^2 I\right] \sin^2 I\right\} \cos 4\omega.$$

For the specific case of Encke's Comet, assuming

$a' = 5.2028$ AU (Jupiter)

$a = 2.217$ AU

$e = 0.847$

$\mu' = Gm' = 0.28253 \times 10^{-6}$ (AU)^3 d^{-2}

$I = 13°.9$,

it follows that the basic parameters of the problem have the values

$A = 9.2976$ AU

$\rho = -25.8374$ (AU)^2

$\rho/A^2 = -0.29889$

$\mu'/A = 0.30388 \times 10^{-7}$ (AU)^2 d^{-2}

$a/A = 0.23845$

$a'/A = 0.55959$

$\sin I = 0.2402$

$\cos I = 0.9707$.

The maximum powers kept in the previous development correspond to reasonably small values:

$$(a/A)^8 = 1.0451 \times 10^{-5}$$

$$(a^3 a'/A^4)^2 = 5.7561 \times 10^{-5}$$

$$(aa'/A^2)^4 = 3.1698 \times 10^{-4},$$

as compared with the main term, which is equal to unity. Of course, the real advantage of the new method is for orbits of high eccentricity which lie partly inside and partly outside the orbit of a 'primary'. The only method available up to now to solve this problem was that of 'partial anomalies' introduced by Hansen.

4. Variation of Elements: Many Planets

The equations for the long-period and secular variations of the elements to be considered are

$$\dot{I} = \frac{\cot I}{na^2(1 - e^2)^{1/2}} \frac{\partial R}{\partial \omega}$$

$$\dot{\omega} = -\frac{\cot I}{na^2(1 - e^2)^{1/2}} \frac{\partial R}{\partial I} + \frac{(1 - e^2)^{1/2}}{na^2 e} \frac{\partial R}{\partial e},$$

while e is obtained through the integral $\sqrt{(1 - e^2)} \cos I = \hat{C}$, and I and Ω can be obtained by direct integration of the respective equations – since they are not present in R. The disturbing function R is defined by the secular part of U, for all planets under consideration, so that it may be written

$$R = \sum_{i=1}^{N} \frac{Gm}{A_i} \sum_{j=0}^{J} M_{ij}(e, I, a_i', a, \rho_i, A_i) \cos 2j\omega,$$

where i refers to the particular planet, N is the number of planets considered and the maximum value J of j is $2[k/2]$, twice the integer part of half the maximum power reached by A_i^{-1} in the development. Evidently this maximum power does not need to be the same for all planets. Also, in this simple formulation, we have assumed all planets in the same plane, a condition which can easily be removed.

Introducing $\hat{C}$ instead of e, we find

$$\dot{I} = \frac{\cos^2 I}{\sin I} \frac{1}{\hat{C}na^2} \frac{\partial R}{\partial \omega}$$

$$\dot{\omega} = -\frac{\cos^2 I}{\sin I} \frac{1}{\hat{C}na^2} \frac{\partial R}{\partial I} + \frac{\hat{C}}{na^2 \cos I} \frac{1}{e} \frac{\partial R}{\partial e},$$

where, after differentiation, e has to be substituted by

$$e = \{1 - \hat{C}^2 \sec^2 I\}^{1/2}.$$

The geometrical behavior of the orbit can be obtained by considering the fact that R, as a function of e and ω with parameters $\hat{C}$ and a, is constant along an orbit (except for short periodic variations), so that one can find $R = $const. curves in the (ω, e) plane. The time dependence is obtained by integrating the above system, and owing to the very complex form of R this is, in general, best done numerically.

5. The Eccentricities of the Planets

It has been shown by several authors that the eccentricity of the orbit of the planet plays an important role in the long-term behavior of cometary orbits. In order to modify the disturbing function accordingly, it is necessary to substitute r' for a'; that is, the ring has to be considered elliptical. Again, one can proceed by considering the identity

$$\Delta^2 = A^2 - (A^2 - \Delta^2),$$

where now

$$A = a(1 + e) + a'(1 + e') \tag{19}$$

and

$$\cos \psi = \cos \delta \cos (\alpha' - \alpha),$$

with the difference that now the integral over the ring is not circular but elliptic. Since

$$\Delta^2 = r^2 + r'^2 - 2rr' \cos \psi,$$

it follows that

$$\Delta^{-1} = A^{-1} \sum_{k=0}^{\infty} \sum_{j=0}^{k} \sum_{v=0}^{j} P_{kjp} A^{-2k}(A^2 - r^2)^{k-j} \cos^p \delta$$

$$\times r'^{2j-p} \cos^p (\alpha' - \alpha), \tag{20}$$

where

$$\alpha' - \alpha = f' - \theta,$$

f' being the true anomaly of the point P' on the ring,

$$\theta = \Omega - \omega' + u,$$

with ω' the longitude of perihelion of the ring and $u = \alpha - \Omega$,

$$r' = \frac{a'(1 - e'^2)}{1 + e' \cos f'},$$

and

$$P_{kjp} = \binom{k}{j} \binom{j}{p}(-2)^p \binom{-\frac{1}{2}}{k}(-1)^k. \tag{21}$$

The integral over the elliptic ring can be performed using the eccentric anomaly E'; that is, introducing the transformation

$$\frac{df'}{dE'} = \frac{a'\sqrt{1 - e'^2}}{r'},$$

so that, for $j \neq 0$,

$$I_{jp} = \frac{1}{2\pi} \int_0^{2\pi} r'^{2j-p} \cos^p (f' - \theta) \, df'$$

$$= \frac{a'\sqrt{1 - e'^2}}{2\pi} \int_0^{2\pi} r'^{2j-p-1} \cos^p (f' - \theta) \, dE'.$$

On the other hand,

$$\cos^p (f' - \theta) = \sum_{q=0}^{p} \binom{p}{q} \cos^q \theta \sin^{p-q} \theta \cos^q f' \sin^{p-q} f',$$

and therefore

$$I_{jp} = \frac{a'\sqrt{1 - e'^2}}{2\pi} \sum_{q=0}^{p} \binom{p}{q} \cos^q \theta \sin^{p-q} \theta$$

$$\times \int_0^{2\pi} x'^q y'^{p-q} r'^{2j-2p-1} dE'. \tag{22}$$

The integration can be easily performed for $j \neq 0$, $p = 0, 1, \ldots, j-1$, for in this case $2j - 2p - 1 > 0$.

For $j = 0$ $(p = 0)$,

$$I_{00} = \frac{1}{2\pi} \int_0^{2\pi} df' = 1.$$

For $j \neq 0$, $p = j$,

$$\frac{1}{2\pi} \int_0^{2\pi} x'^q y'^{j-q} \frac{dE'}{r'} = \frac{a'^{j-1} \eta'^{j-q}}{2\pi} \int_0^{2\pi} \frac{(\cos E' - e')^q \sin^{j-q} E'}{1 - e' \cos E'} dE',$$

which is zero for $j - q$ odd. For $j - q$ even, let $j - q = 2s$. Then, since

$$\cos^p x = 2^{1-p} \sum_{n=0}^{[k/2]} \binom{p}{n} \varepsilon_n \cos [(p - 2n)x]$$

$$\varepsilon_n = 1 \quad (p \text{ odd})$$

$$\varepsilon_n = 1, \quad n = 0, 1, \ldots, \frac{p}{2} - 1 \quad (p \text{ even})$$

$$\varepsilon_{p/2} = \frac{1}{2} \quad (p \text{ even})$$

and

$$\frac{1}{2\pi} \int_0^{2\pi} \frac{\cos nx}{1 - e \cos x} dx = \frac{(\sqrt{1 - e^2} - 1)^n}{(-e)^n \sqrt{(1 - e^2)}},$$

we obtain

$$\frac{1}{2\pi} \int_0^{2\pi} \frac{(\cos E' - e')^q \sin^{j-q} E'}{1 - e' \cos E'} dE'$$

$$= \sum_{\alpha=0}^{q} \sum_{\beta=0}^{s} \sum_{\gamma=0}^{[(\alpha+2\beta)/2]} \binom{q}{\alpha} \binom{s}{\beta} \binom{\alpha + 2\beta}{\gamma} \varepsilon_\gamma (-1)^{q+\beta-\alpha} 2^{1-\alpha-2\beta}$$

$$\times \eta'^{-1} e'^{q-2\alpha-2\beta+2\gamma} (\eta' - 1)^{\alpha+2\beta-2\gamma},$$

so that

$$I_{jj} = a'' \sum_{s=0}^{[j/2]} \binom{j}{j-2s} \cos^{j-2s}\theta \sin^{2s}\theta \, \eta'^{2s}$$

$$\times \sum_{\alpha=0}^{j-2s} \sum_{\beta=0}^{s} \sum_{\gamma}^{[(\alpha+2\beta)/2]} \binom{j-2s}{\alpha}\binom{s}{\beta}\binom{\alpha+2\beta}{\gamma} \varepsilon_\gamma (-1)^{j+\beta} 2^{1-\alpha-2\beta}$$

$$\times e'^{j-2(s+\alpha+\beta-\gamma)}(1-\eta')^{\alpha+2(\beta-\gamma)}. \tag{23}$$

For $j \neq 0$, $p = 0, 1, \ldots, j-1$,

$$I_{jp} = a'^{2j-p}\eta' \sum_{q=0}^{p} \binom{p}{q} \cos^q\theta \sin^{p-q}\theta \, \eta'^{p-q}$$

$$\times \frac{1}{2\pi} \int_0^{2\pi} (\cos E' - e')^q \sin^{p-q} E'(1 - e'\cos E')^{2j-2p-1} \, dE'.$$

Again, this is zero if $p-q$ is odd, so that, setting $p-q=2s$,

$$I_{jp} = a'^{2j-p}\eta' \sum_{s=0}^{[k/2]} \binom{p}{p-2s} \cos^{p-2s}\theta \sin^{2s}\theta \, \eta'^{2s}$$

$$\times \frac{1}{2\pi} \int_0^{2\pi} (\cos E' - e')^{p-2s}(1 - e'\cos E')^{2j-2p-1}(1 - \cos^2 E')^s \, dE',$$

$$\tag{24}$$

which we write as

$$I_{jp} = a'^{2j-p} \sum_{s=0}^{[k/2]} G_{jps}(e') \cos^{p-2s}\theta \sin^{2s}\theta$$

$$G_{jjs} = \binom{j}{j-2s} \eta'^{2s} \sum_{s=0}^{j-2s} \sum_{\beta=0}^{s} \sum_{\gamma=0}^{[\alpha/2]+\beta} \binom{j-2s}{\alpha}\binom{s}{\beta}\binom{\alpha+2\beta}{\gamma} \varepsilon_\gamma$$

$$\times (-1)^{j+\beta} 2^{1-\alpha-2\beta} e'^{j-2(s+\alpha+\beta-\gamma)}(1-\eta')^{\alpha+2(\beta-\gamma)}$$

$$G_{000} = 1, \tag{25}$$

and for $p \neq j$ ($j \neq 0$),

$$G_{jps} = \eta'^{2s+1} \binom{p}{p-2s}$$

$$\times \frac{1}{2\pi} \int_0^{2\pi} (\cos E' - e')^{p-2s}(1 - e'\cos E')^{2j-2p-1}(1 - \cos^2 E')^s \, dE'$$

$$= \binom{p}{p-2s} \eta'^{2s+1} \sum_{\alpha=0}^{p-2s} \sum_{\gamma=0}^{s} \sum_{\beta=[(\alpha+1)/2]}^{[(\alpha-1)/2]+j-p} \binom{p-2s}{\alpha}\binom{2j-2p-1}{2\beta-\alpha}$$

$$\times \binom{s}{\gamma}(-1)^{\gamma+\beta+p} \frac{(2\beta+2\gamma-1)!!}{2^{\beta+\gamma}(\beta+\gamma)!} e'^{k+p-2(\alpha+s)}. \tag{26}$$

We thus arrive at the force function for the elliptical ring:

$$U = \frac{Gm'}{A} \sum_{k=0}^{\infty} \sum_{j=0}^{k} \sum_{p=0}^{j} P_{kjp} A^{-2k} (A^2 - r^2)^{k-j} \cos^p \delta$$

$$\times a^{2j-p} \sum_{s=0}^{[p/2]} G_{jps}(e') \cos^{p-2s} \theta \sin^{2s} \theta. \tag{27}$$

But now,

$$\theta = (\Omega - \omega') + u$$

where

$$\cos \delta \cos u = \cos (\omega + f)$$

$$\cos \delta \sin u = \sin (\omega + f) \cos I.$$

We have that

$$\cos^p \delta \cos^{p-2s} \theta \sin^{2s} \theta$$

$$= \sum_{\alpha=0}^{p-2s} \sum_{q=\alpha}^{\alpha+2s} \binom{p-2s}{\alpha} \binom{2s}{q-\alpha} (-1)^\alpha \cos^{p+q-2(s+\alpha)}(\Omega - \omega')$$

$$\times \sin^{2(s+\alpha)-q}(\Omega - \omega') \cos^{p-q} \delta \cos^{p-q} u \cos^q \delta \sin^q u,$$

and therefore

$$\cos^p \delta \cos^{p-2s} \theta \sin^{2s} \theta = \sum_{\alpha=0}^{p-2s} \sum_{q=\alpha}^{\alpha+2s} \binom{p-2s}{\alpha} \binom{2s}{q-\alpha}(-1)^\alpha$$

$$\times \cos^{p+q-2(s+\alpha)}(\Omega - \omega') \sin^{2(s+\alpha)-q}(\Omega - \omega')$$

$$\times \cos^q I \cos^{p-q}(\omega + f) \sin^q (\omega + f).$$

The contribution to the secular part of U comes evidently from even values of q, so that we keep only those by writing

$$[\cos^p \delta \cos^{p-2s} \theta \sin^{2s} \theta]_{\text{even}}$$

$$= \sum_{\alpha=0}^{p-2s} \sum_{q=[(\alpha+1)/2]}^{[(\alpha+2s)/2]} \binom{p-2s}{\alpha} \binom{2s}{2q-\alpha}(-1)^\alpha \cos^{p+2(q-s-\alpha)}(\Omega - \omega')$$

$$\times \sin^{2(s+\alpha-q)}(\Omega - \omega') \cos^{2q} I \cos^{p-2q}(\omega + f) \sin^{2q}(\omega + f).$$

On the other hand,

$$\cos^{p-2q}(\omega + f) \sin^{2q}(\omega + f) = \sum_{\beta=0}^{q} \binom{q}{\beta}(-1)^{q-\beta} \cos^{p-2\beta}(\omega + f)$$

and

$$(A^2 - r^2)^{k-j} = \sum_{\gamma=0}^{k-j} A^{2(k-j-\gamma)} \binom{k-j}{\gamma}(-1)^\gamma r^{2\gamma},$$

so that, defining as before

$$Q_{2\gamma, p-2\beta} = \frac{1}{2\pi} \int_0^{2\pi} r^{2\gamma} \cos^{p-2\beta}(\omega + f)\, dl,$$

we have, for an elliptic ring,

$$U_s = \frac{Gm'}{A} \sum_{k=0}^{\infty} \sum_{j=0}^{k} \sum_{p=0}^{j} \sum_{s=0}^{[p/2]} P_{kjp} A^{-2k} a^{2j-p} G_{lps}(e')$$

$$\times \sum_{\gamma=0}^{k-} A^{2(k-j-\gamma)} \binom{k-j}{\gamma}(-1)^{\gamma} \sum_{\alpha=0}^{p-2s} \sum_{q=[(\alpha+1)/2]}^{[(\alpha+2s)/2]} \binom{p-2s}{\alpha}\binom{2s}{2q-\alpha}$$

$$\times (-1)^{\alpha} \cos^{2q} I \cos^{p+2(q-s-\alpha)}(\Omega - \omega') \sin^{2(s+\alpha-q)}(\Omega - \omega')$$

$$\times \sum_{\beta=0}^{q} \binom{q}{\beta}(-1)^{q-\beta} Q_{2\gamma, p-2\beta}. \qquad (28)$$

It remains to determine $Q_{2\gamma, p-2\beta}$ for p odd, since for p even ($=2\nu$) we have, from Equation (15),

$$Q_{2\gamma, 2(\nu-\beta)} = a^{2\gamma} \sum_{\sigma=0}^{\nu-\beta} M_{2\gamma, 2(\nu-\beta)}^{\sigma} \cos 2\sigma\omega.$$

Consider therefore $p = 2\nu + 1$, then we require $Q_{2\gamma, 2(\nu-\beta)+1}$; we obtain

$$Q_{2q, 2s+1} = \frac{1}{2\pi} \int_0^{2\pi} r^{2q} \cos^{2s+1}(\omega + f)\, dl,$$

assuming, as is actually the case, that $0 \le s \le q$. Again, we introduce the eccentric anomaly as independent variable so that

$$Q_{2q, 2s+1} = \frac{a^{2q}}{2\pi} \int_0^{2\pi} \left(\frac{r}{a}\right)^{2q+1} \cos^{2s+1}(\omega + f)\, dE.$$

As in Equation (13), we obtain

$$\left(\frac{r}{a}\right)^{2q+1} \cos^{2s+1}(\omega + f)$$

$$= \frac{1}{2^{2s+1}} \sum_{j=0}^{2s+1} \sum_{k=0}^{2s+1-2j} \sum_{\gamma=k}^{k+2q+1} \sum_{\alpha=0}^{k} \sum_{\beta=0}^{\gamma-k} L_{j,k,\gamma,\alpha,\beta}^{2q, 2s+1}$$

$$\times e^{2(s-k-j)+\gamma+1}(1-\eta)^{k-\alpha}(1+\eta)^{\alpha}$$

$$\times \exp[-i(2s+1-2j)\omega]e^{\gamma-2(\alpha+\beta)}.$$

Isolating the even values of γ,

$$\left[\left(\frac{r}{a}\right)^{2q+1} \cos^{2s+1}(f+\omega)\right]_{\text{even}}$$

$$= \frac{1}{2^{2s+1}} \sum_{j=0}^{2s+1} \sum_{k=0}^{2s+1-2j} \sum_{\gamma=[(k+1)/2]}^{[(k+1+2q)/2]}$$

$$\times \sum_{\alpha=0}^{k} \sum_{\beta=0}^{2\gamma-k} L_{j,k,2\gamma,\alpha,\beta}^{2q,2s+1} e^{2(s-k-j+\gamma)+1}(1-\eta)^{k-\alpha}(1+\eta)^{\alpha}$$

$$\times \exp\left[-i(2s+1-2j)\omega\right]e^{2[\gamma-(\alpha+\beta)]}.$$

Finally one finds

$$Q_{2q,2s+1} = a^{2q} \sum_{\nu=1}^{s} M_{2q,2s+1}^{\nu} \cos\left[(2\nu+1)\omega\right], \tag{29}$$

where

$$M_{2q,2s+1}^{\nu} = \frac{1}{2^{2s}} \sum_{k=0}^{2\nu+1} \sum_{\gamma=[(k+1)/2]}^{[(k+2q+1)/2]} \sum_{\alpha=0}^{k} L_{s-\nu,k,2\gamma,\alpha,\gamma-\alpha}^{2q,2s+1}$$

$$\times e^{2(\nu-k+\gamma)+1}(1-\eta)^{k-\alpha}(1+\eta)^{\alpha}, \tag{30}$$

which shows that the definition of the coefficients M is the same as in the previous (even) case; see Equation (16). As was to be expected, both even and odd multiples of ω are now present in the U_s, together with terms in $\Omega - \omega'$. The general argument is of the form

$$2k\omega + j(\omega + \Omega - \omega'),$$

with k, j any integers, positive, negative or zero. Evidently, the system has two degrees of freedom, and there is no simple way of describing its behavior except by numerical means. For a realistic model, the perihelion ω' of the perturbing planet should not be constant but move according to the theory of planetary secular perturbations.

6. Conclusions

The above derivations show that secular perturbations among bodies with arbitrary eccentricities and inclinations can be developed by analytical methods, provided that the possibility of collisions, or more realistically, of close approaches, is excluded. One can consider the secular perturbations by all the planets on the motion of a comet. An approximate calculation for the perturbations by the Earth on P/Encke shows that in order to keep terms (in the disturbing function) of the order of $0''01$, it is necessary to manipulate series of about 400 terms. Of course, the motions of comets are also influenced by nongravitational forces, but since the precise form of these forces seems to be somewhat obscure, it is not practicable to include them in this analysis.

Acknowledgments

This research was partially supported by Office of Naval Research Contract N00014-67-A-0126-0013. We are grateful to P. E. Nacozy and B. G. Marsden for information and several discussions which added to the completeness of the work.

References

Brouwer, D.: 1947, *Astron. J.* **52**, 190.
Whipple, F. L.: 1940, *Proc. Am. Phil. Soc.* **83**, 711.

B. NUMERICAL METHODS

15. A NUMERICAL METHOD OF INTEGRATION BY MEANS OF TAYLOR-STEFFENSEN SERIES AND ITS POSSIBLE USE IN THE STUDY OF THE MOTIONS OF COMETS AND MINOR PLANETS

V. F. MYACHIN and O. A. SIZOVA

Institute for Theoretical Astronomy, Leningrad, U.S.S.R.

Abstract. The Taylor formula is used directly in a method of numerical integration of the n-body problem of celestial mechanics; the derivatives in the expansion of the coordinates are calculated successively at each integration step according to the generalized Steffensen rule. The proposed method is the most precise of all numerical methods based on the predetermined part of the Taylor series. The method is used with a variable number of derivatives at each integration step and also with a variable step. The cumulative error in the coordinates increases more slowly in our method than in any other. We can apply the method to the study of the motion of a comet or minor planet, taking into account the perturbations by eight major planets; the method allows for the simultaneous integration of a great number of objects of zero mass.

All the methods of numerical integration used at present in celestial mechanics, whether they be ones utilizing differences or those of the Runge-Kutta type, are ultimately based on part of a Taylor series in which the derivatives of the right-hand sides of the integrated differential equations are replaced by linear combinations of the right-hand sides themselves, which greatly decreases the precision. We propose a method of numerical integration in which the Taylor series itself is used, with direct computation of all derivatives of the required coordinates by means of equations up to a specified order; the derivatives are computed successively according to the generalized Steffensen rule. We call this process the Taylor-Steffensen method. In so far as this method involves no interpolation, it is the most precise of all numerical methods based on Taylor series. Our method is applicable to any problem of n bodies in which only the forces of mutual Newtonian attraction are acting; all the equations of the system are integrated together in rectangular helio-centric coordinates. Unfortunately, in its present form the method cannot take care of nongravitational forces of a random or intermittent nature, since expansion in Taylor series of all functions included in the integrated system of equations is necessary.

The type of numerical method described here was first suggested by Deprit and Price (1965) for a problem of three bodies with a constant number of derivatives and invariable integration step; our method was developed independently and in a somewhat different form.

The principal features of the method are briefly described below.

(1) In all known numerical methods the coefficients of the differences diminish very slowly (they decrease the most rapidly in Cowell's method – in geometrical progression by factors of 1/4), which results, for each such method, in an optimum number of differences for obtaining (with a fixed step) the greatest accuracy possible.

Chebotarev et al. (eds.), The Motion, Evolution of Orbits, and Origin of Comets, 83–85. All Rights Reserved.
Copyright © 1972 by the IAU.

In the Taylor-Steffensen method the coefficients of the derivatives diminish as $1/n!$, which gives an unlimited increase of precision with the increase in the number of derivatives used. We therefore use the method with a variable number of derivatives at each step, care being taken that the last computed member of the Taylor series is smaller than a given number ε. In practice this is done very simply, since each member of the Taylor series is computed from the preceding ones in a recurrent manner. We may simultaneously provide for a variable step, which can be increased or diminished without additional difficulties any number of times and at any point in the integration interval. This allows us to maintain the required precision of the computation formula without excessive division into more steps, and we can thus decrease the cumulative error in the computed coordinates and velocities.

(2) It follows from the above, in particular, that when comparing the results of integration with the observations at any moment, it is not necessary to resort to interpolation; it will suffice to adjust the integration step to the moment of observation.

(3) Our method allows for greater precision in determining the initial velocities of objects for which two adjacent coordinate values are supplied. The velocities can be obtained without resorting to interpolation by using a special algorithm. This allows us to obtain velocities with the precision of the given coordinates, which is impossible in the conventional computation of initial velocities with the aid of interpolation formulae; and this is highly important, since errors in the initial velocities have an important effect on the total cumulative error in the computed coordinates.

(4) Because of the high precision of the method and the large number of significant figures used in the computer, the total cumulative error in the coordinates over large intervals of time is almost entirely determined by the errors in the initial coordinates; after n integration steps the expected magnitude of the error will be $O(n)$, instead of $O(n^{3/2})$ that follows from Brouwer's (1937) law. Thus the error in the coordinates accumulates much more slowly in our method than in other numerical methods.

We have a computer programme for applying the method to the motions of comets, with account taken of the perturbations by eight major planets, omitting Mercury or Pluto according as to the part of its orbit in which the comet is located. It is also possible to make a simultaneous integration of a large number (50 or more) of objects of zero mass.

Investigation of the motion of comets and minor planets by the Taylor-Steffensen method requires reliable initial data (coordinates and velocities), not only for the object studied but also for the perturbing bodies. Obtaining the latter with sufficient accuracy for any epoch t_0 is a very labour-consuming task. It was for this purpose that combined integration was undertaken of the equations of motion of eight major planets (except Mercury) for the period JD 2428000.5 to 2431820.5. Perturbations caused by Mercury were partially allowed for by the addition of its mass to that of the Sun. For the initial epoch we chose $t_0 = 2430000.5$, for which Schubart and Stumpff (1966) had obtained the values of coordinates and velocities with ten significant figures.

The principal arithmetical operations were made with double precision. The integration step was taken as a constant 10 days. With the accuracy adopted (10^{-15}) in the development of coordinates, we took into account most of the derivatives up to the twelfth order. Heliocentric coordinates and velocities were printed at each step and punched on cards after every other step; these can be used directly as initial data for the major planets in the solution of any cometary problem, as well as for a check, if necessary. The results of the integration by means of our programme were compared with the published ephemerides of the planets; for Venus the departure never exceeded 4×10^{-6}, this being mainly attributable to the perturbations by Mercury.

Integration over an interval of 100–150 yr is now in progress.

References

Brouwer, D.: 1937, *Astron. J.* **46**, 149.
Deprit, A. and Price, J. F.: 1965, *Astron. J.* **70**, 836.
Schubart, J. and Stumpff, P.: 1966, *Veroeffentl. Astron. Rechen-Inst. Heidelberg* No. 18.

16. A LIBRARY OF STANDARD PROGRAMMES FOR CONSTRUCTING NUMERICAL THEORIES FOR STUDYING THE MOTION AND EVOLUTION OF THE ORBITS OF THE MINOR BODIES OF THE SOLAR SYSTEM

N. A. BOKHAN

Institute for Theoretical Astronomy, Leningrad, U.S.S.R.

Abstract. At the Institute for Theoretical Astronomy we have formed a library containing about 120 standard computer programmes. They include calculation of rectangular coordinates and velocities from elements, Lagrangian interpolation, and orbit improvement by the Eckert-Brouwer method. For the investigation of the motions of the minor bodies of the solar system we have constructed a programme for integration with a variable step and making allowance for perturbations by all the major planets and for nongravitational effects. The calculation of the perturbations is carried out using Herrick's vector parameters by the method of variation of arbitrary constants. The programme has been used for studying the motion of P/Encke.

1. The Library of Standard Programmes

When solving astronomical problems that require cumbersome computations it is important to proceed systematically. Consequently, over the years various tables have been made, as the tools of computation have advanced from logarithms to desk calculators. Because of primitive means of computing, the solution of complicated problems in the past took great amounts of time: Encke spent 40 years investigating the motion of the comet named after him; and Backlund studied the same comet for 38 years. Desk calculators considerably facilitated and accelerated the computation process.

The advent of electronic computers in the 1950's resulted in drastic changes: it became possible to pose and solve astronomical problems which were almost insoluble with desk calculators, and all the more so with logarithms. Increasingly difficult problems require more time spent on preparing programmes for the computer. As a result, it is necessary to simplify programming, to use the computer systematically and, particularly, to spend less time debugging the programme.

There are two possibilities for accomplishing this:

(1) writing the programme in ALGOL; (2) forming a library of standard astronomical programmes, as is done for mathematical problems.

When constructing numerical theories for the motion and investigation of the evolution of orbits of the minor bodies of the solar system with the use of computers, it is necessary to have a complex set of programmes. A considerable part of the computer's memory is used for the coordinates of the major planets and other constants. The memory of the computer is used more rationally when programmes are written in machine language rather than in the ALGOL. That is why we have preferred to follow the second possibility above and provide a library of standard programmes.

Rather than store tables of various functions in the computer's memory it is convenient to have special programmes that compute values of these functions according to particular algorithms. These auxiliary programmes are called standard programmes (SP), because they are constructed so as to make them suitable for use in any other programme and with any values of the arguments within the limits given in the instructions for their use. For every computer there is a set of SP's for the calculation of elementary functions and many of the methods of linear algebra, mathematical analysis, and probability theory.

The library set up at the Institute for Theoretical Astronomy has SP's for many common astronomical calculations, such as the solution of Kepler's equation, calculation of rectangular coordinates and velocities from elements, calculation of the coordinates of the Sun from Newcomb's theory, interpolation in tables by the Lagrange method (with a fixed and variable step). More complicated SP's, such as comparison of the observed and computed coordinates of any object, have been constructed by adding to the simple programmes already prepared.

The use of the library of SP's for the solution of complicated astronomical problems with a computer permits economical employment of the computer's memory, because the SP's required at a given stage of calculation are merely brought in as needed.

The library at the Institute contains about 120 SP's. The author has composed more than 50 of them. Together with Makover she has constructed a set of programmes for the improvement of elliptical orbits with six as well as eight unknowns.

The following files may be considered as necessary supplements to the SP's:

(1) rectangular coordinates of the major planets Venus–Pluto at 20-day intervals with allowance for all mutual perturbations during 1660–2060 (Kazimirchak-Polonskaya);

(2) coordinates of Mercury 1900–2000 at 5-day intervals, using the tables by Duncombe *et al.* (1965), with the coordinates thoroughly checked and stored on punched cards and magnetic tape (Bokhan);

(3) coordinates of observatories on punched cards (Belyaev and Bokhan).

There are many methods for the numerical integration of differential equations, such as those of Encke, Cowell, Numerov, and Subbotin. Perturbations for any object can be computed and also checked by using two or more such methods each of which has advantages and disadvantages. At the Institute standard programmes have been constructed for integration using various methods:

(1) integration with double precision, automatic change of step, allowance for perturbations by all major planets and nongravitational effects (Kazimirchak-Polonskaya);

(2) integration by Cowell's method with automatic selection of step and consideration of perturbations by all major planets (Belyaev);

(3) integration of *n* bodies by Myachin's method (Sizova);

(4) integration with a variable step and consideration of perturbations by all

major planets and of nongravitational effects by the method of variation of arbitrary constants using Herrick's vector parameters (Bokhan).

This is all part of a general programme which includes the simultaneous comparison with observations and the computation of ephemerides.

2. Numerical Integration Programme

With the development of modern electronic computers significant changes have been made in the classical methods of the variation of arbitrary constants. To make the process more suitable and systematic for programming, vector elements were used. Introduced by Milankovich (1941), vector elements were developed by Bilimovich (1943) and Popovich (1950). Such methods were not popular until computers were used for astronomical calculations.

Herrick (1948) and Musen (1954) obtained equations for the perturbations of vector elements in a form suitable for computers. Herrick (1953) applied his method to the minor planet Icarus. Vector elements can be successfully used for perturbations on comets of large or moderate orbital eccentricity; the method is especially convenient for objects of small perihelion distance, when Cowell's method requires an extremely small integration step.

The method of quadratures has been used for calculating the perturbations from derivatives, differences up to the fourth order being considered. For automatic selection of the integration interval Kazimirchak-Polonskaya's (1967) criterion was used, depending on the distance from the Sun and perturbing planets.

Nongravitational effects were considered on the assumption that the secular acceleration is in the form of an impulse acting exclusively at perihelion (Makover, 1955). Let t_0 be the osculation epoch and t a given moment of time. If a comet passes through perihelion at time t' in the interval $t_0 - t$, we can write

$$M = M_0 + \mu_0(t - t_0) + \mu'(t - t') + \text{perturbations}$$

$$\mu = \mu_0 + \mu' + \text{perturbations}$$

$$\varphi = \varphi_0 + \varphi' + \text{perturbations},$$

where $\mu' = 2\kappa/P$, P is the period of a comet (in days), and μ' and φ' are the secular variations in the mean daily motion and eccentric angle.

The programme described above has been used for the integration of the equations of motion of P/Encke during 1967–1971, using a variable step and taking into account perturbations by the eight major planets Mercury-Neptune and nongravitational effects. A search ephemeris was calculated for 1970–1971 (Bokhan, 1970).

During the next five years we intend to study the motion of P/Encke over the total interval since discovery. We have started by considering the little-studied period 1911–1921, but the results obtained up to now are preliminary and hence not published.

Acknowledgments

In conclusion I wish to express my cordial thanks to E. I. Kazimirchak-Polonskaya for her constant attention to this work, her valuable comments, for supplying her standard programmes, and for check calculations by her method and integration programme. I am equally grateful to N. A. Belyaev for supplying programmes and for his active help in the investigation. Others who participated in setting up the programme library were S. G. Makover, N. S. Kochina, L. E. Nikonova, V. A. Izvekov, M. Ya. Shmakova, T. I. Podunova and T. K. Nikol'skaya.

References

Bilimovich, A.: 1943, *Astron. Nachr.* **273**, 161.
Bokhan, N. A.: 1970, *Kiev. Komet. Tsirk.* No. 103.
Duncombe, R. L., Tufekcioglu, Z., and Larson, G.: 1965, *U.S. Naval Obs. Circ.* No. 106.
Herrick, S.: 1948, *Publ. Astron. Soc. Pacific* **60**, 321.
Herrick, S.: 1953, *Astron. J.* **58**, 156.
Kazimirchak-Polonskaya, E. I.: 1967, *Trudy Inst. Teor. Astron.* **12**, 24.
Makover, S. G.: 1955, *Trudy Inst. Teor. Astron.* **4**, 133.
Milankovich, M.: 1941, *Kanon der Erdbestrahlung*, Belgrade, p. 78.
Musen, P.: 1954, *Astron. J.* **59**, 262.
Popovich, B.: 1950, *Compt. Rend. Belgrade* **198**, 129.

Discussion

G. Sitarski: It is particularly difficult to link two apparitions of a comet separated by a close approach to Jupiter and absolutely necessary to allow for perturbations in the calculation of the differential coefficients. Do you allow for these?

N. A. Bokhan: Not yet, but work is in progress on this problem.

17. THE SOLUTION OF PROBLEMS OF COMETARY ASTRONOMY ON ELECTRONIC COMPUTERS

N. A. BELYAEV

Institute for Theoretical Astronomy, Leningrad, U.S.S.R.

Abstract. A series of standard programmes has been developed for numerical integration by Cowell's method of the differential equations of motion of minor bodies. A variable step is used, and perturbations by eight major planets and nongravitational effects are taken into consideration. Further programmes have been constructed as part of a general attempt at ITA to produce numerical theories of cometary motion. They include the reduction of observations, the comparison of the observations with theory and the improvement of orbits. The programmes make it possible to calculate $(O-C)$ residuals of up to 2000 observations simultaneously.

The study of the motions of comets and other minor bodies of the solar system is one of the major problems investigated at the Institute for Theoretical Astronomy of the Academy of Sciences of the U.S.S.R. Accordingly, great attention is paid at the Institute to developing various methods of integration and constructing the sets of programmes needed for solving a wide range of problems in cometary and meteor astronomy. Each of these sets has its own method of integration and manner of allowing for perturbations (Herrick, Cowell, Numerov, Myachin, etc.). The programmes that have been developed supplement one another, and this helps to ensure that the results obtained are reliable to the accuracy required. The possibility of choosing a method according to the orbital characteristics, accuracy of the starting data and other factors leads to the greatest compatibility between the method of investigation and the objective.

Following Bokhan's (1972) initiative, in order to reduce the time needed for programming we have introduced a process of standardization of astronomical calculations and have compiled at the Institute a large library of standard programmes (SP's) on punched cards, this being intended for use in the solution of a wide variety of problems. This method of programming appreciably facilitated the construction of large programmes, for any complicated programme is now to some degree the result of mutual activity.

Our set of BESM-4 computer programmes devised for solving the problems of cometary astronomy includes programmes for processing observations, comparison of observations with orbits, improvement of elliptical orbits, computation of ephemerides, and study of the evolution of cometary orbits.

The numerical integration is done by Cowell's method, using a variable step and allowing for up to the fourth differences. The perturbations are computed using a series of osculating elements of the eight major planets Venus–Pluto. The frequency of the epochs of osculation was chosen separately for each planet and such that coordinates calculated from the formulae for undisturbed motion could give the required accuracy at any instant during the 400-year interval 1660 to 2060. The whole series of elements was thoroughly checked.

Chebotarev et al. (eds.), The Motion, Evolution of Orbits, and Origin of Comets, 90–94. All Rights Reserved.
Copyright © 1972 by the IAU.

For checking the integration step we apply Kulikov's (1964) criterion, namely, that the fourth difference $\bar{f}_0^4$ satisfies

$$2^{p+3}\varepsilon_0 < |\bar{f}_0^4| < 2^{p+9}\varepsilon_0, \tag{1}$$

where p gives the order of the first sum (in binary), and ε_0 is the permissible relative error of the calculations. If the lower limit of this inequality does not hold, the integration step is doubled, and if the upper limit does not hold, it is halved.

Table I shows the variation of the integration step for P/Giacobini-Zinner during 1933–1940. The total number of steps for one revolution is 194, so that the 'mean' step is 12 days. A comet with the same period but a perihelion distance of 1.5 AU or more would have a mean step of 15 days. The maximum step for minor bodies of the solar system is 40 days.

TABLE I

Variation with heliocentric distance r of the integration steps $\bar{\omega}$ for P/Giacobini-Zinner during the interval 1933–1940

$\bar{\omega}$ (days)	r (AU)	No. of steps
	1.46	
5		23
	2.50	
10		22
	4.03	
20		18
	5.48	
40	(6.03)	25
	5.05	
20		16
	3.44	
10		16
	2.19	
5		20
	1.27	
2.5		54
	1.00	
5		—
	1.47	

$e = 0.6$, $q = 1.0$ AU, $Q = 6.0$ AU,
$P \sim 2400$ days.

It is well known that when integrating with a variable step it is desirable to store in the computer's memory the precise coordinates of the body for up to 12 or 13 integration points. It is then possible to recalculate the coordinates without loss of accuracy at any point in the integration table. Our programme has been so constructed that the operative memory of the computer stores the precise coordinates for as few as seven integration points, about half the usual number. This was achieved by means of a new method for the recalculation of the integration table. The method involves a new table constructed from osculating elements that are obtained from the precise coordinates (for seven points) given in the previous table. The perturbations are allowed for by means of the formulae by Kazimirchak-Polonskaya (1967). These formulae are intended for calculating the planetary perturbations in rectangular co-ordinates for five initial integration points, provided that the epoch of osculation T_0 and integration step $\bar{\omega}$ are taken to be arbitrary. The perturbations are extrapolated for the two extreme points. The calculation involves successive approximations until the differences between two consecutive iterations are less than the relative accuracy required.

If necessary, the nongravitational effects are taken into consideration by the formulae (Dubyago, 1949)

$$\varphi = \varphi_0 + \varphi'(t - t_0) + \text{perturbations}$$

$$\mu = \mu_0 + 2\mu'(t - t_0) + \text{perturbations} \qquad (2)$$

$$M = M_0 + \mu_0(t - t_0) + \mu'(t - t_0)^2 + \text{perturbations},$$

and changing the step near perihelion (heliocentric distance less than 3 AU), as a rule from $\bar{\omega} = 10$ to $\bar{\omega} = 5$ and vice versa.

In view of the great scope of future studies in cometary and meteor astronomy particular thought was given to the form of the computer input and output. A special standard programme was constructed so that the input data could be in any of about 30 different styles, which means that there is practically no need for any preliminary processing of information taken from the literature. Another programme causes the following quantities to be printed out:

T_0 (civil date), T_0 (Julian date), M_0, ω, Ω, i, φ, μ, e, a, $q = a(1 - e)$, $Q = a(1 + e)$, P (revolution period), $\pi = \omega + \Omega$, r (heliocentric distance), v_0 (true anomaly), $l_0 = \pi + v_0$, C_0 (Jacobian constant), Σ (number of steps), $\bar{\omega}$ (step).

These values, required for additional analysis, plotting, etc., are printed out at perihelion passage, for a number of preselected dates, at encounters with major planets and, if desired, at each integration step. The information output at approaches of comets to the major planets is controlled by a special programme that analyses the results for each separate planet before printing out.

Three programmes intended for comparison with observations, ephemeris computation and investigation of the evolution of cometary orbits were constructed on the basis of the integration programme described above. About 90% of these large programmes consist of the same standard programmes.

The solution of problems of cometary astronomy on electronic computers inevitably poses the problem of an increase in accuracy at all stages of the computations and, above all, in the starting data. We have therefore embarked at ITA on a project of constructing numerical theories of cometary motion, consideration being given to the entire range of observational material. This involves collecting and processing a great number of observations obtained during the last 100 to 120 years. To facilitate this work a scheme was devised for extracting observations from primary sources and cataloguing them, and a special programme was constructed for processing observations. A rubber stamp was made for printing the scheme on the back side of computer cards. Experience at ITA and other institutions has shown that writing down each observation on a separate computer card is very convenient when compiling large catalogues of cometary observations.

The processing programme converts the time of observation to Ephemeris Time, refers α_0 and δ_0 to the standard equinox 1950.0, calculates the precise rectangular coordinates of the Sun and correction for parallax. The coordinates of the Sun are computed from Newcomb's expansions using a standard programme by Podunova. To allow for parallax, a catalogue of 500 observatories has been compiled from all the references at our disposal. The correction for aberration is taken into account in calculating the $(O-C)$ residuals. Up to 2000 observations of a comet may be processed and compared at a time.

A programme for improving elliptical orbits by use of Eckert-Brouwer coefficients has also been constructed, starting from standard programmes by Bokhan, Makover and Nikol'skaya. Up to 200 observations or normal places may be utilized for the improvement. If the accuracy proves to be inadequate, we make use of a procedure based on the variation of elements and seven-fold integration.

Integration of the motion of a comet of the Jupiter family by Cowell's method, with a variable step and considering perturbations by the eight major planets, can be done on the BESM-4 computer at a rate of about 60 yr per hour. For a Trojan minor planet the rate is about 400 yr per hour. Full processing of the observations, including the computation of the solar coordinates, is done at the rate of 14 observations per minute.

The entire set of programmes has been thoroughly checked for proper operation and adequate accuracy. The accumulation of rounding errors has been estimated; it was shown that the rounding error will not build up to the seventh decimal place after 1000 integration steps and will be equal to several units in the sixth decimal place after 10 000 steps on the BESM-4 computer.

As already stated, the mean integration step for comets with a perihelion distance of 1.5 AU or more is about 15 days. This amounts to 25 steps per year (or 1000 steps in 40 yr and 10 000 steps in 400 yr). Our programmes may therefore be used for constructing accurate numerical theories for the motion of a short-period comet for some 30 to 40 yr on each side of the initial osculation epoch; i.e., for a total interval of 70 to 80 yr, provided that there are no close encounters with Jupiter during that interval. General studies of orbital evolution may be extended over several centuries. In cases where our programmes cannot be utilized (either for the problem as a whole or at particular stages), the double-precision programme by Kazimirchak-Polonskaya should be applied. In order to make a thorough check of the operation of the two programmes, we have integrated the equations of motion of P/Daniel from 1909 to 1944 and compared the integration results; see Table II. There was only a single encounter with Jupiter, to about 0.65 AU in 1912.

The whole range of our programmes has been widely utilized in the studies of the orbital evolution of a large number of asteroids (this work being still in progress); and we have calculated ephemerides, constructed numerical theories, and investigated the motions of several comets, notably P/Faye, P/Giacobini-Zinner, P/Daniel, P/-Borrelly, P/Tempel-Tuttle and P/Ashbrook-Jackson.

TABLE II

Comparison of the results of integration of the equations of motion for P/Daniel from 1909 to the epoch 1944 January 31.0 UT

	Kazimirchak-Polonskaya	Belyaev	Difference
M_0	$10°07941$	$10°07874$	$+0''11$
ω	6.13719	6.13715	$+0.14$
Ω	70.43119	70.43119	0.00
i	19.84852	19.84852	0.00
φ	35.07173	35.07174	-0.04
μ	0.14492861	0.14492852	$+0.0003$

$e = 0.6$, $q = 1.4$ AU, $Q = 5.6$ AU, $P = 6.5$ yr.

Acknowledgments

It is my pleasant duty to thank E. I. Kazimirchak-Polonskaya and N. A. Bokhan for their continuous interest and help in my work as well as for useful discussions preparatory to the Symposium. My special thanks are also due to N. A. Bokhan, T. I. Podunova and M. Ya. Shmakova, who were kind enough to contribute their standard programmes.

References

Bokhan, N. A.: 1972, this Symposium, p. 86.
Dubyago, A. D.: 1949, *Opredelenie Orbit*, Moscow. (English translation: 1961, *The Determination of Orbits*, Macmillan, New York.)
Kazimirchak-Polonskaya, E. I.: 1967, *Trudy Inst. Teor. Astron.* **12**, 37.
Kulikov, D. K.: 1964, *Byull. Inst. Teor. Astron.* **7**, 782.

18. A METHOD OF INTEGRATING THE EQUATIONS OF MOTION IN SPECIAL COORDINATES AND THE ELIMINATION OF A DISCONTINUITY IN THE THEORY OF THE MOTION OF PERIODIC COMET WOLF

E. I. KAZIMIRCHAK-POLONSKAYA

Institute for Theoretical Astronomy, Leningrad, U.S.S.R.

Abstract. From the integration formulae of Numerov and Subbotin we have developed and programmed for an electronic computer a particular method for integrating the differential equations of cometary motion in special rectangular coordinates, with a variable step and allowing for all planetary perturbations and nongravitational effects over a time interval of 400 yr. Application of this method and our set of programmes to the investigation of the motion of P/Wolf permits us to eliminate the discontinuity that has hitherto existed in the theory on account of the comet's close approach to Jupiter in 1922.

1. Introduction

Owing to the rapid development of electronic computers and their application to astronomical calculations, cometary astronomy, utilizing the methods of celestial mechanics, is coming to a new stage in its development. Large-scale problems may now be formulated and solved by new methods and using electronic computers. We shall point out here only four of the more important problems.

(1) Construction of continuous numerical theories for the motions of comets over the whole period spanned by their observations, with consideration given to the perturbations by the planets and to nongravitational effects.

(2) Determination of the mass of Jupiter from the large perturbations induced in the orbits of comets that pass through Jupiter's sphere of action.

(3) Investigation of the evolution of cometary orbits over a time scale of centuries and clarification of the role of the giant planets (Jupiter–Neptune) in defining this evolution.

(4) Study of the motion of meteor streams, particularly of the large transformations of their orbits in the sphere of action of Jupiter, and the determination of the effects of planetary perturbations upon the structure of these streams and on the conditions of visibility of the meteor showers.

Our aim is the development of a rigorous procedure for the investigation and solution of these problems, including the compilation of a set of constants and standard programmes.

This paper describes the procedure for investigation and the solution of the first of the above-mentioned problems. The other three problems are discussed in other papers (Kazimirchak-Polonskaya, 1972a, 1972b; Kazimirchak-Polonskaya *et al.*, 1972).

We have prepared punched cards (mainly in binary) containing the coordinates of the eight major planets, with full allowance for their mutual perturbations, and

Chebotarev et al. (eds.), The Motion, Evolution of Orbits, and Origin of Comets, 95–102. All Rights Reserved.

they are given at 20-day intervals for a period of 400 yr (1660–2060). These coordinates are based both on classical and more recent theories for the motions of the major planets (Newcomb, 1895a, 1895b; Clemence, 1943, 1949, 1954, 1961; Duncombe, 1958; Eckert et al., 1951; Morgan, 1945). The computations were performed using programmes by Subbotina (1965) and by the author.

2. Method of Integration

In developing a method for integrating the differential equations of motion for a minor body in the solar system:

$$w^2 \frac{d^2x}{dt^2} = -\frac{k^2 w^2 x}{r^3} + R_x, \qquad x \to y, z, \tag{1}$$

where

$$R_x = k^2 w^2 \sum_{i=2}^{l} m_i \left(\frac{x_i - x}{\Delta_i^3} - \frac{x_i}{r_i^3} \right); \qquad l = 6, 7, 8, 9, \tag{2}$$

we took as our basis the theory proposed by Myachin (1962), which provides an estimate of the total error in the calculations, and also two methods of integration devised by Numerov (1923) and Subbotin (1927, 1928).

According to Myachin, the total linearized error of the integration of Equations (1) is, after n steps, the sum of the following:

(1) an error, proportional to n, arising from errors in the initial data;

(2) a truncation error depending on the number of derivatives or differences taken into account;

(3) errors due to multiple iterations at each integration step;

(4) an error, proportional to $n^{3/2}$, arising from the rounding at each step.

We have aimed at developing a method where each of the partial errors, and hence the total error, would be reduced to a minimum.

Let us consider each of the points in detail.

(1) On the one hand, the error arising from errors in the initial data is dependent upon the accuracy of observations. It is thus of primary importance to increase the accuracy of cometary observations. On the other hand, the error is also dependent upon the theoretical development of the method. This error may be reduced in two ways:

(a) Firstly, considering that the error increases directly as the number of integration steps, we utilize, following Numerov and Subbotin, both the ordinary coordinates x, y, z, and special coordinates $\bar{x}$, $\bar{y}$, $\bar{z}$, that are related to the ordinary ones by

$$\bar{x}_k = x_k \left(1 + \frac{k^2 w^2}{12 r_k^3} \right)$$

$$\bar{r}_k^2 = \bar{x}_k^2 + \bar{y}_k^2 + \bar{z}_k^2. \tag{3}$$

By introducing auxiliary functions λ and σ, defined by

$$\lambda_k = \frac{k^2 w^2}{\bar{r}_k^3}; \qquad \sigma_k = \lambda_k + \tfrac{1}{6}\lambda_k^2 + \tfrac{7}{144}\lambda_k^3 + \tfrac{5}{288}\lambda_k^4 + \cdots \qquad (4)$$

(where it is usually sufficient to ignore further terms in σ_k), Subbotin arrived at the following formulae for integrating Equations (1) in the special coordinates:

$$\bar{x}_k = f_{x_k}^{-2} + \overline{\mathrm{Red}}_{x_k} \qquad (5)$$

$$\overline{\mathrm{Red}}_{x_k} = \tfrac{1}{12}R_{x_k} - \tfrac{1}{240}f_{x_k}^2 + \tfrac{31}{60480}f_{x_k}^4 - \tfrac{289}{3628800}f_{x_k}^6 + \cdots, \qquad (6)$$

where the function f_{x_k} denotes, as in Cowell's method, the right-hand members of Equations (1),

$$f_{x_k} = -\frac{k^2 w^2 x_k}{r_k^3} + R_{x_k} = -\sigma_k \bar{x}_k + R_{x_k}. \qquad (7)$$

The functions $f_{x_k}^{-2}, f_{x_k}^2, f_{x_k}^4, f_{x_k}^6, \ldots$ stand for the second sum and even differences of f_{x_k}. For computing Equations (2) the special coordinates are changed to the ordinary ones by the very simple formula

$$x_k = \bar{x}_k - \bar{x}_k \sigma_k / 12. \qquad (8)$$

Comparing Equations (5) and (6) with the well-known formulae for integration by Cowell's method,

$$x_k = f_{x_k}^{-2} + \mathrm{Red}_{x_k} \qquad (9)$$

$$\mathrm{Red}_{x_k} = \tfrac{1}{12}f_{x_k} - \tfrac{1}{240}f_{x_k}^2 + \tfrac{31}{60480}f_{x_k}^4 - \tfrac{289}{3628800}f_{x_k}^6 + \cdots, \qquad (10)$$

we see, that while the first terms in Equations (6) and (10) are appreciably different, the rest of the terms are identical. Near perihelion, the first term in Cowell's method becomes large and changes rapidly, making it necessary to reduce the integration step. In contrast, the corresponding term in Subbotin's method remains small and changes slowly. Therefore, for any specified accuracy, one may use a larger step-size in Subbotin's method than in Cowell's.

Introduction of special coordinates thus reduces the number of integration steps, and this reduces the error arising from errors in the initial data.

(b) Secondly, we may appreciably increase the accuracy of the initial integration table. Ever since it became customary to refer the tables of the coordinates of the major planets to standard dates, astronomers have imposed the requirement that the epoch of osculation in the initial integration table should also be a standard date. Nevertheless, the initial orbital elements for a comet are frequently referred to arbitrary epochs; this is particularly true for the earlier studies, when the epochs might have been reckoned in Greenwich Mean Time, Berlin Mean Time, or Paris Mean Time, etc. When adopting a new standard epoch of osculation it was usual to adjust only the mean anomaly, the other elements being considered invariable since the perturbations by the planets are small.

In order to refine the initial data we derived and programmed formulae enabling the rectangular coordinates of the body to be obtained, with significantly higher accuracy, for seven initial standard dates (Kazimirchak-Polonskaya, 1967, p. 35). The initial epoch of osculation could have been any date, standard or nonstandard, and we allowed for all planetary perturbations (and nongravitational effects, if necessary). The initial integration table is computed from these coordinates in double precision, thus providing an appreciable reduction in the errors of the initial data.

(2) In order to reduce the truncation error we use Subbotin's quadrature method and perform the integration in double precision (18 significant figures) using Equations (1)–(8). Owing to the rapid decrease of the coefficients, Equation (6) may be restricted to the sixth differences. If the eighth differences become significant, within the limits of the specified accuracy, the step-size is immediately decreased.

On the basis of our theoretical investigations a new criterion has been formulated for automatic choice of the integration step, depending on the specified accuracy and on the distance of the body under study from the Sun and all disturbing planets at any time (Kazimirchak-Polonskaya, 1967). It is interesting to note that the standard programme for automatic choice of the integration step is based exclusively on logical operations.

In connection with this criterion, we have found the most efficient way of doing the computation, with some operations performed in single and others in double precision; as a result, we attain a high degree of accuracy of the coordinates of the body under investigation, even on a time scale of centuries, and this requires a minimum of operation time.

(3) The third partial error arises after multiple iterations at each integration step, the number of iterations being considerably increased when we use double precision.

As is well known, the necessity for successive approximations arises because the expressions in the right-hand members of Equations (6) or (10) are initially unknown. It can be seen from comparison of Equations (9) and (10) with Equations (3)–(6) that the number of iterations in Cowell's method is greater than in Subbotin's method, specifically because of the different first terms in Equations (6) and (10).

We suggest a new way of applying Subbotin's method directly, without approximations. For this purpose we write Equation (6) in the form

$$\overline{\mathrm{Red}}_{x_k} = \tfrac{1}{12}R_{x_k} + a_{-3}f_{x_{k-3}} + a_{-2}f_{x_{k-2}} + a_{-1}f_{x_{k-1}} + a_0 f_{x_k}$$
$$+ a_1 f_{x_{k+1}} + a_2 f_{x_{k+2}} + a_3 f_{x_{k+3}}, \tag{11}$$

where the coefficients $a_{-3}, a_{-2}, \ldots, a_3$ are calculated to the required accuracy. Values of the functions $f_{x_{k-3}}, f_{x_{k-2}}, f_{x_{k-1}}$ in Equation (11) are known at the kth step to a high degree of accuracy. To compute the values for $f_{x_k}, f_{x_{k+1}}, f_{x_{k+2}}, f_{x_{k+3}}$ we proceed from the integration formula in the method of differences by Numerov:

$$\bar{x}_{k+1} = (2 - \sigma_k)\bar{x}_k - \bar{x}_{k-1} + R_{x_k} + \overline{\mathrm{red}}_{x_k} \tag{12}$$

$$\overline{\mathrm{red}}_{x_k} = \tfrac{1}{12}\Delta^2 R_{x_k} - \tfrac{1}{240}f_{x_k}^4 + \tfrac{31}{60480}f_{x_k}^6 - \cdots, \tag{13}$$

where R_{x_k}, σ_k and f_{x_k} are calculated as in Subbotin's method by Equations (2), (4) and (7).

Since Equation (6) in Subbotin's method always involves a very restricted number of significant figures, it is permissible to ignore Equation (13) and obtain the f-functions for dates t_k, t_{k+1}, t_{k+2}, t_{k+3} to appropriate accuracy by Equations (2), (3), (4), (7) and (8) and by Numerov's reduced formula

$$\bar{x}_{k+1} = (2 - \sigma_k)\bar{x}_k - \bar{x}_{k-1} + R_{x_k}; \tag{14}$$

this requires no iteration. The values obtained for f_{x_k}, $f_{x_{k+1}}$, $f_{x_{k+2}}$, $f_{x_{k+3}}$ are then substituted into Equation (11), enabling Subbotin's method to be applied to full accuracy without iteration, and thereby reducing the iteration error to zero.

(4) Since the integration is performed in double precision the rounding error does not become significant until after 10 000 steps or more.

As a result of the expressions and procedures developed by us the total error of our integration method is reduced to a minimum, and this is a necessary condition for solving the four problems defined in the Introduction.

3. Application to the Study of the Motion of P/Wolf

All astronomers who work on the motions of comets know how difficult it is to construct an appropriately accurate numerical theory linking the motion of a comet over a large number of apparitions covering a long interval of time. In addition, if the comet passes well inside Jupiter's sphere of action within this time interval, the calculation of the great changes in its orbit becomes so increasingly complicated that up to now it has been considered impossible to link all the apparitions before and after the close approach to Jupiter. Therefore, the investigators had recourse to an artificial solution: they excluded the critical revolution containing the comet's passage through the sphere of action of Jupiter and constructed numerical theories for separate time intervals on the two sides of the encounter. There was inevitably a discontinuity in the theory of the motion and this was considered irremovable. Elimination of such discontinuities and the construction of continuous numerical theories of the motions of comets for the whole period covered by observations is an urgent necessity and our first aim.

Such a discontinuity existed even in the very skillful theory, constructed by Kamieński (1959), of the motion of P/Wolf. The theory comprises two isolated series of linked apparitions of the comet. The first series covers its motion from discovery to the last apparition before the close approach to Jupiter in 1922, i.e., the interval 1884–1918, and it represents 50 normal places (about 2000 observations) with the very low mean error $\varepsilon = \pm 1''.77$ (Kamieński, 1933). The perturbations by six planets, Venus through Uranus, were considered, and the effect of nongravitational forces was taken into account, to a high degree of accuracy, in the form of secular deceleration in the motion, using the formulae

$$\Delta\mu_t = -0''.0000\ 0042\ (t - t_0)$$

$$\Delta M_t = -0''.0000\ 0021\ (t - t_0)^2,$$

where $\Delta\mu_t$ and ΔM_t are the secular variations in the mean daily motion and mean anomaly over the time interval (measured in days) from the initial epoch t_0 to the running time t.

In the second series Kamieński (1948) linked three apparitions over the interval 1925–1942, after the comet's passage through the sphere of action of Jupiter in 1922, with a still higher degree of accuracy ($\varepsilon = \pm 1''21$). Allowance was again made for the perturbations by the same six planets and nongravitational effects, the only difference being that the coefficients for the secular deceleration were diminished to

$$\Delta\mu_t = -0''0000 \quad 00067 \, (t - t_0)$$
$$\Delta M_t = -0''0000 \quad 00034 \, (t - t_0)^2.$$

Kamieński believes that these changes in the nongravitational effects result from the fact that the comet, through the large perturbations induced by Jupiter, was deflected into a new orbit with considerably changed parameters, particularly the perihelion distance (increasing from 1.6 to 2.4 AU).

Using the second series of elements Kamieński extended his integration of the equations of the motion of P/Wolf to the present time. The comet was observed in 1950–1951, 1959 and 1967, and the agreement between the observations and calculations was quite satisfactory.

Thus there exists a numerical theory of the motion of P/Wolf for two prolonged but isolated time intervals, 1884–1918 and 1925–1967. The critical revolution of 1918–1925 was excluded since Kamieński and Bielicki (1935, 1936) could not get a good representation of the 1925 observations in terms of the elements for 1918 (obtained in the first series). The maximum $(O - C)$ residuals in 1925 were as high as 5^s2 in right ascension and $4''8$ in declination. Until now the residuals have not been reduced by any of the methods employed (Kamieński, 1959).

At Kamieński's request we endeavoured to eliminate the discontinuity in the theory of motion of P/Wolf during the revolution 1918–1925. Having thoroughly analysed the causes that prevented the solution of the problem, we developed the above method of integration in special coordinates and elaborated with great care heliocentric and jovicentric methods for computing the large perturbations on cometary orbits in the sphere of action of Jupiter (Kazimirchak-Polonskaya, 1961).

The application of our procedure and set of programmes enabled us to solve this difficult problem. In spite of very large perturbations in the orbital elements of P/Wolf in Jupiter's sphere of action in 1922, amounting to more than $20°$ in mean anomaly, about $13°5$ in longitude of perihelion, about $10°$ in eccentricity angle, and more than $96''$ in mean daily motion, we obtained a quite satisfactory representation of the normal places of the comet in 1925. The results are given in Table I and compared with those of Kamieński and Bielicki (1936, p. 11).

It should be noted that we have allowed for the perturbations by the six planets Venus-Uranus and the secular deceleration, together with nongravitational effects in other orbital elements, which were detected by Kamieński (1933) but not taken into account by him. They have been calculated by the formulae

$$\Delta\Omega_t = -0''000345 \, (t - t_0)$$
$$\Delta\omega_t = -0.000612 \, (t - t_0),$$

where the coefficients have been determined according to Kamieński's (1933, p. 44) method.

In order to obtain an additional check on our results, we have compared our set of elements for P/Wolf at the epoch of osculation 1925 July 12.5 UT with the set of elements derived by Kamieński for the same epoch from linking the 1925–1942

TABLE I

Representation of normal places of P/Wolf in 1925

1925 UT	Kamieński and Bielicki		Kazimirchak-Polonskaya	
	$\Delta\alpha \cos \delta$	$\Delta\delta$	$\Delta\alpha \cos \delta$	$\Delta\delta$
July 18.0	$-3\overset{s}{.}88$	$+2\overset{''}{.}0$	$-0\overset{s}{.}03$	$+0\overset{''}{.}7$
Aug. 19.0	-4.90	0.0	-0.34	$+1.2$
Sept. 14.0	-5.16	-3.3	-0.30	$+0.4$
Oct. 12.0	-5.04	-3.1	-0.33	-0.5
Nov. 11.0	-4.34	$+3.5$	-0.14	$+2.1$
Dec. 19.0	-3.43	$+4.8$	$+0.19$	$+0.9$
m_J^{-1}	1047.400 (de Sitter)		1047.355 (Hill)	
Nongravitational effects included	$\Delta\mu, \Delta M$		$\Delta\mu, \Delta M, \Delta\Omega, \Delta\omega$	
ε	$\pm 67\overset{''}{.}6$		$\pm 3\overset{''}{.}9$	

apparitions. The differences, in the sense Kamieński minus Kazimirchak-Polonskaya are very small:

$$\Delta T = -0\overset{d}{.}0152, \qquad \Delta\Omega = -1\overset{''}{.}31, \qquad \Delta\varphi = -1\overset{''}{.}17$$

$$\Delta\omega = +3\overset{''}{.}83, \qquad \Delta i = +2\overset{''}{.}23, \qquad \Delta\mu = +0\overset{''}{.}0015.$$

Our orbital elements give a good representation of observations of the comet at subsequent apparitions considerably separated from the approach to Jupiter.

We can conclude from Table I that the discontinuity in the theory of motion of P/Wolf has been eliminated and the revolution 1918–1925 covered by the numerical theory of the motion of comet. Thus the theory has become continuous for the whole period of observations of P/Wolf, from discovery in 1884 up to the present time. However, we still consider our results to be preliminary and hope to arrive at a some-what better agreement between the observations and calculations when solving the second problem, that of determining the mass of Jupiter from the large perturbations on P/Wolf in 1922 (Kazimirchak-Polonskaya, 1972a).

Acknowledgments

We wish to express our deep and sincerest thanks to N. A. Bokhan, who constructed the numerous double-precision standard programmes (Bokhan, 1972) that have been utilized in our set of programmes. We also thank N. A. Belyaev for providing his programme for comparison of calculations and observations (Belyaev, 1972).

References

Belyaev, N. A: 1972, this Symposium, p. 90.
Bokhan, N. A: 1972, this Symposium, p. 86.
Clemence, G. M.: 1943, *Astron. Pap. Washington* **11**, part 1.
Clemence, G. M.: 1949, *Astron. Pap. Washington* **11**, part 2.
Clemence, G. M.: 1954, *Astron. Pap. Washington* **13**, part 5.
Clemence, G. M.: 1961, *Astron. Pap. Washington* **16**, part 2.
Duncombe, R. L.: 1958, *Astron. Pap. Washington* **16**, part 1.
Eckert, W. J., Brouwer, D., and Clemence, G. M.: 1951, *Astron. Pap. Washington* **12**.
Kamieński, M.: 1933, *Acta Astron. Ser. a.* **3**, 1.
Kamieński, M.: 1948, *Bull. Acad. Polon. Sci. Lettres Ser. A* 1.
Kamieński, M.: 1959, *Acta Astron.* **9**, 53.
Kamieński, M. and Bielicki, M.: 1935, *Repr. Astron. Obs. Warsaw Univ.* **30**, 270.
Kamieński, M. and Bielicki, M.: 1936, *Repr. Astron. Obs. Warsaw Univ.* **32**, 1.
Kazimirchak-Polonskaya, E. I.: 1961, *Trudy Inst. Teor. Astron.* **7**, 191.
Kazimirchak-Polonskaya, E. I.: 1967, *Trudy Inst. Teor. Astron.* **12**, 24.
Kazimirchak-Polonskaya, E. I.: 1972a, this Symposium, p. 227.
Kazimirchak-Polonskaya, E. I.: 1972b, this Symposium, p. 373.
Kazimirchak-Polonskaya, E. I., Belyaev, N. A., and Terent'eva, A. K.: 1972, this Symposium, p. 462.
Morgan, H. R.: 1945, *Astron. J.* **51**, 127.
Myachin, V. F.: 1962, *Byull. Inst. Teor. Astron.* **8**, 537.
Newcomb, S.: 1895a, *Astron. Pap. Washington* **6**, part 1, part 3.
Newcomb, S.: 1895b, *The Elements of the Four Inner Planets and the Fundamental Constants of Astronomy*, Washington.
Numerov, B. V.: 1923, *Trudy Gl. Ross. Astrofiz. Obs.* **2**, 188.
Subbotin, M. F.: 1927, *Byull. Sredne-Aziat. Univ. Tashkent* **16**, 273.
Subbotin, M. F.: 1928, *Byull. Sredne-Aziat. Univ. Tashkent* **17**, 21.
Subbotina, N. S.: 1965, *Byull. Inst. Teor. Astron.* **10**, 143.

Discussion

M. Bielicki: What are the causes of the differences between your results and those by Kamieński and myself? In order to obtain the most accurate elements for P/Wolf in 1918 (before the approach to Jupiter) I developed a method of linking two consecutive apparitions with the aim of determining nongravitational effects. These consecutive links enabled me to obtain changes in all the orbital elements as a function of time. According to the tradition then existing I took into consideration the nongravitational secular changes only in the mean motion and mean anomaly. I then calculated the perturbations (by hand, of course) in Jupiter's sphere of action three times, twice using the heliocentric method and once by the more precise jovicentric method. Very great difficulties arose in the determination of accurate initial values of the position and velocity components in the jovicentric method. In the heliocentric calculations, I allowed for the perturbations from Venus through Uranus, and in the jovicentric calculations I allowed for those by the Sun and Saturn only.

E. I. Kazimirchak-Polonskaya: Firstly, the value you adopted for the mass of Jupiter (1/1047.400) is not consistent with the coordinates of Jupiter, which you took from the *Berliner Astronomisches Jahrbuch* and which correspond to Newcomb's value of 1/1047.35. I used Hill's value for the mass of Jupiter (1/1047.355) and the highly accurate coordinates in the *Astronomical Papers*, Volume 12 (plus the corrections in Volume 13) which correspond to Hill's value. Secondly, we developed a more accurate procedure for the calculation of the large perturbations experienced by comets in Jupiter's sphere of action; further, the calculations were performed in double precision and the total integration error reduced to a minimum. Finally, I allowed also for the nongravitational effects in the longitude of the ascending node and longitude of perihelion.

19. THE USE OF THE ELECTRONIC COMPUTER FOR THE URGENT PUBLICATION OF ASTRONOMICAL MATERIAL

V. A. IVAKIN

Institute for Theoretical Astronomy, Leningrad, U.S.S.R.

Abstract. A system is described for processing scientific information with automatic type-setting of tabular text.

Modern astronomy would be very different if it were not for electronic computers. Automatic programming methods, universal algorithmic languages, and the wide availability of electronic computers have simplified as much as possible the process of preparing a programme for a particular problem and carrying out the necessary calculations. Electronic computers are used to provide ephemerides for the members of the solar system, to study the orbital evolution of comets, minor planets, and meteor streams over time spans of centuries and more, etc.

After solving such problems it is necessary to present the results in printed form to observers and research workers. However, a considerable time elapses between completion of the calculations and publication of the results, mainly on account of the type-setting that is necessary. Monotype and linotype equipment also have the drawback that preparation of the punched paper tape that carries type-setting information requires visual inspection of the original text, and errors inevitably appear.

Modern technology makes it possible to set information in type directly from the computers. However, malfunctions in the computer or computer-controlled typewriter can result in errors. Proofreading is still necessary, and considering the great amount of data in such annual publications as *Astronomicheskij Ezhegodnik* and *Efemeridy Malykh Planet*, this is a laborious and time-consuming job.

This is why it is impossible in practice to print urgent information, such as ephemerides or lists of observations of newly discovered bodies at the time it is needed.

At the Computing Centre of the Institute for Theoretical Astronomy we have developed a system for processing scientific information with automatic type-setting of tabular text.

The system comprises the following:

(1) The BESM-4 electronic computer with 8192 words (of 45 bits) of highspeed access memory, 16 384 × 8 words of buffer memory on magnetic drums, and 2 000 000 words of buffer memory on magnetic tapes;

(2) an electronic 'conversion' unit;

(3) monotype type-setting equipment, with controlled programme. Additional memory on magnetic discs may also be incorporated.

The initial information is introduced into the electronic computer in some standard way, i.e., from punched cards, punched paper tape or magnetic tape, or from another computer. The first function of the computer is to reduce the input information to a form in which it may be fed into the programming system utilized. The second

Chebotarev et al. (eds.), *The Motion, Evolution of Orbits, and Origin of Comets*, 103–104. *All Rights Reserved.*

function is to print out the astronomical results in the normal manner. This print-out is used for any checking we may wish to make of the final printed material.

The third function of the computer is to prepare the output information. This information contains the astronomical results, any additional literal or numerical information that is required, commands for controlling the operation of the type-setting equipment, and it is carried on standard punched paper tape.

The 'conversion' unit is an independent electronic logical system. Input information for this consists of the output information from the main computer and some auxiliary commands. An electronic device controls the reading of this input, and it stops in the case of malfunction or detection of an error.

The preliminary information processing unit then removes the auxiliary commands (which are required only for the reading and checking), and forms and stores the remainder. The information converting unit then converts from the computer code to a special code used by the units that operate the type-setting equipment.

The newly coded information is punched on 31-channel paper tape. It is compared line by line with the input information, and the machine stops if an error is detected. When any necessary corrections have been made the 31-channel tape is used for controlling the operation of the type-setting equipment. Final checking of the pre-liminary print-out from the computer can be made if desired.

C. DETERMINATION OF ORBITS

20. A NUMERICAL INTERPRETATION OF THE HOMOGENIZATION OF OBSERVATIONAL MATERIAL FOR ONE-APPARITION COMETS

G. SITARSKI

Astronomical Institute, Polish Academy of Sciences, Warsaw, Poland

Abstract. The formulae developed by Bielicki for the objective weighting of observations are combined with an orbital improvement method in which the effects of perturbations are included in the differential coefficients. The procedure is applied to the orbit of comet 1953 I.

Production of a new catalogue of the orbits of one-apparition comets requires the recomputation of all the cometary orbits in order to make the material homogeneous. Recomputation of the orbits can be realized if the following problems are elaborated mathematically:

(1) weighting of the observations and rejection of doubtful ones (the method should be purely mathematical and quite objective);

(2) improvement of the orbital elements with allowance for all perturbations in the comet's motion and even in the differential coefficients;

(3) taking into account the nongravitational effects in the comet's motion after such effects have been detected by some mathematical process.

The first two of these problems have been solved mathematically and programmed for a digital computer. Formulae and a method for the computation of weights of individual observations, as well as a criterion for rejecting the 'erroneous' observations, have been developed by Bielicki (1972). The formulae were derived on the basis of statistics and probability theory.

Formulae for the correction of the orbital elements, including the perturbations in the differential coefficients, have been derived by the present author (Sitarski, 1971). The differential coefficients are obtained by numerical integration of the differential equation for the correction $\Delta\mathbf{r}$ to the comet's position vector:

$$\frac{d^2\Delta\mathbf{r}}{dt^2} + k^2\frac{\Delta\mathbf{r}}{r^3} - 3k^2\frac{\mathbf{r}}{r^5}(\mathbf{r}\cdot\Delta\mathbf{r}) = \text{grad}\,(\text{grad}\,\Omega\cdot\Delta\mathbf{r}),$$

where k is the Gaussian constant and Ω the disturbing function. Substituting

$$\Delta\mathbf{r} = \sum_{i=1}^{6} C_i\mathbf{G}_i,$$

where the C_i are arbitrary constants, we transform the differential equation for $\Delta\mathbf{r}$ into six differential equations for the $\mathbf{G}_i$. The numerical integration of these equations gives the values of the differential coefficients in the observational equations, from which the constants C_i are obtained by the method of least squares. The appropriate

G. SITARSKI

TABLE I
Residuals for Comet Harrington 1953 I (Epoch 1953 Jan. 23.0 ET, Equinox 1950.0)

No.	$t_i - t_1$	Initial		Without weights		After weighting		$\sqrt{w}$
		$\Delta\alpha \cos \delta$	$\Delta\delta$	$\Delta\alpha \cos \delta$	$\Delta\delta$	$\Delta\alpha \cos \delta$	$\Delta\delta$	
1	0.318	− 33″62	+126″78*	− 34″00*	+126″73*	− 33″27*	+126″51*	3.17
2	2.400	+ 0.32	− 0.48	− 0.05	− 0.52	+ 0.69	− 0.75	3.17
3	3.472	+ 0.71	+ 0.38	+ 0.34	+ 0.34	+ 1.08	+ 0.11	3.17
4	3.487	+ 1.45	+ 0.06	+ 1.09	+ 0.02	+ 1.83	− 0.21	3.17
5	4.949	− 33.91	− 2.40	− 34.27*	− 2.44	− 33.52*	− 2.67	3.17
6	6.837	− 1.42	− 5.21	− 1.76	− 5.24	− 1.01	− 5.48*	3.17
7	7.826	+ 0.42	+ 1.09	+ 0.08	+ 1.06	+ 0.84	+ 0.82	3.17
8	8.793	+ 3.84	+ 1.10	+ 3.51	+ 1.07	+ 4.27*	+ 0.83	3.17
9	10.348	− 0.18	− 0.56	− 0.51	− 0.59	+ 0.26	− 0.83	3.17
10	10.374	− 0.30	+ 1.51	− 0.62	+ 1.49	+ 0.15	+ 1.24	3.17
11	21.152	+ 0.46	+ 1.00	+ 0.20	+ 0.98	+ 1.03	+ 0.74	3.17
12	21.185	+ 0.60	+ 1.11	+ 0.33	+ 1.09	+ 1.16	+ 0.84	3.17
13	21.720	− 1.46	+ 2.48	− 1.72	+ 2.46	− 0.89	+ 2.21	3.17
14	23.746	− 0.29	+ 0.56	− 0.54	+ 0.54	+ 0.31	+ 0.30	3.17
15	24.770	− 3.54	− 0.67	− 3.78	− 0.69	− 2.92	− 0.93	3.17
16	32.864	− 3.71	− 2.52	− 3.89	− 2.54	− 2.98	− 2.76	3.30
17	35.735	− 1.63	− 4.66	− 1.79	− 4.69	− 0.86	− 4.89*	3.72
18	53.903	+ 1.46	+ 32.96	+ 1.50	+ 32.89*	+ 2.51	+ 32.79*	0.96
19	58.901	+ 2.36	− 0.20	+ 2.47	− 0.30	+ 3.49	− 0.36	0.93
20	59.829	− 1.48	+ 5.41	− 1.36	+ 5.31	− 0.33	+ 5.25	0.94
21	63.794	− 0.88	+ 6.47	− 0.69	+ 6.34	+ 0.34	+ 6.31	1.00
22	72.648	+ 1.10	− 1.54	+ 1.45	− 1.76	+ 2.46	− 1.72	0.98
23	81.494	− 0.40	− 3.25	+ 0.13	− 3.58	+ 1.08	− 3.48	0.93
24	82.477	+ 20.00	+ 0.84	+ 20.56*	+ 0.51	+ 21.50*	+ 0.62	0.93
25	84.441	+ 2.26	+ 8.53	+ 2.86	+ 8.17	+ 3.78	+ 8.29	0.90
26	85.648	− 0.03	− 5.17	+ 0.60	− 5.55	+ 1.50	− 5.42	0.88
27	86.639	− 8.08	+ 3.26	− 7.43	+ 2.87	− 6.54	+ 3.01	0.87
28	87.694	+ 0.24	− 0.18	+ 0.90	− 0.59	+ 1.78	− 0.45	0.86
29	88.724	+ 1.19	− 0.36	+ 1.87	− 0.78	+ 2.74	− 0.63	0.86
30	89.660	+ 1.88	+ 0.20	+ 2.58	− 0.22	+ 3.44	− 0.07	0.86
31	90.693	+ 3.90	+ 0.60	+ 4.62	+ 0.16	+ 5.46	+ 0.32	0.86
32	91.527	− 0.66	+ 5.81	+ 0.08	+ 5.36	+ 0.91	+ 5.52	0.86
33	91.756	+ 3.48	+ 29.14	+ 4.22	+ 28.69*	+ 5.05	+ 28.85*	0.87
34	92.524	+ 7.22	+ 0.45	+ 7.98	− 0.01	+ 8.79	+ 0.16	0.87
35	95.527	+ 2.71	− 1.25	+ 3.52	− 1.74	+ 4.30	− 1.56	0.82
36	98.655	− 6.54	− 3.40	− 5.67	− 3.91	− 4.95	− 3.73	0.78
37	108.755	− 207.04*	− 21.09	− 206.03*	− 21.63*	− 205.48*	− 21.45*	1.50
38	109.405	− 1.15	+ 0.98	− 0.13	+ 0.44	+ 0.41	+ 0.62	1.40
39	109.713	− 0.59	− 0.25	+ 0.43	− 0.80	+ 0.97	− 0.62	1.36
40	113.722	− 1.67	− 13.54	− 0.60	− 14.08*	− 0.13	− 13.90*	1.07
41	118.619	− 1.93	+ 2.15	− 0.82	+ 1.63	− 0.43	+ 1.79	0.99
42	118.706	+ 0.94	+ 0.05	+ 2.05	− 0.47	+ 2.43	− 0.31	0.99
43	120.587	− 133.91*	− 121.22*	− 132.79*	− 121.74*	− 132.43*	− 121.58*	0.95
44	121.742	− 11.14	− 0.53	− 10.01	− 1.03	− 9.67	− 0.89	0.95
45	122.478	− 2.39	− 1.53	− 1.25	− 2.03	− 0.93	− 1.88	0.95
46	123.587	− 15.07	− 272.03*	− 13.93*	− 272.52*	− 13.62*	− 272.38*	0.98

TABLE I (continued)

No.	$t_i - t_1$	Initial		Without weights		After weighting		$\sqrt{w}$
		$\Delta\alpha \cos \delta$	$\Delta\delta$	$\Delta\alpha \cos \delta$	$\Delta\delta$	$\Delta\alpha \cos \delta$	$\Delta\delta$	
47	123.700	− 0″48	− 0″86	+ 0″66	− 1″35	+ 0″97	− 1″21	0.98
48	124.587	+158.98*	− 367.63*	+160.13*	− 368.11*	+160.43*	− 367.98*	1.02
49	125.700	− 1.35	− 1.75	− 0.20	− 2.23	+ 0.08	− 2.10	1.06
50	126.597	− 3.75	− 0.74	− 2.59	− 1.21	− 2.32	− 1.08	1.06
51	136.636	+ 2.52	− 0.02	+ 3.70	− 0.35	+ 3.84	− 0.27	0.70
52	138.685	− 10.79	− 22.18	− 9.61	− 22.53*	− 9.49	− 22.46*	0.79
53	140.684	− 9.67	+ 1.08	− 8.48	+ 0.75	− 8.39	+ 0.81	0.87
54	141.623	− 15.92	+ 8.63	− 14.73*	+ 8.31	− 14.66*	+ 8.36	0.89
55	143.496	− 4.57	+ 0.13	− 3.38	− 0.16	− 3.32	− 0.12	0.90
56	144.682	− 0.53	− 1.56	+ 0.66	− 1.84	+ 0.70	− 1.80	0.85
57	146.609	− 1.15	− 3.95	+ 0.04	− 4.21	+ 0.06	− 4.19	0.79
58	147.597	− 7.23	− 2.75	− 6.04	− 3.00	− 6.03	− 2.98	0.75
59	149.737	− 294.35*	− 266.95*	− 293.17*	− 267.17*	− 293.17*	− 267.17*	0.72
60	150.634	+ 175.88*	− 153.86*	+177.06*	− 154.07*	+177.04*	− 154.07*	0.72
61	150.755	− 219.17*	+ 10.39	− 217.99*	+ 10.18	− 218.01*	+ 10.18	0.72
62	152.613	− 0.51	− 0.51	+ 0.67	− 0.71	+ 0.64	− 0.72	0.75
63	153.462	+ 19.37	− 7.79	+ 20.55*	− 7.98	+ 20.51*	− 7.99	0.78
64	153.770	− 1.68	− 8.28	− 0.50	− 8.46	− 0.54	− 8.48	0.79
65	168.631	− 0.30	+ 2.08	+ 0.84	+ 2.07	+ 0.68	+ 1.98	0.86
66	168.658	− 3.97	+ 35.39	− 2.83	+ 35.38*	− 2.99	+ 35.29*	0.86
67	169.638	+ 4.02	− 0.52	+ 5.15	− 0.52	+ 4.99	− 0.62	0.84
68	170.630	+ 0.54	+ 1.89	+ 1.68	+ 1.90	+ 1.50	+ 1.80	0.82
69	174.432	+ 3.98	− 1.16	+ 5.10	− 1.11	+ 4.90	− 1.23	0.82
70	174.651	+ 1.19	+ 2.17	+ 2.31	+ 2.23	+ 2.11	+ 2.10	0.82
71	174.661	+ 3.19	+ 4.10	+ 4.31	+ 4.16	+ 4.11	+ 4.03	0.82
72	178.659	+ 0.72	− 1.00	+ 1.82	− 0.90	+ 1.59	− 1.05	0.74
73	180.671	+ 0.28	+ 1.28	+ 1.37	+ 1.40	+ 1.13	+ 1.25	0.70
74	181.437	+ 1.85	+ 3.48	+ 2.93	+ 3.60	+ 2.69	+ 3.45	0.70
75	181.669	+ 6.43	− 1.54	+ 7.51	− 1.41	+ 7.27	− 1.57	0.70
76	184.674	− 3.68	+ 6.55	− 2.61	+ 6.71	− 2.87	+ 6.53	0.70
77	199.758	+ 1.89	− 1.13	+ 2.85	− 0.82	+ 2.52	− 1.07	0.70
78	202.682	− 8.81	− 35.48	− 7.87	− 35.15*	− 8.22	− 35.41*	0.70
79	203.678	− 22.39	− 2.83	− 21.45*	− 2.49	− 21.81*	− 2.76	0.70
80	207.744	+ 1.60	− 0.40	+ 2.50	− 0.02	+ 2.13	− 0.31	0.70
81	214.710	− 4.47	− 4.58	− 3.63	− 4.14	− 4.04	− 4.46	0.70
82	231.735	+ 0.05	+ 0.47	+ 0.70	+ 1.02	+ 0.22	+ 0.62	0.70
83	236.705	− 3.00	+ 0.92	− 2.40	+ 1.50	− 2.90	+ 1.08	0.70

$T = $ 1953 Jan. 5.41783	Jan. 5.41874 ± 0.00139	Jan. 5.41888 ± 0.00101
$q = $ 1.66498366	1.66498073 ± 0.00001656	1.66499216 ± 0.00001564
$e = $ 0.99594159	0.99593281 ± 0.00005744	0.99592389 ± 0.00005168
$\omega = $ 191°37′52″53	191°37′53″97 ± 3″14	191°37′54″30 ± 2″33
$\Omega = $ 220 41 25.32	220 41 25.04 ± 1.12	220 41 25.22 ± 1.17
$i = $ 59 07 11.86	59 07 12.77 ± 1.89	59 07 11.80 ± 2.14
$1/a = $ +0.00243752	+0.00244280 ± 0.00003452	+0.00244816 ± 0.00003106
a (AU) = 410.25	409.36 ± 5.81	408.46 ± 5.23
P (yr) = 8309.6	8282.6 ± 176.5	8255.2 ± 158.8

choice of the initial data for starting the integration of the equations for G_i allows us to obtain the coefficients in the observational equations for the corrections to the conventional orbital elements: v, q, e, ω, Ω, i (v being the true anomaly). The formulae derived for these elements appear to be independent of the kind of orbit (elliptical, hyperbolic or parabolic).

The process of improvement of the orbit is realized by the GIER computer in the following manner. At first the orbit of a comet is corrected by application of the traditional method (rejecting the obviously erroneous observations only and regarding all the observations as equivalent). Starting from these preliminary improved orbital elements the $(O-C)$ residuals are computed, all the perturbations in the comet's motion being included; and from the unperturbed Keplerian orbit of the comet an ephemeris is computed for the moments of observation. The $(O-C)$ residuals are added to this unperturbed ephemeris and thus 'artificial observations' are obtained to correct the Keplerian orbit of the comet.

The Keplerian orbit is corrected iteratively, the 'erroneous' observations being rejected by means of Chauvenet's criterion. The third differences of the $(O-C)$ residuals that result from this orbit improvement then serve as a basis for weighting the observations as follows. The local mean error of one observation for the mean moment of 20 successive observations is computed using Bielicki's formulae. If the number of observations is N, then the mean errors are obtained for $N-20$ mean moments. Thus, we have a numerical representation of the mean error function with respect to time. This function is smoothed by applying the Woolhouse parabolic method, as modified by Bielicki. Then the mean error of each individual observation is computed, and hence the weights of the observations are calculated by Bielicki's method. The process of weighting the observations is iterated three times.

The Keplerian orbit is improved once more by means of the weighted observations, the 'erroneous' observations now being rejected according to Bielicki's criterion. Finally the orbit is corrected with the perturbations included in the differential coefficients. In this single process the weighted observations are used and the doubtful observations rejected previously are again omitted.

The method described has been applied to the improvement of the orbit of the long-period comet 1953 I (Harrington). Table I contains the $(O-C)$ residuals. The asterisked residuals indicate the observations rejected in the process of orbit improvement. The initial residuals were obtained after a preliminary correction of the orbital elements. It is obvious that some observations can be regarded as erroneous and therefore they were rejected at once. Then the orbit was improved, first by giving all the observations unit weight, then by assigning the 'real' weights. In both cases the observations were subjected to selection by application of Bielicki's criterion.

The last column in Table I contains the values of square roots of the weights as computed by Bielicki's method. The rapid change in the run of weights after the seventeenth observation is due to the lack of observations during the next 20 days or so (probably because of the Full Moon). We can see that the observations of the comet were more exact before than after perihelion (perihelion time falling around the middle of the table); the ratio of the weights of the observations at the beginning and at the end of the observation interval amounts to 20:1. Hence we can hardly regard all the observations as equivalent when improving the orbit.

The orbital elements and their mean errors, as resulting from the improvement of the orbit without and with weights, are given below the table. The differences of the two improved sets are small and are contained within the limits of the mean errors of the elements. The question arises as to whether the use of this rather complicated process, described above, is reasonable in practical computations. However, the mathematical selection of observations may be applied only to equivalent observations. Furthermore, the nongravitational effects in the comet's motion can be detected by analysing the (O−C) residuals of the weighted observations. Therefore, it seems that the weighting of the observations should not be neglected in the process of orbit improvement if we really want to determine the 'definitive' orbit of the comet.

References

Bielicki, M.: 1972, this Symposium, p. 112.
Sitarski, G.: 1971, *Acta Astron.* **21**, 87.

21. THE PROBLEM OF ELABORATION AND CLASSIFICATION OF OBSERVATIONAL MATERIAL FOR ONE-APPARITION COMETS

M. BIELICKI

Astronomical Observatory, Warsaw University, Warsaw, Poland

Abstract. A new, purely objective criterion is developed for the rejection of observations with nonaccidental errors. A mathematically rigorous procedure is devised for the determination of the weights of individual observations.

For the investigation of orbits of one-apparition comets from a statistical point of view it is necessary to discuss all the comets in the same manner, using the same astronomical constants and the same methods for reducing the observations and utilizing the observational material.

The first problem will then be to introduce systematic corrections to the observations: corrections due to differences between the star catalogues (including the constants of precession, nutation and aberration, and the proper motions of the stars) and between methods of reduction, time scales, etc.

The other important problem in the homogenization of the observations is the quite objective, purely mathematical question as to the best use of the observational material. It is known that observations of the celestial coordinates of a comet at specified moments of time are just the particular values of the observed variables arising from the general functional dependence between the elements of the orbit and the position in the sky at these times. But the observational data of position and time are influenced by errors of various types and are of different weights. We may simplify the problem arbitrarily by putting it in a fictitious form in which we consider the observed times to be completely free from error but introduce the errors that necessarily exist in the times into further errors in the positions. We consider therefore the positions as particular values of a random variable and the mean square error of the positions as the dispersion of this random variable.

Summing up, we take into consideration two properties of an observation: its weight and its correctness of performance and elaboration. These two problems have been investigated analytically and numerically and solved with the help of statistical methods. We present here a brief account of these investigations and their results.

First let us consider briefly the classification of observations from the point of view of their doubtfulness, i.e., the conventional classification as a set of observations with accidental errors only. This implies that the observations should have been obtained and reduced correctly within the possibilities of the instruments used and the methods of observation. There are many methods used for this classification – mainly the ones for rejecting observations with nonaccidental errors – from the naïve guess or primitive criterion such as 3σ to procedures that are very complicated both analytically and numerically and well-nigh impossible to apply.

Chebotarev et al. (eds.), The Motion, Evolution of Orbits, and Origin of Comets, 112–117. All Rights Reserved.
Copyright © 1972 by the IAU.

Among these methods one draws particular attention because it is fairly objective and easy to handle numerically. This method, based on Chauvenet's (1891) reasoning, derives from a simple and sound principle: the probability that an error will exceed a given magnitude $\sigma\theta$ is $1 - P(\theta)$, where σ denotes the dispersion, θ the limiting coefficient, and we have

$$P(\theta) = \frac{2}{\pi} \int_0^{2^{-1/2}\theta} e^{-t^2}\,dt. \tag{1}$$

When N observations are made the probable number of those surpassing $\sigma\theta$ is $N[1 - P(\theta)]$. If $N[1 - P(\theta)] = 1/2$, then

$$P(\theta) = (2N - 1)/2N. \tag{2}$$

Let ε be an error such that if $|\varepsilon| > \sigma\theta$, the corresponding observation is to be rejected on the ground that it will have a greater probability against it than for it. The numerical application is thus based on the consecutive formulae

$$P(\theta_{Cr\,1/2}) = 1 - \frac{1}{2N} \tag{3}$$

$$P(\theta_{Cr\,1/2}) = \frac{2}{\pi} \int_0^{2^{-1/2}\theta_{Cr\,1/2}} e^{-t^2}\,dt \tag{4}$$

$$\sigma_{Cr\,1/2} = \sigma\theta_{Cr\,1/2}, \tag{5}$$

where we define this criterion as Cr 1/2.

The coefficient $\theta_{Cr\,1/2}$ is a function only of N and can be tabulated. The reasoning is based on the fact that the probable number of residuals (a residual has to replace the deviation of a measurement from the truth) larger than the limit $\sigma_{Cr\,1/2}$ is smaller than 1/2, which means that measurements with such residuals should not exist: if any do, they apparently suffer from an abnormal source of error and should be rejected. The process should of course be repeated to the limit, where none of the measurements retained has a residual exceeding the corresponding value $\sigma_{Cr\,1/2}$.

In principle, the criterion Cr 1/2 seems to be reasonably correct, and what is more important, it is quite objective. It has, however, a drawback resulting from the fact that the factor σ in the calculation of the limit $\sigma_{Cr\,1/2}$ is a random variable as well. Therefore in actual applications this factor leads to uncertainty in classifying measurements into those with accidental and those with nonaccidental errors.

The random variable σ has a distribution in which the probable error can be estimated. This error is equal to

$$\frac{0.4769363\ldots}{\sqrt{N}}\,\sigma; \tag{6}$$

hence the condition

$$_{theor}\sigma\left(1 - \frac{0.47\ldots}{\sqrt{N}}\right) < {}_{real}\sigma < {}_{theor}\sigma\left(1 + \frac{0.47\ldots}{\sqrt{N}}\right) \tag{7}$$

is realized with probability 1/2, and so is the condition

$$_{real}\sigma \left(1 + \frac{0.47\ldots}{\sqrt{N}}\right)^{-1} < {}_{theor}\sigma < {}_{real}\sigma \left(1 - \frac{0.47\ldots}{\sqrt{N}}\right)^{-1}. \tag{8}$$

In our case the criterion Cr 1/2 is based on the mean value, i.e., on the theoretically accepted $_{theor}\sigma = {}_{real}\sigma$. We obtain the limiting condition for $_{theor}\sigma$ when we accept that this criterion should work when the probability of existence of a class of non-accidental errors in the range from a vanishing state to the theoretical mean state attains a value of not more than 1/2. This is the condition when the probability of existence of $_{theor}\sigma$ reaches the value 1/2 in an increasing direction from the mean value.

According to the above reasoning, taking into account the uncertainty of the expression $_{theor}\sigma$, we accept as our basis the formula

$$\sigma_{Cr\,1/2,1/2} = \sigma \left(1 - \frac{0.47\ldots}{\sqrt{N}}\right)^{-1} \theta_{Cr\,1/2}, \tag{9}$$

where Cr 1/2, 1/2 represents the new criterion and $\sigma_{Cr\,1/2,\,1/2}$ gives us the limiting residual for the rejection of measurements with nonaccidental errors. This new criterion takes into consideration the fact that the number of measurements with residuals larger than the limit should be less than 1/2, with the probability of existence of these measurements also smaller than 1/2 because of the uncertainty of the observational mean square error. It is easy to see that the criterion Cr 1/2, 1/2 is more liberal than the Cr 1/2 criterion in classifying the observations as correct, i.e., with accidental errors only. This feature is particularly important when the number of measurements is not too large. On the other hand, the number of measurements cannot be too small, or the statistical methods used for developing the criterion would not be applicable. In practice the minimum value of N is about 20, the accuracy of the determination of $\sigma_{Cr\,1/2,\,1/2}$ being about 10%. In actual examples we have found that application of this criterion causes the elimination of about 10 to 15% of the observations. Perhaps 60 to 70% of the observations eliminated can be corrected and recovered; there remains about 3 to 5% of the whole that is completely lost. Owing to the increasing use of automation in the measurement and reduction of observations we lose a smaller percentage of modern observations. The criteria Cr 1/2 and Cr 1/2, 1/2 can be included in computer programmes using tables or suitable functional models. It must be remembered that the necessary condition for application of this criterion is that the observations have equal weights. To investigate the effects of the criteria in sets of observations a large number of numerical tests have been devised and performed on digital computers. The results of these tests and investigations lead us to accept the above criterion Cr 1/2, 1/2 as correct and applicable in the classification of observational material.

Let us take now into consideration the other property of the observations – their different weights. Attention should be drawn to the fact that for the homogenization of the observations it is necessary to consider weights, both when solving the observational equations by the method of least squares and when applying the criterion of rejection for observations with nonaccidental errors. When the weights have been determined numerical application is easy.

The weights should be determined in a purely objective way and they must in principle result from the same observational material. There are many methods for giving weights to observations. One can quote from Plummer (1939): "In any case the decision should be based on the intrinsic circumstances which affect the observation: the fact that it may give a result not agreeing with expectation or with the results of other observers, is a most dangerous guide in assessing its value." Indeed, all methods based on this type of expectation are applied incorrectly.

Only the mathematical determination of the local mean error at different points in a series of observations enables us to obtain local weights and hence the whole set of weights for the series. The determination of the local mean square error (i.e., the local dispersion of the observation function) can be carried out fairly easily under certain suppositions. One way of doing it is to consider a subinterval surrounding a point in the series and containing a sufficient number (which cannot be too small) of observations, and to represent this subinterval with some particular function according, say, to the method of least squares. The next step is to determine the dispersion, i.e., the mean square error of a single observation. We suppose that all observations have the same accuracy in this subinterval – it cannot be too long – and that they have only accidental errors. It is advisable to take into consideration only observations that are quite certainly reliable, i.e., to base the result on a conventional 'nucleus of accuracy' of observations, defined identically in all the subintervals investigated. This can be achieved if we apply a rather sharp criterion for the local rejection of doubtful observations, separately in each subinterval. It is known from practice that the Chauvenet criterion of Cr 1/2 is sufficient for this. This is permissible, since in the determination of sets of weights it is not the values of the mean errors themselves, but the ratios of these errors, that are important. Repetition of this process at different points of the observation interval gives us local mean errors of the 'nucleus of accuracy' type at these points. These local mean errors correspond to definite values of the time and hence they can be interpolated and extrapolated for all points in the series of observations. By choosing a conventional mean square error for the unit of weight we may determine the whole set of observational weights.

The basic problem is the form of the function representing the residuals in the subintervals. The set of residuals depends on this form, and hence the elimination process of doubtful measurements does as well. So does the mean square error of the observations in each subinterval, and hence the weights of the observations.

Two particular cases can be quoted here: the *a posteriori* type of mean error, resulting from the application of the same functional form as for solving the whole problem; and the *a priori* type of mean error, demonstrating the observational inaccuracy of a single observation. Both of these types of mean error are of great importance in assigning sets of weights, which therefore can be of different types also.

Accepting as the basis of the solution of a problem the determination of the most probable orbit according to Newtonian gravitation only, i.e., using *a posteriori* mean

errors, we shall consider the nongravitational effects as additional influences augmenting the accidental errors of the observations. Then the elements of the orbit
will be the best according to Newtonian gravitation and suitable for the detectability
of nongravitational effects – their existence and influence on the residuals.

On the other hand, when we use *a priori* mean errors, the orbit will be the best
mean result for both pure Newtonian gravitation and various nongravitational
effects. Such an orbit will be particularly suitable for the definitive analysis and
determination of such nongravitational effects and the Newtonian orbit of the
comet.

Numerical mean errors of the *a posteriori* type, i.e., after the application of the
Newtonian mathematical model, can be obtained immediately. The mean errors of
the *a priori* type must be determined in another way: there does exist a purely mathematical manner for determining them, utilizing variance properties of ordinary and
divided differences; see Bielicki (1958, 1967). There are many interesting details of
application which will not be mentioned here.

It is easy to see that the two problems – the calculation of a set of weights and the
elimination of observations with abnormal sources of errors – have some common
points, and therefore they are numerically solved together as follows: First we eliminate the observations that are obviously wrong (e.g., the preliminary residuals are
more than 50″). Then, by means of the subinterval procedure, we determine the local
mean square 'nucleus of accuracy' errors (either *a priori* or *a posteriori* or some other
type) for separate points on the observational interval. Next we find mean errors of
the adopted type for particular observations using methods of smoothing, interpolation, and extrapolation, and hence we establish the weights. Since the local mean
error depends on the set of residuals, and this set depends on the weights of the
observations, the process of weight determination must be an iterative one. It is
also connected with the elimination process. The procedure is rapidly convergent,
however, and because the weights are not required with very great accuracy (e.g.,
to about 10% only), two, three or four approximations already give a good result.
After obtaining the final weights we apply them to the observations and then subject
the whole observational material to the new classification criterion (Cr 1/2, 1/2),
which eliminates almost all the observations affected by some abnormal source of
error.

The problem of weighting and the elimination of doubtful observations has been
solved for the case where the total number of observations is more than 20. When
this number is less there are other, simplified methods for application that make use
of less information.

We have in general two or more possibilities for the solution, depending on which
type of weights we assign to the observations. Each of these possibilities has its importance for further investigation of the motion of the comet.

In connection with the above reasoning about weights, elimination processes, etc.,
we have performed a large number of investigations and tests on computers. As an
example we considered 83 observations of comet 1953 I. A series of experiments
was carried out in which we considered different divisions into subintervals, different

methods and criteria for rejecting doubtful observations, different mathematical models for representing the residuals, etc. Finally we investigated the iterative process for determining mean square errors of the *a priori* and *a posteriori* types and the sets of weights that follow from these errors. Figure 1 shows that the *a priori* mean error increases rather suddenly from 1″ to 3″ about a hundred days before the comet passes perihelion; then it increases more slowly to 4″ at the end of the observational interval. The sudden change of this mean error is accompanied by an increase in the

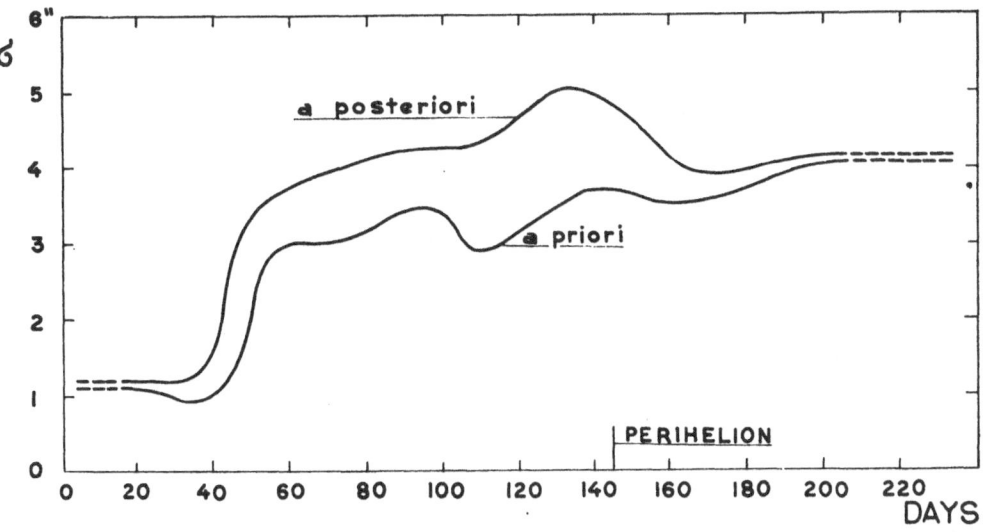

Fig. 1.   The local mean errors for comet 1953 I (Harrington).

nongravitational effects, which in turn cause the great increase to 5″ in the *a posteriori* mean error. The nongravitational influence decreases slowly after perihelion, and both mean errors end up near 4″. It follows from the changes in the mean errors that the weights of the observations vary considerably during the observational interval, by a factor of 15 for the *a priori* case and 20 for the *a posteriori* weights. This has an important effect on determination of the orbit. Further details concerning the application to comet 1953 I are given by Sitarski (1972).

Acknowledgment

I should like to express my thanks to G. Sitarski for his collaboration, discussion, and assistance with the computing.

References

Bielicki, M.: 1958, *Acta Astron.* **8**, 131.
Bielicki, M.: 1967, *Acta Astron.* **17**, 409.
Chauvenet, W.: 1891, *A Manual of Spherical and Practical Astronomy*, 5th ed., Philadelphia, p. 564.
Plummer, H. C.: 1939, *Probability and Frequency*, London.
Sitarski, G.: 1972, this Symposium, p. 107.

22. THE INFLUENCE OF PROPERTIES OF A SET OF OBSERVATIONS ON THE WEIGHTS OF DETERMINATION OF THE ORBITAL ELEMENTS OF A ONE-APPARITION COMET

M. BIELICKI

Astronomical Observatory, Warsaw University, Warsaw, Poland

Abstract. An expression is developed for the dependence of the weights of the determination of the elements of the orbit of a one-apparition comet on the number of observations, the length of the observational interval and the apparent motion of the comet during that interval.

The determination of the definitive orbit of a celestial body consists in finding those values of the orbital elements for which the discrepancies between the observed and calculated positions are smallest. This is generally achieved by means of the method of least squares. This method satisfies the real distribution of random errors, provided that the measurements are of equal weight and free from nonaccidental errors. The determination of weights and the rejection of observations with nonaccidental errors involve only the quality of the measurements and can frequently be treated quite objectively (Bielicki, 1972).

We are interested here, however, in certain properties of the observational material as a whole and in the influence of these properties on the accuracy of the determination of the orbital elements. In the case of a one-apparition comet it is intuitively obvious that this accuracy is affected by three factors:

(1) the number M of observational equations;

(2) the interval of time T covered by the observations;

(3) a quantity K that depends on the apparent motion of the comet during the observational interval.

By application of Cauchy's theorem Jacobi obtained the following formula connecting the solutions of particular groups of observational equations with the general solution of all the observational equations by the method of least squares:

$$E_j^m - E_j^c = \frac{\sum_{(r)=1}^{\binom{M}{p}} (D_{(r)})^2 (E_j^m - E_j^c)_{(r)}}{\sum_{(r)=1}^{\binom{M}{p}} (D_{(r)})^2}, \tag{1}$$

where $E_j^m - E_j^c$ is the most probable correction to the parameter E_j^c, with $j = 1, 2, \ldots, p$; $D_{(r)}$ is the determinant of the coefficients of the combination (r) of observational equations, selected from the total of M observational equations in $\binom{M}{p}$ discrete ways; and $(E_j^m - E_j^c)_{(r)}$ is the rigorous solution of the combination (r) of observational equations. The above formula permits only a qualitative discussion on the accuracy

Chebotarev et al. (eds.), The Motion, Evolution of Orbits, and Origin of Comets, 118–122. All Rights Reserved.
Copyright © 1972 by the IAU.

of the results, and such general discussions have been described by various authors (Whittaker and Robinson, 1924; Plummer, 1939; etc.).

This prompted us to find a formula that would directly relate the coefficients in the observational equations with the weights Q_{jj}^{-1} of the parameters j of the orbit. It is as follows:

$$Q_{jj}^{-1} = \frac{\sum_{(r)=1}^{\binom{M}{p}} (D_{(r)})^2}{\sum_{(r)=1}^{\binom{M}{p-1}} (D_{(r)}^{jj})^2}, \tag{2}$$

where $D_{(r)}^{jj}$ is the determinant of the coefficients of the combination (r) of observational equations, selected in $\binom{M}{p-1}$ discrete ways, and in which the column of coefficients corresponding to the unknown j has been removed. A summary of the reasoning follows.

Let $(D_{(r)})^2$ be a random variable with the same probability density as an element of the general population. Then its expected value is

$$E[(D_{(r)})^2] = \binom{M}{p}^{-1} \sum_{(r)=1}^{\binom{M}{p}} D_{((r))}^2.$$

Taking a sample of m elements $(D'_{(r)})^2$, we have

$$E[(D'_{(r)})^2] = E[(D_{(r)})^2].$$

When we add this sample to the total population we have $M+m$ elements $(D''_{(r)})^2$ and also

$$E[(D''_{(r)})^2] = E[(D_{(r)})^2].$$

From the above there results

$$\binom{M+m}{p}^{-1} \sum_{(r)=1}^{\binom{M+m}{p}} (D''_{(r)})^2 = \binom{M}{p}^{-1} \sum_{(r)=1}^{\binom{M}{p}} (D_{(r)})^2,$$

and, with analogous reasoning for the determinants $D_{(r)}^{jj}$, we have for $M+m$ and M observations

$$(Q_{jj}^{-1})_{M+m} = \frac{\sum_{(r)=1}^{\binom{M+m}{p}} (D''_{(r)})^2}{\sum_{(r)=1}^{\binom{M+m}{p-1}} (D''_{(r)}^{jj})^2} = \frac{M+m-p+1}{M-p+1} = (Q_{jj}^{-1})_M;$$

Since the limit M^{-1} tends to zero, we have

$$Q_{jj}^{-1} \sim M. \tag{3}$$

Let two consecutive values $c^j_{(r)i}$ and $c^j_{(r)i+1}$ of the differential coefficient $c^j_{(r)}$ for the parameter j in the combination (r) of observational equations be of the form

$$c^j_{(r)i+1} = c^j_{(r)i} + k^j_{(r)i}(w_{(r)i+1} - w_{(r)i}) + \cdots .$$

We suppose that the independent variable w is chosen so that quadratic and higher terms may be ignored. Then

$$
D_{(r)} = \begin{vmatrix}
c^1_{(r)1} & c^2_{(r)1} & \cdots & c^p_{(r)1} \\
c^1_{(r)2} & c^2_{(r)2} & \cdots & c^p_{(r)2} \\
\cdot & \cdot & \cdot & \cdot \\
c^1_{(r)p} & c^2_{(r)p} & \cdots & c^p_{(r)p}
\end{vmatrix}
$$

$$
= \begin{vmatrix}
c^1_{(r)1} & & c^2_{(r)1} & & \cdots \\
c^1_{(r)1} + k^1_{(r)1}(w_{(r)2} - w_{(r)1}) & & c^2_{(r)1} + k^2_{(r)1}(w_{(r)2} - w_{(r)1}) & & \cdots \\
c^1_{(r)2} + k^1_{(r)2}(w_{(r)3} - w_{(r)2}) & & c^2_{(r)2} + k^2_{(r)2}(w_{(r)3} - w_{(r)2}) & & \cdots \\
\cdot & \cdot & \cdot & \cdot & \cdot \cdot \cdot
\end{vmatrix}
$$

$$
= \begin{vmatrix}
c^1_{(r)1} & c^2_{(r)1} & \cdots \\
k^1_{(r)1} & k^2_{(r)1} & \cdots \\
k^1_{(r)2} & k^2_{(r)2} & \cdots \\
\cdot & \cdot & \cdot
\end{vmatrix} \times \prod_{i=1}^{p-1} (w_{(r)i+1} - w_{(r)i})
$$

$$
= D^{c,k}_{(r)} \prod_{i=1}^{p-1} (w_{(r)i+1} - w_{(r)i}).
$$

Now, $D^{c,k}_{(r)} = c^1_{(r)1} D^{k1}_{(r)} - c^2_{(r)1} D^{k2}_{(r)} + \cdots$, where $D^{k1}_{(r)}$, $D^{k2}_{(r)}$, are the minors of the determinant $D^{c,k}_{(r)}$. Then,

$$
\frac{d}{dw} D^{c,k}_{(r)} = \frac{d}{dw} c^1_{(r)1} D^{k1}_{(r)} + c^1_{(r)1} \frac{d}{dw} D^{k1}_{(r)} - \cdots
$$

$$
= k^1_{(r)1} D^{k1}_{(r)} - k^2_{(r)1} D^{k2}_{(r)} + \cdots
$$

But the determinant $D^{k1}_{(r)}$ is of the form

$$
D^{k1}_{(r)} = \begin{vmatrix}
k^2_1 & k^3_1 & \cdots & k^p_1 \\
k^2_2 & k^3_2 & \cdots & k^p_2 \\
\cdot & \cdot & \cdot & \cdot
\end{vmatrix} = \text{const} \prod_{i \times 1}^{p-2} (w_{(r)i+1} - w_{(r)i}),
$$

and so are $D^{k2}_{(r)}$, etc. Thus

$$
\frac{d}{dw} D^{c,k}_{(r)} \sim \prod_{i=1}^{p-2} (w_{(r)i+1} - w_{(r)i}).
$$

TABLE I

Influence of the number of observations (M), the length of the observational interval (T) and the apparent motion factor (K) on the relative weights of the elements determined for the orbit of comet 1953 I

Observations included	No. of observations	Observational interval (days)	$\bar{r}$ (AU)	$\bar{\Delta}$ (AU)	M	T^4	K^4	theor Q_{jj}^{-1}	obs Q_{jj}^{-1}
Sensitivity to M									
1–83	83	236	2.1	1.7	1.00	1.00	1.00	1.00	1.00
1, 3, 5, …, 81, 83	42	236	2.1	1.7	0.51	1.00	1.00	0.51	0.53
1, 4, 7, …, 79, 82	28	236	2.1	1.7	0.34	1.00	1.00	0.34	0.34
Sensitivity to T									
23–82	60	150	1.9	1.4	0.72	0.16	1.46	0.17	0.19
33–72	40	87	1.8	1.2	0.48	0.018	2.18	0.019	0.023
43–62	20	32	1.6	1.5	0.24	0.00034	0.56	0.000045	0.000040
Sensitivity to K									
1–20	20	59	2.4	1.6	0.24	0.0039	2.17	0.0020	0.0017
21–40	20	50	1.8	1.0	0.24	0.0020	4.51	0.0022	0.0022
41–60	20	32	1.6	1.4	0.24	0.00034	0.73	0.000060	0.000053
61–80	20	57	1.8	1.9	0.24	0.0034	0.34	0.00028	0.00023

The figures given as obs Q_{jj}^{-1} are the geometric means of the determinations of the components of position and velocity of the comet.

In which case,

$$D_{(r)}^{c,k} \sim \prod_{i=1}^{p-1} (w_{(r)i+1} - w_{(r)i}),$$

$$D_{(r)} \sim \prod_{i=1}^{p-1} (w_{(r)i+1} - w_{(r)i})^2.$$

Analogously, when p is replaced by $p-1$,

$$D_{(r)}^{jj} \sim \prod_{i=1}^{p-2} (w_{(r)i+1} - w_{(r)i})^2.$$

If the observations are uniformly distributed with respect to w, then $w_{(r)i+1} - w_{(r)i} \sim W$, the total length of the observational interval. Hence $D_{(r)} \sim (W^{p-1})^2$ and $D_{(r)}^{jj} \sim (W^{p-2})^2$, and from Equation (2) we obtain

$$Q_{jj}^{-1} \sim \frac{(W^{p-1})^4}{(W^{p-2})^4} = W^4.$$

We now define the quantity K, the measure of the comet's apparent motion, to be the average value of dw/dt during the interval of observation T (which corresponds to W). Hence

$$Q_{jj}^{-1} \sim K^4 T^4,$$

and combining this with Equation (3), we find the complete dependence of the weights on the properties of the observations to be

$$Q_{jj}^{-1} \sim M K^4 T^4. \tag{4}$$

In practice, e.g., when we determine the components of position and velocity as the cometary orbit, K is the average value of the ratio of the heliocentric distance r and the geocentric distance Δ during the observation interval:

$$K = \langle r/\Delta \rangle. \tag{5}$$

The results in Table I show the effects of different selections of observations on the relative weights of the orbital elements of comet 1953 I, using Equations (4) and (5); see also Sitarski (1972).

References

Bielicki, M.: 1972, this Symposium, p. 112.
Plummer, H. C.: 1939, *Probability and Frequency*, London, p. 170.
Sitarski, G.: 1972, this Symposium, p. 107.
Whittaker, E. A. and Robinson, G.: 1944, *The Calculus of Observations*, 4th ed., London and Glasgow, p. 251.

23. ON THE DIFFERENTIAL CORRECTION OF NEARLY PARABOLIC ORBITS

P. HERGET

University of Cincinnati Observatory, Cincinnati, Ohio, U.S.A.

The differential correction of nearly parabolic orbits was discussed by the author (Herget, 1939) in the era of lead pencil computing. The Gauss-Marth method is the best one to use whenever the appropriate conditions exist, i.e., $|E| < 64°$ and e nearly unity. The crucial point in the above-cited discussion is the use of the first differences from the Gauss-Marth tables in order to simplify the computation of the partial differential coefficients, namely dB/dA, dC/dA, and dD/dA.

In the present era of computation by electronic calculators, the Gauss-Marth tables have been replaced by series expansions (Benima *et al.*, 1969), which are more appropriate to the use of high-speed electronic storage devices. The purpose of this note is simply to call attention to the efficacy of a subroutine which will evaluate the derivatives of these series, and then the same formulae may be used for the partial differential coefficients as were cited above. The arrangement of formulae in the form of matrix multiplications is still a useful device in modern programming methods.

At the Cincinnati Observatory we have attained what is probably the ultimate in the differential correction of orbits by expanding the widely used N-body integration program of Schubart and Stumpff (1966). The partial differential coefficients are each obtained by a separate, implicit, numerical double integration in exactly the same way as was presented by the writer (Herget, 1968) for the outer satellites of Jupiter. This is equivalent to the simultaneous integration of the trajectory and its variational equation, and this leaves little more to be desired, especially if there are excessive perturbations involved. This program has been used with great success in our study of the motion of P/Pons-Brooks during two complete revolutions (Herget and Carr, 1972).

References

Benima, B., Cherniack, J. R., Marsden, B. G., and Porter, J. G.: 1969, *Publ. Astron. Soc. Pacific* **81**, 121.
Herget, P.: 1939, *Astron. J.* **48**, 105.
Herget, P.: 1968, *Astron. J.* **73**, 737.
Herget, P. and Carr, H. J.: 1972, this Symposium, p. 195.
Schubart, J. and Stumpff, P.: 1966, *Veroeffentl. Astron. Rechen-Inst. Heidelberg* No. 18.

24. STANDARDIZATION OF THE CALCULATION OF NEARLY PARABOLIC COMETARY ORBITS

L. E. NIKONOVA
Kazan University, Kazan, U.S.S.R.

and

N. A. BOKHAN
Institute for Theoretical Astronomy, Leningrad, U.S.S.R.

Abstract. A complex of standard programmes has been developed for computing nearly parabolic orbits. The true anomaly is related to the time using Herrick's formulae. Perturbations are calculated by Encke's method.

A library of standard programmes (SP) has been compiled in the department of minor planets and comets of the Institute for Theoretical Astronomy. These programmes, written for electronic computers of the type M-20, are intended for serving many of the problems frequently encountered in astronomical calculations, and they are continuously being supplemented.

The standardization of calculations on electronic computers is a new stage in the development of computational techniques, and it enables us to formulate and solve new topical problems in cometary astronomy. Among these problems is the systematic calculation of orbits for all nonperiodic comets, and it has been our purpose to provide a complex of standard programmes for computing definitive, nearly parabolic orbits.

This complex of SP's may be divided into three groups. One group is used for computing perturbations by the major planets, another is for comparison with observations, and the third enables one to carry out a multiple orbit improvement.

The SP's for calculating perturbations are based upon a set of coordinates, stored on magnetic tape, of the eight major planets Venus to Pluto over a period of 400 years (Kazimirchak-Polonskaya, 1967). For calculating the perturbations Encke's method is used, our programme consisting of fourteen SP's written by different authors.

A standard programme by Nikonova causes the coordinates of the major planets during the period of the comet's visibility to be read from the magnetic tape into the computer's high-speed memory. This is used in conjunction with two SP's by Bokhan, which

(1) identify the zone number on the tape and determine the corresponding initial date for the coordinates of the major planets on the magnetic drum and in high-speed memory;

(2) read out the coordinates of the planets from the magnetic drum forward and backward according to the sign of the integration step. The coordinates are interpolated for intervening dates by Lagrange's formula.

To determine the rectangular heliocentric coordinates of the comet the true anomaly is first calculated using Herrick's (1960) formula:

$$\frac{k(1 + e)^2(t - T)}{2p^{3/2}} = \frac{\sigma}{1 + \lambda\sigma^2} + \frac{\sigma^3}{1 + e}$$

$$\times \left[\frac{1}{1 + \lambda\sigma^2} - \frac{1}{3} + \sum_{n=1}^{\infty} \frac{(-1)^{n+1}(\lambda\sigma^2)^n}{2n + 3} \right],$$

where

$$\sigma = \tan\frac{v}{2}, \qquad \lambda = \frac{1 - e}{1 + e},$$

k being the Gaussian constant, v the true anomaly, e the eccentricity, T the time of perihelion passage, p the parameter, and t a given moment. The unperturbed rectangular coordinates of the comet are obtained from the usual formulae.

The total disturbing acceleration on the comet is calculated by means of an SP by Kazimirchak-Polonskaya, and SP's are then used for integrating the perturbations on the comet in rectangular coordinates. The initial integration table is constructed for seven starting moments, using an SP for obtaining the first and second differences. The orbit is then integrated in groups of three consecutive steps forward (or backward), and the final perturbed rectangular coordinates of the comet are obtained for the total visibility period (Nikonova).

For comparison of the observed and calculated spherical coordinates of the comet the observations are first processed using the programme by N. A. Belyaev, which reduces the coordinates to a standard equinox and calculates the topocentric coordinates of the Sun. The disturbed rectangular coordinates of the comet are interpolated for each observation (the time being corrected for aberration, with the distance from the Earth calculated by successive approximations). The comet's calculated right ascension and declination are obtained from the rectangular coordinates of the comet and the Sun, and then the (O−C) residuals are printed out and punched onto cards (Bokhan).

The programme for the orbit improvement combines SP's for

(1) the calculation of auxiliary values;

(2) formation of the Eckert-Brouwer differential coefficients using the formulae modified by Dubyago (1949) for $e \simeq 1$ (Nikonova);

(3) formulation of the equations of condition (S. G. Makover, Bokhan);

(4) application of the method of least squares (Makover).

The programme allows the simultaneous treatment of up to 190 observations or normal places. The improved elements, the mean square errors of the unknowns, the mean square error of unit weight and the final residuals are then printed out and punched. The programme has been applied to comet 1963 V (Nikonova, 1968).

References

Dubyago, A. D.: 1949, *Opredelenie Orbit*, Moscow, p. 293

Herrick, C. E.: 1960, *Details of Computational Methods: Space Trajectories*, p. 135.
Kazimirchak-Polonskaya, E. I.: 1967, *Trudy Inst. Teor. Astron.* **12**, 24.
Nikonova, L. E.: 1968, *Trudy Astron. Obs. Kazan No.* 35, 169.

Discussion

G. Sitarski: What principle do you use for the rejection of observations?

L. E. Nikonova: The observations are rejected by the 3σ principle.

M. Bielicki: Did you take into account the weights of the observations in solving the normal equations?

L. E. Nikonova: No.

25. DÉTERMINATION D'ORBITES PARABOLIQUES À PARTIR DE *N* OBSERVATIONS AU MOYEN DE L'ORDINATEUR ÉLECTRONIQUE

H. DEBEHOGNE

Observatoire Royal de Belgique, Bruxelles, Belgique

Abstract. The Olbers-Banachiewicz method for computing parabolic orbits has been programmed. A least-square improvement of the orbit is obtained by varying the first and last geocentric distances.

1. Introduction

Dans le cas des astéroïdes et comètes, l'étude des écarts entre l'observation d'une part et d'autre part le calcul, dérivant des éléments d'une première orbite, permet de sélectionner parmi toutes les observations les trois qui serviront de base à l'établissement d'une orbite meilleure, sans perdre de vue pour autant que cette sélection reste soumise à l'examen qualitatif des clichés.

A l'Observatoire Royal de Belgique, c'est la méthode de Olbers-Banachiewicz (Stracke, 1929, p. 171) qui a été utilisée par Arend et ses collaborateurs.

2. La première approximation

L'étape initiale consiste à rechercher, au moyen de la règle de fausse position, la distance Δ_1 de la comète à la Terre à l'instant correspondant à la première observation de base. On sait que cette méthode utilise au départ deux valeurs approchées de Δ_1. Banachiewicz obtient ces deux valeurs à partir d'une valeur approchée de Δ_1 à moins de 0,01 à laquelle il ajoute et retranche 0,01. Les calculs doivent conduire à obtenir l'égalité entre le déplacement héliocentrique géométrique S_g et le déplacement héliocentrique dynamique S_d de la comète.

Ce déplacement dynamique est obtenu grâce à la formule tirée du théorème de Lambert (Stracke, 1929, p. 35). Banachiewicz trouve par exemple $S_g - S_d = +0{,}000\,01$ à partir de $\Delta_1 = 1{,}712\,52$ (comète 1925c). L'ordinateur effectue les calculs avec 16 chiffres significatifs et il donne pour $\Delta_1 = 1{,}711\,94$ une valeur $S_g - S_d$ vérifiant $S_g - S_d < 0{,}000\,001\,003$. Cet écart important entre les 2 valeurs de Δ_1 a été déjà rencontré entre les 2 valeurs de la solution de l'équation de Képler en sinus hyperbolique. La première avait été obtenue en utilisant 8 chiffres significatifs (comme le permettent seulement les tables du sinus hyperbolique). La deuxième l'avait été en utilisant 16 chiffres significatifs comme dans notre programme de calcul des résidus pour une orbite hyperbolique (Debehogne, 1968). Comme, en fin de calcul, nos résidus relatifs à la seconde observation sont moitié moindres ($+0{,}''3$ et $-0{,}''8$) que ceux de Banachiewicz on peut considérer que notre valeur de Δ_1 est meilleure que celle de l'exemple exposé par Stracke. Il convient de noter que cet écart porte déjà sur le quatrième chiffre décimal alors que l'auteur travaille constamment avec cinq chiffres décimaux.

Chebotarev et al. (eds.), The Motion, Evolution of Orbits, and Origin of Comets, 127–129. All Rights Reserved.

3. Valeur approchée de Δ_1

L'obtention de la première valeur approchée de Δ_1 (1,71) exige la connaissance d'une valeur approchée de Δ_2 (distance comète–Terre à l'instant correspondant à la deuxième observation de base). Cette valeur doit vérifier l'équation $(CR)^2CS - c = 0$, où c est égal à $2d^2R_2^2$, R_2 étant la distance Terre-Soleil à l'instant correspondant à la deuxième observation de base et d valant $g(1+M)/2h$, M étant fonction des instants des 3 observations de base, des coordonnées du Soleil à l'instant de la seconde observation et des coordonnées géocentriques de la comète, g et h étant liés respectivement au déplacement de la Terre et de la comète.

Quant à CR et CS on voit leur signification sur le graphique (Stracke, 1929, p. 183) qui servait à Banachiewicz pour déterminer Δ_2 à moins de 0,01.

Nous introduisons, par une carte contenant certaines données, la première valeur approchée de Δ_2 d'où l'ordinateur tire les valeurs de CS, RC et donc de $(CR)^2CS - c$. Nous augmentons successivement Δ_2 de $0,001 \times n$ ($n = 1, 2, 3, \ldots$). Quand le signe de $(CR)^2CS - c$ change nous retournons à l'avant-dernière valeur de Δ_2 que nous augmentons successivement de $0,001 \times n/2$ et ainsi de suite jusqu'à la détermination de Δ_2 avec 2 décimales.

4. Détermination des éléments

Pour ne pas dépasser la capacité de l'ordinateur nous avons découpé le programme en programmes partiels, afin d'obtenir l'angle voulu puisque l'ordinateur donne pour $x = $ arc tg y des valeurs comprises entre $-\pi/2$ et $+\pi/2$.

Nous obtenons ainsi sur cartes les éléments suivants: l'instant de passage T au périhélie, le noeud Ω, la distance ω du Ω au périhélie, l'inclinaison i du plan de l'orbite sur l'écliptique et la distance périhélique q.

5. Les résidus

Le but poursuivi au cours de ce travail est le calcul des 108 résidus correspondant aux observations de la comète 1957 III (Arend-Roland), effectuées à Uccle.

On peut recourir à la méthode décrite plus haut à propos de la résolution de $(CR)^2CS - c = 0$ ou utiliser la méthode de fausse position comme ci-dessus pour le calcul de Δ_1 ou encore programmer la formule de Lemaître suivant le procédé exposé par Arend (1966). Nous avons préféré la méthode suivante qui nous paraissait la plus rapide: soit à résoudre l'équation

$$x + \frac{x^3}{3} = \frac{k}{\sqrt{2}} \frac{t - T}{q^{3/2}},$$

ou $x = $ tg $v_1/2$, $k = 0,017\ 202\ 098\ 95$, et $t = $ instant d'observation. Si on pose successivement

$$x = 2 \cotg 2y$$
$$y = \sqrt[3]{\operatorname{tg} \frac{z}{2}},$$

l'équation s'écrit

$$\text{tg } z = \frac{2\sqrt{2}}{3k} \frac{1}{M}.$$

6. Conclusion

On calcule d'abord une orbite préliminaire à partir de trois observations de base et les résidus de toutes les observations pour lesquelles on dispose d'un résultat précis: on calcule d'abord la distance Terre-comète Δ à partir de $t-A$, A étant la constante de lumière; Δ est donc supposé égal à l'unité, puis on corrige t de $-A\Delta$, d'où en réalité deux calculs, alors qu'auparavant on se contentait d'une règle de trois à partir de $A\Delta_1$, $A\Delta_2$, $A\Delta_3$, où Δ_1, Δ_2, Δ_3 sont les distances astre-Soleil pour les trois observations de base. Les résultats varient peu pour $\varepsilon = \varepsilon + \Delta\varepsilon$, avec $\Delta\varepsilon$ valant environ vingt secondes d'arc (ε = inclinaison de l'équateur sur l'écliptique).

Enfin, on améliore l'orbite par la méthode de variation des distances Δ_1 et Δ_3 et application des moindres carrés.

Réferences

Arend, S.: 1966, *Commun. Astron. Obs. Roy. Belg.* No. 243.
Debehogne, H.: 1968, *Acad. Roy. Belg. Bull. Cl. Sci.* **54**, 941.
Stracke, G.: 1929, *Bahnbestimmung der Planeten und Kometen*, Springer, Berlin.

PART III

MOTIONS OF THE SHORT-PERIOD COMETS

A. PLANETARY PERTURBATIONS AND NONGRAVITATIONAL EFFECTS

26. NONGRAVITATIONAL EFFECTS ON COMETS: THE CURRENT STATUS

B. G. MARSDEN

Smithsonian Astrophysical Observatory, Cambridge, Mass., U.S.A.

Abstract. A method for allowing for the effects of nongravitational forces on the motions of comets is summarized. Study of the motions of specific comets indicates that these forces act essentially continuously but have a high inverse dependence on heliocentric distance; there is also evidence for secular changes. The equations of motion employed are discussed in terms of the Whipple icy-conglomerate model. Nongravitational parameters are tabulated for all 46 comets observed at three or more perihelion passages. We point out the particular problems that still exist for certain comets and suggest directions for future research.

Few astronomers would deny that comets are subjected to forces of a nongravitational nature: one has merely to consider the many instances where comets have split or have exhibited great surges in brightness. There has been severe disagreement, however, as to whether nongravitational forces have detectable effects on the *motions* of comets. Controversy has raged ever since Encke first claimed, a century and a half ago, that such effects were present in the motion of the comet that bears his name, and similar claims made during the second half of the nineteenth century and the first half of the twentieth with respect to a number of other comets have failed to clarify the situation.

The conservative element has pointed out, with considerable justification, that conclusions regarding the orbits of comets can be influenced, not only by possible systematic errors in the observations of comets, but also – and particularly – by errors in the computations themselves (Roemer, 1961). Even in the most sophisticated computations approximations have been introduced, and there was no way of knowing how the resulting errors would propagate when a comet was moderately perturbed by Jupiter, for example. The partial derivatives used for the differential correction of the orbit were invariably calculated from the traditional Keplerian expressions (e.g., Eckert and Brouwer, 1937), whereas they should be the derivatives of the instantaneous, perturbed coordinates with respect to the osculating elements at some specified epoch. The easiest and surest way of obtaining correct partial derivatives is to calculate the perturbations, not only on the preliminary orbit, but also on the six orbits obtained by varying each element one by one by a small amount; the derivatives are then taken to be the differences between the residuals from the preliminary orbit and from each variation orbit in turn. The process should be iterated until it has converged to the requisite number of figures.

In the days of logarithms and desk-calculators it was clearly impractical to try to fit rigorously the gravitational orbit of a periodic comet to observations at several apparitions. Now that high-speed computers can be used this is no longer the case, and the rigorous calculations made during the last few years for a number of comets,

Chebotarev et al. (eds.), The Motion, Evolution of Orbits, and Origin of Comets, 135–143. All Rights Reserved.
Copyright © 1972 by the IAU.

including the perturbations by all the principal planets, show quite conclusively that, as a general rule, unacceptable systematic trends remain in the residuals (Marsden, 1968). These systematic trends become larger the longer the interval of time covered by the observations used, and they sometimes amount to several minutes of arc over only three apparitions. Since the possible departure of the center of mass of a comet from the center of light is not more than a few seconds of arc – for modern photographic observations with long-focus reflectors at any rate – it is quite clear that additional *forces* are involved. The failure to fit a gravitational orbit is most extreme for a comet such as P/d'Arrest or P/Honda-Mrkos-Pajdušáková, where any condensation is hard to detect in the midst of the large coma; on the other hand, comets like P/Arend-Rigaux and P/Neujmin 1, which have nearly always been indistinguishable from minor planets in appearance, are found to conform to purely gravitational theory.

Another controversy, and one not so easily resolved, concerns whether the nongravitational forces take the form of discrete impulses or whether they act more or less continuously. The most readily detectable nongravitational effect is, of course, the progressive advance or delay of a comet at successive returns to perihelion, and the standard way to allow for this has been to postulate a secular variation in the mean motion. Studies made in this manner – affected though they might have been by approximations – indicated that for a particular comet this variation was surprisingly regular, which is a strong point in favor of the hypothesis that the forces act continuously. And attempts to relate nongravitational effects in the motion of a comet to direct observations of the influence of nongravitational forces on the comet's physical appearance have failed miserably: according to the calculations by Cunningham (1968) and by Herget (1968) P/Schwassmann-Wachmann 1 – a comet that quite frequently throws off shells of matter and suddenly increases in brightness a hundredfold – shows in its motion no nongravitational effects whatsoever.

Accordingly, we have supposed the nongravitational forces to act continuously and have included in the cometary equations of motion additional acceleration components F_1, F_2 and F_3, where F_1 is directed outward along the radius vector, F_2 is directed parallel to the line from the Sun to the point in the orbit with true anomaly 90° ahead of the comet (i.e., approximately along the velocity vector in the case of low orbital eccentricity), and F_3 is directed perpendicular to the orbit plane and such that one has a right-handed system (i.e., it is toward the north pole of the orbit). It is reasonable that the F_i should depend on the comet's heliocentric distance r, and while it is not possible to determine this dependence precisely, we have found that in general it is considerably more extreme than an inverse square or inverse cube. For definiteness, but certainly not uniqueness, we have generally adopted the form

$$F_i = G_i r^{-3} \exp(-r^2/2), \tag{1}$$

where r is measured in AU.

The orbit of a comet can thus be differentially corrected, not only for the usual six elements, but also for the three nongravitational parameters G_1, G_2 and G_3, assumed constant. The additional partial derivatives may be formed numerically in precisely the manner indicated for the others. Whenever we have solved for G_3 we have found it to be determined to less than its mean error, so the solution may be limited to the nongravitational components in the comet's orbit plane. Such solutions have been made to date for 16 periodic comets (e.g., Marsden, 1969, 1970). The transverse component G_2 is related rather closely to the secular variation in the comet's mean motion and is sometimes determined to several hundred times its mean error. However, the radial component G_1 can often be determined surprisingly well too, occasionally to several tens of times its mean error.

In terms of Whipple's (1950) icy-conglomerate model for a cometary nucleus (assumed spherical) the G_i are:

$$G_1 = \xi[\cos \lambda(1 - \tfrac{1}{2} \sin^2 I)$$

$$+ \tfrac{1}{2} \sin^2 I - \tfrac{1}{2}(1 - \cos \lambda) \sin^2 I \cos (2\varphi + 2f)]$$

$$G_2 = \xi[\sin \lambda \cos I + \tfrac{1}{2}(1 - \cos \lambda) \sin^2 I \sin (2\varphi + 2f)] \tag{2}$$

$$G_3 = \xi \sin I[\sin \lambda \cos (\varphi + f) - (1 - \cos \lambda) \cos I \sin (\varphi + f)],$$

where f is the comet's true anomaly, φ the longitude of the meridian of the comet facing the Sun at the comet's passage through perihelion, I the inclination of the comet's equator to its orbit, λ ($\geqslant 0$) the lag-angle of the direction of maximum mass ejection behind the subsolar meridian, and ξ ($\geqslant 0$) gives the magnitude of the reactive force on the comet.

If we make the usual assumption that $\sin I = 0$, Equations (2) become, simply,

$$G_1 = \xi \cos \lambda$$

$$G_2 = \pm \xi \sin \lambda \tag{3}$$

$$G_3 = 0,$$

the choice of sign depending on whether $I = 0$ or $180°$. When G_1 is well determined we have found it to be positive, showing that the radial component of the force acts away from the Sun, and particularly for comets of perihelion distance q greater than about 1.4 AU G_1 is an order of magnitude greater than G_2. This suggests that, in general, λ is a small and thus approximately constant angle in the first quadrant and that the sign of G_2 corresponds directly to the choice $I = 0$ or $180°$. This is perhaps what one would expect for a 'new' comet, a 'clean snowball' that has only recently been perturbed by Jupiter into an orbit of relatively small q. On the other hand, although the solutions are not completely satisfactory, there are indications that G_1 may be slightly negative for P/Encke and P/Pons-Winnecke. But we know that these comets are 'old' and that they have existed for some time with q less than 0.8 AU (only 0.3 AU for P/Encke); their nuclear constructions must now be extremely complex, and we cannot expect the simple Equations (3), or even Equations (2), to apply.

Solutions for constant G_1 and G_2, as well as the six orbital constants, greatly extend the interval over which one can obtain a satisfactory representation of the orbit of a periodic comet. However, as discovered a century ago by Asten and Back-lund in the case of P/Encke, the nongravitational effects do not completely reproduce themselves every revolution. Asten, Backlund, and more recently Makover and others at the Institute for Theoretical Astronomy here in Leningrad have found that the secular variation in the mean motion of P/Encke has been decreasing. This sug-gests that when solutions for constant G_1 and G_2 cease to be satisfactory we might consider these quantities to vary in some regular way with the time. In order to avoid the possibility that G_2 for P/Encke would change sign in the near future and that it would then rapidly increase in magnitude – which does not seem physically to be very probable – we have adopted the exponential variation

$$G_i = A_i \exp(-B_i \tau), \tag{4}$$

where the A_i and B_i are now constants, and τ is the time from an initial epoch (meas-ured, for convenience, in units of 10^4 days, or approximately 27.4 yr).

We have not found it useful to solve separately for the radial variation B_1 (for this does not significantly affect the residuals) and have generally taken it to be zero. And since A_2 and B_2 are rather highly correlated we have solved for B_2 only if it produces a substantial improvement in the residuals; even then the results for B_2 should be regarded with a certain amount of caution. A solution for P/Encke, fitted to observa-tions over the interval 1927–1967, gave $B_2 = +0.8$ (implying that G_2 decreases with a half-life of 36 yr). Extrapolation back to the early nineteenth century, however, would make the nongravitational force much larger than was observed, and we must suppose that B_2 has steadily increased from the value then of $+0.3$. Four other comets, notably the well-observed P/Schwassmann-Wachmann 2, have given positive values of B_2, and indeed this is to be expected if the nongravitational effect eventually all but vanishes, and we are left with a comet like P/Arend-Rigaux and P/Neujmin 1 that appears to have lost practically all its volatile material.

With the use of the additional parameters A_1 and A_2, and if necessary also B_2, it is sometimes possible to represent the observations of a short-period comet at as many as six apparitions and to predict the comet's position with moderate accuracy at several more. Occasionally, however, a comet will appear far from its expected place. The most celebrated example of this is P/Perrine-Mrkos at its return in 1968. Sitarski (1968) had provided a careful, but purely Newtonian, prediction based on the observations in 1955 and 1961–1962. The comet had also been observed in 1909 and 1896–1897, and extrapolation back suggested that nongravitational effects would advance T in 1968 by 0.10 day, a relatively large correction but by no means unusual. In actual fact T was advanced by an additional 0.7 day, and since the comet was rather near the Earth, the error in the geocentric position was well over $2°$; the error along the comet's orbit was almost 2 000 000 km. We surmised that the reason for the discrepancy stemmed from the fact that the comet passed only 0.4 AU from Jupi-ter in 1959, the resulting orbital change altering the pattern of solar radiation on the comet, which in turn affected the nongravitational parameters. The anomaly could be considered as a sudden decrease of 3.5 m s^{-1} in velocity when the comet was near Jupiter.

There is also evidence that the same phenomenon, though to a lesser extent, occurred in the case of P/Schaumasse around the time of its approach to Jupiter (also to 0.4 AU) in 1937. On the other hand several more comets, even comets on which the non-gravitational effects are normally large, have passed near Jupiter and subsequently been perfectly predictable. A noteworthy example is P/d'Arrest, the orbit fitted to observations over 1923–1964 requiring ΔT corrections of only some 0.02 day in 1910 and in 1970, this in spite of approaches to Jupiter of less than 0.5 AU in 1920 and 1968. And we don't know for sure that the troubles with P/Perrine-Mrkos and P/Schaumasse arose during the revolutions in which there were approaches to Jupiter – they might have arisen during the neighboring revolutions. Yeomans (1972) has conclusively shown that a large anomaly occurred in the motion of P/Giacobini-Zinner between 1959 and 1965, an interval that did not involve an approach to Jupiter.

Even if we ignore the most recent apparition Yeomans' results on P/Giacobini-Zinner show quite definitely that B_2 is negative. A somewhat less certain, but possibly more negative, B_2 has now been obtained for P/Honda-Mrkos-Pajdušáková. So the simple picture of the nongravitational effects slowly decreasing as a comet loses its volatiles cannot be correct. Furthermore, solutions for A_1 and A_2 over discrete arcs of observations of P/Pons-Winnecke and P/Faye have shown that G_2 can change sign. The change of sign for P/Pons-Winnecke between the nineteenth century and the present is very definite, but in the meantime the comet repeatedly made close approaches to Jupiter, and among the orbital changes there was a net increase of 50% in q. The nongravitational effects on P/Faye are smaller and not so easily detectable, although the change of sign of G_2 also seems to be established; there have also been two moderately close approaches to Jupiter, but any modifications in the comet's orbit were minor.

Nevertheless, the values of G_2 for P/Pons-Winnecke and P/Faye are nowadays numerically smaller than they were, and while the change of sign means that we must abandon the straightforward exponential variation of Equation (4), it may be appropriate to modulate it with a periodic term; i.e., to adopt

$$G_i = \exp(-B_i \tau)[A_i + D_i \cos(\beta_i \tau + \gamma_i)]. \tag{5}$$

This form of damped oscillation was in fact adopted by Michielsen (1968) in a preliminary study of the secular acceleration of P/Encke. One could certainly assume that $D_1 = 0$, and solutions for all the quantities A_1, A_2, B_2, D_2, β_2, and γ_2 would only rarely be practical.

In modifying the equations that correspond to the Whipple model we made the assumption that $\sin I = 0$. It followed from the observations that λ is generally small. As an alternative, we could assume from the start that λ is small but make no assumption about I. To an appropriate degree of approximation it follows that

$$G_1 = \xi \tag{6}$$

$$G_2 = \xi \lambda \cos I.$$

G_3 is not now zero but would exhibit periodic variations with amplitude of order $\xi \lambda$. We certainly cannot exclude observationally the possibility that G_3 is of the same order as G_2, for G_3 is by no means as well determined as G_2; but by the same token it follows that G_3 can be ignored.

The important difference between Equations (3) and Equations (6) is the presence of the factor cos I in the latter. From a physical point of view nothing is known about the variation of cos I with time, but if there is a single dominant long-term variation, it is not unreasonable to associate this variation with the D_2 term in Equation (5). The exp $(-B_i\tau)$ factor would be associated with the variation of ξ; and we should adopt $B_1 = B_2$ and expect to find it positive. Gehrels (1970) has suggested that as a comet ages there may be a tendency for I to stabilize at 90°, so oscillations of I about 90° and hence changes of sign of G_2 would not be uncommon in a comet's dying stages. There is a need for more theoretical, and if possible even experimental, work on the variations of I. It could be that large and sudden changes occur in I, and these may be responsible for the peculiar anomalies observed in the motions of P/Perrine-Mrkos, P/Schaumasse and P/Giacobini-Zinner.

In order to form the actual nongravitational acceleration components F_i the quantities G_i must be multiplied by some function of r. We found it convenient to define the relationship between the F_i and G_i by Equation (1). From studies of the periodic comets we know that an inverse square law is inappropriate for the transverse component, but it is not clear that an inverse square law is unsatisfactory in the case of the radial component. We could perhaps adopt an inverse square law for both components and then assume that the excess variation with heliocentric distance of F_2 is associated with the factor λ in Equations (6); further theoretical study of the lag-angle λ is also most desirable.

There is some evidence that the radial component may indeed vary according to an inverse square law. The results discussed until now have been concerned with the short-period comets, and observations have to be made at three apparitions before the nongravitational effects show up. There are two recent long-period, single-apparition comets where it also appears to be possible to detect in their motions the effects of nongravitational forces. These are comets 1957 III (Arend-Roland) and 1960 II (Burnham), and a Newtonian orbit solution is particularly unsatisfactory in the latter case, even though the comet was under observation for only six months. For these comets it is the radial nongravitational component that is the better determined, and while there is some improvement if one defines it by Equation (1), the improvement is significantly greater if one supposes F_1 to vary simply according to r^{-2}. The repulsive force amounts to about 7×10^{-5} that of solar gravitational attraction for comet 1957 III and to 20×10^{-5} for comet 1960 II. These values are particularly high. Hamid and Whipple (1953) attempted to determine the radial nongravitational forces on 64 long-period comets by modifying the definitive orbit determinations. Most of their individual results are probably suspect, but their average result, a repulsion of about 1×10^{-5} that of gravitational attraction by the Sun, could well be meaningful.

Our computations on the short-period comets suggest too that the radial repulsive force amounts to about 10^{-5} that of solar attraction. Equation (1) was adopted, but for both components the figures may be *roughly* converted to an effective inverse square law by dividing them by $q \exp (q^2/2)$. The results so modified, and designated

TABLE I

Nongravitational parameters for the 46 comets of three or more appearances

Comet	$\bar{q}$	$\bar{P}$	N	Arc	ΔP	$10^5 A_1'$	$10^6 A_2'$	B_2	Note
Encke	0.34	3ʸ3	12	1927–1967	−0ᵈ02	−0.1?	−0.4	+0.8	(1)
Honda-Mrkos-Pajdušáková	0.56	5.2	4	1948–1969	−0.15	+0.7	−2.2	−0.4	
Halley	0.59	76	3	1758–1911	+4		(+1)		
Brorsen	0.60	5.5	3	1857–1873	+0.2?		(+3?)		
Crommelin	0.75	28	4	1818–1956	+2?		(+2?)		
Pons-Brooks	0.78	72	3	1812–1954	−4		(−1?)		
Grigg-Skjellerup	0.86	4.9	4	1947–1962	−0.005		(−0.1)		
Biela	0.89	6.7	4	1805–1846	−0.25		(−4)		
Tempel-Tuttle	0.97	33	3	1699–1965	+0.4		(+0.4)		
Giacobini-Zinner	0.98	6.5	6	1913–1960	+0.08	+0.9	+1.6	−0.2	(2)
Tuttle	1.03	14	3	1926–1967	+0.09		(+1)		
Finlay	1.07	6.9	3	1953–1967	+0.06		(+1)		(3)
Tempel-Swift	1.09	5.5	4	1869–1908	−0.13	+0.4?	−4.6	+0.2	
Tuttle-Giacobini-Kresák	1.12	5.5	3	1907–1962	+0.07		(+2)		
Pons-Winnecke	1.16	6.2	4	1939–1964	+0.002	−0.1?	+0.06		(4)
Schaumasse	1.20	8.2	3	1944–1960	−0.08	+1.5	−1.6		(5)
Olbers	1.20	72	3	1815–1956	+5		(+1)		
Perrine-Mrkos	1.20	6.5	3	1909–1962	−0.1?		(−3?)		(6)
Tempel 2	1.37	5.2	6	1930–1967	+0.001	<0.05?	+0.03		(7)
d'Arrest	1.37	6.7	4	1923–1964	+0.12	<0.5?	+3.7	+0.1	
de Vico-Swift	1.40	5.9	3	1844–1965	+0.04		(+2)		(8)
Arend-Rigaux	1.40	6.7	3	1951–1963	0.00		0		
Borrelly	1.41	6.9	8	1904–1968	−0.04	+0.5	−1.5		(9)
Kopff	1.54	6.4	3	1958–1970	−0.08?		(−2?)		
Neujmin 1	1.54	18	4	1913–1966	0.00		0		
Forbes	1.54	6.4	4	1929–1961	+0.05	+1.2	+1.9	+1.2	
Daniel	1.55	6.8	4	1937–1964	+0.06	+3.6	+2.2		(10)
Wolf-Harrington	1.61	6.5	3	1951–1965	−0.04		(−2)		
Wirtanen	1.62	6.7	4	1948–1967	−0.07	<4?	−2.5		
Faye	1.64	7.4	6	1932–1970	−0.001	+0.9	−0.03		(11)
Tempel 1	1.70	5.9	3	1867–1879	0.00		0		(12)
Väisälä 1	1.75	11	3	1939–1960	0.00		0		
Comas Solá	1.77	8.5	5	1926–1962	+0.01		(+0.2)		
Harrington-Abell	1.78	7.2	3	1955–1969	0.00		0		
Arend	1.82	7.8	3	1951–1967	−0.02	<2?	−0.5		
Brooks 2	1.87	6.9	5	1925–1954	−0.11		(−4)		
Reinmuth 2	1.90	6.7	4	1947–1967	0.00		0		
Reinmuth 1	2.0	7.6	3	1949–1965	0.00		0		
Schwassmann-Wachmann 2	2.1	6.5	7	1929–1968	−0.05	+1.6	−2.0	+0.3	
Holmes	2.2	7.0	4	1892–1965	+0.02		(+0.4)		(13)
Johnson	2.3	6.9	3	1949–1964	0.00		0		
Ashbrook-Jackson	2.3	7.4	3	1948–1965	0.00		0		
Whipple	2.5	7.5	5	1933–1964	−0.01	+0.2	−0.2		
Wolf	2.5	8.4	3	1942–1960	+0.002		(+0.1)		
Oterma	3.4	7.9	(3)	1942–1962	0.00		0		
Schwassmann-Wachmann 1	5.5	16	(4)	1902–1965	0.00		0		

(1) B_2 is evidently increasing and was only +0.3 in the early nineteenth century.
(2) In 1966 an additional correction $\Delta T = +0ᵈ3$ was required.
(3) Approach to 0.60 AU of Jupiter in 1957.
(4) Five approaches within 0.7 AU of Jupiter occurred every alternate revolution between 1882 and 1942. Before that ΔP (and A_2') had the opposite sign. During 1858–1875 (when $\bar{q} = 0.79$, $\bar{P} = 5.6$) $\Delta P = -0.02$, $10^5 A_2' = +0.6$, $10^6 A_2' = -0.4$.
(5) Approach to 0.37 AU of Jupiter in 1937. In 1927 an additional correction $\Delta T = -0ᵈ4$ was required.
(6) Approach to 0.38 AU of Jupiter in 1959. In 1968 an additional correction $\Delta T = -0ᵈ7$ was required.
(7) Approach to 0.63 AU of Jupiter in 1943.
(8) Approaches to 0.60 AU of Jupiter in 1885 and to 0.44 AU in 1897.
(9) Approach to 0.54 AU of Jupiter in 1936.
(10) Approach to 0.53 AU of Jupiter in 1959.
(11) ΔP (and A_2') formerly had the opposite sign. During 1888–1926 (when $\bar{q} = 1.69$, $\bar{P} = 7.5$) $\Delta P = +0.01$, $10^5 A_1' = +0.9$, $10^6 A_2' + 0.3$. Approaches to 0.51 AU of Jupiter in 1899 and to 0.60 AU in 1959.
(12) Approach to 0.36 AU of Jupiter in 1870.
(13) Approach to 0.54 AU of Jupiter in 1908.

This table is based on calculations by K. Aksnes, J. L. Brady, M. P. Candy, H. J. Carr, A. D. Dubyago, J. Hepperger, P. Herget, M. Kamieński, L. Kresák, B. G. Marsden, J. G. Porter, E. K. Rabe, H. Q. Rasmusen, G. Schrutka, J. Schubart, L. R. Schulze, Z. Sekanina, G. Sitarski, P. Stumpff and D. K. Yeomans.

by A'_1 and A'_2, are given in Table I, together with the values of B_2 and the effective period-change ΔP, in days per period, due to the nongravitational forces. Rigorous computations of our nongravitational parameters have been made in relatively few cases, but ΔP values, though uncertain in many instances, can be estimated for all the comets of three or more apparitions. These ΔP values can be approximately converted into A'_2 values, the results so obtained being given in parentheses in the table. The column N gives the number of apparitions considered, and 'Arc' shows the actual span of the observations. The epoch for B_2 is the middle of this span. $\bar{P}$ denotes the mean period, and the comets are listed according to mean perihelion distance $\bar{q}$.

For only eleven of the entries are nongravitational forces completely undetectable, while positive and negative values of ΔP (and hence A'_2) are equally numerous. The comets that show no detectable effects may perhaps be expected to show them when observations are available over longer intervals, but it should be noted that the two of smallest $\bar{q}$ are the two 'asteroidal' comets. Among the comets of $\bar{q}$ greater than about 1.9 AU small nongravitational effects prevail, and only P/Schwassmann-Wachmann 2 quite definitely shows them at all in fewer than four apparitions. Particularly noteworthy is the failure to detect nongravitational effects for the two comets of $\bar{q}$ greater than 2.5 AU, for low eccentricities have made it possible to observe these comets regularly even near aphelion, so their orbits are very well determined. It is perhaps to be expected that the motion of an icy comet would completely cease to be subject to nongravitational effects beyond a heliocentric distance of, say, 2.5 to 3.0 AU. Our continuous equations of motion, with the variation with r given by Equation (1), simulate this but continue to give a nonzero contribution beyond the suggested cutoff. There would presumably be a more serious discrepancy if we were to adopt an inverse square law for F_1. P/Schwassmann-Wachmann 2 would thus not have been subject to nongravitational effects prior to its close approach to Jupiter in 1926, q previously having been 3.5 AU. P/Brooks 2, another comet with large nongravitational effects, considering its present value of q, had a q of 5.5 AU until 1886. Long dormant, these comets have recently and suddenly become very active.

Although we have had a moderate amount of success in applying the procedures discussed here, it is by no means clear that we are really tackling the problem of nongravitational effects on cometary motions in the right manner. There is much room for further experimentation by a variety of methods, and the parentheses, blank spaces and question marks in Table I show that there are very many more comets left to be studied.

References

Cunningham, L. E.: 1968, private communication.
Eckert, W. J. and Brouwer, D.: 1937, *Astron. J.* **46**, 125.
Gehrels, T.: 1970, private communication.
Hamid, S. E. and Whipple, F. L.: 1953, *Astron. J.* **58**, 100.
Herget, P.: 1968, *Astron. J.* **73**, 729.
Marsden, B. G.: 1968, *Astron. J.* **73**, 367.

Marsden, B. G.: 1969, *Astron. J.* **74**, 720.
Marsden, B. G.: 1970, *Astron. J.* **75**, 75.
Michielson, H. F.: 1968, private communication.
Roemer, E.: 1961, *Astron. J.* **66**, 368.
Sitarski, G.: 1968, *Acta Astron.* **18**, 423.
Whipple, F. L.: 1950, *Astrophys. J.* **111**, 375.
Yeomans, D. K.: 1972, this Symposium, p. 181.

Discussion

S. K. Vsekhsvyatskij: You speak of the continuous action of nongravitational forces. What is your opinion of the cases, studied by Sekanina, where cometary nuclei have split, the forces involved directly exhibiting their impulsive character?

B. G. Marsden: I am talking about a phenomenon that affects the motions of most, if not all, comets; and analysis of the orbit solutions suggests that it is continuous, inasmuch as anything is continuous. Of course, impulsive forces might yield observable effects in the case of a comet that has split.

L. Kresák: What is the period of the periodic term in your Equation (5) and how many cycles are covered by the observations of P/Encke?

B. G. Marsden: Michielsen obtained a period of something like 150 to 200 yr for P/Encke – barely a complete cycle since discovery. Calculations by Sekanina and myself suggest that the period may be similar in the case of P/Faye.

L. M. Shul'man: Outbursts in brightness are connected with symmetrical explosions in the nucleus. But the nucleus acquires almost zero momentum in this process, and we cannot expect any nongravitational forces.

B. G. Marsden: This is a good point, but are the outbursts of P/Schwassmann-Wachmann 1 sufficiently symmetrical for the forces to be cancelled out?

27. ON THE DETERMINATION OF NONGRAVITATIONAL
FORCES ACTING ON COMETS

P. E. ZADUNAISKY

University of La Plata and Instituto T. Di Tella, Buenos Aires, Argentina

Abstract. Some recent investigations on the existence and nature of nongravitational forces acting on comets have required the application of a good deal of high precision numerical procedures. In this paper these procedures are examined critically; special attention is given to cases where difficulties may appear when a comet makes a close approach to a planet or the Sun.

1. Introduction

In recent times important advances have been made in the study of irregularities in the motions of comets. The results, especially those obtained by Marsden (1968, 1969, 1970), seem to correspond very closely to Whipple's (1950) theory on the physical nature of cometary nuclei and give evidence for the existence of nongravitational forces acting on comets. In Marsden's work the equations of motion for a comet in a system of heliocentric cartesian coordinates are adopted in the following form:

$$\ddot{x} = -\mu x r^{-3} + \partial R/\partial x + F_1 x r^{-1} + F_2(r\dot{x} - \dot{r}x)h^{-1} + F_3(y\dot{z} - z\dot{y})h^{-1},$$

where $x \to y, z$; $h^2 = (y\dot{z} - z\dot{y})^2 + (z\dot{x} - x\dot{z})^2 + (x\dot{y} - y\dot{x})^2$; $r^2 = x^2 + y^2 + z^2$; μ is the gravitational constant, R the planetary disturbing function, and F_1, F_2 and F_3 are orthogonal components of a nongravitational force to be determined. Further, the F's are assumed to be of the form

$$F_i = A_i \exp(-B_i\tau) \exp(-r^2/C)r^{-\alpha}; \qquad i = 1, 2, 3;$$

where C, α, A_i, B_i are constants and τ is the time from the initial osculation epoch. The constants C and α are more or less arbitrarily set at 2 and 3, respectively. On the other hand, a selection of the constants A_i and B_i, together with the six Keplerian elements of the orbit, are determined by a process of successive differential corrections.

To establish the equations of condition it is necessary to integrate the equations of motion given above. In these calculations certain difficulties arise when a comet makes a close approach to Jupiter. Particularly in the cases of P/Schaumasse and P/Perrine-Mrkos it seems to be very difficult to establish a set of Keplerian and nongravitational parameters in order to link several returns without the appearance of systematic trends in the residuals. It has been considered that in these cases the nature of the trouble may be more mathematical than physical. We agree with such an assumption, and the purpose of the present communication is to point out possible sources of error in the numerical calculations and to indicate possible ways of solving this type of problem.

We believe that one of the difficulties stems from truncation errors accumulated in the numerical integration of the equations of motion when the comet makes a

Chebotarev et al. (eds.), The Motion, Evolution of Orbits, and Origin of Comets, 144–151. All Rights Reserved.

close approach to a planet or to the Sun. These errors are then reflected and can be magnified in the process of differential corrections of the parameters. In the following sections we examine in order those numerical processes.

2. The Numerical Integration of the Equations of Motion

Let us first consider the following example, which may be a typical case of the motion of a comet that periodically makes close approaches to the Sun. To simplify matters, we have considered that the heliocentric motion of the comet is perturbed only by Jupiter, and that the motion occurs in the orbital plane of the planet – which is assumed in turn to describe a circular orbit around the Sun. The motion can then be referred to synodic coordinates and described by the equations of the restricted problem of three bodies; the unknowns are in this case the coordinates (x, y) of the comet and their derivatives $(\dot{x}, \dot{y})$. We have chosen the initial conditions corresponding to a periodic orbit of a type described by Rabe (1961); it has been demonstrated by Deprit and Palmore (1966) that this type of orbit is stable. The orbit is defined by the following elements: $a = 5.20$ AU $(P = 11.86$ yr$)$, $e = 0.91$ $(q = 0.45$ AU$)$.

For the numerical integration we used the Runge-Kutta-Gill method; during the whole computation the step-size was controlled in order to keep the local truncation error under a certain tolerance limit, given as an input parameter. To estimate the local truncation error we used the well-known method of advancing the computation for two steps (of size h), then repeating it in one step of double size $(2h)$ and comparing the results. We found also a good estimate of the *total* errors accumulated after n steps of integration by applying a method developed by ourselves that can be outlined as follows:

Let us consider an 'original problem' of ordinary differential equations that, without loss of generality, may be written in the form:

$$dx/dt = f(t, x), \qquad x(0) = x_0.$$

By any numerical process we obtain numerical results $\tilde{x}_n$, and we want an estimate of the error $\xi_n = x(t_n) - \tilde{x}_n$ after n steps of integration. For that purpose we may determine first an empirical function $P(t)$, which can be a polynomial or any other simple function involving coefficients that are adjusted to represent, in the best possible manner, the numerical values $\tilde{x}_n$ for a certain interval of t. Now we can establish a 'pseudo-problem' of the form

$$dz/dt = f(t, z) + P'(t) - f(t, P(t)), \qquad z(0) = x_0.$$

The exact solution is evidently $z = P(t)$. If we apply the same numerical process to this pseudo-problem we shall obtain numbers $\tilde{z}$, and the error after n steps will be exactly $\zeta_n = P(t_n) - \tilde{z}_n$. Under certain conditions this error ζ_n is also a good estimate of the error ξ_n in the original problem. These conditions are based on the asymptotic theory of error propagation; for a discussion see Zadunaisky (1964), and for practical applications see Zadunaisky (1969).

We first performed the computation to a moderate degree of accuracy by carrying nine significant digits and controlling the step size so as to keep the local truncation errors under 10^{-8}. The computation was extended to three orbital periods, and the results are shown in Figure 1, where the accumulated error x and $\dot{y}$, obtained by our method outlined above, are plotted on a semilogarithmic scale as a function of time.

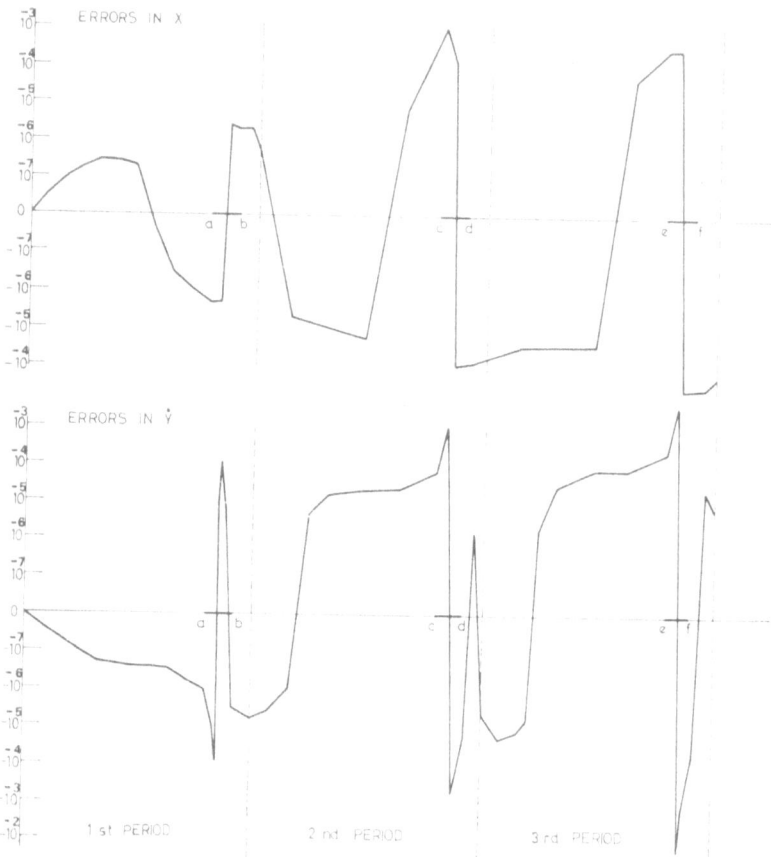

Fig. 1.   The accumulated errors in x and $\dot{y}$ during three orbital periods. The local roundoff error is 10^{-9} and the tolerated local truncation error 10^{-8}. In the intervals (a, b), (c, d), and (e, f) the heliocentric distance of the comet is smaller than 3 AU.

It is clearly seen how the errors follow an oscillating pattern as a consequence of the fact that both the orbit and the numerical process are stable. However, when the comet comes closer to the Sun and the heliocentric distance becomes smaller than three astronomical units there is a sudden increase in the size of the accumulated errors, which become much larger than the tolerated local error. The reason for this is that the estimation of the local errors is based on the implicit assumption that the higher derivatives of the unknowns do not become too large, which is true only while the comet is not close to one of the primaries.

We have also performed the same calculation with higher standards of accuracy, namely carrying 16 significant digits and setting the tolerance limit for the local error at 10^{-13}. The behaviour of the accumulated error followed more or less the same oscillatory pattern as before but, of course, its size was much smaller, reaching a maximum limit of the order of 10^{-8}.

In the ordinary process for the correction of just the six Keplerian elements of an osculating orbit, errors of that size should not bother us too much. However, if one adds as unknowns the parameters defining nongravitational forces, the order of magnitude may be that of the errors, and their determination can be substantially affected. The point of interest here is that nonnegligible errors in the calculation may occur precisely in that region of the orbit where most of the observations are made. If one considers a linear system of equations of condition $Ax=b$, those errors appear as perturbations δA in the matrix A and δb in the vector b of residuals. As we shall show in the following sections, the effect of those perturbations can be considerably magnified in the solution x due to the instability that characterizes the process of least squares solution.

3. The Process of Successive Differential Corrections

A. CONVERGENCE AND ESTIMATION OF ERRORS

The standard method of differential correction, as used in practical applications, can be described as follows. Let

$$f_i(a_1, a_2, \ldots, a_k) = y_i; \qquad i = 1, 2, \ldots, n, \tag{1}$$

be a system of equations of condition, where f_i are given functions, in general not linear, of certain parameters $a_1, a_2, \ldots, a_k$, and y_i are observed quantities. Assuming $n > k$, it is proposed to solve the system for the unknowns $a_1, a_2, \ldots, a_k$ in the least squares sense. By using the vectorial notation $A = (a_1, a_2, \ldots, a_k)$, $B = (y_1, y_2, \ldots, y_n)$, $F(A) = (f_1, f_2, \ldots, f_n)$, and assuming that an approximate solution A_r is known, one wants to obtain a further approximation $A_{r+1} = A_r + \Delta A_r$, where the correction ΔA_r is so small that its square and higher powers are supposed negligible.

Such a correction is obtained as the least squares solution of the linear system

$$M \times \Delta A_r = \bar{B}(A_r), \tag{2}$$

where M is the Jacobian matrix $(\partial f_i/\partial a_j)$; $i = 1, 2, \ldots, n$; $j = 1, 2, \ldots, k$; and $\bar{B}(A_r) = B - F(A_r)$ is the vector of residuals. The least squares solution is given by

$$\Delta A_r = [(M^T M)^{-1} M^T]\bar{B}(A_r). \tag{3}$$

The procedure can be repeated, and we have the iteration formula

$$A_{r+1} = \varphi(A_r), \tag{4}$$

where

$$\varphi(A) = A + N^{-1}(A)M^T(A)\bar{B}(A), \tag{5}$$

$N = M^T M$ being the normal matrix of the system.

In the language of functional analysis the process of successive differential corrections is represented by the equivalent problem of finding a 'fixed point' of the function $\varphi(A)$. On the other hand, the operator φ is said to be a contraction operator when there exists a factor α such that

$$\|\varphi(A_{r+1}) - \varphi(A_r)\| \leq \alpha \|A_{r+1} - A_r\|, \qquad 0 < \alpha < 1, \tag{6}$$

for any pair of vectors A_{r+1} and A_r. Under these conditions the set of approximations A_r converges to a unique solution A^*. Furthermore, an upper bound of the error of the rth approximation is given by

$$\|A^* - A_r\| \leq \frac{\alpha}{1 - \alpha} \|A_r - A_{r-1}\|. \tag{7}$$

Applying some known results of the theory (Zadunaisky and Pereyra, 1965; Pereyra, 1967), we have found an upper bound for the factor α to be

$$\alpha \leq \|N^{-1}(A_r)\| \times \|\Omega(A_r)\| \times \|\bar{B}(A_r)\|, \tag{8}$$

where $\Omega(A_r)$ is a matrix whose elements are defined by

$$\Omega_{pq} = \max_i \frac{\partial^2 f_i}{\partial a_p \partial a_q}. \tag{9}$$

This upper bound of α evidently has, by Equation (6), an important effect on the precision and speed of convergence of the iterated least squares process. According to Equation (8), it depends on three factors, each of them having a special meaning, as shown below.

The factor $\|\Omega(A_r)\|$ evidently reduces to zero in a linear problem; in the general case it measures the influence of the nonlinear terms neglected in the process.

The third factor is the sum of the squares of the residuals (if one adopts the Euclidean norm for vectors), and it depends, of course, on both an adequate choice of the functions f_i and the good quality of the observations.

B. NUMERICAL STABILITY OF THE PROCESS

Now let us turn our attention to the factor $\|N^{-1}(A_r)\|$ in Equation (8). In our calculations we have assumed implicitly that in the linear system, Equation (2), the $n \times k$ matrix M has a rank $r = k$; i.e., all the columns are linearly independent. In that case the normal matrix N is nonsingular and the least squares solution is given by Equation (3), where the expression in brackets is called a pseudo-inverse of the matrix M and is usually indicated by the notation

$$M^+ = (M^T M)^{-1} M^T. \tag{10}$$

When $r < k$ the normal matrix is singular, but it is still possible to obtain two different types of least squares solutions in a way that can be outlined as follows (Rosen, 1964; Pereyra and Rosen, 1964).

To simplify the notation let us write the linear system, Equation (2), in the form

$$Ax = b, \tag{11}$$

where A is an $n \times k$ matrix, and x and b are vectors of k and n dimensions, respectively. In A there are r linearly independent columns, and without loss of generality the matrix A can be partitioned in the form

$$A = (B, \underline{B}),$$

where B is an $n \times r$ matrix of independent columns, and $\underline{B}$ is an $n \times (k-r)$ matrix formed by the rest of the columns of A. The normal matrix $(B^T B)$ is then nonsingular, and the pseudo-inverse of B is

$$B^+ = (B^T B)^{-1} B^T.$$

Then a pseudo-inverse of A is given by the formula

$$A^+ = C^T (CC^T) B^+,$$

where

$$C = B^+ A.$$

It is possible to show that the vector

$$x_m = A^+ b$$

satisfies the least squares condition and x_m has minimum modulus. On the other hand, if we form the matrix

$$A^\# = \begin{pmatrix} B^+ \\ \vdots \\ 0 \end{pmatrix},$$

where the first r rows consist of the matrix B^+ and the remaining rows are zero, the vector

$$x_b = A^\# b$$

also satisfies the least squares condition, and it has at most r nonzero components. x_m is called a *minimum approximate solution* and x_b is a *basic approximate solution*.

The matrices B and $(B^T B)^{-1}$ may be determined by the following recursive algorithm. Let a_q be the qth column of A and B_q a submatrix of A formed by its first q columns. Assuming $(B_q^T B_q)^{-1}$ to be known, one obtains

$$(B_{q+1}^T B_{q+1})^{-1} = \left(\begin{array}{c|c} (B_q^T B_q)^{-1} + \alpha_{q+1}^{-1} u_q u_q^T & -\alpha_{q+1}^{-1} u_q \\ \hline -\alpha_{q+1}^{-1} u_q^T & \alpha_{q+1}^{-1} \end{array} \right),$$

where

$$u_q = B_q^+ a_{q+1}$$
$$\alpha_{q+1} = \| P_q a_{q+1} \|^2,$$

and

$$P_q = (I - B_q (B_q^T B_q)^{-1} B_q^T).$$

The process is initiated with the first column of A, obtaining $(B_1^T B_1)^{-1} = (a_1^T a_1)$ and then adding one column at a time. The column a_{q+1} is linearly independent of those in B_q if $\alpha_{q+1} > 0$, because P_q is a projection matrix that takes the vector a_{q+1} into the space orthogonal to that spanned by B_q. In an actual computation one should never obtain a value of α_{q+1} exactly equal to zero, so that one gives a properly chosen parameter γ, and a column a_{q+1} is considered as linearly independent of those of B_q when $\alpha_{q+1} > \gamma$.

The whole process can be performed by rows instead of by columns. If the system $Ax = b$ represents a linear model of a physical process, the analysis by columns may give an indication of how well the parameters have been selected in the sense that strong correlations do not exist among them. On the other hand, the analysis by rows should show the effects of the successive observations on the model.

So far we have assumed that the rank of A can be well determined and x computed without difficulties. But when the rank of A is not well determined and the normal matrix becomes nearly singular or ill-conditioned, serious difficulties may arise.

In the case that A is a square nonsingular matrix it is known that a perturbation δb in the right-hand member of $Ax = b$, or a perturbation δA in A, may produce changes δx in the solution such that

$$\frac{\|\delta x\|}{\|x\|} \leq \|A\| \times \|A^{-1}\| \frac{\|\delta b\|}{\|b\|}$$

and

$$\frac{\|\delta x\|}{\|x + \delta x\|} \leq \|A\| \times \|A^{-1}\| \frac{\|\delta A\|}{\|A\|},$$

respectively.

The product $\|A\| \times \|A^{-1}\|$ is the 'condition number' of A, and when A is nearly singular it becomes large, and the effects of the perturbations δb and δA can be greatly magnified.

When the matrix A is rectangular, similar, although more complicated, results may be obtained by introducing the 'pseudo-condition' number $\|A\| \times \|A^+\|$. We have described before a method of selection of the successive columns of A, in order to form the submatrix B, by checking the degree of correlations among the variables. The procedure may be completed so as to obtain a pseudo-inverse that produces the smallest possible magnification of the errors introduced by the perturbations δb and δA. The details of these procedures fall beyond the limits of this report; see Pereyra (1969).

4. Final Remarks

We have shown in Section 2 how the numerical integration of the equations of motion may introduce in the equations of condition perturbations, which may be small but not negligible. In Section 3 we have shown how these perturbations can be magnified in the least squares solution of the equations of condition. We think that this can be a possible explanation of the anomalies observed in those comets that make close approaches to Jupiter or the Sun.

These difficulties are not unavoidable; the standard procedures for the numerical integration of the equations of motion can be applied and completed with the method of error estimation described in Section 2. On the other hand, the resolution of the equations of condition can be performed by the methods outlined in Section 3, with all the precautions indicated there for avoiding the troublesome effects that result from their instability.

We intend to perform a series of numerical experiments on typical cometary orbits by applying this type of technique.

Acknowledgments

The participation of the writer in the Symposium has been made possible through a grant from the National Council of Geo-Heliophysics Research of Argentina.

References

Deprit, A. and Palmore, J.: 1966, *Astron. J.* **71**, 94.
Marsden, B. G.: 1968, *Astron. J.* **73**, 367.
Marsden, B. G.: 1969, *Astron. J.* **74**, 720.
Marsden, B. G.: 1970, *Astron. J.* **75**, 75.
Pereyra, V.: 1967, *Soc. Indust. Appl. Math. J. Num. Anal.* **4**, 27.
Pereyra, V.: 1969, *Aequationes Mathematicae* **2**, 194.
Pereyra, V. and Rosen, J. B.: 1964, *Stanford Univ. Tech. Rept.* CS 13.
Rabe, E.: 1961, *Astron. J.* **66**, 500.
Rosen, J. B.: 1964, *J. Soc. Indust. Appl. Math.* **12**, 156.
Whipple, F. L.: 1950, *Astrophys. J.* **111**, 278.
Zadunaisky, P. E.: 1964, in G. Contopoulos (ed.), 'The Theory of Orbits in the Solar System and in Stellar Systems', *IAU Symp.* **25**, 281.
Zadunaisky, P. E.: 1969, in G. E. O. Giacaglia (ed.), *Periodic Orbits, Stability and Resonances*, Reidel, Dordrecht, p. 216.
Zadunaisky, P. E. and Pereyra, V.: 1965, *Proceedings of the Int. Fed. Inf. Processing Congress*, New York, Vol. 2, p. 488.

Discussion

B. G. Marsden: I appreciate the difficulties you have mentioned, but I think Sitarski and Yeomans will agree with me that these strange anomalies in the motions of comets are sometimes very large indeed. Furthermore, they consistently occur for the same comets, and around the same time. In particular, a large anomaly appears in the case of P/Perrine-Mrkos whether one fits the 1955 and 1962 observations and extrapolates forward to 1968, or whether one fits the 1962 and 1968 observations and extrapolates back to 1955; and calculations have been made, by Sitarski and myself, using completely independent procedures.

P. E. Zadunaisky: I should like to make experiments with this method and see what happens. I am in a position now to obtain quantitative results about the dependence and the upper bounds of the errors we may expect.

28. A SEARCH FOR ENCKE'S COMET IN ANCIENT CHINESE RECORDS: A PROGRESS REPORT

F. L. WHIPPLE and S. E. HAMID

Smithsonian Astrophysical Observatory, Cambridge, Mass., U.S.A.

Abstract. The catalog by Ho Peng Yoke lists ancient Chinese observations of 581 transient astronomical objects of which some 300 might conceivably have been P/Encke, between the years -156 and $+1600$. Using the Gauss-Hill method it was possible to establish roughly the plane of the orbit and the direction of perihelion for P/Encke over this interval of time. All but about 40 of the objects could be eliminated by comparing the observed locations with possible locations of P/Encke. Several more could be eliminated on the basis of their calculated magnitudes and general observability. Large uncertainties in the non-Newtonian motion as yet prevent our certain identification of P/Encke among the remaining possibilities.

The extended and torroidal character of the Taurid-Arietid meteor complex in space (Whipple and Hamid, 1952) and its dynamical relationship with Encke's Comet (Whipple, 1940) demonstrate that this periodic comet has been active in contributing meteoroids for at least hundreds and probably for thousands of revolutions. Observations, general logic, and specifically the icy comet model (Whipple, 1950) support the concept that the intrinsic brightness of a comet should progressively decrease with time. Hence P/Encke, which has probably moved in much its present orbit throughout all historic time, should generally have been brighter in earlier centuries than during recent apparitions. The rate at which comets systematically change in brightness with age is not established satisfactorily on either an observational or a theoretical basis. Hence it is highly desirable to establish whether P/Encke, frequently a naked-eye object during the past 180 years, was sufficiently bright in ancient times to attract the attention of the systematic Chinese observers. Identification of P/Encke in the ancient records could also give us valuable information about its past nongravitational motion and perhaps a clearer idea of how it attained its present unusual orbit. The extremely valuable *Ancient and Medieval Observations of Comets and Novae in Chinese Sources* by Ho Peng Yoke (1962) provides an unparalleled source in which to search for possible observations of P/Encke. For most of the 581 observations listed, Ho Peng Yoke includes a date, a direction in the sky and frequently detailed information on the appearance of the object plus additional observations of its motion.

The long intervals between probable observations of P/Encke in the past and its known variable period (see e.g., Marsden, 1969) preclude the possibility of meaningful predictions of the comet's position for apparitions predating its discovery in 1786. On the other hand, the position of the orbital plane and the line of apsides can be predicted approximately over long periods of time by computation of the secular perturbations, even though the phase relationship of the comet in its orbit has become completely uncertain. Thus for any given date in the past twenty centuries one can ascertain with some confidence whether an observed comet in the sky at a given date

Chebotarev et al. (eds.), The Motion, Evolution of Orbits, and Origin of Comets, 152–154. All Rights Reserved.
Copyright © 1972 by the IAU.

could reasonably be attributed to P/Encke at some position in its orbit as seen from the Earth.

Before beginning such calculations we eliminated some 280 objects from Ho Peng Yoke's catalog because of their previous identification with Halley's Comet, with novae, with fireballs, or because of lack of specific information on the date or position in the sky. Many of the sky positions were rather vague because of their location by means of Chinese constellations, some of which extend for over great distances on the sky.

We calculated the orientation of the orbit of Encke's Comet in space by numerical integration of the equations of motion defining the secular perturbations through the Gauss-Hill method. The positions of the planets in the past were obtained from the Brouwer-van Woerkom (1950) theory. The small errors in numerical results obtained from this theory as noted by Sharaf and Budnikova (1967) are trivial with respect to the much larger observational uncertainties in the comet positions.

At 100-yr intervals, from the calculated position of the orbit of Encke's Comet, we derived the directions as seen from the Earth at 10-day intervals for all positions around the Earth's orbit and around the comet's orbit from 140 days before the perihelion to 130 days after. For each recorded cometary observation we searched in the ephemeris of the orbital plane nearest in time to ascertain whether or not P/Encke could have been seen in the stated direction at the given time, errors in direction up to 20° being allowable. For each allowable observation, an approximate date of perihelion could be determined. More than 40 cometary observations remain from Ho Peng Yoke's catalog in the interval -156 to $+1600$ after the application of this geometrical exclusion.

A number of these possible observations of Encke's Comet can be eliminated by the criterion that $r^{-4}\Delta^{-2} > 16$, where r and Δ (in AU) are the solar and geocentric distances at the time of observation. This three-magnitude enhancement of the brightness from that at unit distance from the Sun and Earth might be relaxed as a criterion among the earlier observations should clear evidence be found for an intrinsic brightening of Encke's Comet in ancient times. Some observations can be questioned because of close proximity of the observed position to the Sun. Furthermore, some of the Chinese comets were too bright or lasted too long to have been Encke's Comet.

A few observations might be added tentatively to the list of possibilities should a clear-cut sequence of identification of Encke's Comet be established in the abridged list. The process of searching for possible identifications is being continued, although the sequences so far found appear to be the result of chance. The number of revolutions between widely separated possible observations is, of course, unknown, depending on the nongravitational effects.

References

Brouwer, D. and van Woerkom, A. J. J.: 1950, *Astron. Pap. Washington* **13**, part 2.
Ho Peng Yoke: 1962, *Vistas Astron*—. **5**, 127.

Marsden, B. G.: 1969, *Astron. J.* **74**, 720.
Sharaf, S. G. and Budnikova, N. A.: 1967, *Byull. Inst. Teor. Astron.* **11**, 231.
Whipple, F. L.: 1940, *Proc. Am. Phil. Soc.* **83**, 711.
Whipple, F. L.: 1950, *Astrophys. J.* **111**, 375.
Whipple, F. L. and Hamid, S. E.: 1952, *Bull. Roy. Obs. Helwan* No. 41.

Discussion

G. A. Chebotarev: Can one be sure that the orbit of P/Encke has not undergone large changes during this time?

F. L. Whipple: The general turning of the orbital plane is rather well established, but there is uncertainty in the semimajor axis and period, of the order of a few percent, because of the nongravitational forces.

L. Kresák: The present mean value of the semimajor axes of the shower meteors associated with P/Encke is definitely smaller than that of the comet. Do you think that this difference is due to the past history of the comet's orbit, or to the nongravitational effects acting on the meteoroids?

F. L. Whipple: I believe that unknown forces act to reduce the semimajor axes of meteor orbits and therefore that the semimajor axes of associated meteor streams do not give useful information about the past history of the comet orbit.

29. THE MOTION OF HALLEY'S COMET FROM 837 TO 1910

J. L. BRADY

Lawrence Radiation Laboratory, University of California, Livermore, Calif., U.S.A.

Abstract. Numerical experiments have been made in an attempt to remove the residuals of P/Halley and link the seven apparitions from 1456 to 1910. All efforts to link more than two apparitions using Newtonian equations have invariably failed. However, by the addition of a secular term to the equations of motion, the four apparitions from 1910 back to 1682 can be linked by a numerical integration which represents the observations to contemporary accuracy. When this integration is continued, the apparitions of 1607, 1531, and 1456 show residuals of less than one day in the time of perihelion passage. Prior to 1456 the residuals begin to run off but, with the exception of 1222 and 1066, the apparitions back to 837 show residuals no greater than four days in the time of perihelion passage. The residuals of 20 days in 1222 and 7 days in 1066 appear anomalous but can be made reasonable if the Chinese records are adopted in preference to the European records.

30. A NUMERICAL ANALYSIS OF THE MOTION OF PERIODIC COMET BROOKS 2

P. STUMPFF

Max Planck-Institut für Radioastronomie, Bonn, F.R.G.

Abstract. Various sets of osculating elements of P/Brooks 2, derived by Dubyago, are introduced into an *N*-body integration programme and run from 1686 to 1976. Attempts are made to find a system of elements which links the apparitions before and after the close approach to Jupiter in 1922. The propagation of differential perturbations, and also nongravitational effects, is examined.

It is unavoidable that a set of osculating elements of a comet, determined from observations, contains finite errors. Consequently our knowledge of the orbit's evolution in phase space is represented by a 'tube' rather than by a 'thin line'. The coordinates of the centre of the tube, as a function of time, vary due to the combined effect of all planetary perturbations. The variation in shape and size of the tube cross-section is caused by differential perturbations among a group of massless bodies whose initial conditions correspond to the original orbital errors. If the error tube representing a first apparition, at the time of a second one, overlaps with the error tube of the latter to a certain extent, we would probably feel certain we have allowed for all the forces acting. However, if we find systematic separations between the tubes, we have to conclude that the motion of the comet was affected by additional forces. Therefore, the question of whether or not nongravitational forces influence a particular periodic comet depends very much on, among other things, its history of differential perturbations. A close approach to a planet, between two apparitions, not only changes the osculating elements of the comet but may also change the shape and size of the cross-section of the tube that represents the osculating elements (and their errors) of the first apparition. Without knowledge of these changes it would be difficult to compare the elements of the two apparitions.

I want to demonstrate these relations in the case of a comet whose motion is suitable for that purpose; P/Brooks 2, discovered in 1889, is such a case. Its period is seven years. Osculating elements have been derived from observations in 1889, 1896, 1903, 1911, 1925, 1932, 1939, and 1946 by Dubyago (1950, 1956), and these elements are classified as of highest quality in the catalogue of cometary orbits by Porter (1961). We know of four approaches to Jupiter; for brevity we denote them by $\mathrm{2\!\!\!\!|}_1$, $\mathrm{2\!\!\!\!|}_2$, $\mathrm{2\!\!\!\!|}_3$ and $\mathrm{2\!\!\!\!|}_4$. The first one took place three years before discovery, in 1886 (JD 2410108), with a minimum distance of less than 0.001 AU. The second one occurred in 1922 (JD 2423077), with a minimum distance of 0.086 AU. It is particularly the $\mathrm{2\!\!\!\!|}_2$ approach which makes P/Brooks 2 such a suitable object for our purpose since there is probably no other comet known where two sets of observed and carefully evaluated apparitions are interrupted by such a close Jupiter encounter. The approaches $\mathrm{2\!\!\!\!|}_3$ (1958, JD 2436280, minimum distance about 1.3 AU) and $\mathrm{2\!\!\!\!|}_4$ (1969, JD 2440345, minimum distance about 1.6 AU) are of minor importance, compared with the first

Chebotarev et al. (eds.), *The Motion, Evolution of Orbits, and Origin of Comets*, 156–166. *All Rights Reserved.*
Copyright © 1972 by the IAU.

two spectacular ones. Dubyago found it impossible to represent the apparitions by one gravitational orbit both before and after $\mathcal{2}_2$. Therefore, for each series of apparitions, he included additional (nongravitational) terms in the orbit determination. His procedure, as seen from today's knowledge, may be somewhat doubtful, because he did not actually introduce nongravitational forces into his equations of motion. His results are nevertheless striking because they show a rather systematic nongravitational variation of all elements in periods both before and after $\mathcal{2}_2$. Assuming that Dubyago's elements are at least a reasonable first approach to the problem of the true motion of the comet, my numerical analysis may be valuable for a future orbit determination which includes a more modern treatment of the nongravitational forces. In addition, the analysis of a comet with both nongravitational effects and close Jupiter approaches might throw some light on the difficulties which, for instance, have been reported by Marsden (1969).

All perturbation calculations were carried out with the Heidelberg N-body programme (Schubart and Stumpff, 1966); the planets Venus to Neptune were taken into account throughout the whole investigation using the initial values in the original publication of the programme (1.c., Table VII). The results of the N-body calculations were then converted into heliocentric ecliptical osculating elements by a special programme. Since Dubyago uses the mean equinox 1890.0 for the definition of his first set of nongravitational effects, I have chosen the same equinox for the presentation of my results.

The osculating elements for the period 1889–1946 (Dubyago 1950, p. 25; 1956, pp. 26–27) were introduced into the N-body programme in the form of eight massless bodies and were integrated over the entire time interval 1887–1976. The steplength normally was two days, except during the time of the approach $\mathcal{2}_2$ when it had to be reduced to 0.5 days. I have attempted to find an orbit which gravitationally 'links' the two observed periods before and after $\mathcal{2}_2$, using a trial-and-error method based on the matrix of partial derivatives of the elements in the neighbourhood of the Jupiter approach. The conditions for the linking orbit were set so that a forward extrapolation of the nongravitational effects observed before $\mathcal{2}_2$, and a backward extrapolation of these effects observed after $\mathcal{2}_2$, would lead to approximately the same orbit. The forward extrapolation in itself was problematic because no elements were available for the 1918 apparition. A unique solution of the link problem is impossible with the methods I have applied. However, two of the solutions which I found may serve as a first approximation; they are represented by two bodies denoted by $L1$ and $L2$, respectively, $L1$ being somewhat better than $L2$.

In order to demonstrate the total perturbations caused by the planets, four plots are given in Figure 1 for the four osculating elements, $\tilde{\omega} = \Omega + \omega$,* i, φ, and n. During the approach, all elements oscillate heavily and reach maximum and minimum values which are not visible in Figure 1. To the left of $\mathcal{2}_2$, the thick line represents the 'tube' in phase space which contains the bodies 1889, 1896, 1903, and 1911. The

---

* This was chosen instead of a single representation for both Ω and ω, because these two elements are each changed due to the Jupiter perturbations by about 180° – an effect which, of course, is only a formal consequence of the usual convention for the orbital inclination.

strong differential perturbations during $\mathcal{2}_2$ let them appear separately to the right of $\mathcal{2}_2$. The direct numerical results indicate that the 'magnification factor' for the tube diameter is of the order of 30–300. To the right of $\mathcal{2}_2$, the thick line represents the tube containing the bodies 1925, 1932, 1939 and 1946, and the differential perturbations (if we follow a backward calculation) split them into four different curves which are distinguishable in the left half of the plots; the magnification factors are here of the order of 80–600. That the bodies $L1$ and $L2$ are linking the two observational periods is clearly indicated by the fact that they are the only bodies which are contained in the thick line on both sides of $\mathcal{2}_2$.

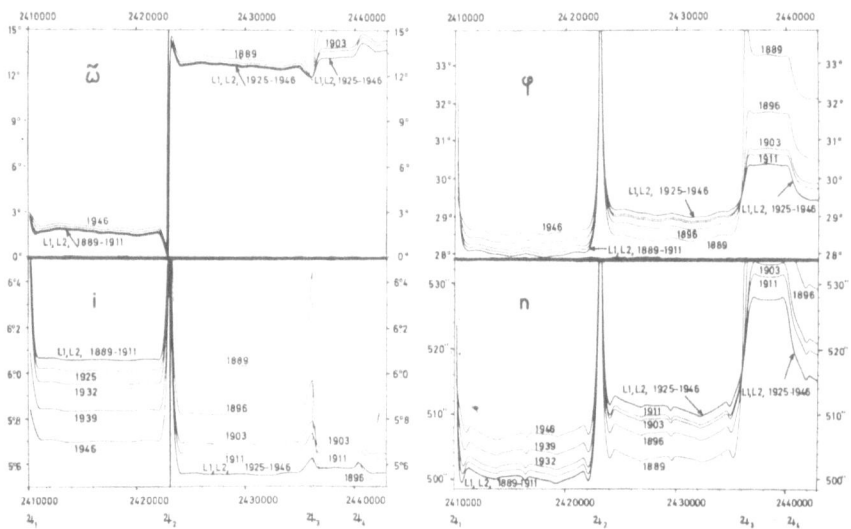

Fig. 1.   Osculating elements between 1887 and 1976. Abscissa is the Julian Date. The curves are marked by the year of apparition. The times of Jupiter approaches are indicated by $\mathcal{2}_1 - \mathcal{2}_4$ on the bottom. $L1$ and $L2$ link the observations before and after $\mathcal{2}_2$. Note that they coincide before $\mathcal{2}_2$ with the apparitions 1889–1911 and after $\mathcal{2}_2$ with the apparitions 1925–1946. Mean equinox 1890.0.

On the left edge of the plots, all the curves approach the singularity zone which corresponds to $\mathcal{2}_1$; this will be discussed later. In the right half of the diagrams, one can see the effect of the $\mathcal{2}_3$ and $\mathcal{2}_4$ approaches. One should note that the order of magnitude of the total perturbations is similar for $\mathcal{2}_2$, $\mathcal{2}_3$ and $\mathcal{2}_4$, whereas in the case of differential perturbations, $\mathcal{2}_2$ has an effect which is tremendously large compared to $\mathcal{2}_3$ and $\mathcal{2}_4$.

Total perturbations by all planets are rather weak during the long time intervals from one close Jupiter approach to the next, and differential perturbations are negligible. The sharp minima in n near JD 2430000 and JD 2416000, and the corresponding maxima in φ, are caused by Jupiter (distance 3 AU).

We will now look in more detail at the differential perturbations and the nongravitational effects. In Figures 2a and 2b, the differences of the osculating elements of the observed bodies relative to the osculating elements of our linking body, $L1$, are plotted over the entire time interval 1887–1976. Body $L1$ is represented by the zero line; note that by definition, this line crosses the $2\!\!\downarrow_2$ singularity without any disturbances. The epochs of the apparitions are marked by arrows on the curves; the arrow immediately to the left of $2\!\!\downarrow_2$ corresponds to the unobserved 1918 apparition. Let us particularly look at the plot of Δi, because its behaviour is ideally suited for the demonstration of nongravitational effects. Within both observing periods, all curves are parallel to the zero line and do not contain any disturbances. This means

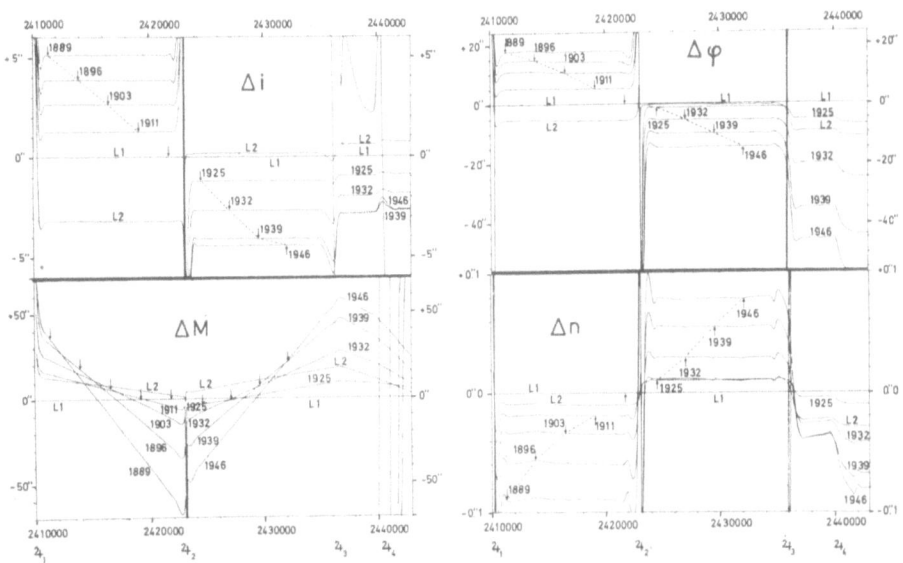

Fig. 2a. Differential evolution of inclination, eccentricity, mean daily motion, and mean anomaly between 1887 and 1976. Abscissa is the Julian Date. For each body, the difference of the elements relative to body $L1$ is plotted. The epochs of introduction of the bodies into the N-body programme (i.e., osculation epochs of Dubyago's elements) are marked by arrows on the curves. The arrow on the zero line left of $2\!\!\downarrow_2$ corresponds to the perihelion time of the missing apparition. The broken line connecting the observed apparitions corresponds to the assumption that nongravitational effects are acting along the entire orbit.

that the orbital inclination was not affected by differential perturbations. The distance between each pair of consecutive curves corresponds to the total nongravitational effect per revolution. The complete absence of differential perturbations makes it possible to explain the action of nongravitational forces in many different ways. Firstly, these forces could have acted continuously along the entire orbit; in this case their effect on the inclination would then be represented by the broken line which connects the arrows. Secondly, they could have acted only during small time intervals near the perihelia, in which case a step function, jumping from one curve to the next, would demonstrate the effect in the orbital element. Finally, the distances between the curves could have been produced by a number of discrete actions along the orbit. This latter model is not very probable because it would be difficult to understand why a series of discrete events produces always the same total change per revolution.

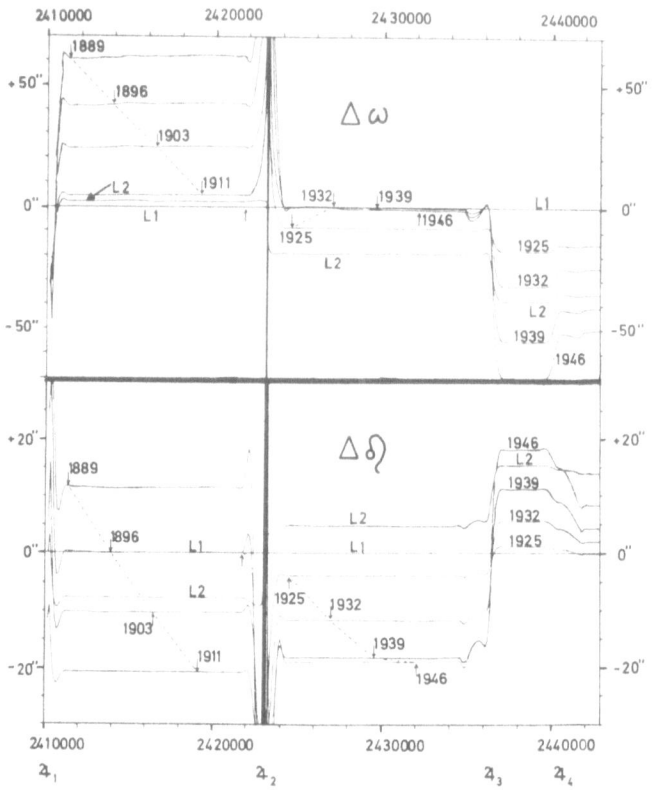

Fig. 2b.   Differential evolution of longitudes of perihelion and ascending node between
1887 and 1976.

The diagrams for $\Delta\varphi$ and Δn are similar to the one for Δi, although we find here some differential perturbations caused by Earth and Mars at the times of the first three apparitions. In all three elements, our linking body allows us to approximate the nongravitational effects over the entire period by a straight line which is not disturbed by the $2\!\!\!\downarrow_2$ singularity.

The small curvature in Δn before $2\!\!\!\downarrow_2$ corresponds to a second-order term introduced by Dubyago in his evaluation of the observations; looking at the ΔM diagram, one finds it possible to approximate the envelope of the arrows by a parabola (which would correspond to a linear Δn), but it is questionable whether one can apply a more refined interpretation.

Differential perturbations are also absent in the case of $\Delta\Omega$ and $\Delta\omega$, as Figure 2b shows. However, the general behaviour of $\Delta\Omega$ and $\Delta\omega$ appears to be different from the behaviour of the other elements. In the case of $\Delta\omega$, this certainly reflects features of Dubyago's osculating elements which have nothing to do with the $2\!\!\!\downarrow_2$ singularity or with the method I have used to analyse his data. In the case of $\Delta\Omega$, where on both sides of $2\!\!\!\downarrow_2$ the nongravitational effect decreases linearly with a similar slope, I assume that it is my definition of the linking body, $L1$, which produces the strong discontinuity observed in Figure 2b. It is easy to understand that we would obtain a $\Delta\Omega$ representation similar to the diagrams in Figure 2a, if we were to use a linking body other than $L1$. Indeed, in my attempts to link the two observing periods, I have found such solutions. However, they always produced discontinuities in all other elements. One possible explanation for this difficulty would be that the observations were not treated correctly by Dubyago. The other possibility is that the nongravitational forces act along the entire orbit (and, therefore, also within the $\mathrm{2\!\!\!\!\!+}_2$ singularity zone). As I said before, this hypothesis cannot be proven in the observing periods which were free of differential perturbations. In Figures 2a and 2b, the curves enter the singularity zone at heliocentric distances of 3.3 AU (JD 2422000) and 3.7 AU (JD 2424000). Both these distances are outside the range of cometary activity proposed by Whipple (1950) in his comet model. If we assume that the nongravitational forces exist even at large distances from the Sun, the following will happen: changes will be continuously produced in the elements and will be continuously magnified by the differential perturbations – a feature not contained in my calculations but which might be extremely helpful in order to study possible mechanisms of cometary activity. If the numerical integration included those forces which correspond to the nongravitational effects observed in the osculating elements outside the singularity zone, a solution, free of discontinuities, for the linking body, might be found.

In the present investigation, it was not possible to carry out such an analysis. Therefore, I chose arbitrarily a solution for the linking body which left a discontinuity in $\Delta\Omega$ only but which permitted the nongravitational effects to be represented satisfactorily in all other elements. Even in this case, the discontinuity is only of the order of 40″, which is not very much if one takes into account the fact that one missing apparition and one very close Jupiter approach had to be linked.

Body $L1$ may be a useful starting point for future investigations. In particular, one could possibly try to use it to search for the comet during the missing 1918 apparition, if photographic plates should still exist somewhere. Table I gives a set of osculating elements defining body $L1$.

TABLE I

Osculating elements of Body
$L1$ at Epoch JD 2421600.5

$$M = 354°49'11\rlap{.}''3$$
$$\left.\begin{array}{l} \omega = 343\ 36\ 25.7 \\ \Omega = 17\ 53\ 15.6 \\ i = 6\ 03\ 41.7 \end{array}\right\} \quad 1890.0$$
$$\varphi = 28\ 02\ 03.9$$
$$n = 501\rlap{.}''267769$$
$$T = \text{JD } 2421637.7$$

As was mentioned before, $L2$ is not as good as $L1$; the Figures 2a and 2b illustrate that it causes discontinuities in all elements except the mean daily motion.

The remaining features of Figure 2 are almost self-explanatory. The vertical lines near $\mathrm{2\!\!\!\!\!+}_3$ and $\mathrm{2\!\!\!\!\!+}_4$ correspond to some of the bodies 1889–1911 which, due to differential perturbations, cross the plots. The curve entering the plot of Δi from above, and reaching two minima near $♃_3$ and $♃_4$, belongs to body 1896.

From the rather systematic behaviour of the nongravitational effects within the entire interval 1889–1946, as indicated in Figures 2, I gained some confidence in the possibility of their backward extrapolation. This is indeed extremely interesting because the orbit before $♃_1$ must have been completely different from the present orbit.

Dubyago (1950) himself had made such backward calculations, using his elements A (l.c, p. 25) and B (l.c., p. 26) as starting orbits, and following their development through the $♃_1$ approach back until 1883. The comparison between these two bodies does not give a realistic picture of the influence of differential perturbations on the orbit, because their definition is not a measure for the uncertainties of the backward extrapolation of the nongravitational effects. In my own calculations, I am considering a group of eight bodies. The first one is body 1889 (Dubyago's elements A). Its backward integration was started at JD 2411284; at the date JD 2411000, a new body was introduced which corresponds to body 1889 plus the nongravitational effects accumulated from 2411284 until 2411000. This body is denoted here by $B1$. Similarly, at the date JD 2410800, a body $B2$ was introduced by adding to $B1$ the nongravitational effects accumulated since JD 2411000. In this way, a series of five bodies, $B1$–$B5$, was created which might be considered an approximation for all nongravitational effects back to JD 2410214, a date which is only 106 days away from $♃_1$ and where the distances to the Sun and Jupiter were 5.3 and 0.3 AU, respectively. For this date, Dubyago gives osculating elements also, from his own calculations, and I have introduced these into my calculations for comparisons; the corresponding body is denoted here by D. The last body which I have considered is obtained by adding to body 1889, at the starting epoch JD 2411284, the total nongravitational effects to be expected for the interval between JD 2410214 and the starting epoch. This body is denoted by T; it demonstrates a model where half of the nongravitational effects per revolution are added to the osculating elements at perihelion. Since only the time from the starting epoch to JD 2410800 is completely free of differential perturbations, and since at the latter date the heliocentric distance was about 4 AU, we may also consider the body T as an approximation to Whipple's model where the nongravitational forces are acting continuously within 4 AU of the Sun. On the other hand, body $B5$ corresponds to a model where the forces have continuously acted within 5.3 AU of the Sun.

For Jupiter's flattened potential field, I have used a first-order term with the same coefficient, $J=0.022273$, that Dubyago used. Dubyago did not include the perturbations by the other planets during the approach and treated the passage in jovicentric coordinates, whereas in my calculations heliocentric coordinates were kept (smallest steplength 0.001 day), and the influence of all planets was taken into account. These differences are probably responsible for the fact that bodies $B5$ and D, as will be seen below, are not identical. As far as the above definition of the eight test bodies is concerned, I feel that bodies $B5$, D and T are a reasonable approximation to the motion of the actual comet before and at $♃_1$.

The results of the calculations are shown in Figure 3. Here, the differences of the osculating elements of all test bodies are plotted relative to the osculating elements of body 1889; the latter is then by definition represented by the zero line. The ratio between the 'tube' in phase space before and after $2\!\!\downarrow_1$, as can be seen directly from Figure 3, is not as large as one might expect in the case of such a close approach to

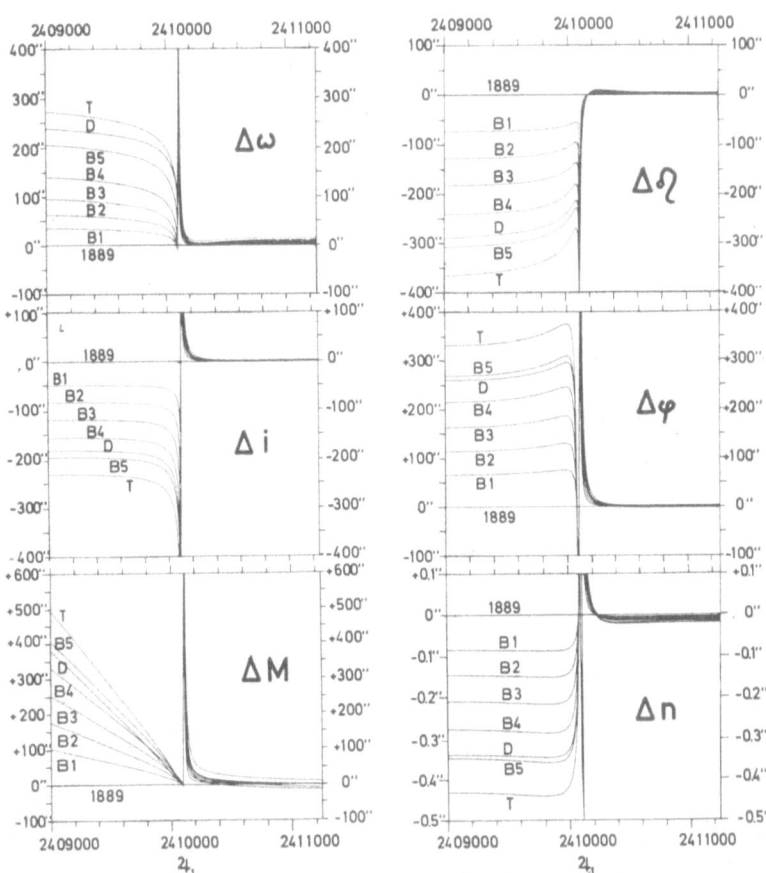

Fig. 3.   Differential evolution of orbital elements near the $2\!\!\downarrow_1$ approach. Abscissa is the Julian Date (1883–1889). For each of eight test bodies the element difference relative to body 1889 is plotted. The diameter of the 'tubes' near the right edge of the plots corresponds roughly to one half of the nongravitational effects per revolution. For the definition of the test bodies, see explanation in the text. The plot of Δn may be looked on as an illustration of celestial mechanics energy splitting due to strong perturbations.

Jupiter. The magnification factors for the various elements vary between 20 and 115, as the direct numerical results show.

The dates of closest approach to Jupiter, for the eight bodies considered here, are all within JD 2410108.51 and 2410108.78. The minimum distances from Jupiter's surface, in units of its equatorial radius (4.8×10^{-4} AU) are shown in Table II. The sequence of these numbers indicates that smaller distances would be obtained if

TABLE II

Jupiter Approach 1886 ($\mathfrak{2}_1$)

Body	Minimum distance from surface
1889	1.055
$B1$	1.040
$B2$	1.028
$B3$	1.016
$B4$	1.008
$B5$	0.995
D	1.001
T	0.987

stronger nongravitational effects were assumed. However, a more precise description of the potential field of Jupiter would also change these results, and Dubyago's remarks concerning the negligible influence of the satellites would have to be carefully studied before one can go any further with these speculations.

In order to investigate the history of the comet prior to $\mathfrak{2}_1$, I have integrated the eight test bodies backward in time, ignoring nongravitational effects, to 1686. The results of this calculation are summarized in Table III. Between the first two dates

TABLE III

Osculating elements of bodies $B5$, D, and T before $\mathfrak{2}_1$

Body	JD	1890.0			φ	n	q (AU)	Q (AU)
		ω	Ω	i				
$B5$	2409000	2°21	186°43	6°51	26°83	112″55	5.48	14.48
	2357000	1.80	187.35	6.53	26.10	118.52	5.40	13.88
	2353000	354.91	190.50	6.83	22.30	140.93	5.33	11.85
	2337000	348.66	195.90	6.33	25.69	120.81	5.39	13.64
D	2409000	2.22	186.44	6.51	26.82	112.56	5.49	14.48
	2357000	1.82	187.35	6.54	26.10	118.49	5.40	13.89
	2353000	354.65	190.63	6.85	22.19	141.64	5.33	11.80
	2337000	342.66	201.14	5.78	27.13	113.26	5.41	14.47
T	2409000	2.23	186.42	6.50	26.84	112.47	5.48	14.49
	2357000	1.74	187.37	6.53	26.04	118.87	5.40	13.85
	2353000	357.51	189.30	6.68	23.38	133.98	5.36	12.41
	2337000	358.64	189.93	6.68	23.32	134.74	5.35	12.36

tabulated, all three orbits remain close to each other, and the mean orbit is quite stationary. Between the second and third dates, a perturbation by Jupiter occurs; the dates of the approach vary from JD 2354266 to JD 2354324 (Aug.–Oct. 1733) and the minimum distances are between 0.81 and 1.10 AU. All eight test bodies participate in this event, and it might be mentioned that body 1889 comes closest to Jupiter (0.31 AU). The orbit of body T remains rather stable from this encounter to the end of the calculations. The two other bodies, however, again pass close to Jupiter on JD 2344600 (March 1707), the minimum distances being 0.90 and 0.64 AU, respectively.

The mean anomalies at JD 2337000 are 95° ($B5$), 110° (D), and 79° (T). Thus the uncertainty in the position of the comet at the end of the calculations has become so large that it would be meaningless to follow the motion further back in the past. Just for curiosity it should be mentioned that body $B2$, which may not represent the actual comet, passes Saturn on JD 2343866 (1705) within 0.7 AU.

Acknowledgments

This investigation is based in part on an academic post-doctoral thesis (P. Stumpff, Habilitationsschrift Universität Heidelberg, 1965). It was carried out during my stay at the National Radio Astronomy Observatory (operated by Associated Universities, Inc., under contract with the National Science Foundation), Green Bank, West Virginia, U.S.A. The computations were made with the IBM 360/50 of the N.R.A.O. I am grateful to Mrs E. Litman and Mrs S. Huang for programming assistance. I wish to thank B. G. Marsden for his comments which made it clear to me that a final solution of the motion of P/Brooks 2 would have to be based on a re-evaluation of observations.

References

Dubyago, A. D.: 1950, *Trudy Astron. Obs. Kazan* No. 31.
Dubyago, A. D.: 1956, *Byull. Astron. Obs. V.P. Engel'gardta* No. 32.
Marsden, B. G.: 1969, *Astron. J.* **74**, 720.
Porter, J. G.: 1961, *Mem. Br. Astron. Assoc.* **39**, No. 3.
Schubart, J. and Stumpff, P.: 1966, *Veroeffentl. Astron. Rechen-Inst. Heidelberg* No. 18.
Whipple, F. L.: 1950, *Astrophys. J.* **111**, 375.

Discussion

Yu. V. Evdokimov: Is there a nongravitational effect perpendicular to the orbital plane of P/Brooks 2?

P. Stumpff: If one believes Dubyago's orbit determinations, components in all directions must exist. If these forces were limited to the orbital plane, one would not expect the changes in Ω and i. However, I may not have made clear enough that in the case of Ω and ω the situation is more uncertain than in the case of the other elements, possibly due to the fact that i is so small.

S. K. Vsekhsvyatskij: P/Brooks 2 was observed at its first apparition in 1889 with a large number of satellites, indicating the comet's decay. Can one really therefore maintain that the comet made approaches to Jupiter in the eighteenth century?

P. Stumpff: After the discovery there was speculation that the comet was identical with P/Lexell, which had been observed in 1770, was known to have subsequently passed near Jupiter, but was never detected again. Dubyago had shown that the time between the approaches of the two comets to Jupiter was inconsistent with the revolution periods. I became interested in the pre-1886 orbit because it goes beyond Saturn's orbit, and I wanted to discover the role of Saturn in the comet's history. The backward calculations, showing possible earlier approaches to Jupiter and Saturn, are a reasonable extrapolation of the orbital information available to us, but of course no definite proof that these events actually took place.

E. I. Kazimirchak-Polonskaya: Did you consider that an inaccurate value for the mass of Jupiter could have affected the representation of the observations before and after the approach to Jupiter in 1922, or have you attributed the residuals entirely to nongravitational effects? Have you tried to use the close approach for determining the mass of Jupiter?

P. Stumpff: I wanted to define a strictly gravitational orbit – a sort of interpolation orbit, so to speak – that gives reasonable agreement with Dubyago's elements both before and after the approach. I found that varying Jupiter's mass by about $\pm 0.02\%$ within an interval of about 2000 days around the approach would produce differential effects in the elements which have a magnitude similar to the observed nongravitational effects. So the interpolation orbit will certainly depend on the value assumed for Jupiter's mass, but I doubt that one can in this case come to any conclusions about Jupiter's mass because the whole picture is so heavily disturbed by the existence of the nongravitational effects – or if you wish, by the existence of residuals which have not been explained properly.

31. LINKAGE OF SEVEN APPARITIONS OF PERIODIC COMET FAYE 1925–1970 AND INVESTIGATION OF THE ORBITAL EVOLUTION DURING 1660–2060

N. A. BELYAEV and F. B. KHANINA

Institute for Theoretical Astronomy, Leningrad, U.S.S.R.

Abstract. Seven apparitions of this comet, from 1925 to 1970, have been linked, with account taken of the perturbations by Venus to Neptune; 202 observations have been represented with mean square error $\pm 2''.15$. The resulting orbit was integrated from 1947 back to 1843 and compared with Möller's results. The result of the comparison after 104 years was found to be very good. The comet's orbital evolution was studied altogether over a period of 400 yr (1660 to 2060). There were nine approaches within 1.5 AU of Jupiter, the comet on one occasion passing through Jupiter's sphere of action. The orbital evolution was free from catastrophic changes, the comet remaining in the Jupiter family throughout the interval. The problem of the comet's discovery was considered.

Periodic comet Faye, under observation for more than 125 yr, is one of the few short-period comets with well-studied motion. Such comets are of particular interest for cometary astronomy. On the one hand, there is a wealth of observational material on which to build a reliable orbital theory; 15 or more apparitions can be linked and the results checked over many decades. On the other hand, the changes in their outward appearance, as well as unexpected variations in and systematic reduction of their brightness, render such comets highly valuable for exploring the problem of cometary ageing and disintegration.

Such comets have the added interest of offering a possibility for checking integration methods and computer programmes. All our calculations have been done on a BESM-4 computer, using the complex of programmes described by Belyaev (1972).

Extensive accounts of the history of P/Faye and its motion have been given by Khanina and Barteneva (1959, 1961), Khanina (1966) and Kazimirchak-Polonskaya (1961). After its discovery in 1843, the comet has returned to perihelion as many as 18 times, escaping observation but twice. It was last at perihelion 1969 October 7; the comet was recovered on 1969 May 17 and observed at several observatories until 1970 March 8. Forty precise positions at this apparition have been published to date.

Many scientists have investigated the motion of P/Faye. Its motion between 1843 and 1881 was studied in great detail by Möller, who devoted 40 years of his life to this comet. The period from 1881 to 1932 was studied by Kreutz, Strömgren, Fayet, Ristenpart, Prager, and others. Studies of the comet's motion since 1932 have been conducted at the Institute for Theoretical Astronomy, Leningrad (Zheverzheev, 1952; Khanina and Barteneva, 1959, 1961; Khanina, 1966; Khanina and Belyaev, 1968). The comet's motion prior to discovery was investigated by Möller (1873) and Zhdanov (1885a, 1885b).

Our own objective was to do the following:

(1) Link all the apparitions between 1925 and 1970, so that the last seven apparitions of the comet might be represented, if possible, by one set of elements;

Chebotarev et al. (eds.), *The Motion, Evolution of Orbits, and Origin of Comets,* 167–172. *All Rights Reserved.*
Copyright © 1972 by the IAU.

(2) Check to see how the new set of elements represents all the other apparitions of the comet, as far back as 1843;

(3) If this representation proves to be satisfactory, study the comet's motion back to 1660 and forward to 2060, so that a picture of the comet's orbital evolution over 400 years might be obtained.

We started from Khanina's set of elements for 1947, obtained by linking the five apparitions of the comet 1932–1962. It represented 21 normal places with a mean square error of $\pm 2\overset{''}{.}3$. Linkage of the seven apparitions 1925–1970 was accomplished by varying the elements of the cometary orbit (by $0\overset{\circ}{.}000001$ for μ and $0\overset{\circ}{.}1$ for the other elements) and by successive sevenfold integration, taking account of the perturbations by the seven planets Venus to Neptune. This method was necessary because in 1959 P/Faye approached Jupiter within 0.60 AU, which resulted in large perturbations, as much as 7° in the line of nodes. This improvement and all our subsequent investigations were made without consideration of the effects of nongravitational forces.

The new set of elements below represents 202 observations 1925–1970 with mean square error $\pm 2\overset{''}{.}15$. Initially, 286 observations were assembled, but 84 were rejected as erroneous. In the future the observations will be revised.

$$
\begin{aligned}
\text{Epoch} &= 1947 \text{ October } 22.0 \text{ ET} & T &= 1947 \text{ September } 28.4072 \text{ ET} \\
M_0 &= 3\overset{\circ}{.}1238 & \omega &= 200\overset{\circ}{.}5266 \\
\varphi &= 34.3144 & \Omega &= 206.3504 \quad\Big\} \quad 1950.0 \\
\mu &= 0.132400 & i &= 10.5337
\end{aligned}
$$

For the latest apparition this system gives $T = 1969$ October 7.5918 ET, compared with the published value of 1969 October 7.639 ET (Khanina and Belyaev, 1968). For the next return we obtain $T = 1977$ February 27.8418 ET.

Starting from the above elements the equations of motion were integrated back for 104 yr from 1947, again taking into account perturbations by Venus to Neptune. This was done to enable comparison with the observations and calculations by earlier investigators. Cometary astronomy offers very few opportunities for such comparisons, and this chance was used for checking both the starting elements and the operation of the programme over considerable time periods, and also for revealing possible nongravitational effects.

Table I compares the perihelion passages given by observation (Porter, 1961) and our calculations for the apparitions between 1910 and 1843. It is to be noted that ΔT exceeds $0\overset{d}{.}35$ only once; and in any case, the observed perihelion times from 1881 onwards are somewhat uncertain.

Table II compares our integrated elements for an epoch in 1843 with Möller's elements (Porter, 1961), the latter having been obtained by linking the four apparitions 1843–1866. The largest discrepancy is $0\overset{\circ}{.}017$ in ω; in Ω and i the discrepancies are several seconds of arc, while in q and e they are practically zero. We remark that in 1899 the comet approached Jupiter to within 0.51 AU and that the total number of integration steps from 1947 to 1843 was 2350. The results obtained testify to the high

TABLE I

Comparison of observed and calculated perihelion times

Year	T_{obs}	T_{calc}	$\Delta T = T_{obs} - T_{calc}$
1910	Nov. 1.96	Nov. 2.19	$-0\overset{d}{.}23$
1896	Mar. 19.76	Mar. 19.30	$+0.46$
1881	Jan. 23.17	Jan. 22.96	$+0.21$
1873	July 18.99	July 18.74	$+0.25$
1866	Feb. 14.47	Feb. 14.22	$+0.25$
1858	Sept. 13.37	Sept. 13.11	$+0.26$
1851	Apr. 2.44	Apr. 2.14	$+0.30$
1843	Oct. 17.63	Oct. 17.30	$+0.33$

TABLE II

Comparison of Möller's elements for 1843 with the results of our integration

	Möller	Belyaev-Khanina	Difference
T (UT)	1843 Oct. 17.63	1843 Oct. 17.30	$+0\overset{d}{.}33$
ω	$200°024$	$200°007$	$+0°017$
Ω } 1950.0	211.013	211.012	$+0.001$
i	11.364	11.367	-0.003
q (AU)	1.69223	1.69222	$+0.00001$
e	0.55583	0.55582	$+0.00001$
P (yr)	7.4364	7.4366	-0.0002

quality of the orbital elements obtained by Möller for 1843 and by ourselves for 1947. Besides, they confirm the possibility of utilizing our programme for this type of investigation and justify the supposition that all 16 apparitions of the comet might be linked without taking into account nongravitational effects, although this latter question will be settled finally only after a careful analysis of all the observational material.

The satisfactory results back to 1843 made it worthwhile to run the integration further back to 1660 and ahead to 2060. The evolution of the orbit of P/Faye is presented in Table III, which also lists the comet's nine approaches within 1.5 AU of Jupiter during these 400 yr. The approach in 1816 was inside the sphere of action. (For the known approaches during 1816–1959 the present work yielded more accurate data than has previously been published.) The effect of these approaches is a continuous retrograde rotation of the line of nodes through 50°6 in the 400 yr, and since the argument of perihelion advanced by 39°5, the total motion of the line of apsides was $\Delta \pi = -11°1$. The orbital inclination varied from 6°5 to 13°1. No significant changes occurred in the shape or size of the orbit, so the comet quite definitely remains in the Jupiter family during the entire period in question.

We have also integrated the motion of the comet from 1843 to 1813, starting from Möller's elements in Table II, and again taking into account the perturbations by Venus to Neptune. This was done in order to examine how the evolution of the comet's orbit through the approaches to Jupiter in 1841 and 1816 would be affected

TABLE III

Evolution of the orbit of P/Faye during 1660–2060

Epoch, t_J	Δ_J (AU)	ω	Ω	$\pi = \omega + \Omega$	i	e	q (AU)	Q (AU)	P (yr)
1660 Feb. 28		168°9	242°1	51°0	6°5	0.554	1.74	6.05	7.69
1673 Aug. 10.71	0.57								
1679 Feb. 28		180.1	232.6	52.7	8.1	0.528	1.91	6.18	8.13
1734 Oct. 14.07	1.34								
1736 Aug. 19		182.9	231.4	54.3	8.1	0.533	1.87	6.13	8.01
1758 Aug. 8.78	0.66								
1760 July 13		190.7	226.1	56.9	7.5	0.559	1.69	5.99	7.53
1816 Apr. 5.21	0.249								
1820 Nov. 16		193.4	219.2	52.5	13.1	0.534	1.81	5.94	7.63
1841 Mar. 5.58	0.63								
1843 Aug. 28		200.0	211.0	51.5	11.4	0.556	1.69	5.93	7.44
1875 July 7.49	1.49								
1880 Nov. 21		201.2	210.6	51.8	11.3	0.549	1.74	5.97	7.57
1899 June 26.78	0.51								
1903 Mar. 27		199.4	206.8	46.3	10.6	0.566	1.65	5.94	7.39
1959 Feb. 17.15	0.60								
1962 June 25		203.6	199.1	42.7	9.1	0.576	1.61	5.97	7.38
2018 Mar. 7.32	0.63								
2021 Sept. 23		207.0	192.2	39.2	8.0	0.577	1.62	6.03	7.48
2059 Nov. 3		208.4	191.5	39.9	8.0	0.579	1.60	6.02	7.44

The short lines give the dates t_J of approaches to Jupiter and the minimum separations Δ_J.
The other lines give the comet's orbital elements for various epochs, for the beginning and end of the integration interval and for dates near the perihelion dates following the approaches to Jupiter.

by errors in the initial data (which are particularly large in T). The circumstances of the approaches to Jupiter, as given by these two integrations, are compared in Table IV, the final line giving Möller's own results, obtained by an approximate method. The two integrations do not differ significantly, either in time or minimum distance. Consequently, the elements for the 1813 perihelion passage differ but slightly. Notwithstanding the comet's passage through Jupiter's sphere of action and the discrepancy of 0.33 day in the 1843 perihelion times, the difference in perihelion times in 1813 increased only to 1.07 days.

TABLE IV

Circumstances of the approaches of P/Faye to Jupiter in 1841 and 1816

	t_J	Δ_J (AU)	t_J	Δ_J (AU)
Integration 1947–1813	1841 Mar. 5.58	0.63	1816 Apr. 4.92	0.246
Integration 1843–1813 from Möller's elements	1841 Mar. 5.68	0.63	1816 Apr. 5.21	0.249
Möller	1841 Mar. 5	0.6	1816 Mar.	0.3

Of particular interest in connection with orbital evolution studies is the question of comet discovery. It is necessary to trace the changes in perihelion distance prior to discovery and investigate the comet's encounters with the Earth, considering that the most favourable conditions for discovery occur when the distances of the comet from the Sun and the Earth are minimal and the date of perihelion passage close to that of the comet's approach to the Earth. Table III shows that after the 1841 approach to Jupiter the comet's perihelion distance decreased from 1.81 to 1.69 AU. When the comet was next at perihelion (1843 October 17) it was 0.92 AU away from the Earth. On 1843 November 24 it approached the Earth within 0.79 AU; and it was discovered two days before.

Table V lists all the instances between 1660 and discovery when the comet came within 1.1 AU of the Earth or the time interval between closest approach to the Earth and perihelion passage was less than three months. It shows that the comet was discovered under the most favourable conditions since 1660, although such conditions are not the only ones responsible for the comet's discovery in 1843.

TABLE V

Approaches of P/Faye to the Earth between 1660 and 1843

t_E	Δ_E (AU)	T	q (AU)	Δ_E^T (AU)	$T-t_E$ (dy)
1703 Nov. 19	0.98	1703 Oct. 28	1.95	1.02	-22
1736 Dec. 13	1.10	1736 Sept. 12	1.87	1.61	-91
1744 Dec. 7	1.04	1744 Sept. 24	1.88	1.45	-74
1752 Dec. 4	1.00	1752 Oct. 1	1.87	1.31	-64
1775 Dec. 14	0.92	1775 Sept. 30	1.70	1.28	-75
1790 Dec. 4	0.80	1790 Oct. 17	1.68	0.96	-48
1805 Oct. 18	0.96	1806 Jan. 11	1.74	1.40	$+86$
1843 Nov. 25	0.79	1843 Oct. 17	1.69	0.92	-38

t_E are the dates of closest approach to the Earth, the separation being Δ_E.
Δ_E^T are the separations at perihelion times T.

We plan to link all the apparitions of P/Faye, in order that a continuous numerical theory may be developed to account for the comet's motion from 1843 to 1970. On completion of that, we can obtain a more precise picture of the comet's orbital evolution over 1660–2060. We hope that the definitive version will not differ considerably from that described in this paper, considering that the orbit obtained to date forms a very reliable foundation for this kind of study.

References

Belyaev, N. A.: 1972, this Symposium, p. 90.
Kazimirchak-Polonskaya, E. I.: 1961, *Trudy Inst. Teor. Astron.* **7**, 131.
Khanina, F. B.: 1966, *Problemy Dvizheniya Malykh Tel Solnechnoj Sistemy*, Baku, p. 69.
Khanina, F. B. and Barteneva, O. N.: 1959, *Byull. Inst. Teor. Astron.* **7**, 466.
Khanina, F. B. and Barteneva, O. N.: 1961, *Byull. Inst. Teor. Astron.* **8**, 229.
Khanina, F. B. and Belyaev, N. A.: 1968, *IAU Circ.* No. 2062.

Möller, A.: 1873, *Ofvers. Kongl. Vetensk. Akad. Stockholm Förhandl.* **1**, 3.
Porter, J. G.: 1961, *Mem. Br. Astron. Assoc.* **39**, No. 3.
Zhdanov, A.: 1885a, *Mem. Acad. Imp. Sci. St. Petersb.* **33**, 1.
Zhdanov, A.: 1885b, *Astron. Nachr.* **111**, 113.
Zheverzheev, V. F.: 1952, *Byull. Inst. Teor. Astron.* **5**, 97.

Discussion

S. K. Vsekhsvyatskij: P/Faye is famous for its rapid secular decrease in brightness. Have you tried to identify it with any comets of the seventeenth or eighteenth centuries?

N. A. Belyaev: Not yet, but we plan to do so in the near future.

W. J. Klepczynski: In his study, Möller determined a correction to the mass of Jupiter. Do you plan to solve for such a correction?

N. A. Belyaev: No. Requirements for comets that may be utilized for determining corrections to the mass of Jupiter have been formulated and will be reported by Kazimirchak-Polonskaya in Paper 40. P/Faye does not meet those requirements.

32. INVESTIGATION OF THE MOTION OF PERIODIC COMET GIACOBINI-ZINNER AND THE ORIGIN OF THE DRACONID METEOR SHOWERS OF 1926, 1933 AND 1946

YU. V. EVDOKIMOV

Kazan University, Kazan, U.S.S.R.

Abstract. Orbital elements of P/Giacobini-Zinner have been obtained from 577 observations at the eight apparitions 1900 to 1965 by linking apparitions in pairs. By this technique we established that the comet has a nongravitational secular deceleration amounting, on the average, to 0.081 day/(period)^2. As a result of the comet's approach to Jupiter in 1969, a return of the Draconid meteor shower is possible on 1972 October $8^d15^h45^m$ UT. The 1946 meteor shower was due to meteoroids ejected forward along the orbit of the comet in 1940 with velocities of 14 m s^{-1}. The meteor showers of 1933 and 1926 were apparently produced by meteoroids ejected in 1900 with velocities of 15.0 m s^{-1} forward and 14.5 m s^{-1} backward, respectively.

Because of its association with the very strong meteor showers of 1933 and 1946 periodic comet Giacobini-Zinner is of particular interest. The comet was first discovered in 1900 by Giacobini, and having a period of approximately 6.5 yr, it was observed also in 1913, 1926, 1933, 1939, 1946, 1959, and 1965; it was missed at its returns in 1907, 1920, and 1953.

A total of 577 observations, made at 47 observatories, have been reported. Attempts to link successive apparitions (Evdokimov, 1963) convinced us that the motion of this comet is not governed by pure gravitational theory. The elements of the orbit vary irregularly, and it is impossible to represent with reasonable accuracy all these changes for the whole interval 1900 to 1965.

The same conclusion was made by Dubyago in the case of P/Brooks 2. In his last work (Dubyago, 1956) he abandoned his method for determining variations in the mean daily motion and eccentricity from equations. Instead, he calculated orbital elements from each pair of successive apparitions and then compared, for some osculation epoch during each apparition, the elements obtained by linking that apparition with the preceding one and with the following one. This technique roughly simulates, in the form of a jump, the accumulated nongravitational change in the elements over a comparatively small arc near perihelion.

We have made use of the same technique in this investigation of the motion of P/Giacobini-Zinner. First we linked the four apparitions 1926–46 in pairs, with the result

Epoch and osculation 1933 July 27.0 ET

	1926–1933		1933–1939	
$M_0 =$	$1°77099$	$\pm 0°00011$	$1°76944$	$\pm 0°00006$
$\omega =$	171.77039	± 0.00070	171.76983	± 0.00026
$\Omega =$	196.24331	± 0.00050	196.24323	± 0.00046
$i =$	30.68403	± 0.00040	30.68393	± 0.00017
$e =$	0.7159891	± 0.0000004	0.7160021	± 0.0000007
$a =$	3.5193712	± 0.0000030	3.5194500	± 0.0000016 AU

(the ω, Ω, i values for 1933–1939 are referred to 1950.0)

Chebotarev et al. (eds.), The Motion, Evolution of Orbits, and Origin of Comets, 173–180. All Rights Reserved.

Epoch and osculation 1939 October 14.0 ET

	1933–1939		1939–1946		
$M_0 =$	341°.11596	± 0°.00007	341°.11575	± 0°.00032	
$\omega =$	171.79101	± 0.00026	171.79169	± 0.00025⎫	
$\Omega =$	196.24993	± 0.00046	196.24946	± 0.00020⎬ 1950.0	
$i =$	30.74130	± 0.00017	30.74123	± 0.00013⎭	
$e =$	0.7166606	± 0.0000007	0.7166605	± 0.0000003	
$a =$	3.5138522	± 0.0000016	3.5139385	± 0.0000019 AU.	

Of the total of 173 observations made at these four apparitions, 30 were erroneous or of low precision and therefore rejected. The remainder were combined into normal places. The mean errors corresponding to the three orbits were ±1".85, ±0".75 and ±1".42 respectively, and the maximum residuals were about ±2".5, ±1".5 and ±3".

Integration of the 1926–1933 orbit back to 1913 gave residuals of 1200", while integration of the 1939–1946 orbit forward to 1959 gave residuals of 1400".

The differences between the above pairs of elements are listed in Table I in the columns headed 1933 and 1939, respectively, and also designated (a) and (b). Column (c) is essentially a mean, and these variations were applied to the 1926 elements before integrating back to 1920. The elements so obtained for an epoch near perihelion in 1920 were corrected using the same mean variations and the integration then continued to 1913. The resulting (O − C) residuals in 1913 were of the order 2" to 20", suggesting that the mean variations of the elements conform to reality.

After an orbit improvement using the 1913 observations we obtained the following comparison of the elements for an epoch in 1926. The mean errors are ±2".42 and ±2".30, respectively, and the differences are given in Table I as column (d).

Epoch and osculation 1926 December 11.0 ET

	1913–1926		1926–1933		
$M_0 =$	359°.89405	± 0°.00011	359°.89395	± 0°.00011	
$\omega =$	171.75871	± 0.00080	171.75292	± 0.00070⎫	
$\Omega =$	196.24336	± 0.00050	196.24273	± 0.00050⎬ 1950.0	
$i =$	30.73808	± 0.00040	30.73870	± 0.00040⎭	
$e =$	0.7170158	± 0.0000003	0.7170151	± 0.0000004	
$a =$	3.5115892	± 0.0000026	3.5116703	± 0.0000025 AU.	

For the integration from 1913 back to 1900 we first adjusted the 1913–1926 elements by the mean of variations (d) and (c) and integrated to 1907. We then corrected the 1907 elements similarly and continued the integration back to 1900. The resulting (O − C) residuals obtained for 1900 ranged from 1".7 to 15". The 1900 and 1913 observations were then used to correct this orbit.

The two sets of elements for an epoch in 1913 are given below. The mean residuals were ±3".06 and ±2".42, respectively, and the differences are listed in Table I as column (e).

TABLE I

Variations in the elements

Elements	1913 (e)	1926 (d)	1933 (a)	1939 (b)	1946 (f)	1959 (g)	Average 1933–39 (c)	Average 1913–39	Average 1913–59
$10^5 \Delta M_0$ (deg)	+ 15 ± 15	− 10 ± 15	− 156 ± 12	− 21 ± 33	+ 10 ± 30	+ 72 ± 9	− 150 ± 17	− 60 ± 44	− 7 ± 38
$10^5 \Delta\omega$ (deg)	+ 155 ± 120	− 580 ± 100	− 56 ± 27	+ 68 ± 26	− 8 ± 80	− 130 ± 80	+ 6 ± 20	− 26 ± 62	− 15 ± 56
$10^5 \Delta\Omega$ (deg)	+ 9 ± 80	− 63 ± 90	− 8 ± 46	− 47 ± 36	+ 90 ± 60	− 88 ± 70	− 21 ± 22	− 29 ± 14	− 17 ± 22
$10^5 \Delta i$ (deg)	+ 26 ± 55	− 62 ± 60	− 10 ± 17	− 7 ± 50	+ 18 ± 30	− 2 ± 34	− 10 ± 17	− 19 ± 11	− 5 ± 8
$10^7 \Delta e$	+ 56 ± 12	− 7 ± 5	+ 130 ± 8	− 1 ± 8	+ 73 ± 8	+ 193 ± 10	+ 64 ± 6	+ 28 ± 35	+ 51 ± 30
$10^7 \Delta a$ (AU)	+ 757 ± 35	+ 811 ± 35	+ 778 ± 16	+ 863 ± 25	+ 1013 ± 20	+ 2062 ± 60	+ 808 ± 5	+ 815 ± 6	+ 819 ± 17

Epoch and osculation 1913 August 21.0 ET

	1900–1913		1913–1926	
$M_0 =$	348°89657	$\pm$ 0°00010	348°89762	$\pm$ 0°00011
$\omega =$	171.48398	$\pm$ 0.00090	171.48553	$\pm$ 0.00080
$\Omega =$	196.36857	$\pm$ 0.00060	196.36866	$\pm$ 0.00050
$i =$	30.74613	$\pm$ 0.00036	30.74639	$\pm$ 0.00040
$e =$	0.7206350	$\pm$ 0.0000012	0.7206406	$\pm$ 0.0000003
$a =$	3.4936561	$\pm$ 0.0000020	3.4937318	$\pm$ 0.0000030 AU.

(The ω, Ω, i values for 1913–1926 are bracketed as 1950.0)

Linking the 1946 and 1959 apparitions was more difficult, probably due to the comet's approach to Jupiter in 1958. The final improved elements for 1946 are:

Epoch and osculation 1946 September 17.0 ET

	1939–1946		1946–1959	
$M_0 =$	359°77761	$\pm$ 0°00032	359°77781	$\pm$ 0°00080
$\omega =$	171.81091	$\pm$ 0.00025	171.81074	$\pm$ 0.00072
$\Omega =$	196.29317	$\pm$ 0.00020	196.29497	$\pm$ 0.00060
$i =$	30.72683	$\pm$ 0.00013	30.72720	$\pm$ 0.00030
$e =$	0.7166747	$\pm$ 0.0000003	0.7166894	$\pm$ 0.0000008
$a =$	3.5143590	$\pm$ 0.0000019	3.5145616	$\pm$ 0.0000004 AU,

(The ω, Ω, i values for 1946–1959 are bracketed as 1950.0)

and on running the integration forward to 1965 we obtained the following improved sets of elements for 1959:

Epoch and osculation 1959 September 29.0 ET

	1946–1959		1959–1965	
$M_0 =$	355°71004	$\pm$ 0°00008	355°71149	$\pm$ 0°00004
$\omega =$	172.84615	$\pm$ 0.00072	172.84356	$\pm$ 0.00033
$\Omega =$	196.03171	$\pm$ 0.00060	196.02994	$\pm$ 0.00030
$i =$	30.90435	$\pm$ 0.00030	30.90440	$\pm$ 0.00016
$e =$	0.7289063	$\pm$ 0.0000008	0.7289449	$\pm$ 0.0000006
$a =$	3.4525903	$\pm$ 0.0000004	3.4530026	$\pm$ 0.0000060 AU.

(The ω, Ω, i values for 1959–1965 are bracketed as 1950.0)

The mean errors were $\pm 1''42$ and $\pm 5''91$ in 1946 and $\pm 5''90$ and $\pm 2''20$ in 1959. Since the comet made two revolutions between 1946 and 1959, the variations are halved, the results being given in Table I as columns (f) and (g).

The large residuals for 1946–1959 do not allow us to consider these elements as final, but they are suitable for the determination of the variation in the daily mean motion with fair reliability.

In the course of our analysis we obtained the following results:

(1) Before being discovered in 1900, the comet had rather a close approach to Jupiter. Our elements in 1900 were:

Epoch and osculation 1900 September 18.0 ET

$M_0 = 349°\!.10035$
$\omega = 171.05303$
$\Omega = 197.43596$ } 1950.0
$i = 29.83385$
$e = 0.7315889$
$a = 3.4704377$ AU
$q = 0.931504$ AU.

On 1898 October 27.92798 the comet was at its minimum distance of 0.1887 AU from Jupiter. Some nine months earlier, when the comet had receded to 1.465 AU, its orbital elements were:

Epoch and osculation 1898 January 11.0 ET

$M_0 = 202°\!.93232$
$\omega = 166.99162$
$\Omega = 198.90023$ } 1950.0
$i = 33.76276$
$e = 0.6641633$
$a = 3.6382879$ AU
$q = 1.22187$ AU.

The decrease in perihelion distance from 1.22 to 0.93 AU must have had a profound influence on the comet. On passing through perihelion in 1900 there would have been a significant change in the physical condition of the nucleus, and this apparently led to the ejection of meteoric particles.

From the variation of the elements, we deduce that the comet has a secular deceleration, amounting to about 0.081 day/(period)^2; this increases the comet's velocity, from which we conclude that most of the meteoric material was ejected back along the orbit. Having observed in 1933 and 1946 a rather dense swarm of meteoroids behind the comet, we can expect a denser swarm of such bodies ahead of the comet (or, at least, the more massive meteoroids; there is evidence of this in the large bolide observed at 53 locations on 1926 October 9).

(2) The comet approached Jupiter again on 1910 February 3.709, passing it at a minimum distance of 1.0049 AU. This encounter changed the orbit of the comet, bringing it closer to the orbit of the Earth ($q=0.976$), and the subsequent cumulative action of planetary perturbations brought the orbits (at the comet's descending node) even closer. The uniform variation in the comet's daily mean motion indicates that ejection of meteors took place in a similar manner at every perihelion.

(3) The third approach to Jupiter during the period under investigation took

place on 1958 January 19.5, the minimum distance being 0.93 AU. This approach caused the comet's perihelion distance to decrease to 0.936 AU. It is likely that this new approach brought about vigorous activity in the comet's nucleus and a further ejection of particles. The larger change in the elements (particularly the semimajor axis) from the 1946–1959 to 1959–1965 representations is consistent with this.

(4) Belyaev and the author have obtained predicted elements and an ephemeris for the 1972 return. This was based on 127 observations in 1959 and 1965. An approach of the comet to within 0.577 AU of Jupiter on 1969 September 23 caused the perihelion distance to increase to 0.994 AU. This means that the orbits of the comet and the Earth will pass within 0.0006 AU of each other, suggesting that there might be a return of the meteor showers, if not in 1972, then at the following returns of the comet. The Earth will pass near the descending node of the comet's orbit on October $8^d15^h45^m$ UT. The comet will then be 58.5 days ahead of the Earth. This situation will be similar to that on 1933 October 9, the comet then passing the point of approach 80 days before the Earth.

The precise orbital elements we have obtained for P/Giacobini-Zinner enable us to calculate orbits for Draconid meteors. Keeping in mind that the Earth passes the point of approach 58.5 days after the comet, let us imagine a meteor moving at the same distance from the comet, not in 1972 but in 1965, with the same orbital elements as the comet. Integration of the motion of this imaginary meteor gave an approach to Jupiter on 1969 August 16 (more than a month before the comet) to a distance of 0.710 AU. The resulting change in the meteor's orbit is very different from that in the comet's orbit, and the meteor will pass 0.0636 AU from the Earth in 1972. On the strength of this calculation we suggest that all other meteors moving at approximately the same distance from the comet in 1965 will be diverted by Jupiter into very different orbits, and their subsequent encounters with the Earth will be very unlikely.

Observations in 1933 and 1946 indicated that meteoroids are ejected from the comet with velocities of the order of 13 to 14 m s^{-1}. In order to encounter the Earth in October 1972 they must be 58.5 days behind the comet, which means that they should have made four revolutions around the Sun since leaving the comet. We have made calculations for four imaginary meteors ejected from the comet's nucleus near perihelion in 1946. We suppose that the meteors were ejected forward along the orbit of the comet at a velocity of 13.4 m s^{-1}, when the comet's true anomaly was $+30°$, $+45°$, 0 and $-30°$, respectively. The subsequent elements of the orbits of the meteors are shown in Table II. It is clear that none of these meteors can encounter the Earth in 1972. We obtained a similar result when we traced up to 1972 the orbit of a meteor ejected in 1940. An intense shower like those of 1933 and 1946 is very unlikely in 1972. It is more likely that the shower will be poor or missing completely.

From these calculations we see that at every approach to Jupiter the meteor swarm is considerably dispersed (up to 0.1 AU at perihelion for one approach). This supports our conclusions (Evdokimov, 1955) about the youth of the Draconid showers of 1933 and 1946. As further confirmation, we considered the ejection of meteors by the comet in 1940 with a moderate velocity (13 to 14 m s^{-1}) and calculated orbits for

TABLE II

Orbital elements of meteors leaving the comet in 1946

True anomaly of meteor at ejection	1946			1959		
	a (AU)	e	q (AU)	a (AU)	e	q (AU)
+30°	3.528562	0.717801	0.995757	3.455725	0.740067	0.898258
+45	3.527917	0.717730	0.995837	3.452743	0.750017	0.863128
0	3.528465	0.717808	0.995706	3.456468	0.732909	0.923192
−30	3.528562	0.717801	0.995757	3.452074	0.740930	0.894330

True anomaly of meteor at ejection	1965			1972		
	a (AU)	e	q (AU)	a (AU)	e	q (AU)
+30	3.452468	0.740449	0.896092	3.482198	0.728500	0.945415
+45	3.449419	0.750400	0.860976	3.477261	0.739697	0.905142
0	3.453476	0.733301	0.921040	3.484520	0.720305	0.974460
−30	3.449106	0.741320	0.892216	3.478642	0.729328	0.941570

several meteors ejected when the comet was at perihelion (1940 February 17). All the meteors with forward ejection velocities of 14 m s^{-1} would pass very close to the Earth (0.001 AU) on 1946 October 10 around 3^h–4^h UT, i.e., at the maximum of the meteor shower. This is proof that the meteoroids which became the meteors in 1946 were ejected from the comet in 1940.

We were able to explain the 1933 meteor shower by a similar ejection from the comet in 1900. A meteor ejected forward from the comet in 1900 with a velocity of 15.0 m s^{-1} passed only 0.0004 AU from the Earth on 1933 October 9. A meteor ejected at the same moment back along the orbit with a velocity of 14.5 m s^{-1} passed 0.02 AU from the Earth in 1926, indicating that the meteor showers of both 1926 and 1933 were formed by meteoroids ejected in 1900.

Acknowledgments

The calculations linking the apparitions of 1900, 1913, 1946, and 1959 as well as all the calculations on imaginary meteors could be made only using the excellent computer programmes available now at the department of minor planets and comets of the Institute for Theoretical Astronomy. The programme by Belyaev (1972) was used for the integration – taking into account the perturbations by seven major planets – and the calculation of residuals. The programme by Bokhan (1972) was used for the orbit improvements. The author thanks N. A. Belyaev, N. A. Bokhan, and E. I. Kazimirchak-Polonskaya for the possibility of using these programmes and for suggestions made in the course of this investigation. My sincere thanks are due to the administration of the Institute for Theoretical Astronomy for providing an opportunity for personnel from the Kazan State University to work at the Institute.

References

Belyaev, N. A.: 1972, this Symposium, p. 90.
Bokhan, N. A.: 1972, this Symposium, p. 86.
Dubyago, A. D.: 1956, *Byull. Astron. Obs. V. P. Engel'gardta* No. 32.
Evdokimov, Yu. V.: 1955, *Astron. Tsirk.* No. 159.
Evdokimov, Yu. V.: 1963, *Astron. Zh.* **40**, 544.

Discussion

L. A. Katasev: What is the basis for your statement that a great cluster of large meteoroids moves ahead of the nucleus of P/Giacobini-Zinner?

Yu. V. Evdokimov: I omitted to say at the beginning of my report that I agree with the widely approved model that a comet is a conglomerate of ices and solid particles. My work is based on the fact that the particles are liberated when the comet comes close to the Sun. Since this comet has a secular deceleration most of the meteoric material is ejected back from the nucleus and moves ahead of the comet.

S. K. Vsekhsvyatskij: Have you made an attempt to estimate the mass of the comet's nucleus through consideration of the nongravitational forces and from the data on the Draconid meteor shower?

Yu. V. Evdokimov: I have determined a rate of mass loss, rather than the total mass of the comet's nucleus. In 1963 I showed that the mass loss is as great as about 10% per revolution. I came to this conclusion from the displacement of the orbital planes of the meteors from that of the comet. Using the programme by Belyaev, I have now been able to determine the displacement of the orbital planes of meteoroids ejected in 1940 and 1946. I now find that the comet ejects about 5% of its mass per revolution, so that it will cease to exist after 50 to 100 revolutions.

33. NONGRAVITATIONAL FORCES AND PERIODIC COMET GIACOBINI-ZINNER

D. K. YEOMANS

Astronomy Program, University of Maryland, College Park, Md., U.S.A.

Abstract. Observations of P/Giacobini-Zinner could not be successfully represented for more than two successive apparitions using Newtonian equations of motion. The comet's motion has a secular deceleration approximately equal to 0.08 day per (period)^2. After extending the differential correction to include nongravitational terms in the equations of motion the observations were successfully represented over the time intervals 1900–1946, 1913–1959, and 1939–1965. There is an indication that the magnitude of the nongravitational forces increases with time and that the motion of the comet was discontinuous between 1959 and 1965. A Giacobinid meteor shower prediction has been provided for the 1972 return.

Short-period comet Giacobini-Zinner was discovered by M. Giacobini at Nice in 1900 and accidentally rediscovered in 1913 by E. Zinner. It has been observed at eight passages through perihelion and missed altogether at three. Debris from P/Giacobini-Zinner is believed responsible for the Giacobinid (Draconid) meteor showers. An investigation of this comet's motion was undertaken to determine the role of nongravitational forces.

The numerical integration employed is a predictor-corrector technique running at one-day time steps throughout the comet's orbit. Planetary and solar coordinates are read from magnetic tapes. The differential correction is a 'variation of orbits' procedure and is similar to that used by Marsden (1968). The numerical integration, formation of residuals, and differential correction procedure was programmed for the IBM 360/95 computer at Goddard Space Flight Center.

The first step in determining an orbit for this comet was an attempt to link the successive 1933, 1939 and 1946 apparitions using strictly Newtonian equations of motion. However, after the differential correction routine had converged, the mean residual was nearly 2″.12 and the two 1939 observations had residuals nearly equal to 27″. Attempting to link the 1946, 1959, and 1965 apparitions met with even less success. The mean residual was 11″.48 and the five 1965 observations had residuals close to 190″. Newtonian orbits were then successfully computed for pairs of consecutive apparitions, and a comparison of the predicted and observed times of perihelion passage revealed a secular deceleration in this comet's mean motion approximately equal to 0.08 day per (period)^2. As a check upon the present work, it was noted that the Newtonian orbits computed for the 1933, 1939 and 1946 apparitions agree with the earlier results of Evdokimov (1963).

In an attempt to represent the observations for more than two apparitions, three nongravitational terms were added to the Newtonian equations of motion. The form of these orthogonal terms is that used by Marsden (1969, 1972). The differential correction scheme was extended to include the unknowns A_i and B_i. Generally A_1

Chebotarev et al. (eds.), The Motion, Evolution of Orbits, and Origin of Comets, 181–186. All Rights Reserved.

and A_2 were sufficient to provide a good representation of the observations for four or five apparitions, while the inclusion of B_2 allowed a satisfactory representation for up to six apparitions. Inclusion of a fourth nongravitational parameter (A_3 or B_1) had a negligible effect upon the results. Ultimately, it was possible to represent successfully the observations over the intervals 1913–1959, 1900–1946 and 1939–1965; see Table I.

TABLE I

Osculating elements and nongravitational parameters (with probable errors)

Interval	1913–1959		1900–1946		1939–1965	
Mean resid.	1″22		1″48		0″94	
Epoch (ET)	1913 Oct. 10.0		1900 Dec. 7.0		1939 Aug. 25.0	
T (ET)	1913 Nov. 2.56654		1900 Nov. 28.49641		1940 Feb. 17.21648	
		± 0.00013		± 0.00023		± 0.00071
e	0.7206115	± 0.0000001	0.7315675	± 0.0000002	0.7166567	± 0.0000001
q (AU)	0.9760088	± 0.0000003	0.9315211	± 0.0000005	0.9956677	± 0.0000003
$\Omega°$ (1950.0)	196.36633	± 0.00006	197.43252	± 0.00008	196.24967	± 0.00006
$\omega°$ (1950.0)	171.48630	± 0.00017	171.05086	± 0.00028	171.79083	± 0.00011
$i°$ (1950.0)	30.74715	± 0.00003	29.83658	± 0.00006	30.74135	± 0.00003
$A_1 \times 10^5$	+0.6584	± 0.0175	+0.5043	± 0.0267	+0.8226	± 0.0232
$A_2 \times 10^6$	+1.0911	± 0.0013	+1.1340	± 0.0022	+0.3567	± 0.0042
B_2	−0.1544	± 0.0012	−0.0533	± 0.0018	−3.6586	± 0.0188

Table II represents successful orbits using different nongravitational parameters. Each orbit was differentially corrected over the time interval indicated, solving for the given nongravitational parameters and their probable errors. The mean residual ('Resid.') for each orbit is listed above the number of observations employed ('Obs.'), and the epoch of osculation appears under the observed interval. Each orbit was integrated forward and backward in time to determine the maximum residuals (in seconds of arc) for the apparitions not included in the differential correction. The units associated with the accelerations are AU per (40 ephemeris days)². Orbits No. 8, 5, and 12 are the ones listed in Table I.

Assuming an icy-conglomerate model for the cometary nucleus (Whipple, 1950), a number of results are evident from Table II:

(1) The radial component of the nongravitational force is directed away from the Sun ($A_1 > 0$). The transverse component is directed 90° forward of the radial component ($A_2 > 0$), implying a direct rotation of the nucleus.

(2) There is no evidence for a significant nongravitational force component normal to the orbit plane.

(3) The ratio A_2/A_1 implies a small lag angle ($< 15°$) between the subsolar meridian and the direction of maximum mass ejection.

(4) The magnitude of the transverse component of the nongravitational acceleration is increasing with time over the observed time interval ($B_2 < 0$).

TABLE II

Nongravitational parameters for P/Giacobini-Zinner

Orbit	Interval Epoch	Resid. Obs.	$A_1 \times 10^5$	$A_2 \times 10^6$	B_2	Maximum residuals							
						1965	1959	1946	1940	1933	1926	1913	1900
1	1933–46 33 Jan. 28.0	0".96 41	+0.4673±0.0717	+1.2830±0.0245	—	139	85	—	—	—	21	165	330
2	1926–46 26 Aug. 25.0	0.93 51	+0.6272±0.0498	+1.2142±0.0018	—	149	224	—	—	—	—	40	122
3	1913–46 13 Oct. 12.0	1.29 86	+0.4817±0.0264	+1.2021±0.0006	—	153	252	—	—	—	—	—	59
4	1913–46 13 Oct. 12.0	1.24 86	+0.5338±0.0259	+1.1579±0.0049	-0.0597±0.0069	143	143	—	—	—	—	—	9
5	1900–46 00 Dec. 7.0	1.48 99	+0.5043±0.0267	+1.1340±0.0022	-0.0533±0.0018	144	152	—	—	—	—	—	—
6	1933–59 33 Jan. 28.0	0.90 81	+0.7108±0.0175	+1.3156±0.0004	—	131	—	—	—	—	19	267	493
7	1926–59 26 Aug. 25.0	0.90 91	+0.7461±0.0167	+1.1267±0.0030	-0.2219±0.0040	125	—	—	—	—	—	83	257
8	1913–59 13 Oct. 12.0	1.22 126	+0.6584±0.0175	+1.0911±0.0013	-0.1544±0.0012	128	—	—	—	—	—	—	81
9	1913–59 13 Oct. 12.0	1.18 126	+0.3951±0.0403	+1.0897±0.0013	-0.1515±0.0014 $B_1 = -0.4800\pm0.0910$	127	—	—	—	—	—	—	88
10	1913–59 13 Oct. 12.0	1.19 126	+0.6842±0.0177	+1.0887±0.0013	-0.1562±0.0012 $A_3 = (-1.2933\pm 0.2339)\times10^{-6}$	128	—	—	—	—	—	—	81
11	1946–65 46 May 12.0	1.46 68	+1.2090±0.0339	+4.2638±0.0158	—	—	—	—	182	2242	4537	17700	20791
12	1939–65 39 Aug. 25.0	0.94 71	+0.8226±0.0232	+0.3567±0.0042	-3.6586±0.0188	—	—	—	—	156	561	3738	5274
13	1900–26 00 Aug. 25.0	1.76 60	+0.3172±0.1278	+1.1537±0.0026	—	—	—	—	—	—	—	—	—

(5) Orbit No. 5 (1900–1946) and orbit No. 8 (1913–1959) both spanned six apparitions, whereas orbit No. 12 (1939–1965) successfully represented the observations of only four apparitions.

The last two points require some elaboration. If B_2 is negative, P/Giacobini-Zinner would be the first short-period comet to exhibit an increase in the transverse component of the nongravitational acceleration with time. Independent evidence that this is indeed the case comes from a comparison of the increasing values of A_2 from orbits No. 13 and 6 in Table II. Although the sign of B_2 is likely to be negative in all orbits in which it was satisfactorily determined, the magnitude of B_2 in orbit No. 12 is quite suspicious. It is only with the inclusion of the 1965 observations that the magnitude of B_2 increases 24 times compared to orbit No. 8.

In terms of the icy-conglomerate model, two processes come to mind for a secular time increase in the transverse component of nongravitational acceleration. First, if the comet nucleus had an ice mantle, the effective nongravitational force would increase with time as the comet radius decreased. Secondly, one could imagine the magnitude of the outgassing to remain constant while the inclination of the rotation axis with respect to the orbit plane slowly 'straightened up' from an initially inclined position.

The negative value of B_2 for P/Giacobini-Zinner is not the only surprise concerning this comet's behavior. The motion of the comet between 1959 and 1965 appears to be discontinuous. The apparent discontinuous behavior is suggested by the results presented in Table II. Orbit No. 8, which successfully represented the observations in 1913, 1926, 1933, 1939, 1946 and 1959, produced residuals of approximately 2' in 1965. The comet was nearly 3 AU from the Earth when observed in 1965, and orbit No. 8 required a correction to the perihelion passage time of $\Delta T = +0\overset{d}{.}3$. There is no obvious reason why such a large ΔT correction should be necessary in 1965. The required ΔT correction in 1965 is apparently due to an enhancement of the nongravitational forces between 1959 and 1965. The effect could be due to a sudden increase in the comet's orbital velocity. Marsden (1969, 1970) has noted that some comets appear to have discontinuous changes in their nongravitational forces following a close approach to Jupiter. Although P/Giacobini-Zinner had a moderately close approach (0.93 AU) to Jupiter in 1958, there was no close approach between 1959 and 1965. A case could be made for the splitting of this comet between 1959 and 1965. The 1959 observations showed no evidence of splitting, while the 1965 observations were made near the limit of visibility. The brightest and faintest 1965 observed magnitudes were 19.7 and 20.5, so that a slightly fainter fragment could easily have been missed. For example, the discrepancy in the 1965 perihelion passage time could be explained if the comet split while approaching aphelion ($r = 5$ AU) in 1962 and the larger fragment received a sudden increase in velocity of 1.4 m s^{-1}. Table III lists osculating elements for epochs near the perihelion times 1900–1966.

Finally, a word about the Giacobinid meteor showers is in order. The Giacobinid meteor showers have probably been the most spectacular meteor displays of the present century. The showers were particularly strong in 1933 and 1946, with lesser showers being observed in 1926 and 1952. The duration of the showers is remarkably

TABLE III

Osculating elements for epochs near passages through perihelion

1900 Dec. 7.0	1907 May 25.0	1913 Nov. 9.0	1920 May 16.0
1900 Nov. 28.49641	1907 May 19.50678	1913 Nov. 2.56656	1920 May 18.89607
0.7315675	0.7317000	0.7206140	0.7199385
0.9315211	0.9308234	0.9760089	0.9799927
197.43252	197.36299	196.36625	196.31669
171.05086	171.12828	171.48637	171.54205
29.83658	29.85905	30.74721	30.71705
1926 Dec. 31.0	1933 July 27.0	1940 Feb. 21.0	1946 Sept. 17.0
1926 Dec. 11.70840	1933 July 15.14737	1940 Feb. 17.21100	1946 Sept. 18.48600
0.7170052	0.7159931	0.7167041	0.7166791
0.9937348	0.9995269	0.9956051	0.9957057
196.24237	196.24296	196.24976	196.29323
171.75492	171.77146	171.79206	171.81054
30.73865	30.68389	30.74136	30.72686
1953 April 13.0	1959 Nov. 8.0	1966 Mar. 16.0	
1953 April 16.91679	1959 Oct. 26.91926	1966 Mar. 28.29078	
0.7178923	0.7289474	0.7293999	
0.9886907	0.9359554	0.9335067	
196.23984	196.02990	195.96510	
171.88762	172.84433	172.91988	
30.82435	30.90379	30.94417	

For each set of elements, the consecutive lines represent the osculation epoch, time of perihelion passage (ET), eccentricity, perihelion distance (AU), the longitude of the ascending node, argument of perihelion, and the inclination. The three angular elements are in degrees and are referred to the ecliptic and mean equinox of 1950.0.

The elements for 1900 and 1907 are from orbit No. 5, and those for 1913 to 1959 are from orbit No. 8. The 1966 elements are from an orbit fitted to observations in 1959 and 1965 but adopting the nongravitational parameters of orbit No. 8.

short, with most of the meteor activity taking place within 1 h. In 1972, the Earth will pass only 0.00074 AU outside the comet's orbit at the descending node. The Earth will follow the comet to the node by 58.5 days, and the prospect of a meteor shower in 1972 Oct. 8.65 UT is quite good.

References

Evdokimov, Yu. V.: 1963, *Astron. Zh.* **40**, 544.
Marsden, B. G.: 1968, *Astron. J.* **73**, 367.
Marsden, B. G.: 1969, *Astron. J.* **74**, 720.
Marsden, B. G.: 1970, *Astron. J.* **75**, 75.
Marsden, B. G.: 1972, this Symposium, p. 135.
Whipple, F. L.: 1950, *Astrophys. J.* **111**, 375.

Discussion

F. L. Whipple: The evidence for an increasing nongravitational effect in P/Giacobini-Zinner and the suggestion of a transition in the nucleus from ices on the outside toward denser earthy material –

providing a greater time lag in heat transfer – are most intriguing. Confirming evidence may be found in the fact that the related meteor stream contains the extreme examples of low density and fragile materials. On the other hand, the comet may be uniform throughout and near its final dissolution, which suggests that the comet was never a very large one with an asteroidal core.

L. Kresák: Isn't the observed increase in the nongravitational effects only an effect of long-period variations?

D. K. Yeomans: Yes, I rather think so.

34. A NON-NEWTONIAN ORBIT FOR PERIODIC COMET BORRELLY

D. K. YEOMANS

Astronomy Program, University of Maryland, College Park, Md., U.S.A.

Abstract. A non-Newtonian orbit has been determined for P/Borrelly which successfully represents all eight apparitions. Although this comet has a substantial secular acceleration in its mean motion, the transverse nongravitational force is time independent over the 64-yr observed interval.

Alphonse Borrelly discovered the comet that bears his name at Marseilles in December 1904. P/Borrelly has been observed at eight apparitions and missed at its returns in 1939 and 1946. Using Newtonian equations of motion, Marsden (1968) noted a secular acceleration in the motion of this comet approximately equal to 0.05 day per (period)^2.

The computational procedure used in the present work is similar to that employed by the author in his investigation of P/Giacobini-Zinner (Yeomans, 1972). Three nongravitational terms were added to the standard Newtonian equations of motion and the differential correction yielded solutions for selected nongravitational parameters as well as the osculating orbital elements.

Table I displays the various successful non-Newtonian orbits for P/Borrelly. For each orbit, the observed interval is given above the epoch of osculation, and the mean residual is placed above the number of observations employed in the differential correction. The given nongravitational parameters are a result of the differential correction, and the units of acceleration are AU per (40 ephemeris days)^2. Orbit

TABLE I

Nongravitational parameters for P/Borrelly

Orbit	Interval Epoch	Mean resid. Obs.	$A_1 \times 10^5$	$A_2 \times 10^6$	B_2
1	1918–54 18 July 26.0	1″32 70	$+0.6375 \pm 0.1195$	-2.5211 ± 0.0160	
2	1954–68 53 May 23.0	1.00 20	$+0.7941 \pm 0.5357$	-2.9600 ± 0.0126	
3	1918–68 18 July 26.0	1.23 76	$+1.0876 \pm 0.0638$	-2.6022 ± 0.0062	
4	1918–68 18 July 26.0	1.21 76	$+1.0477 \pm 0.0630$	-2.5374 ± 0.0164	-0.0327 ± 0.0084
5	1904–68 04 Dec. 6.0	1.48 172	$+0.8427 \pm 0.0385$	-2.6470 ± 0.0027	
6	1904–68 04 Dec. 6.0	1.45 172	$+0.8082 \pm 0.0392$	-2.6661 ± 0.0059	$+0.0123 \pm 0.0035$
7	1904–68 04 Dec. 6.0	1.41 164	$+0.8669 \pm 0.0373$	-2.6478 ± 0.0028	

Chebotarev et al. (eds.), The Motion, Evolution of Orbits, and Origin of Comets, 187–189. All Rights Reserved.

No. 7 in Table I is considered the best solution over the entire 64-yr observed interval. The mean residual (1.″41) is quite satisfactory and there are no systematic residual trends. As is evident from Table I, there is no indication of a time dependence in the magnitude of the transverse nongravitational term ($B_2 \simeq 0$) over the entire observed interval. Marsden (1969) has pointed out that the nongravitational forces of some, but not all, comets appear to have been enhanced following a close approach to Jupiter. P/Borrelly made a moderately close approach to Jupiter in March 1936

TABLE II

Nongravitational orbit for P/Borrelly

Epoch (ET)	1904 Dec. 6.0
T (ET)	1905 Jan. 17.29547 $\pm$ 0.00023 (p.e.)
e	0.6152593 $\pm$ 0.0000002
q (AU)	1.3953626 $\pm$ 0.0000006
Ω (1950.0)	77.°38168 $\pm$ 0.00004
ω (1950.0)	352.°35243 $\pm$ 0.00015
i (1950.0)	30.°48415 $\pm$ 0.00003
$A_1 \times 10^8$	+0.8669 $\pm$ 0.0373
$A_2 \times 10^8$	−2.6478 $\pm$ 0.0028

TABLE III

Osculating elements for epochs near passages through perihelion

1905 Jan. 15.0	1911 Dec. 10.0	1918 Nov. 23.0	1925 Oct. 17.0
1905 Jan. 17.29511	1911 Dec. 18.49032	1918 Nov. 17.09841	1925 Oct. 7.52656
0.6152297	0.6140728	0.6150881	0.6164495
1.3953650	1.4026547	1.3957867	1.3881794
77.38117	77.37547	77.37065	77.37869
352.35244	352.37496	352.39845	352.42255
30.48471	30.44170	30.49142	30.51054
1932 Aug. 21.0	1939 June 6.0	1946 June 9.0	1953 June 12.0
1932 Aug. 27.29210	1939 June 8.85191	1946 June 9.44330	1953 June 9.49829
0.6167257	0.6054618	0.6044994	0.6040735
1.3855408	1.4416113	1.4479157	1.4500429
77.30822	76.21477	76.16989	76.17790
352.55089	350.82753	350.93275	350.95574
30.52947	31.09903	31.06621	31.08975
1960 June 15.0	1967 June 19.0		
1960 June 13.24020	1967 June 17.71729		
0.6033964	0.6044628		
1.4541183	1.4465989		
76.19417	76.14186		
350.97387	351.02879		
31.06647	31.11549		

For each set of elements, the consecutive lines represent the osculation epoch, time of perihelion passage (ET), eccentricity, perihelion distance (AU), the longitude of the ascending node, argument of perihelion, and the inclination. The three angular elements are in degrees and are referred to the ecliptic and mean equinox of 1950.0.

(minimum separation 0.54 AU), but there was apparently no noticeable effect upon the nongravitational forces. In addition, it seems unlikely that there will be any discontinuity in the nongravitational forces when P/Borrelly approaches to within 0.61 AU of Jupiter in 1972.

Table II gives the osculating elements for orbit No. 7 of Table I. This orbit was run forward to provide the osculating elements given in Table III for various epochs.

References

Marsden, B. G.: 1968, *Astron. J.* **73**, 367.
Marsden, B. G.: 1969, *Astron. J.* **74**, 720.
Yeomans, D. K.: 1972, this Symposium, p. 181.

35. AN INVESTIGATION OF THE MOTION OF PERIODIC COMET BORRELLY FROM 1904 TO 1967

L. M. BELOUS

Northwestern Extramural Polytechnic Institute, Leningrad, U.S.S.R.

Abstract. The motion of P/Borrelly has been investigated using all precise published positions and taking into account the perturbations by Venus to Pluto. Nongravitational forces have been found to influence the motion of the comet considerably, formerly as a secular deceleration, more recently as a secular acceleration. By linking the seven apparitions 1904–1960, with allowance for the non-gravitational effects, we have obtained a set of elements which represents 802 observations with a mean error of 1″89.

1. Introduction

Periodic comet Borrelly was discovered by Alphonse Borrelly at the Marseilles Observatory on 1904 December 28 as an object of magnitude 10, 1′–2′ in diameter and possessing a small nucleus (Vsekhsvyatskij, 1958). Having a period of 6.9 yr, the comet was observed at eight apparitions 1904–1967; it was missed in 1939 and 1946. The first apparition of the comet was thoroughly investigated by Fayet (1914), who calculated a final orbit from 430 observations in eight normal places, taking into account perturbations by the six planets Mercury to Saturn; his set of elements represents the normal places with a mean error of 1″397. Later (Fayet, 1925), he formed ten normal places from 259 observations at the 1911–1912 apparition and obtained a set of elements that gave a good representation of the first nine normal places; the last normal place, formed from seven observations at the end of the apparition, produced a residual that was considerably larger than the rest. Fayet then linked the apparitions of 1904–1905 and 1911–1912 and derived a predicted set of elements for 1919, taking into account perturbations by the planets Venus, Earth, Jupiter, and Saturn.

2. Investigation of the Comet's Motion under the Influence of Gravitational Forces

We have investigated the motion of the comet from 1904 to 1967, with the help of the BESM-4 computer at the Institute for Theoretical Astronomy. The following programmes have been used: integration with a variable step, in double precision and with allowance for perturbations by all the major planets (Kazimirchak-Polonskaya, 1967), integration by Cowell's method (Belyaev), comparison of observed and computed coordinates (Belyaev and Bokhan), and improvement of elliptical orbits (Bokhan and Makover). There were 105 accurate observations at the 1918–1919 apparition, 82 in 1925–1926, 33 in 1932–1933, 10 in 1954 and 7 in 1960. The 1967 apparition is represented by a single approximate observation. Using modern star catalogues, we have refined the positions of reference stars for the visual observations, proper motions being taken into account.

Chebotarev et al. (eds.), *The Motion, Evolution of Orbits, and Origin of Comets,* 190–194. *All Rights Reserved.*
Copyright © 1972 by the IAU.

We first considered the set of elements (A), obtained by Schaumasse (1931) from an integration of the motion from 1911 to 1919, taking into account the perturbations by the planets Venus-Saturn:

$$
\begin{aligned}
\text{Epoch} &= \text{1919 March 23.0 ET}\\
M_0 &= \ 17°96984\\
\omega &= 352.39656\\
\Omega &= \ \ 77.36882\\
i &= \ \ 30.49100\\
\varphi &= \ \ 37.96048\\
\mu &= \ \ \ 0.1427191.
\end{aligned}
\qquad 1950.0 \qquad\qquad (A)
$$

We have compared the 1918–1919 observations with the set (A) and concluded that these elements do not represent these observations adequately. We therefore made a preliminary orbit correction using 92 observations at the 1918–1919 apparition; normal places were not formed. The improved set of elements (I) represents the observations with a mean error of 2″56:

$$
\begin{aligned}
\text{Epoch} &= \text{1919 March 23.0 ET}\\
M_0 &= \ 17°9742090\\
\omega &= 352.392113\\
\Omega &= \ \ 77.3686643\\
i &= \ \ 30.4899165\\
\varphi &= \ \ 37.9565861\\
\mu &= \ \ \ 0.142752508.
\end{aligned}
\qquad 1950.0 \qquad\qquad (I)
$$

The set (I) was adopted as a basis for further investigations. We integrated the motion up to 1925 and then up to 1932, taking into account perturbations by all the major planets. We then succeeded in linking 87 observations at the returns of 1925 and 1932, obtaining the set of elements (II):

$$
\begin{aligned}
\text{Epoch} &= \text{1919 March 23.0 ET}\\
M_0 &= \ 17°9665203\\
\omega &= 352.391078\\
\Omega &= \ \ 77.3684744\\
i &= \ \ 30.4901315\\
\varphi &= \ \ 37.9569632\\
\mu &= \ \ \ 0.14274929.
\end{aligned}
\qquad 1950.0 \qquad\qquad (II)
$$

The mean error was 2″44. Using 38 observations at the same two returns Marsden (1968) obtained an orbit with mean error 2″5.

Our attempt to link the three apparitions 1918–1919, 1925-1926 and 1932–1933 gave a mean error of 15″4. On attempting to fit the observations at the 1932, 1953 and 1960 returns Marsden found (O–C) residuals as high as 30″. He pointed out that P/Borrelly evidently has a nongravitational secular acceleration, amounting to something like 0^{d}05 per (period)^2.

3. Influence of Nongravitational Forces on the Motion of the Comet

On comparing the observations with the calculated orbit, and taking into account gravitational forces very completely, we confirm that P/Borrelly is affected by non-gravitational forces, these appearing as systematic secular accelerations and decelerations, their values changing from apparition to apparition.

It is probable that this is not the only comet with this kind of motion. Instances where investigators have found only accelerations or only decelerations, or no non-gravitational effects at all, can be explained by the small number of apparitions considered. It is possible that such variations in the action of the nongravitational forces could result in the failure to link particular apparitions of a comet, even when a large number of precise observations is available.

We have investigated the nongravitational variation of the mean anomaly M_0 of P/Borrelly during the whole period covered by observations. This investigation shows that from discovery to its approach to Jupiter (1936 March 26.04717, minimum separation 0.53935 AU) the comet had a secular deceleration; subsequently, the comet had a secular acceleration. The variations of M_0 and the corresponding variations in perihelion time T are given in Table I.

TABLE I

Nongravitational variations in mean anomaly
(ΔM_0) and perihelion time (ΔT)

Apparition	ΔM_0	ΔT
1911	$-0°0310$	$+0°2180$
1918	-0.0237	$+0.1660$
1925	-0.0154	$+0.1076$
1932	-0.0072	$+0.0503$
1953	$+0.0192$	-0.1366
1960	$+0.0204$	-0.1453

4. Improvement of the Elements

The set of elements (III) was obtained by linking 120 observations of the comet at the five returns 1918–1960 and taking into account the variation of the mean anomaly due to the influence of nongravitational forces; the mean error was $1''61$:

$$
\begin{aligned}
\text{Epoch} &= \text{1919 March 23.0 ET} \\
M_0 &= 17°9743901 \quad \pm\ 0°0000294 \\
\omega &= 352.3914940 \quad \pm\ 0.0001514 \\
\Omega &= 77.3684353 \quad \pm\ 0.0000719 \\
i &= 30.4900482 \quad \pm\ 0.0000519 \\
\varphi &= 37.9565787 \quad \pm\ 0.0000256 \\
\mu &= 0.142752542 \pm\ 0.000000005.
\end{aligned}
\quad \left.\begin{array}{}\\ \\ \\ \end{array}\right\} \ \text{1950.0} \qquad \text{(III)}
$$

Finally, we have incorporated the observations at the 1904–1905 and 1911–1912 apparitions, using for this purpose the normal places given by Fayet (1914, 1925), but omitting the tenth normal place at the second apparition. The set of elements (IV) satisfies the observations with a mean error of $1.''89$:

$$
\begin{aligned}
\text{Epoch} &= \text{1919 March 23.0 ET} \\
M_0 &= 17.°9743786 \quad \pm\ 0.°0000300 \\
\omega &= 352.3915210 \quad \pm\ 0.0001439 \\
\Omega &= 77.3684353 \quad \pm\ 0.0000752 \\
i &= 30.4901341 \quad \pm\ 0.0000552 \\
\varphi &= 37.9565464 \quad \pm\ 0.0000234 \\
\mu &= 0.142752536 \pm 0.000000004.
\end{aligned}
$$

$\left. \right\}$ 1950.0   (IV)

The final elements for the seven apparitions considered are given in Table II. The set of elements for 1967 was obtained by using an approximate value for the nongravitational effects.

TABLE II

Improved elements (equinox 1950.0, times in ET)

Epoch	1905 May 15.0	1911 Oct. 31.0	1919 Mar. 23.0	1925 Aug. 3.0
M_0	16.°8055948	353.°101935	17.°9743901	350.°618223
ω	352.350961	352.373229	352.391494	352.424228
Ω	77.3815379	77.3758291	77.3684353	77.3785897
i	30.4852542	30.4414967	30.4900482	30.5108143
φ	37.9604383	37.8823025	37.9565787	38.0532219
μ	0.142776147	0.142261202	0.142752542	0.143173005
T	1905 Jan. 17.2945	1911 Dec. 18.4893	1918 Nov. 17.0979	1925 Oct. 7.5250

Epoch	1932 July 17.0	1953 Dec. 9.0	1960 Aug. 29.0	1967 Sept. 17.0
M_0	354.°078527	25.°6743858	10.°7737019	12.°8830691
ω	352.551043	350.948992	350.974719	351.035042
Ω	77.3083608	76.1780288	76.1942250	76.1406239
i	30.5294665	31.0896158	31.0661465	31.1140951
φ	38.0765135	37.1548362	37.1150968	37.1975837
μ	0.143401646	0.140679086	0.140381401	0.140862085
T	1932 Aug. 27.2931	1953 June 9.5153	1960 June 13.2578	1967 June 17.5385

Acknowledgments

It is my pleasant duty to express my cordial thanks to G. A. Chebotarev, Director of the Institute for Theoretical Astronomy, for the opportunity to carry out these calculations using the computer at the Institute. I also express my sincere thanks to E. I. Kazimirchak-Polonskaya for supplying her integration programme, for her helpful comments and constant attention to the investigation; I also thank N. A.

Belyaev and N. A. Bokhan for their comments and for supplying computer programmes.

References

Fayet, G.: 1914, *Ann. Obs. Paris Mem.* **30**, A1.
Fayet, G.: 1925, *J. Obs.* **8**, 109.
Kazimirchak-Polonskaya, E. I.: 1967, *Trudy Inst. Teor. Astron.* **12**, 39.
Marsden, B. G.: 1968, *Astron. J.* **73**, 367.
Schaumasse, A.: 1931, *J. Obs.* **14**, 147.
Vsekhsvyatskij, S. K.: 1958, *Fizicheskie Kharakteristiki Komet*, Moscow.

36. THE MOTION OF PERIODIC COMET PONS-BROOKS, 1812–1954

P. HERGET and H. J. CARR

University of Cincinnati Observatory, Cincinnati, Ohio, U.S.A.

Abstract. The orbit of P/Pons-Brooks has been studied using the observations of 1812, 1883–1884 and 1953–1954. A curious deviation of the residuals has been established during the last two months of observation in April and May 1884, but no discontinuity in the orbit due to an impulsive action on the comet has been found. The observations covering two complete revolutions cannot be fitted into one single orbit.

In 1939 the observations of P/Pons-Brooks were treated in much the same manner as Cowell and Crommelin had used to predict the return of Halley's Comet in 1909. The observations of 1883–1884 were collected only for those observatories which had made 40 or more, so that systematic errors could be examined, but eventually no corrections on this basis seemed justified. These observations were first collected into normal places which were separated by each Full Moon, but they subsequently were replaced by normal places extending over an interval of about ten days each. All the comparison star positions were examined and compared with the *Geschichte des Fixsternhimmels* in order to provide improved positions whenever possible. The perturbations in rectangular coordinates during the interval of observations were computed by Encke's method. The differential correction process was by Herget's (1939) extension of the Eckert-Brouwer method. The dynamical variables used were dT, de, and da/a. The equations of condition were solved only for five unknowns in terms of the residuals (O − C) and da/a (since the latter was otherwise nearly indeterminate).

The orbit represented by the best preliminary fit to the observations was then integrated backward to 1812 in terms of the barycentric, rectangular coordinates. The normal places which had been provided by Schulhof and Bossert for the observations of 1812 were then used to determine only the value of da/a. In other words, on the assumption that the observations of 1812 were of poorer quality, the values of the five unknowns in the equations of condition for 1812 were eliminated in terms of the solution obtained from 1883–1884, and only da/a remained in the equations. It was now well determined because an entire period had been observed, and thus the finally adopted orbit for 1883–1884 could be derived.

At this point it was apparent that all the observations in the Southern Hemisphere during the last two months of observation were of increasingly poorer quality. The residuals of the individual observations in the last normal place spread over a circle more than 20″ in diameter and they all lay on one side of the finally adopted path. It seemed futile to attempt at that time to determine the effect of any nongravitational action due to the perihelion passage. Therefore the adopted trajectory was integrated forward to 1954, entirely disregarding what was known of the poor fit at the end of

Chebotarev et al. (eds.), *The Motion, Evolution of Orbits, and Origin of Comets*, 195–199. *All Rights Reserved.*
Copyright © 1972 by the IAU.

the observed arc in 1884. After the recovery in 1953, the predicted perihelion passage proved to be in error by 4.5 days.

After the Prague meetings of IAU Commission 20 in 1967, we undertook to determine the magnitude of an assumed discontinuity in the gravitational trajectory which must have taken place shortly after perihelion passage in 1884. It was proposed to determine for Case I an orbit from the observations in 1812 and those before perihelion in 1884, and for Case II a second orbit from those in 1953–1954 and after perihelion in 1884. Then the representation of the observations in 1883–1884 after perihelion as given by Case I, and those before perihelion as given by Case II would be an indication of the magnitude of the discontinuity.

Our efforts, however, produced a different result. Along with the Schubart-Stumpff N-body integration program for the trajectory of the comet, we integrated the variational equations in terms of the starting values, x_0, y_0, z_0, x'_0, y'_0, and z'_0, with the epoch near perihelion at JD 2409200.5. Normal places were abandoned, and each observation was used individually. Both Case I and Case II, separately, showed that the observations in the Southern Hemisphere during the last two months deviated in nearly the same fashion as had been found earlier. Therefore these observations were not used in what follows. Case A is a differential correction based on all the remaining observations of 1883–1884 and 1953–1954, and Case B is a differential correction based on all the remaining observations of 1883–1884 and the six normal places of 1812. The residuals of all the observations of 1883–1884 show an almost identical pattern in both Cases. The unit vectors perpendicular to the two orbit planes differ in direction by less than one second of arc. The starting values of both Cases and the differences of the position and velocity vectors at the epoch are as shown in Table I. Between the two Cases: $(dP/P) = 0.0001702 = 4\overset{d}{.}37$. With this disparity, we have

TABLE I

Position and velocity components for the epoch JD 2409200.5

Case B	Case A	Diff.
+0.132082942748	+0.132082068306	+874442
+0.758828947641	+0.758829039873	−92232
+0.094968656120	+0.094969079510	−423390
(40k) $\bar{v}_0$	(40k) $\bar{v}_0$	
−0.362633594277	−0.362633469038	−125239
+0.166990147726	+0.166988666760	+1480966
−1.016524295006	−1.016523056834	−1238172

found it impossible to combine the observations of all three apparitions into one orbit. The increasingly wide systematic deviations of the observations during the last two months in 1884 also remain unexplained. All the residuals are shown in Figures 1–3, except those of 1812, which are given in Table II. Their systematic character also shows that our results are far from satisfactory. Undoubtedly there must be non-gravitational forces acting, but we feel that the distribution of observations is such that the arbitrary introduction of additional parameters is not justified.

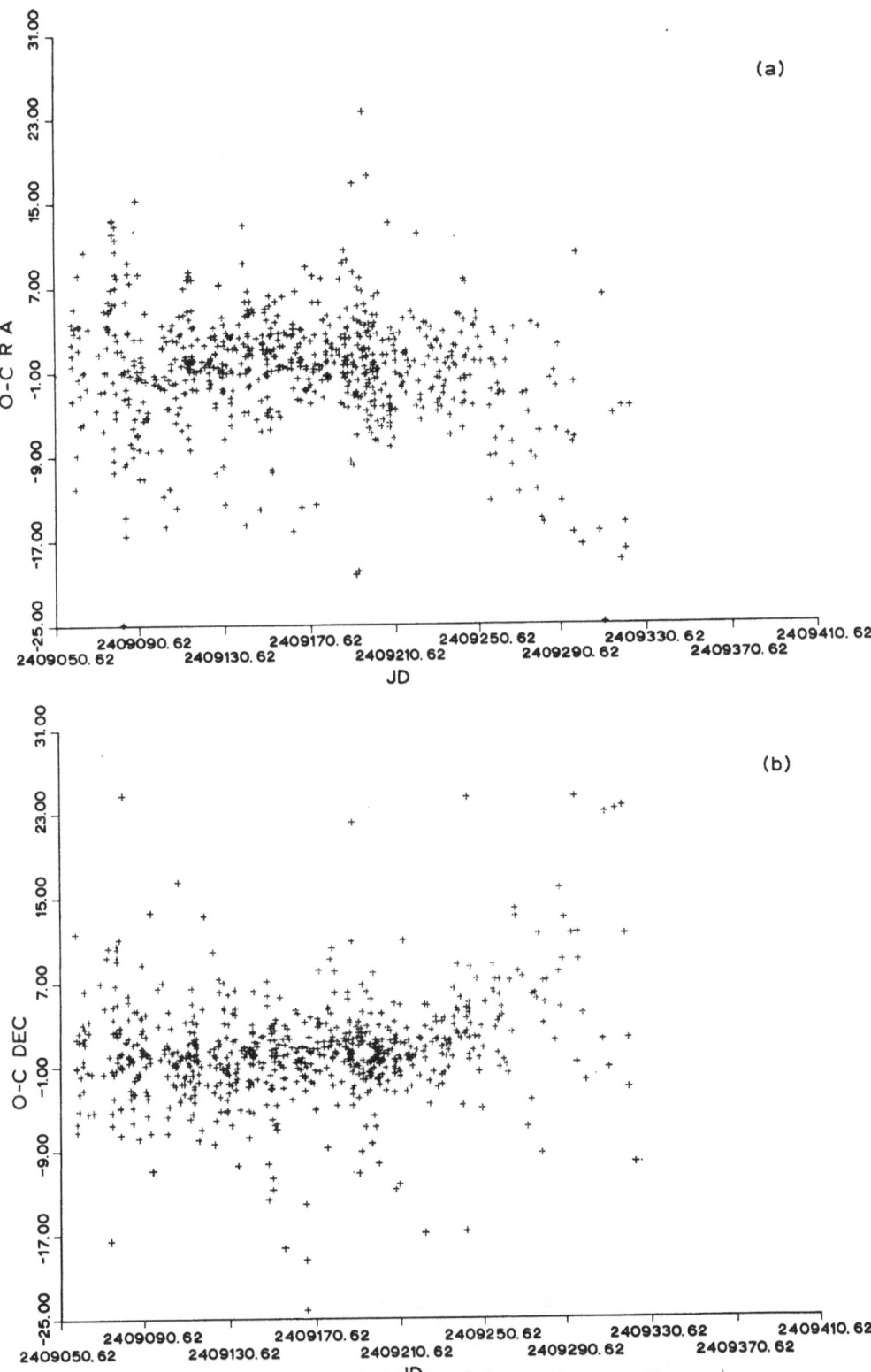

Figs. 1a–b.   (O − C) residuals in 1883–1884; Case A.

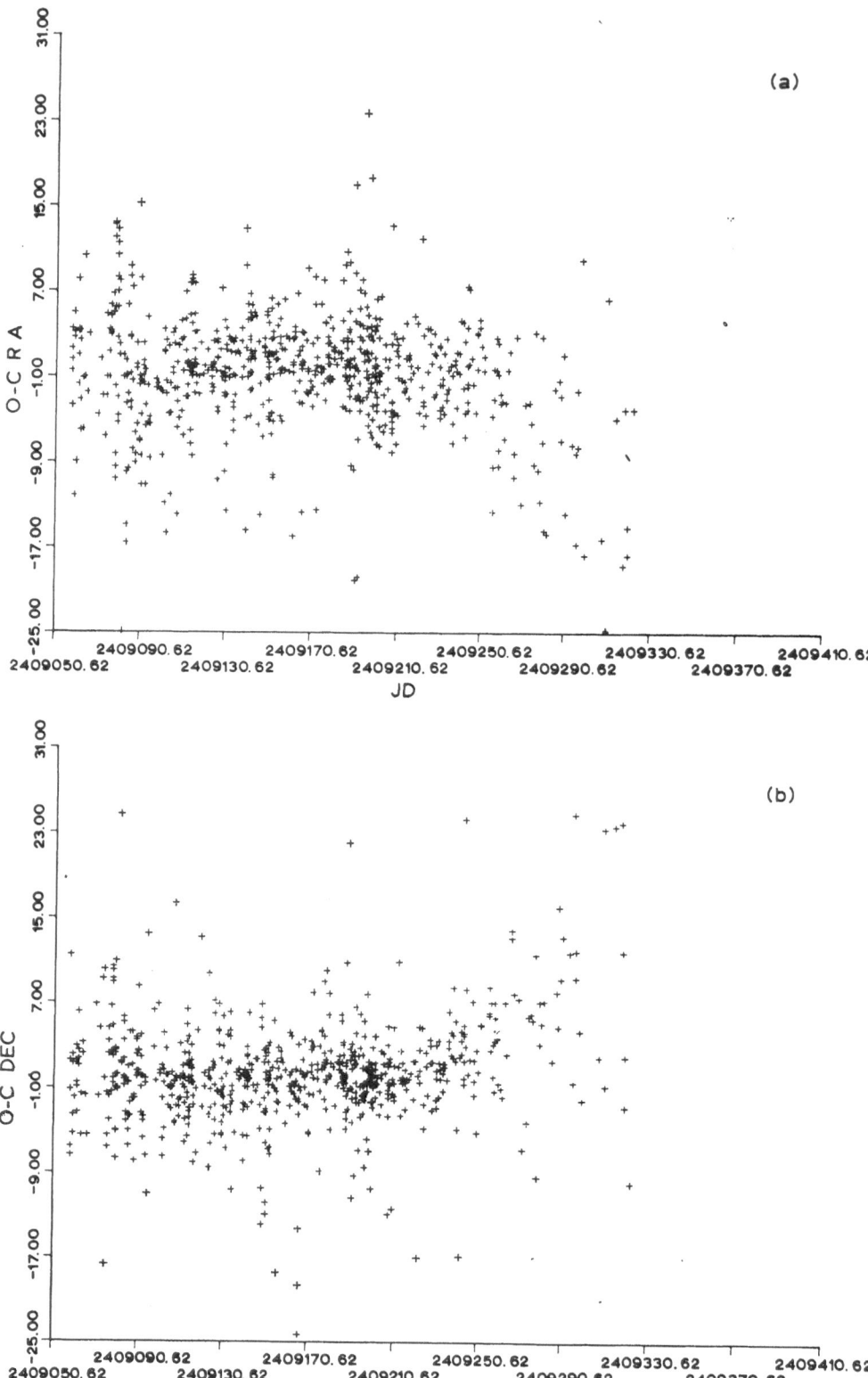

Figs. 2a–b.   (O—C) residuals in 1883–1884; Case B.

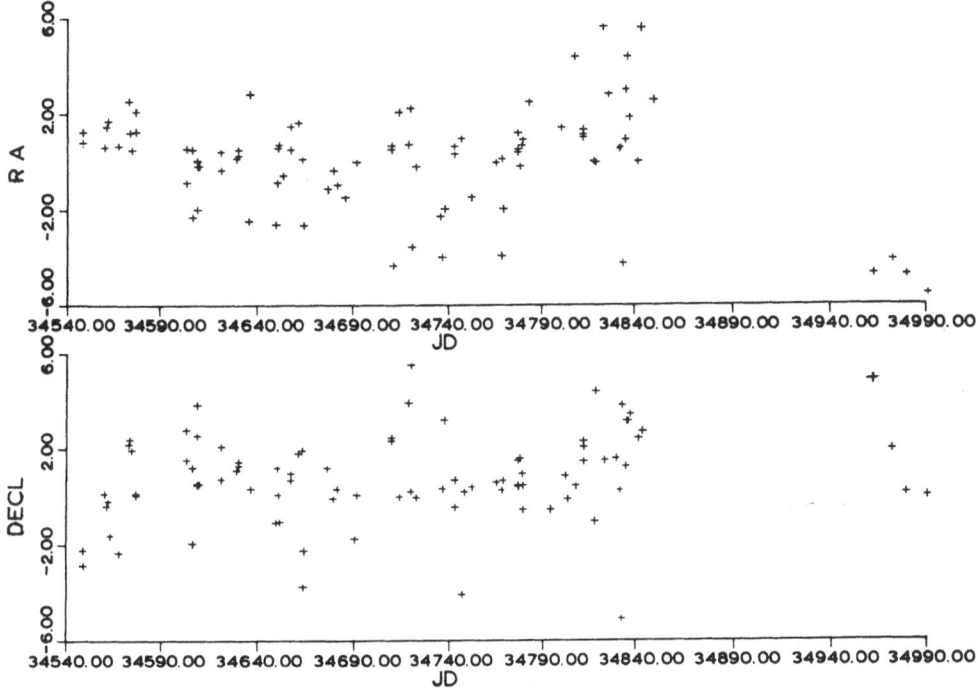

Fig. 3.   (O—C) residuals in 1953–1954.

TABLE II
(O–C) residuals in 1812

1812	$\cos \delta \, \Delta\alpha$	$\Delta\delta$
July 23	− 0″05	− 11″31
Aug. 6	− 4.25	− 9.02
Aug. 22	− 1.17	− 1.17
Sept. 3	+ 3.69	+ 2.43
Sept. 14	+ 3.22	+ 4.10
Sept. 23	+ 1.98	+ 8.58

Reference

Herget, P.: 1939, *Astron. J.* **48**, 105.

37. PERIODIC COMET TEMPEL-TUTTLE AND THE LEONID METEOR SHOWER

E. D. KONDRAT'EVA

Kazan University, Kazan, U.S.S.R.

Abstract. P/Tempel-Tuttle, the comet associated with the Leonid meteor shower, was observed at only two of its last four passages through perihelion, in 1865–1866 and 1965. We have re-reduced the observations in 1865–1866, and with the help of Belyaev's computer programme for numerical integration have linked the two apparitions.

Periodic comet Tempel-Tuttle was discovered on 1865 December 21 and observed up to 1866 February 9. It was a diffuse object of magnitude 12 with a central condensation of diameter 5'. During this interval 118 observations were made, almost the same number before perihelion as after. Oppolzer (1866) computed an orbit but pointed out that the period of revolution obtained was not very certain. Using that orbit Bidschof (1897) calculated a search ephemeris for the comet's 1899 return. The comet was not found then, and neither was it found in 1932. Kanda (1932) calculated the orbit of the comet of 1366, using observations on October 25, 26 and 29, and he suggested that this comet was identical with P/Tempel-Tuttle. Schubart (1965) integrated the equations of motion of P/Tempel-Tuttle, making allowance for the perturbations by the outer planets, and assuming various values for the period. The calculations were made for 15 revolutions, back from 1866 to 1366. Despite the inaccuracy of the ancient observations, but with the help of what appeared to be a single observation of the comet in 1699, he could choose an appropriate value for the period of revolution and derive a prediction for 1965. He then found diffuse, sixteenth magnitude images of the comet on plates obtained on 1965 June 30 and July 1 by M. J. Bester with 25-cm astrograph of the Boyden Observatory. Only four observations were obtained at this apparition, and they agree well with Schubart's prediction, provided that the latter is corrected by $\Delta T = +5$ days.

We have improved Schubart's elements by means of the observations in 1965. After applying the ΔT correction we corrected the elements three times by the Eckert-Brouwer method, with the result:

$$
\left.
\begin{aligned}
\Omega &= 234°32'49\overset{..}{.}85 \\
\omega &= 172\ 58\ 09.21 \\
i &= 162\ 43\ 28.33
\end{aligned}
\right\} \quad 1950.0
$$

$$
\begin{aligned}
a &= 10.259719 \\
e &= 0.9039816 \\
T &= 1965\ \text{April}\ 30.09581\ \text{ET},
\end{aligned}
$$

and the $(O - C)$ residuals:

1965	$\Delta\alpha \cos \delta$	$\Delta\delta$
June 30.1	$-0\overset{s}{.}03$	$+0\overset{"}{.}15$
30.5	$+0.22$	$+1.01$
July 1.1	-0.16	-4.73
26.5	$+0.13$	$+0.13.$

We then integrated these elements back to 1866, using the computer programme kindly supplied by Belyaev, which takes into account the perturbations by Venus to Pluto. The result was:

$$\text{Epoch} = 1866 \text{ February } 8.0 \text{ UT}$$
$$M_0 = \quad 0°36'54\overset{"}{.}31$$
$$\Omega = 232°39'02.27$$
$$\omega = 171 \ 16 \ 24.78 \quad 1950.0$$
$$i = 162 \ 42 \ 26.60$$
$$a = 10.394061$$
$$e = \ 0.9056516.$$

The 118 observations of 1865–1866 have been re-reduced, the positions of the reference stars being taken from the Yale zone catalogues or AGK_2 and corrected to the FK_4 system. Oppolzer's seven normal places were then adjusted, and the following elements were obtained by differential correction of the above elements:

$$\text{Epoch} = 1866 \text{ January } 11.0 \text{ UT}$$
$$M_0 = -0°01'06\overset{"}{.}68718$$
$$\Omega = 232°34'58.50$$
$$\omega = 170 \ 55 \ 56.28 \quad 1950.0$$
$$i = 162 \ 41 \ 13.51$$
$$a = 10.356750$$
$$e = \ 0.9057112;$$

these elements represent the normal places as follows:

1865/66	$\Delta\alpha \cos \delta$	$\Delta\delta$
Dec. 22.0	$+0\overset{s}{.}61$	$+4\overset{"}{.}07$
27.5	$+0.38$	$+4.72$
Jan. 4.5	-1.24	-2.63
9.5	-1.12	-0.31
15.5	-0.57	-6.72
22.5	-1.14	-4.14
Feb. 5.5	$+0.35$	$+1.47.$

The new elements were then reintegrated from 1866 to 1965. It was necessary to make a further correction to the mean anomaly, and a final differential correction gave the result:

$$
\begin{aligned}
\text{Epoch} &= \text{1866 January 11.0 UT} & \text{1965 April 30.0 UT}\\
M_0 &= -67\overset{''}{.}55842 & -10\overset{''}{.}38647\\
\Omega &= 232°34'38\overset{''}{.}13 & 234°33'07\overset{''}{.}19\\
\omega &= 170\ 56\ 18.65 \quad\Big\}\ 1950.0 & 172\ 59\ 09.19 \quad\Big\}\ 1950.0\\
i &= 162\ 41\ 48.33 & 162\ 43\ 40.33\\
a &= 10.356748 & 10.259808\\
e &= 0.9057159 & 0.9039863.
\end{aligned}
$$

The final residuals are:

1865/66	$\Delta\alpha\cos\delta$	$\Delta\delta$
Dec. 22.0	$-0\overset{s}{.}13$	$+0\overset{''}{.}38$
27.5	$+0.70$	-1.25
Jan. 4.5	-0.10	$+3.08$
9.5	$+0.11$	$+2.46$
15.5	$+0.12$	-3.51
22.5	-0.12	-3.09
Feb. 5.5	$+0.16$	-6.55
1965		
June 30.1	-0.30	$+3.45$
30.5	-0.29	$+3.99$
July 1.1	$+0.63$	-0.70
26.5	$+0.08$	$+0.07.$

From the exact orbital elements of P/Tempel-Tuttle we can determine by Dubyago's method the circumstances of the Earth's interception of the Leonid meteor stream. For 1966 the time of maximum was calculated to be November $17^d 10^h 42^m$ UT, which is in good agreement with the observations.

References

Bidschof, F.: 1897, *Astron. Nachr.* **144**, 299.
Kanda, S.: 1932, *Astron. Herald* **25**, 187.
Oppolzer, T.: 1866, *Astron. Nachr.* **68**, 241.
Schubart, J.: 1965, *IAU Circ.* No. 1907.

38. INVESTIGATION OF THE MOTION OF PERIODIC COMET STEPHAN-OTERMA

M. YA. SHMAKOVA

Institute for Theoretical Astronomy, Leningrad, U.S.S.R.

Abstract. The orbit of P/Stephan-Oterma has been determined using 128 observations at the two apparitions (1867 and 1942) and taking into account perturbations by Venus to Neptune. The evolution of the comet's orbit has been studied over the interval 1660–2060.

The short-period comet Stephan-Oterma, with a revolution period of about 40 years, has been observed at two apparitions, in 1867 and 1942. The comet was discovered by Stephan on 1867 January 25 at Marseilles. Orbital elements were obtained by Becker (1891) from 65 post-perihelion observations and taking into account perturbations by Venus to Jupiter by Encke's method. The maximum residual was slightly larger than 1″.

The comet was not observed at its next return to perihelion ($T = 1904$ May 23.931). It was rediscovered accidentally at the end of 1942 by Oterma, Whipple and Tevzadze. Dubyago (1943) determined the orbit from seven post-perihelion observations.

The aim of the present investigation has been to obtain a fairly reliable orbit from observations at the two apparitions. We first determined an orbit from the three observations 1942 Nov. 28, 1943 Jan. 6 and Feb. 8. This was then corrected using 74 observations in 1942–1943 and allowing for the perturbations by Venus to Neptune, with the result:

$$
\begin{aligned}
&\text{Epoch} = \text{1942 Oct. 18.0} \\
&T = \text{1942 Dec. 19.09137} & \varphi &= 59°41888 \\
&M_0 = 358°42444 & n &= 91″34666 \\
&\omega = 358.27211 & q &= 1.59529 \text{ AU} \\
&\Omega = 78.60846 \quad \Big\} \ 1950.0 & Q &= 21.34361 \text{ AU} \\
&i = 17.90115 & P &= 38.84317 \text{ yr.}
\end{aligned}
$$

The mean error is 1″4, with individual residuals rising to 3″5. We can see that the comet is a member of the Uranus family ($Q = 21.3$ AU).

Belyaev's programme was then used to integrate the comet's motion (by Cowell's method) from 1942 back to 1867 on the BESM-4 computer, and again taking into account perturbations by Venus to Neptune. This first integration was made using a variable step (5–40 days). The 1867 observations were precessed to equinox 1950.0, most of the reference stars being reduced to the FK_4 system. The (O – C) residuals in 1867 were in some cases as high as 2°4. As an experiment, the osculating elements at two epochs, 1942 October 18 and 1867 October 6, were improved separately, using the Eckert-Brouwer method and the observations at both apparitions. From the

Chebotarev et al. (eds.), *The Motion, Evolution of Orbits, and Origin of Comets*, 203–205. *All Rights Reserved.*

results obtained one can see that it is more efficient to take for the improvement the elements for an epoch during the apparition which gives the largest residuals (in this case 1867). Since the 1867 residuals are so large the corresponding differential co-efficients must be calculated with great accuracy. After the first improvement the 1867 residuals decreased by a factor of 30, but after integration forward the 1942–1943 residuals were found to be 3 orders higher. Three successive improvements were carried out in this manner, and the maximum residual decreased to $0\overset{''}{.}7$.

In order to converge on a satisfactory result more rapidly, we considered next the effect of a change in the comet's mean motion. We applied a certain increment $(\Delta n = 0\overset{\circ}{.}000003 = 0\overset{''}{.}011)$ to the osculating mean daily motion (n_0) in 1867. This had a negligible effect on the 1867 residuals (which were less than 2″). Integration of this new orbit forward gave quite different residuals in 1942–1943, and by comparison with the residuals for the initial orbit we established that a correction $\Delta n_1 = +0\overset{\circ}{.}00000032$ was required. Numerical integration with a new value of the mean daily motion $(n_1 = n_0 + \Delta n)$ yielded 1942–1943 residuals of only 6″. A further calculation of the same type gave a second correction $\Delta n_2 = -0\overset{\circ}{.}000000014$.

We were thus able to obtain an orbit linking the two apparitions and considering a total of 128 observations (35 in 1867 and 93 in 1942–1943). The maximum residuals (in right ascension) do not exceed 1″ in 1942–1943 and reach 2″ in 1867. The final elements were:

$$
\begin{array}{ll}
\text{Epoch} = \text{1942 Oct. 18.0} & \\
T = \text{1942 Dec. 19.07104} & \varphi = 59\overset{\circ}{.}41721 \\
M_0 = 358\overset{\circ}{.}42478 & n = 91\overset{''}{.}35711 \\
\omega = 358.26892 & q = \ \ 1.59534 \text{ AU} \\
\Omega = \ \ 78.59958 \ \Big\} \ 1950.0 \quad & Q = 21.34181 \text{ AU} \\
i = \ \ 17.89615 & P = \ \ 38.83872 \text{ yr.}
\end{array}
$$

These elements are not greatly different from those obtained from the observations in 1942–1943 alone.

It is of interest to integrate the above orbit over the 400-yr interval 1660–2060 in order to find approaches to the major planets and generally to investigate the evolution of the comet's orbit. During this period P/Stephan-Oterma had approaches to the Earth, Jupiter, Saturn, and Uranus; they are listed in Table I. At its next return $(T = 1980$ Dec. 5.165) the comet will have rather a close approach to the Earth, minimum distance occurring 4 days after perihelion passage. As a result, conditions for observing the comet will be favourable.

The elements of the orbit have changed rather insignificantly during the 400-yr interval, the values at the beginning and end of the interval being given in Table II. Inside the interval the largest changes were as follows:

$$
\begin{array}{l}
36.9 \leq P \leq 38.9 \\
91\overset{''}{.}4 \leq n \leq 96.1 \\
20.6 \leq Q \leq 21.3.
\end{array}
$$

TABLE I

Minimum distances (in AU) between P/Stephan-Oterma
and the major planets

	Earth	Jupiter	Saturn	Uranus
1720 May 12	—	1.954	—	—
1731 Dec. 15	—	—	—	1.270
1819 Aug. 30	—	—	—	2.171
1866 Nov. 6	0.919	—	—	—
1901 Apr. 14	—	—	2.848	—
1903 Jan. 17	—	1.661	—	—
1942 Dec. 7	0.631	—	—	—
1980 Dec. 9	0.593	—	—	—
1984 June 1	—	—	1.424	—

TABLE II

Approximate elements at beginning and end of integration interval

Date	ω	Ω	i	e	P (yr)	q (AU)	Q (AU)	n
1660	357°	81°	18°	0.858	38.07	1.60	21.03	93″19
2060	0	77	18	0.860	38.42	1.59	21.18	92.36

All the calculations were made on the BESM-4 computer using programmes prepared by the staff of the Institute for Theoretical Astronomy: N. A. Belyaev, N. A. Bokhan, S. G. Makover and the author.

References

Becker, L.: 1891, *Monthly Notices Roy. Astron. Soc.* **51**, 475.
Dubyago, A. D.: 1943, *Astron. Tsirk.* No. 17.

B. DETERMINATION OF PLANETARY MASSES

39. DETERMINATION OF PLANETARY MASSES FROM THE MOTIONS OF COMETS

W. J. KLEPCZYNSKI

U.S. Naval Observatory, Washington, D.C., U.S.A.

Abstract. A brief survey is given of past determinations of the masses of the principal planets from analyses of the motions of comets. Some numerical experiments using comets which have close approaches to Jupiter are made. As a result of these experiments, it is concluded that the conventional least squares solution for the correction to the mass of Jupiter is inadequate for comets which have a close approach to Jupiter. It is further concluded that perhaps, in some cases, the apparent presence of nongravitational forces is merely a manifestation of the failure of the conventional orbit correction process to adjust correctly the orbits of objects which undergo very large perturbations, and it also may be a consequence of errors in the adopted planetary masses. It is suggested that the use of partial derivatives obtained through the numerical integration of the variational equations may overcome the difficulties.

1. Introduction

The determination of the mass of a planet is based on (1) an analysis of the motion of a satellite, if it possesses one; (2) an analysis of the secular or periodic perturbations it produces in the motion of another celestial object or; (3) an analysis of the large perturbations induced in the trajectory of an object during an especially close approach to the body whose mass is being sought.

The last case offers many advantages over the other methods, the primary ones being that the observational history of the object whose motion is being studied does not have to be as long as in the other methods, and that the magnitude of this type of perturbation is considerably larger than in the second method.

Comets should be extremely useful objects for determining the masses of the planets because they can come closer to the major planets than any other class of celestial objects. In the case of Jupiter, close approaches by comets are relatively common occurrences, and a careful analysis of their motions would be very useful in improving the knowledge of the mass of this great planet. However, this has not been the case. There have not been very many determinations of planetary masses based on analyses of the motions of comets, and those that have been made give widely varying results.

This investigation was undertaken with the hope of finding a plausible explanation for the disparate results among the various investigations and to suggest possible avenues for future investigations.

2. Mass Determinations from Cometary Motion

Since the late nineteenth century there have been fourteen investigations which have attempted to determine the mass of a planet through perturbations induced in the

Chebotarev et al. (eds.), The Motion, Evolution of Orbits, and Origin of Comets, 209–226. All Rights Reserved.
Copyright © 1972 by the IAU.

W. J. KLEPCZYNSKI

TABLE I

Planetary mass determinations based on analysis of cometary motion

Planet	Reciprocal mass and mean error	Reference	Comet
Mercury	5 669 700 ± 600 000	Haerdtl (1889a)	Encke, 1871–1885
	5 648 600 ± 2 000	Haerdtl (1889a)	Encke, 1819–1868
	5 012 842 ± 697 863	Haerdtl (1889a)	Pons-Winnecke, 1858–1886
	9 697 000	Backlund (1894)	Encke, 1819–1858
	9 745 000	Backlund (1894)	Encke, 1871–1891
	6 280 000 ± 350 000	Makover (1956)	Encke, 1937–1954
	5 980 000 ± 170 000	Makover and Bokhan (1961)	Encke, 1897–1954
Earth-Moon	329 097	Haerdtl (1889b)	Pons-Winnecke, 1858–1886
Jupiter	1 047.788 ± 0.408	Möller (1872)	Faye, 1843–1866
	1 050.478	Haerdtl (1889a)	Encke, 1819–1868
	1 047.175 ± 0.021	Haerdtl (1889a)	Pons-Winnecke, 1858–1886
	1 050.99 ± 0.98	Rasmusen (1967)	Halley, 1759–1911
	1 050.93 ± 0.33	Rasmusen (1967)	Olbers, 1815–1956
Saturn	3 497.6 ± 0.3	Herget (1970)	Schwassmann-Wachmann 1, 1927–1965

motions of comets; see Table I. The determinations primarily analyze large perturbations produced in the motion of the comet caused by a close approach to the planet. An inspection of Table I shows that the results of these investigations are not in agreement with the values determined using other techniques and objects, except for the investigations of Herget (1970) and of Makover and Bokhan (1961).

Attempts have been made to attribute these discrepancies and other deficiencies noted in studies of the motions of comets to errors of observation, inaccurate numerical integrations of the orbits, nongravitational forces, or other unknown cause. True, there are many observational problems associated with the determination of the position of a comet. However, in the case of a close approach to a planet, the effects we are looking for should be an order of magnitude greater than the errors of observation. Furthermore, current procedures of astrometry have greatly reduced the errors associated with cometary positions. The use of electronic computers has all but eliminated the problems commonly associated with the numerical integrations of orbits. The existence of nongravitational forces is generally accepted (Marsden, 1968, 1969, 1970); however, their magnitude and effect on the observed motions of comets have been questioned (Roemer, 1961).

There are two error sources which, I believe, may affect the results of analyses made thus far: first, the system of planetary masses used in the investigations may be in error; second, the approximations used in the differential correction process may not be sufficiently accurate.

3. Comments on Adopted Masses and the Orbit Correction Process

It is an accepted fact that the currently adopted IAU set of planetary masses is not necessarily the best set. Roemer (1961) has pointed out that this might explain some of the irregularities observed in the motions of comets. Since we normally use the technique of successive approximations in our analysis of the motion of celestial objects, inaccurate initial conditions in our numerical integrations and in the differential correction process can drastically affect our results. Marsden (1972) achieved a significant improvement in the orbit of the minor planet 944 Hidalgo by including in his analysis a solution for a correction to the mass of Saturn.

The equations we use to make a differential correction of the initial values of the osculating orbital elements and other constants which affect the motion of a celestial object are also subject to some approximations. The partial derivatives which form the coefficients of the unknowns used in our equations of condition can be formed in several ways. Numerically integrating the variational equations is considered the best method to use in order to obtain accurate partial derivatives. In fact, this is the procedure used by Herget (1970) in his study of the motion of P/Schwassmann-Wachmann 1. Another method of obtaining accurate partial derivatives makes use of a general theory of the motion of a comet. In this case, the only limitation on the accuracy of the partial derivatives is the accuracy of the general theory.

The two approaches just mentioned require extensive computation, which is not practical for a large number of investigators. As a result other methods have been introduced to obtain the partial derivatives. The technique of Eckert and Brouwer (1937) is one approximate method for evaluating the partial derivatives which has gained wide acceptance. Its application implicitly assumes that the real orbit of the object does not differ significantly from the osculating orbit. Another technique frequently used to evaluate a partial derivative is to difference two numerical integrations which use two different values for the unknown in question and identical values for all other parameters. The major drawback of this method is that we do not know for how large a difference of the variable in question this procedure is valid.

In this investigation use is made of the last two techniques: the partial derivatives for the coefficients of the corrections to the orbital elements are obtained using the method of Eckert and Brouwer (1937), while the partial derivative for the coefficient of the correction to the disturbing mass is obtained by differencing two numerical integrations which utilize the same osculating elements but two different values for the mass of Jupiter.

4. Numerical Experiments

In order to determine the effect of an error in the adopted mass of a disturbing body and to test the adequacy of the differential correction procedure just described, the following numerical experiment was performed.

Six comets were selected which had close approaches to Jupiter. They were P/Brooks 2, P/Grigg-Skjellerup, P/Kopff, P/Oterma, P/Pons-Winnecke, and P/Wolf. Elements for these six comets were given by Porter (1961). The dates of perihelion passage were rounded to the nearest half day for computational convenience. Therefore, the elements used in this study are only a reasonable approximation to those of the real objects. For this reason, Table II lists only approximate osculating elements of these objects for the epoch of the integration, JED 2434000.5.

TABLE II

Approximate osculating elements of comets used in this investigation
(epoch JED 2434000.5)

	Brooks 2	Grigg-Skjellerup	Kopff	Oterma	Pons-Winnecke	Wolf
l_0	275°	344°	10°	65°	16°	49°
ω	196°	356°	32°	355°	170°	161°
Ω	178°	215°	253°	155°	94°	204°
i	6°	18°	7°	4°	22°	27°
e	0.487	0.704	0.556	0.143	0.653	0.396
a	3.638	2.887	3.369	3.971	3.347	4.134
n	511	723	574	448	579	422
P	6.938	4.905	6.183	7.912	6.124	8.405
q	1.867	0.856	1.495	3.404	1.161	2.498
Q	5.408	4.918	5.242	4.538	5.534	5.770

l_0 = mean anomaly
ω = argument of perihelion
Ω = longitude of ascending node
i = inclination
e = eccentricity
a = semimajor axis in AU
n = mean motion in seconds of arc per day
P = period in years
q = perihelion distance in AU
Q = aphelion distance in AU

Using the numerical integration program by Schubart and Stumpff (1966), a simultaneous integration of the equations of motion of the planets Venus through Pluto and the six comets was performed at a half-day interval, backward from the epoch of integration for a period of 60 yr. This integration used the IAU values for the planetary masses, the reciprocal mass of Jupiter in this set being 1047.355. The heliocentric rectangular coordinates which resulted from this integration were transformed into heliocentric longitude and latitude. This set of coordinates became the standard of comparison for all subsequent studies and is referred to as our *standard set* of observations.

In order to see the character of the orbits of the six comets used in this investigation, the paths of these objects were plotted in a rotating frame of reference with Jupiter fixed on the abscissa at unit distance and the frame rotating with the same rate as Jupiter (Figure 1). The comet with the minimum distance from Jupiter at the time of closest approach is P/Brooks 2. In increasing minimum distance we next have P/Wolf, followed by P/Oterma, P/Grigg-Skjellerup, P/Pons-Winnecke, and finally P/Kopff.

COMET P/BROOKS 2

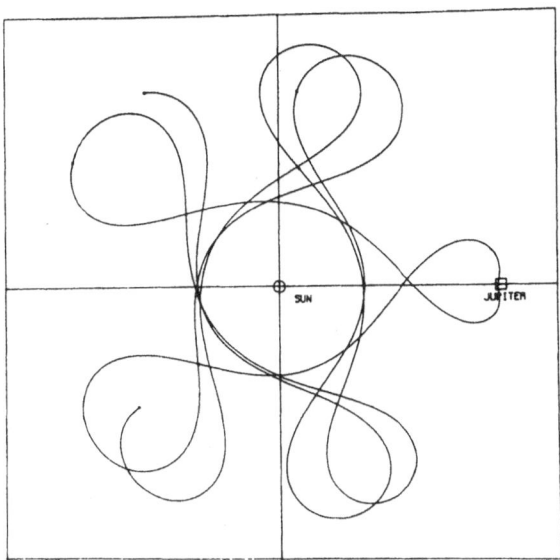

Fig. 1a.

COMET P/GRIGG-SKJELLERUP

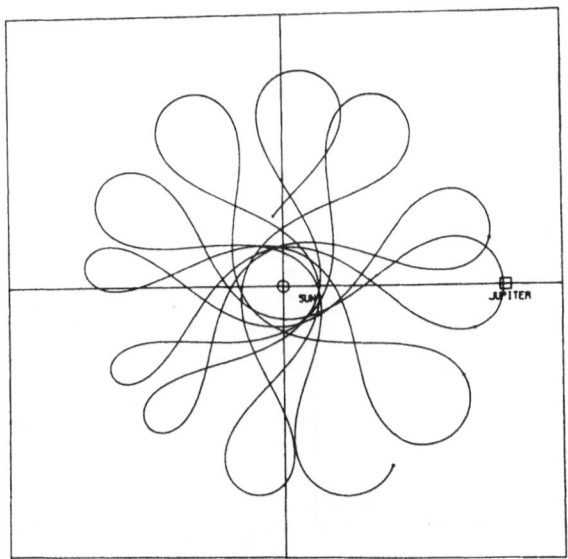

Fig. 1b.

COMET P/KOPFF

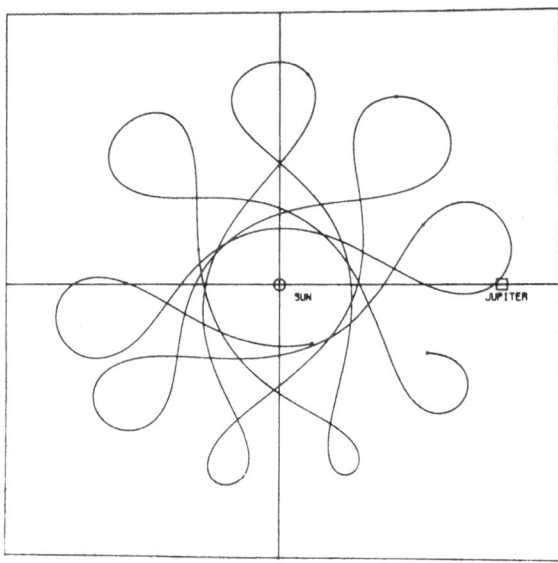

Fig. 1c.

COMET P/OTERMA

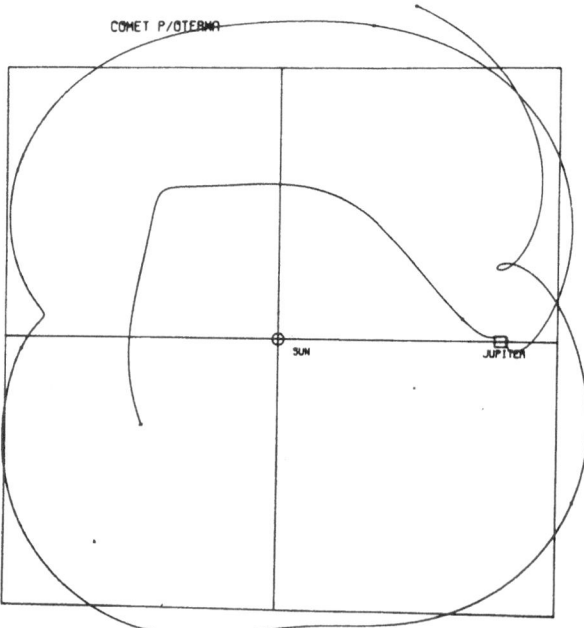

Fig. 1d.

COMET P/PONS-WINNECKE

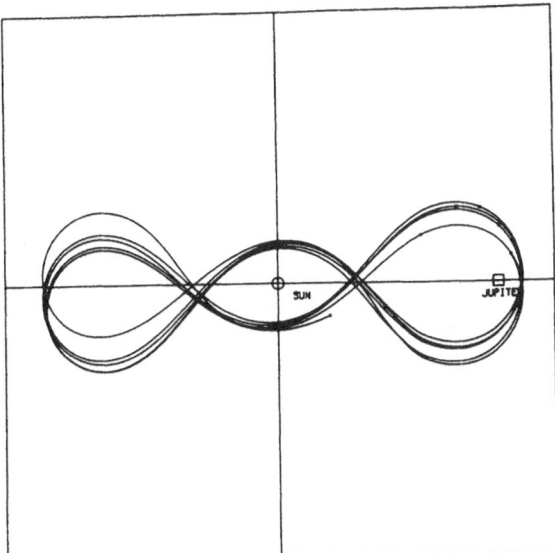

Fig. 1e.

COMET P/WOLF

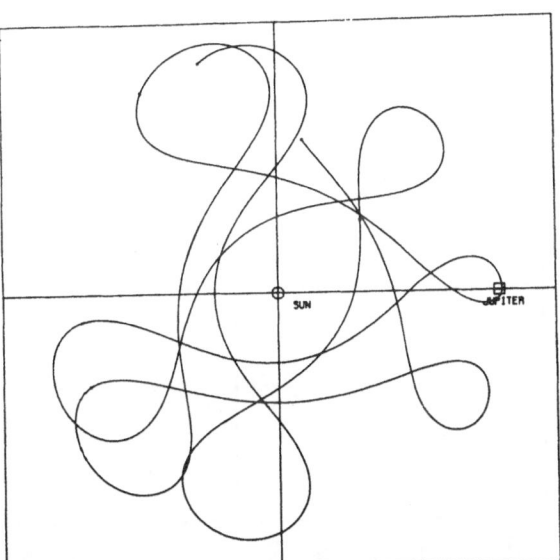

Fig. 1f.

Figs. 1a–f. Orbits of comets used in this investigation plotted in rotating frame of reference. Jupiter is at unit distance and the frame is rotating with the actual velocity of Jupiter.

Next, keeping all other parameters fixed, the disturbing mass of Jupiter was changed to 1047.390, and a new integration was performed. This new integration is sometimes referred to as the 'unfitted' integration. This new value of Jupiter's mass corresponds to a change of 0.035 or 0.003% in the reciprocal mass. The differences in heliocentric longitude for each comet between this integration and the standard integration which used 1047.355 as the reciprocal mass of Jupiter are exhibited in Figure 2. Note that there is one vertical line corresponding to the epoch of integration and one at the point of closest approach of each comet to Jupiter.

An attempt was then made to fit the integration which utilized 1047.390 as the disturbing mass of Jupiter to our standard artificial observations. *Two classes* of solutions were made. In the first class of solutions, the observational subset which went into the differential correction program included those observations for each comet from the epoch of the integration to some date beyond the point of closest approach. In the second class of solution, the differential correction was based on the entire span of our standard artificial observations. In both classes, new integrations based on the new elements were made.

In both classes, after new integrations had been formed, *two types* of solutions were made: one involved corrections to the six orbital elements; the other added a seventh unknown to correct the mass of Jupiter. The six-unknown solution was made to test the adequacy of the previous correction. The coefficient of the six unknowns, the partial derivatives, were determined by the method of Eckert and Brouwer. The unknowns actually used are those designated Set III in Brouwer and Clemence (1961).

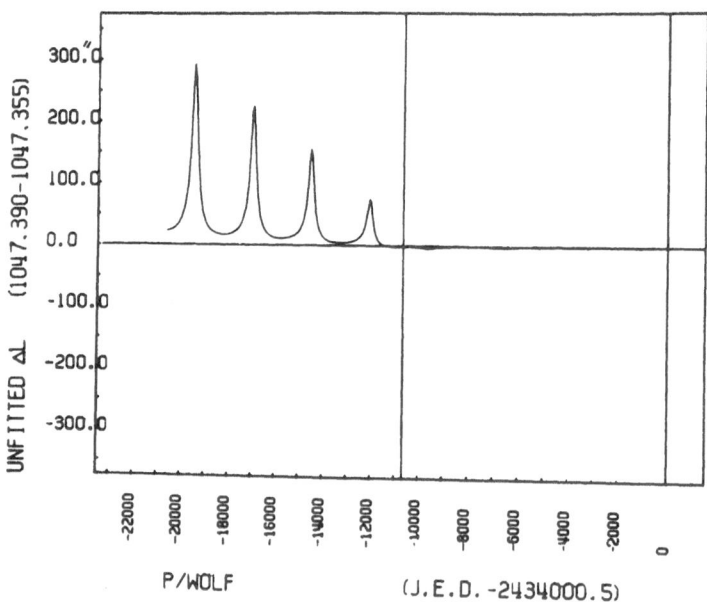

Fig. 2a.

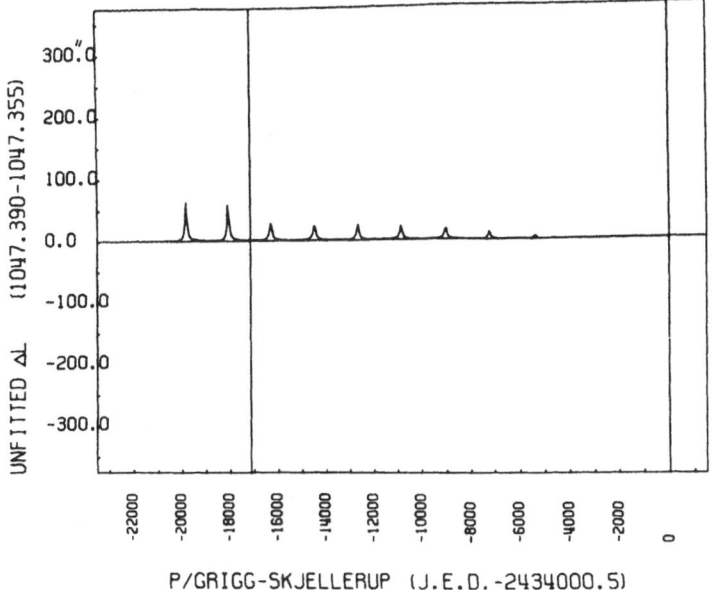

Fig. 2b.

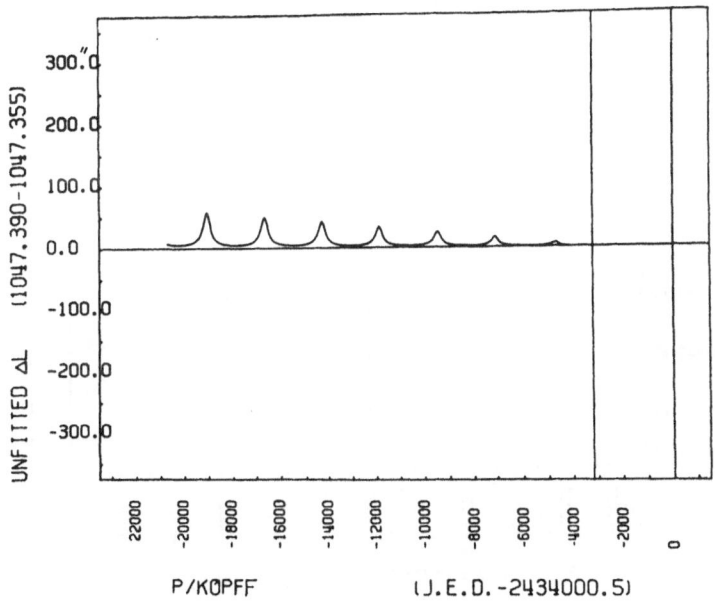

Fig. 2c.

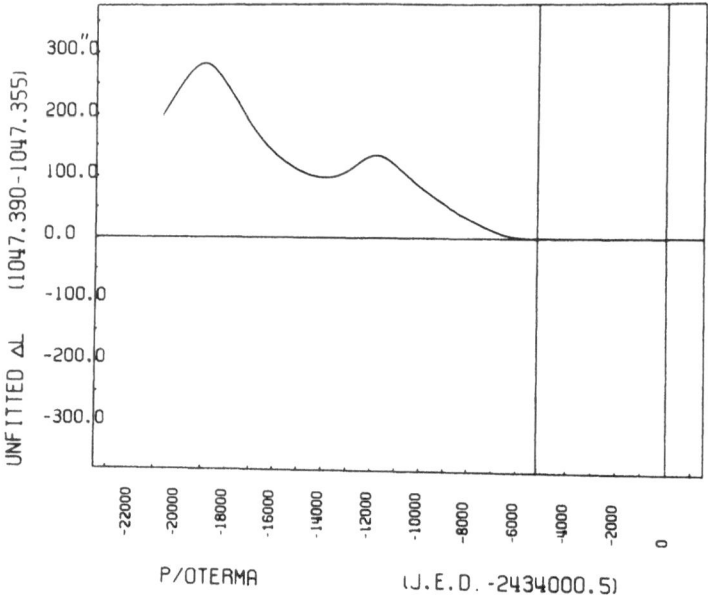

Fig. 2d.

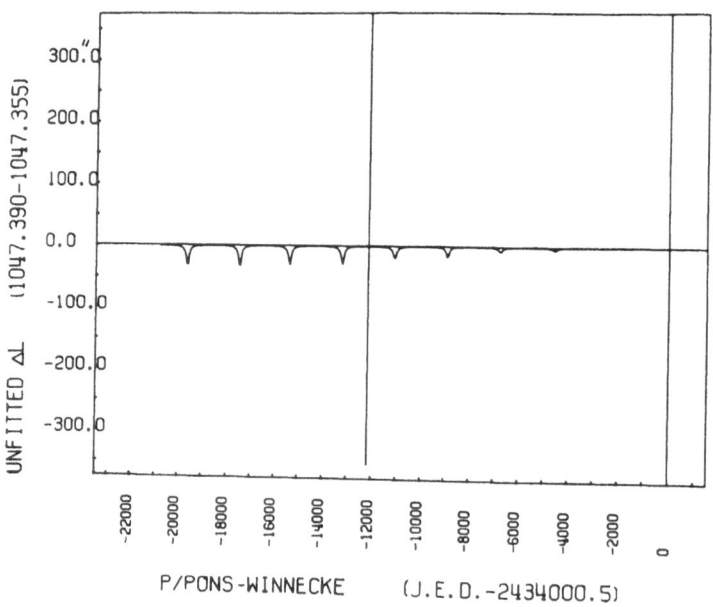

Fig. 2e.

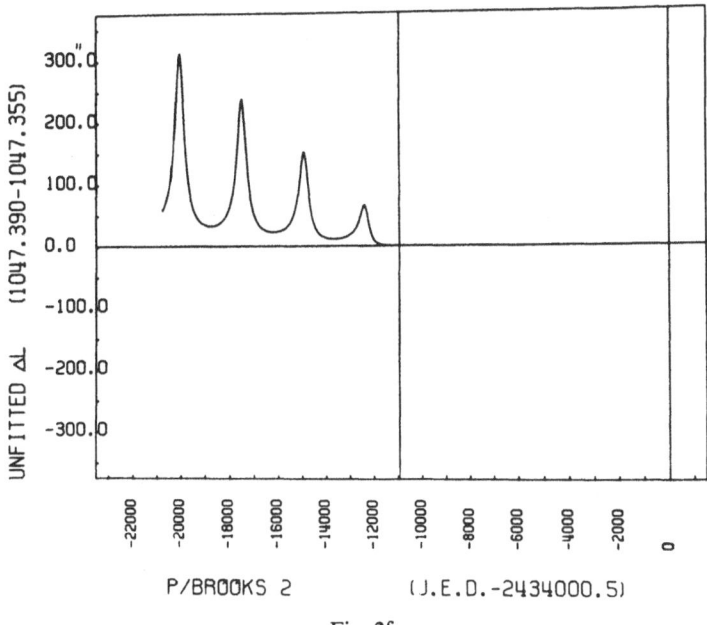

Fig. 2f.

Figs. 2a–f. For the six comets used in this investigation, the differences in heliocentric longitude between the standard numerical integration and one which uses the same osculating elements for each object but 1047.390 for the reciprocal mass of Jupiter are given in seconds of arc.

The coefficient of the seventh unknown in the equations of condition was formed by the differencing techniques already described.

5. Results of Numerical Experiments

The attempts to fit an integration based on 1047.390 for the mass of Jupiter to the standard observations extending from the epoch of the integration to beyond the points of closest approach were extremely successful. Figure 3 is a plot of the differences in orbital longitude for each of the comets between the standard observations and the integration fitted to the close approach data. Since several close approaches of P/Pons-Winnecke had inadvertently been included in the fitting process, not one as had been desired, the results for this comet are not similar to those of the others, and it is omitted from this part of the discussion.

After fitting *the close approach observations* with the six-unknown solutions, a seven-unknown solution was made. For each comet the resulting value for the correction to the mass of Jupiter should be -0.035, yielding a value of 1047.355 for the true mass of Jupiter. The results contained in Table III are listed in a chronological order where time is measured in days after close approach; that is to say, the observations which were used in the differential correction process included close approach plus

W. J. KLEPCZYNSKI

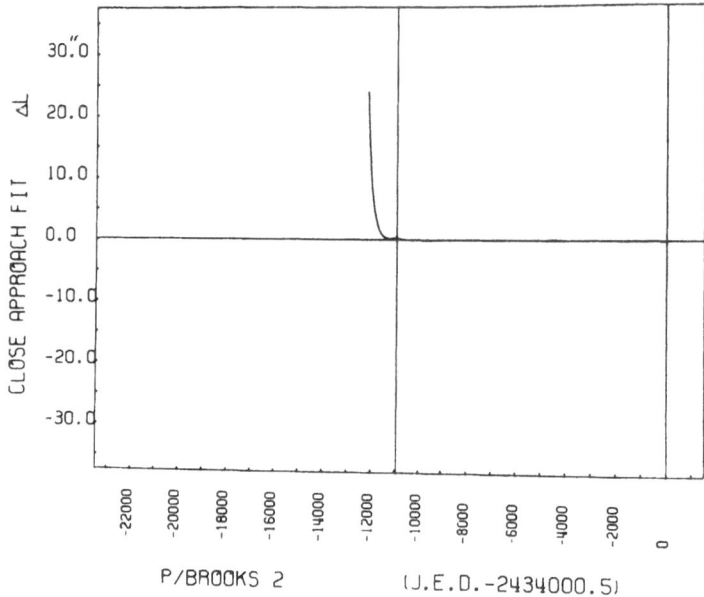

Fig. 3a.

Fig. 3b.

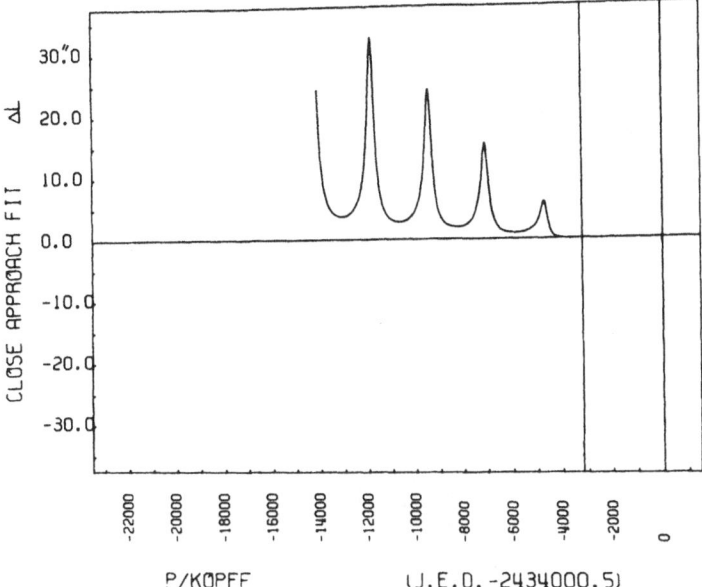

Fig. 3c.

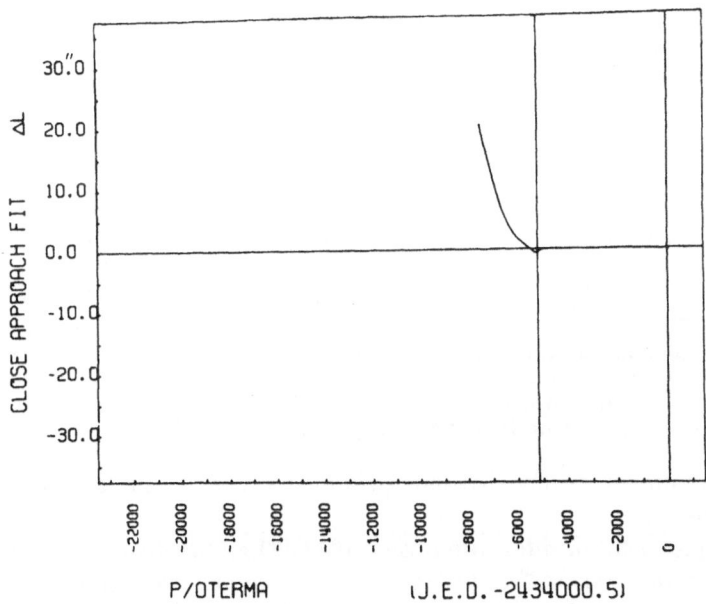

Fig. 3d.

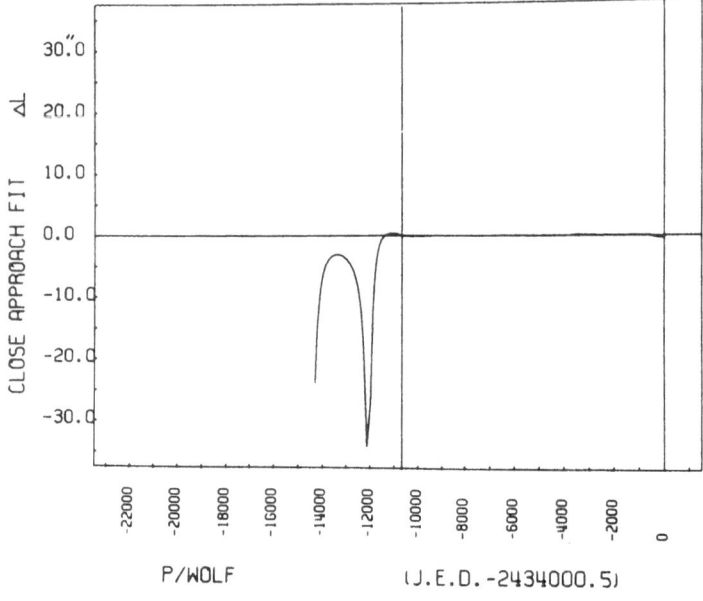

Fig. 3e.

Figs. 3a–e. For five comets used in this investigation, the difference in heliocentric longitude between the standard numerical integration and a numerical integration, which uses 1047.390 for the reciprocal mass of Jupiter and which has been fitted to the close approach observational set, are given in seconds of arc.

TABLE III

Solution for reciprocal mass of Jupiter based on close approach data set

Comet	t	Δm	ρ	T
P/Wolf	25	−0.035	0.119	242 3325
P/Brooks 2	65	−0.034	0.089	242 3065
P/Kopff	93	−0.036	0.577	243 0793
P/Oterma	126	−0.030	0.166	242 8826
P/Grigg-Skjellerup	167	−0.018	0.173	241 6867

t = number of days after close approach
Δm = correction to reciprocal mass of Jupiter
ρ = distance of minimum separation
T = approximate JED of close approach

the indicated number of days beyond. Also included in the table is the distance of minimum separation.

The attempts to fit an integration using 1047.390 as the assumed mass of Jupiter to the *entire period covered by the standard observations* were not very successful. It was only possible to fit three of the comet orbits to the observations. These three

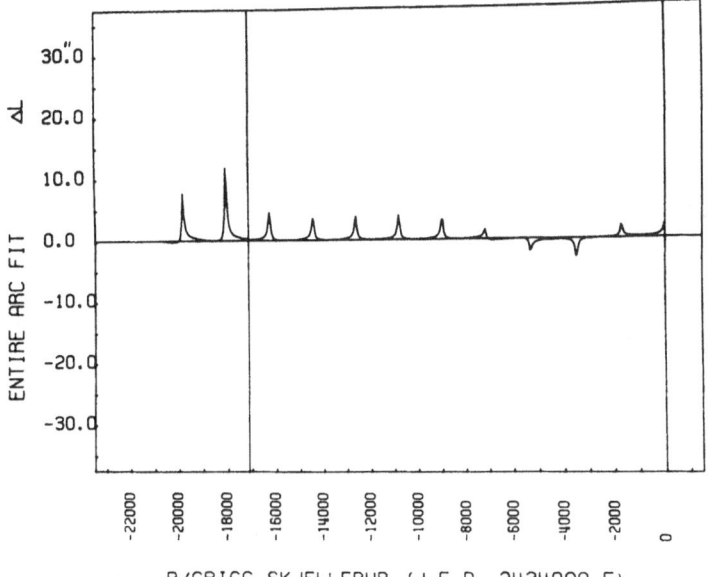

P/GRIGG-SKJELLERUP (J.E.D.-2434000.5)

Fig. 4a.

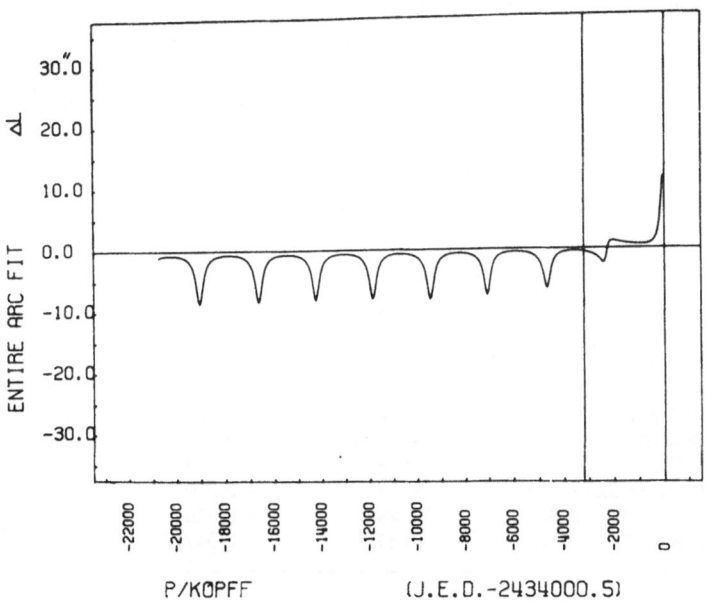

P/KOPFF   (J.E.D.-2434000.5)

Fig. 4b.

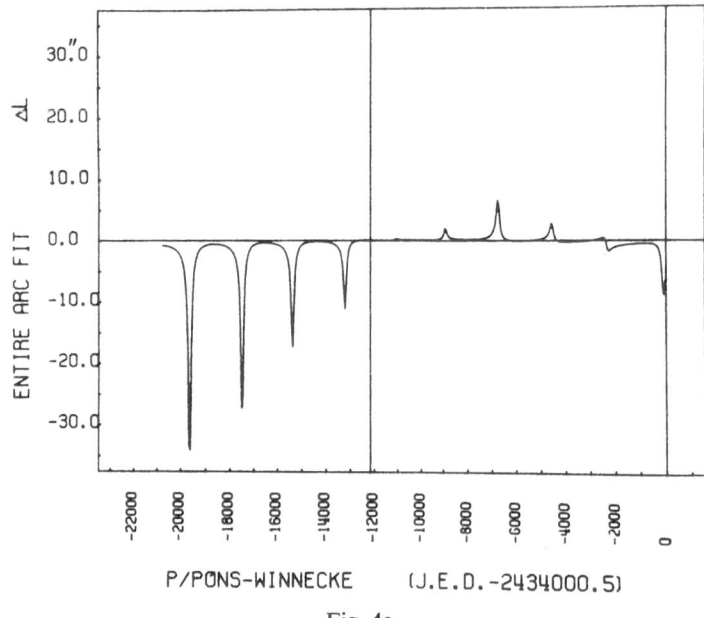

Fig. 4c.

Figs. 4a–c. For three comets used in this investigation, the differences in heliocentric longitude between the standard numerical integration and a numerical integration, which uses 1047.390 for the reciprocal mass of Jupiter and which has been fitted to the entire observational data set, are given in seconds of arc.

were those which had the largest minimum distance from Jupiter. Figure 4 compares the unfitted orbits with those which resulted from the six-unknown fit over the entire period of observations for the comets P/Grigg-Skjellerup, P/Kopff and P/Pons-Winnecke. After adjusting these orbits with six-unknown solutions, a solution for seven unknowns was made. The resulting corrections to the assumed disturbing mass of Jupiter are listed in Table IV. As before, the correct answer should be -0.035. Here the material has been arranged in order of increasing minimum separation from Jupiter.

TABLE IV

Solution for mass of Jupiter based on total data set

Comet	ρ	Δm
P/Brooks 2	0.089	-0.980
P/Wolf	0.119	-0.887
P/Oterma	0.166	-3.642
P/Grigg-Skjellerup	0.173	-0.010
P/Pons-Winnecke	0.494	-0.020
P/Kopff	0.577	-0.047

ρ = distance of minimum separation
Δm = correction to reciprocal mass of Jupiter

6. Discussion

Several important considerations can be deduced from the results as presented. From Table III, we see that the values for the correction to the mass deteriorate, not as a function of minimum distance to Jupiter but as a function of time elapsed from close approach. After a considerable time has elapsed from close approach, then distance to Jupiter plays a significant role.

From Figure 3 we see that an orbit based on an incorrect value for the mass of Jupiter could be forced to fit adequately the standard observations, provided we only used data up to the time around close approach in our fitting process. From our attempt to fit the standard artificial observations over the entire period they covered, we see that the ability to fit was a function of minimum separation at the time of close approach of the comet to Jupiter.

We can now draw the general conclusion that the six-unknown orbit-correction procedure used in this paper can be used to fit an orbit with an arbitrary mass of Jupiter to observations of comets which undergo a close approach to Jupiter if we do not try to include data far beyond the time of close approach. Furthermore, if we restrict ourselves to using observations not beyond 100 days after close approach, then the seven-unknown solution does correctly determine a correction to the mass of Jupiter and the orbit is meaningful. If we try to adjust an orbit using observations which extend considerably beyond close approach, then depending on the character of the comet orbit we might get seemingly meaningful, but nevertheless incorrect results.

The term 'seemingly meaningful' was purposely chosen. If we were working with real comet observations which are nonuniformly distributed and are imperfectly made, then we might be satisfied that we had adequately represented the observations.

Since we do not know the true mass of Jupiter, we hope that the value we use in practice is a sufficiently accurate representation of the true one. But to what should we attribute our inability to represent adequately the motions of some comets? We see that an incorrect value can lead to seemingly meaningful results in some cases and not in others. Perhaps, in those cases where we attribute deficiencies in our ability to represent the motion of comets to nongravitational forces, we might be able more correctly to attribute the deficiencies to inadequacies of our modeling. With respect to Figure 4, an important observation is to be made. It should be pointed out that the effect of an error in the mass of Jupiter occurs about one-half a revolution after a close approach or at perihelion passage of the comet.

From Figure 2 we can see that small changes in the mass of Jupiter can cause significant perturbations in the motions of those comets which come close to Jupiter. It seems that comets which have a close-approach distance less than 0.17 AU should be extremely sensitive indicators to an error in the mass of Jupiter. On further consideration, it is also possible to conclude that the observational history of such a comet does not have to be too long to get meaningful results. Of the objects studied in this investigation, P/Oterma had the most recent approach to Jupiter. This occurred in 1963

(Marsden, 1970); unfortunately, this comet is now extremely faint. Despite this fact, it appears that a careful analysis of existing observations of this comet could give a reliable determination of the mass of Jupiter.

It was not possible to compare the results of this investigation, which used Eckert-Brouwer partial derivatives for the coefficients of six unknowns, with results obtained by using partial derivatives evaluated by numerically integrating the first-order variational equations. It is hoped that this study can be done sometime in the future.

References

Backlund, O.: 1894, *Bull. Astron.* **11**, 473.

Brouwer, D. and Clemence, G. M.: 1961, *Methods of Celestial Mechanics*, Academic Press, New York and London.

Eckert, W. J. and Brouwer, D.: 1937, *Astron. J.* **46**, 125.

Haerdtl, E.: 1889a, *Denk. Acad. Wiss. Wien Math.-Nat. Kl.* **56**, 151.

Haerdtl, E.: 1889b, *Astron. Nachr.* **120**, 257.

Herget, P.: 1970, private communication.

Makover, S. G.: 1956, *Byull. Inst. Teor. Astron.* **6**, 67.

Makover, S. G. and Bokhan, N. A.: 1961, *Trudy Inst. Teor. Astron.* **8**, 135.

Marsden, B. G.: 1968, *Astron. J.* **73**, 367.

Marsden, B. G.: 1969, *Astron. J.* **74**, 720.

Marsden, B. G.: 1970, *Astron. J.* **75**, 75.

Marsden, B. G.: 1972, this Symposium, p. 239.

Möller, A.: 1872, *Viertelj. Astron. Ges.* **7**, 85.

Porter, J. G.: 1961, *Mem. Br. Astron. Assoc.* **39**, No. 3.

Rasmusen, H. Q.: 1967, *Publ. Mind. Medd. Kbh. Obs.* No. 194.

Roemer, E.: 1961, *Astron. J.* **66**, 368.

Schubart, J. and Stumpff, P.: 1966, *Veroeffentl. Astron. Rechen-Inst. Heidelberg* No. 18.

Discussion

E. I. Kazimirchak-Polonskaya: Since you started with very approximate elements and integrated with a half-day step for 60 yr the accumulation of error should be considerable.

W. J. Klepczynski: I performed integration at two-day intervals, one-day intervals, half-day intervals, and quarter-day intervals, and after 60 yr of integration the coordinates all agree. But I should emphasize that I was merely trying to test the feasibility of the method, not make actual determinations of planetary masses, and for that reason it was not necessary for the elements to agree exactly with those of the real comets.

40. THE DETERMINATION OF JUPITER'S MASS FROM LARGE PERTURBATIONS ON COMETARY ORBITS IN JUPITER'S SPHERE OF ACTION

E. I. KAZIMIRCHAK-POLONSKAYA

Institute for Theoretical Astronomy, Leningrad, U.S.S.R.

Abstract. Necessary and sufficient conditions are formulated for determining the mass of Jupiter from large perturbations induced in cometary orbits in the sphere of action of Jupiter. A procedure for the investigation has been developed and programmed for an electronic computer. Comparison of heliocentric and jovicentric computations shows that the perturbations on P/Wolf could be determined with great accuracy when this comet passed through Jupiter's sphere of action in 1922. The first attempt has been made to determine the mass of Jupiter using this passage and the observations of the comet in 1925. The resulting value for the reciprocal mass is 1047.345.

1. Statement of the Problem

There is increasing interest in determining the masses of major planets from the motions of minor bodies of the solar system.

The study of the passage of a comet through Jupiter's sphere of action is advantageous for determining the mass of Jupiter, because the slightest variation in Jupiter's mass produces an appreciable change in the large perturbations on the cometary orbit and significantly affects the representation of the observations of the comet after the encounter (Kazimirchak-Polonskaya, 1961, p. 19).

We select for our study P/Wolf, because in its revolution 1918–1925 it passed within 0.125 AU of Jupiter. For our initial elements we adopt the most accurate set obtained by Kamieński (1959, Table 3) for the last apparition of the comet before the encounter with Jupiter.

We make use of two values for Jupiter's mass, within the possible range of uncertainty, namely: $1/m_1 = 1047.325$, $1/m_2 = 1047.400$ (de Sitter's value). For these values we have performed two identical integrations of the comet's equations of motion, with allowance for the perturbations by the planets Venus to Uranus, and the nongravitational secular deceleration. Accordingly, we have obtained two representations of the comet's normal places in 1925. These are given in Table I, and the fourth and seventh columns show that there exist significant differences in $\Delta\alpha \cos \delta$ and $\Delta\delta$ derived from the two values.

Thus, the formulation of the following problem is justified: to perform a series of integrations of the comet's equations of motion based on the same initial set of elements but varying the mass of Jupiter within definite limits, to obtain from each computation a representation of the comet's normal places after the encounter with Jupiter, and to choose the one that gives the least value for the mean error of one normal place; this representation will correspond to the most probable value for the mass of Jupiter.

Chebotarev et al. (eds.), The Motion, Evolution of Orbits, and Origin of Comets, 227–232. All Rights Reserved.

TABLE I

Representation of normal places of P/Wolf in 1925 for the two values for the mass of Jupiter

1925 UT	$(\Delta\alpha\cos\delta)_1$	$(\Delta\alpha\cos\delta)_2$	$(\Delta\alpha\cos\delta)_1-$ $(\Delta\alpha\cos\delta)_2$	$(\Delta\delta)_1$	$(\Delta\delta)_2$	$(\Delta\delta)_1-(\Delta\delta)_2$
July 18.0	-7.3	-14.9	$+7.6$	$+5.5$	$+3.2$	$+2.3$
Aug. 19.0	-11.7	-21.8	$+10.1$	$+6.7$	$+3.6$	$+3.1$
Sept. 14.0	-11.1	-22.3	$+11.2$	$+4.6$	$+1.6$	$+3.0$
Oct. 12.0	-10.4	-21.5	$+11.1$	$+0.7$	-0.7	$+1.4$
Nov. 11.0	-4.3	-15.0	$+10.7$	$+0.5$	$+1.2$	-0.7
Dec. 19.0	$+5.4$	-5.2	$+10.6$	-2.2	-0.4	-1.8

2. Conditions for Determining the Mass of Jupiter

Encke (1831a, 1831b) was the first to determine a correction to the mass of Jupiter while improving the orbit of a minor body of the solar system. To do this he introduced into the equations of condition both a correction to the mass of Jupiter and corrections to the orbital elements of the body. This procedure was used afterwards by Möller (1872), Asten (1878), and Haerdtl (1889) in their studies of P/Faye, P/Encke, and P/Pons-Winnecke, and it is nowadays widely used when determining corrections to the values of the masses of the major planets.

We have decided to use a substantially different procedure, the basic principle of which is to formulate necessary and sufficient conditions whereby a single unknown, the mass of Jupiter (or some other planet), is left in the problem. The solution will then be unambiguous and the most probable one.

These conditions may be formulated as follows:

(1) The numerical theory of the comet's motion before the approach to Jupiter should be constructed over a large interval of time and to a high degree of accuracy, with allowance made for the perturbations by all the planets and nongravitational effects.

(2) Sufficiently extensive and accurate observational material should be collected both at the comet's last return before the approach and at its first return afterwards.

(3) At least three apparitions of the comet should be linked after the encounter with Jupiter to the same accuracy as before, because otherwise it will be impossible to allow for the nongravitational effects after the encounter.

(4) The variations in the elements due to nongravitational forces should be as small as possible; thus comets with very small perihelion distances should not be chosen because their nuclei are exposed to strong solar radiation.

(5) The approach to Jupiter should be close enough that the effect of a change in Jupiter's mass can be detected in subsequent observations.

Nevertheless, since Jupiter's oblateness is not known to the necessary accuracy, it is not practicable to make use of extremely close encounters, such as that of P/Brooks 2 to Jupiter in 1886, because a second unknown, the oblateness of Jupiter, is introduced, and this would also affect the motion of the comet and the representation of its observations. This complicates the problem and might lead to fictitious results. It is most advantageous to utilize those encounters where the least separation is between 0.01 and 0.20 AU.

P/Wolf is a very suitable object for solving the problem because its theory of motion and its passage through Jupiter's sphere of action in 1922 satisfy all the stated conditions.

3. Method of Investigation

In addition to the method of integration in special coordinates, programmed in double precision on the BESM-4 electronic computer (Kazimirchak-Polonskaya, 1972), we make use of the following:

(1) A highly accurate procedure for computing large perturbations on cometary orbits in Jupiter's sphere of action in both heliocentric and jovicentric form (Kazimirchak-Polonskaya, 1961, p. 191). Table II compares the application of the two forms to P/Wolf in a five-month interval during the comet's passage through Jupiter's sphere of action. Altogether, the gravitational attractions of the Sun and the planets Venus to Uranus have been taken into account.*

TABLE II

Perturbations in the orbital elements of P/Wolf in the sphere of action of Jupiter 1922 July 8.0 to December 15.0

	Jovicentric	Heliocentric	Jovicentric – Heliocentric
δM	$+20°05'52''50$	$+20°05'52''48$	$+0''02$
$\delta \pi$	$-13\ 25\ 34.87$	$-13\ 25\ 34.89$	$+0.02$
$\delta \Omega$	$-\ 0\ 31\ 36.05$	$-\ 0\ 31\ 36.05$	0.00
δi	$+\ 1\ 23\ 46.18$	$+\ 1\ 23\ 46.17$	$+0.01$
$\delta \varphi$	$-\ 9\ 55\ 16.84$	$-\ 9\ 55\ 16.87$	$+0.03$
$\delta \mu$	$-96''36763$	$-96''36769$	$+0''00006$

(2) Extremely precise coordinates of Jupiter during the comet's passage through the sphere of action. These are obtained by integrating Jupiter's orbit (using a step-size of 2 days, 1 day, or even less, nine decimal places, and considering perturbations by all the planets), rather than by interpolating from the magnetic tape that gives coordinates at 20-day intervals. This is exceedingly important when a comet goes deeply into the sphere of action, say, to within 0.10 AU.

(3) Additional procedures for increasing the accuracy of the determination of Jupiter's mass which we do not have the opportunity to describe here.

* The figures given are from manual calculations because only the heliocentric method has so far been programmed. Computations in double precision will undoubtedly yield still higher accuracy.

4. The First Attempt to Determine the Mass of Jupiter from the Passage of P/Wolf through Jupiter's Sphere of Action in 1922

Since Kamieński allowed for the perturbations only by Venus to Uranus (those by the other planets being almost undetectable to the accuracy he used), we also find it practicable to carry out the calculations over the critical revolution 1918–1925 using the same planets.

We have made several accurate integrations from 1918 to 1925 starting from the same set of elements and changing only the value for the mass of Jupiter within the limits

$$1047.325 \leq 1/m_J \leq 1047.400.$$

Kamieński (1933) took into account secular decelerations $\Delta\mu$ and ΔM along with the planetary perturbations. In addition, we have allowed for nongravitational effects in the orientational elements of the orbit (Kazimirchak-Polonskaya, 1972). The nongravitational effects were considered from the epoch of osculation $T_0 = 1918$ December 16.0 Berlin Mean Time to the time of the comet's approach to Jupiter. Precise positions of the comet, in relation to the Sun and the six planets, were thus obtained while the comet was in Jupiter's sphere of action. We also took the nongravitational effects into account until after the last observation in 1925.

For each integration we made a comparison with all the observations and normal places in 1925. It made little difference whether we used observations or normal places, so we adopted the latter. We have given (Kazimirchak-Polonskaya, 1972, Table I) residuals for the normal places in 1925 as found by Kamieński and Bielicki (1936) and as found by us for Hill's value of the mass of Jupiter. We find that the normal places are best represented by the value

$$1/m_J = 1047.345,$$

and the corresponding residuals are shown in Table III.

TABLE III

Representation of normal places for P/Wolf in 1925 for the most probable value for the mass of Jupiter

1925 UT	$\Delta\alpha \cos \delta$	$\Delta\delta$
July 18.0	$+0\overset{s}{.}03$	$+1\overset{''}{.}0$
Aug. 19.0	-0.21	$+1.6$
Sept. 14.0	-0.20	$+0.7$
Oct. 12.0	-0.20	-0.4
Nov. 11.0	-0.05	$+2.1$
Dec. 19.0	$+0.21$	$+0.6$
m_J^{-1}	1047.345	
Nongravitational effects included	$\Delta\mu, \Delta M, \Delta\Omega, \Delta\pi$	
ϵ	$\pm 3\overset{''}{.}3$	

When we compared our elements for the epoch of osculation 1925 July 12.5 UT with those by Kamieński and Bielicki from linking the three apparitions 1925–1942, we found differences of the order of those obtained earlier (Kazimirchak-Polonskaya, 1972).

Our results are of threefold importance:

(1) The elimination of the discontinuity in the theory of motion of P/Wolf has been more accurately shown.

(2) The new value 1047.345 for the reciprocal mass of Jupiter has a fair degree of confidence.

(3) The method is basically correct.

We intend to continue our studies on the problem in the future using a still more accurate procedure. To this end the theory of the motion of P/Wolf should be constructed to a very high degree of accuracy, with allowance for the perturbations from the nine planets (Mercury-Pluto) and the nongravitational effects in all the elements, for two time intervals 1884–1918 and 1925–1967. Other features that have already been developed, such as allowance for the perturbations by Jupiter's Galilean satellites, should be introduced into the study of the motion of the comet in the vicinity of and inside the sphere of action of Jupiter. To our regret, we have no opportunity to concern ourselves with such matters in this paper.

Acknowledgments

We wish to express our sincerest thanks to N. A. Belyaev, who was kind enough to contribute his programmes for comparing the observations and calculations.

References

Asten, E.: 1878, *Mem. Acad. Imp. Sci. St. Petersb.* **26**, No. 2.
Encke, J.: 1831a, *Math. Abh. Akad. Wiss. Berlin* **1**, 93; **2**, 35.
Encke, J.: 1831b, *Astron. Nachr.* **9**, 317.
Haerdtl, E.: 1889, *Denk. Akad. Wiss. Wien Math.-Nat. Kl.* **55**, 215.
Kamieński, M.: 1933, *Acta Astron. Ser. A* **3**, 1.
Kamieński, M.: 1959, *Acta Astron.* **9**, 53.
Kamieński, M. and Bielicki, M.: 1936, *Repr. Astron. Obs. Warsaw Univ.* **32**, 1.
Kazimirchak-Polonskaya, E. I.: 1961, *Trudy Inst. Teor. Astron.* **7**.
Kazimirchak-Polonskaya, E. I.: 1972, this Symposium, p. 95.
Möller, A.: 1872, *Astron. Nachr.* **79**, 119.

Discussion

V. A. Shor: One of the reasons for your being able to link successfully the apparitions of P/Wolf in 1918 and 1925 was that the coordinates of Jupiter were consistent with its adopted mass. You destroyed this agreement because you did not vary the coordinates when varying the mass.

E. I. Kazimirchak-Polonskaya: The correction I introduced to the mass of Jupiter is negligible, and hence the representation of the observations and correction to the mass would be unchanged if I varied Jupiter's coordinates. When I am able to include the perturbations by all the planets and full nongravitational effects, however, I shall carry out a supplementary investigation to preserve the agreement between the coordinates and mass of Jupiter. Eventually I intend to establish the limits by which the mass may be varied without destroying the theories of the motions of the major planets published in the *Astronomical Papers*, and a procedure will be developed for applying appropriate corrections to Jupiter's coordinates.

M. Bielicki: I can confirm that your new value for the mass of Jupiter is so negligibly different from Hill's value that adoption of the new value will not appreciably affect Jupiter's coordinates and hence your results. When Kamieński and I adopted de Sitter's value there was no effect on the coordinates.

41. DETERMINATION OF THE MASS OF JUPITER FROM OBSERVATIONS OF 10 HYGIEA DURING 1932–1969

N. S. CHERNYKH

Crimean Astrophysical Observatory, Crimea, U.S.S.R.

Abstract. An attempt has been made to derive the mass of Jupiter from modern observations of the minor planet 10 Hygiea. The calculations have been made using computer programmes prepared at the Institute for Theoretical Astronomy. From 250 observations 1932 to 1967 we obtained the reciprocal mass 1047.345 ± 0.040 (rms error). After excluding some less reliable observations and allowing for the elliptic aberration terms we obtained 1047.326 ± 0.033. By means of a more sophisticated computer programme and utilizing 100 additional observations, some of them made in 1969, we obtained the value 1047.324 ± 0.023. Our results are compared with recent determinations by other investigators.

1. Introduction

At present the value adopted for the mass of Jupiter is still that obtained by Newcomb and Hill at the end of the last century. Modern authorities are of the opinion that this value for the reciprocal mass, 1047.355, should be increased by about 0.05 (de Sitter, 1938; Clemence, 1949; Kulikov and Subbotina, 1963). Because of the recent overhaul of the whole system of fundamental astronomical constants redetermination of the mass of Jupiter has become a particularly urgent problem. Several astronomers in the United States are following Hill's (1873) plan of improving Jupiter's mass by means of prolonged series of observations of 13 asteroids with mean motions approximately in 2:1 resonance with Jupiter and thus experiencing large long-period perturbations. For some of these asteroids the investigations have already been completed (O'Handley, 1968; Zielenbach, 1969; Klepczynski, 1969; Fiala, 1969).

As was shown by Newcomb (1895), the mass of Jupiter can be reliably determined using less extensive series of observations of some asteroids. Newcomb's well-known value was derived from a discussion of observations of 33 Polyhymnia over a period of 34 yr.

In this investigation an attempt is made to derive a value for the mass of Jupiter using *modern* observations of a minor planet over approximately the same time-span. Modern observations are preferred since they are substantially more accurate than the older, almost exclusively visual, observations, and we have therefore considered only those made during the last four decades.

We wanted to select an asteroid similar to 33 Polyhymnia, i.e., one that is quite bright, has a fairly large eccentricity and small inclination and approaches to within 1.5 AU of Jupiter. We have examined the circumstances of the approaches of several dozen of the brighter minor planets to Jupiter and have checked on the observations available. As a result we found that 10 Hygiea has obvious advantages and therefore selected this planet. It is the brightest object in Hill's list (never fainter than magnitude

Chebotarev et al. (eds.), The Motion, Evolution of Orbits, and Origin of Comets, 233–238. *All Rights Reserved.*

11 at opposition) and has thus been widely observed. During the last four decades 10 Hygiea has made four approaches to Jupiter:

1932 December	2.16 AU
1942 May	1.73 AU
1952 September	1.55 AU
1963 February	1.79 AU.

These approaches, recurring with a period of about 10.5 yr, bring about considerable perturbations in the planet's orbit: during an interval of 35 yr the daily mean motion changes by 6″, the mean longitude by 2° and the mean anomaly by 12°.

2. Observations

We have collected 410 observations obtained at 29 observatories at 32 oppositions from 1932 to 1969. This number includes 71 observations made by the author at the Crimean Astrophysical Observatory at five oppositions from 1963 to 1969. Before 1950 there was still a substantial proportion of visual observations. Most of them were recalculated using new positions for the reference stars. The positions of the planet had been determined by the observers relative to star positions in the AGK, the Yale zones, or the Carte du Ciel. Using the tables by Orel'skaya (1962), we reduced all the observations to the common system of the FK_3, although as a rule these corrections were not large and thus not of great importance. The times of the observations were reduced to Ephemeris Time and all positions were precessed to the standard equinox 1950.0. About 60 erroneous observations were rejected.

The discussion of the observations was carried out twice and will be described in detail below. In the first solution about 250 observations were treated, 56 of them being visual and the rest photographic. The majority of the observations had accuracies in the range 0″.5 to 1″. Later we added about 100 observations of fairly high quality.

3. Numerical Integration. Normal Places for 1932–1967

As the basis for our calculations we used the following set of elements, obtained by Khanina (1965) from observations 1932–1964:

$$
\begin{aligned}
\text{Epoch} &= 1948 \text{ July } 28.0 \text{ ET} = \text{JED } 243\ 2760.5 \\
M_0 &= 257°032358 \\
\omega &= 310.27339 \\
\Omega &= 285.37736 \quad \bigg\} \quad 1950.0 \\
i &= 3.81364 \\
\varphi &= 5.71756 \\
n &= 0.17622832 \\
a &= 3.1507861 \text{ AU.}
\end{aligned}
$$

We then integrated the equations of motion of 10 Hygiea from 1931 to 1968, taking into account the perturbations by Venus to Pluto. This was done with the help of the BESM-2 computer programme by Belyaev (1967), which makes use of a file of osculating elements of these eight planets for 1660–2060. The integration is done by Cowell's method in rectangular coordinates, with 9–10 significant figures and a variable step.

The numerical integration was compared with all the observations 1932 to 1967. Separate observations were then linked into normal places. We formed two normal places for the oppositions of 1957 and 1963 and one for each of the remaining oppositions. Altogether, we formed 28 normal places from 250 observations.

4. Conditional Equations. Solution Using the Observations 1932–1967

In addition to the Jupiter mass correction and the usual six unknowns of the problem of improving the orbit of a minor planet, it is possible to include as unknowns corrections to the elements of the Earth's orbit and to the equinox and equator of the system of star positions. In this general case, the conditional equations have the form:

$$\cos \delta \sum_{i=1}^{6} \frac{\partial \alpha}{\partial E_i} \Delta E_i + \cos \delta \sum_{j=1}^{5} \frac{\partial \alpha}{\partial E'_j} \Delta E'_j + \cos \delta \frac{\partial \alpha}{\partial m_5} \Delta m_5 + \cos \delta \Delta \alpha_0$$

$$= \cos \delta (O - C)_\alpha$$

$$\sum_{i=1}^{6} \frac{\partial \delta}{\partial E_i} \Delta E_i + \sum_{j=1}^{5} \frac{\partial \delta}{\partial E'_j} \Delta E'_j + \frac{\partial \delta}{\partial m_5} \Delta m_5 + \Delta \delta_0 = (O - C)_\delta,$$

where E_i $(i = 1, 2, \ldots, 6)$ are the elements of the asteroid's orbit, E'_j $(j = 1, 2, \ldots, 5)$ are the elements of the Earth's orbit, m_5 is the mass of Jupiter, and $\Delta \alpha_0$ and $\Delta \delta_0$ are the equinox and equator corrections. $(O-C)_\alpha$ and $(O-C)_\delta$ are the residuals of the observed positions (normal places) from the positions calculated from the initial conditions. A pair of such equations is formed for each observation or normal place, and the complete solution is made by the method of least squares.

The equations for the 28 normal places of 10 Hygiea contained all 14 unknowns mentioned. The coefficients $\partial \alpha / \partial E_i$ and $\partial \delta / \partial E_i$ were obtained from the numerical integration, while the same coefficients for the Earth were derived using the rectangular coordinates of the Sun. In order to obtain the coefficients of Δm_5 we made an additional integration using a slightly different value for the mass of Jupiter. All the equations of condition were assigned unit weight.

First of all we just solved for seven of the unknowns: the corrections to the orbital elements of the minor planet and to the mass of Jupiter. The resulting mass of Jupiter was found to be

$$1/m_5 = 1047.345 \pm 0.040 \text{ (rms error)}.$$

The residuals of some of the normal places were as high as 2″, with an rms error of unit weight $\sigma_0 = \pm 0.''88$. To improve the accuracy we revised the normal places, eliminating some of the less reliable observations and introducing the elliptic aberration terms. We also solved for all 14 unknowns, but owing to the smallness of the determinant of the coefficients the result was unsatisfactory. This was to be expected because in order to separate the corrections to the elements of the Earth and the asteroid one requires observations made far from opposition, and these were not available. Removal of the correction to the Earth's longitude improved matters, but the corrections to the equinox and to the Earth's remaining elements were still very poorly determined. We could determine with confidence only eight unknowns: the corrections to the elements of the asteroid, to the mass of Jupiter, and to the system of declination. This solution gave for Jupiter's mass

$$1/m_5 = 1047.326 \pm 0.033.$$

5. New Treatment of the Observations

At this point in our calculations the Institute for Theoretical Astronomy acquired a BESM-4 computer, and two new programmes for numerical integration were written. We decided to perform the calculations anew, using these more accurate programmes and also including a number of additional observations made earlier at the U.S. Naval Observatory and in 1969 at the Crimean Observatory.

Belyaev's new programme has most of the features of the old one, but it can also be used for calculating ephemerides and comparing calculations with observations, and the results are good to one more decimal place. The programme by Kazimirchak-Polonskaya (1967) has also been revised for the new computer; it makes use of a file of rectangular coordinates of the planets Venus to Pluto for 400 yr, and the integration is carried out in double precision using the Numerov-Subbotin method.

Comparison of integrations of the orbit of 10 Hygiea, made using the two new programmes, gave differences in α and δ of only $0\overset{s}{.}01$ and $0\overset{''}{.}1$, respectively, after 19 yr. The new calculation had an insignificant effect on the residuals.

Inclusion of a large number of additional observations from 1956 to 1967 allowed us to form more normal places. The total number of observations was increased to 350, and we formed 37 normal places. With few exceptions, the distribution of observations in the normal places became more uniform.

The coefficients in the conditional equations were calculated as before, and the equations were all given unit weight, but in view of the unsatisfactory results obtained when we solved for corrections to the Earth's orbit, we decided to retain seven unknowns only. The new solution proved to be very similar to the previous one. In particular, the reciprocal mass of Jupiter was

$$1/m_5 = 1047.324 \pm 0.023.$$

The difference between the two values is negligible, and even our preliminary result of 1047.345 is contained within the error limits.

Table I shows several recent determinations of the mass of Jupiter using minor planets and satellites, the results being listed in order of increasing mass. These

TABLE I

Modern determinations of the mass of Jupiter

$1/m_5$ and mean error	Reference	Object	Note
1047.387 ± 0.004	O'Handley (1968)	65 Cybele	
1047.386 ± 0.041	Bec (1969)	Jupiter IX	
1047.381 ± 0.020	Klepczynski (1967)	52 Europa	(1)
1047.372 ± 0.006	Klepczynski (1969)	31 Euphrosyne	
1047.367 ± 0.004	Fiala (1969)	57 Mnemosyne	(2)
1047.359 ± 0.010	Klepczynski (1969)	24 Themis	
1047.356 ± 0.004	Fiala (1969)	57 Mnemosyne	(2)
1047.351 ± 0.006	Klepczynski (1969)	10 Hygiea	
1047.350 ± 0.004	Klepczynski (1969)	57 Mnemosyne	(2)
1047.345 ± 0.040	This investigation	10 Hygiea	(3)
1047.340 ± 0.024	Zielenbach (1969)	48 Doris	(4)
1047.337 ± 0.027	Klepczynski (1969)	52 Europa	(1)
1047.335 ± 0.077	Herget (1968)	Jupiter VIII	
1047.333 ± 0.024	Zielenbach (1969)	48 Doris	(4)
1047.326 ± 0.033	This investigation	10 Hygiea	(3)
1047.324 ± 0.023	This investigation	10 Hygiea	(3)

(1) From 297 and 561 observations, respectively.
(2) The first value is from 986 observations reduced to the FK_4 and the second included additional observations; the third value is also by Fiala, quoted by Klepczynski.
(3) The first value is preliminary.
(4) For the second value the differential coefficients for the orbital corrections were obtained from variational equations.

determinations are all grouped near the Newcomb-Hill value, suggesting that the correction indicated by de Sitter and by Clemence is unwarranted. We may note also that Klepczynski's result using 10 Hygiea is on the same side of the Newcomb-Hill figure as ours.

Acknowledgments

I wish to express my cordial thanks to N. S. Yakhontova for her constant attention and to N. A. Belyaev and E. I. Kazimirchak-Polonskaya for their help with the calculations.

References

Bec, A.: 1969, *Astron. Astrophys.* **2**, 381.
Belyaev, N. A.: 1967, *Byull. Inst. Teor. Astron.* **10**, 696.
Clemence, G. M.: 1949, *Proc. Am. Phil. Soc.* **93**, 7.
de Sitter, W.: 1938, *Bull. Astron. Inst. Neth.* **8**, 307.
Fiala, A. D.: 1969, *Bull. Am. Astron. Soc.* **1**, 342.
Herget, P.: 1968, *Astron. J.* **73**, 737.
Hill, G. W.: 1873, *Collected Mathematical Works*, Vol. I.
Kazimirchak-Polonskaya, E. I.: 1967, *Trudy Inst. Teor. Astron.* **12**.
Khanina, F. B.: 1965, *Byull. Inst. Teor. Astron.* **10**, 424.
Klepczynski, W. J.: 1967, *Astron. J.* **72**, 808.

Klepczynski, W. J.: 1969, *Astron. J.* **74**, 774.
Kulikov, D. K. and Subbotina, N. S.: 1963, in *Problemy Dvizheniya Iskusstvennykh Nebesnykh Tel*, Moscow.
Newcomb, S.: 1895, *Astron. Pap. Washington* **5**, part 5.
O'Handley, D. A.: 1968, *Astron. J.* **73**, 529.
Orel'skaya, V. I.: 1962, *Byull. Inst. Teor. Astron.* **8**, 660.
Zielenbach, J. W.: 1969, *Astron. J.* **74**, 567.

Discussion

W. J. Klepczynski: I wish to congratulate you on this fine work. I agree that Newcomb's mass of Jupiter is about the best so far, and I think that any further improvement will have to come from the close approach of a space probe to Jupiter.

42. THE MOTION OF HIDALGO AND THE MASS OF SATURN

B. G. MARSDEN

Smithsonian Astrophysical Observatory, Cambridge, Mass., U.S.A.

Abstract. The principal features of the motion of Hidalgo over the interval 1400–2900 are described. The possibility that this object is an extinct (or nearly extinct) comet nucleus is discussed. A determination of the mass of Saturn, using observations of Hidalgo during 1920–1964, is presented and compared with other recent determinations.

Unusual though the orbits of many of the minor planets may be, none is so anomalous in so many different ways as that of 944 Hidalgo. In many respects the orbit of Hidalgo represents a compromise among those of the periodic comets Tuttle, Wild, and Neujmin 1, all four objects having their aphelia near the orbit of Saturn and rather high orbital eccentricities and inclinations.

Perhaps the most significant difference between minor planets and short-period comets is that the orbits of the latter are continually being disturbed as the result of passages near Jupiter, while the orbits of the former – except for Hidalgo – are stable. That Hidalgo can pass only 0.4 AU from Jupiter (Belyaev and Chebotarev, 1968) can certainly be regarded as suggestive of its cometary nature. Actually, the orbit of Hidalgo would be relatively stable for a short-period comet, only P/Neujmin 1 and P/Arend-Rigaux having been more successful at avoiding Jupiter in recent centuries (Marsden, 1970). These two comets are unusual in that they are almost invariably asteroidal in appearance, their cometary character having been evident only when they were considerably closer to the Earth than Hidalgo ever comes. It is not unreasonable to conclude that Hidalgo is also a comet, and that the relative stability of their orbits and regular passages within 2 AU of the Sun have, in the course of centuries, caused all three comets to lose almost all their volatiles.

Figure 1 shows some of the results of a long-term integration of the motion of Hidalgo over the interval 1400–2900. Close approaches to Jupiter are indicated by arrows. The very close approaches in 1673 (0.38 AU), 2752 (0.42 AU), and 2883 (0.32 AU) are particularly to be noted. Between 1922 and 2752, when there is no approach within 1.2 AU, small periodic variations are evident, especially in the mean distance a, and these reflect the approximate 6:7 mean motion commensurability with Jupiter (although the situation is also influenced by the 7:8 commensurability); as expected, there is no secular trend in a, and the trends in eccentricity e and inclination i (to the ecliptic) are effectively canceled out in the combination $(1 - e^2)^{1/2} \cos i$. Between 1400 and 2900 the longitude of the ascending node regresses from 26° to 12°, while the argument of perihelion changes from 58° to 54°.

A determination of the orbit of Hidalgo, using 94 reliable observations spanning 1920–1964 and allowing for the gravitational attractions of all nine planets, gave a mean residual of 1″.95, and there were systematic trends of 3″ and more. Suspecting that Hidalgo might be a comet and that these residuals were due to the effects of

240 B. G. MARSDEN

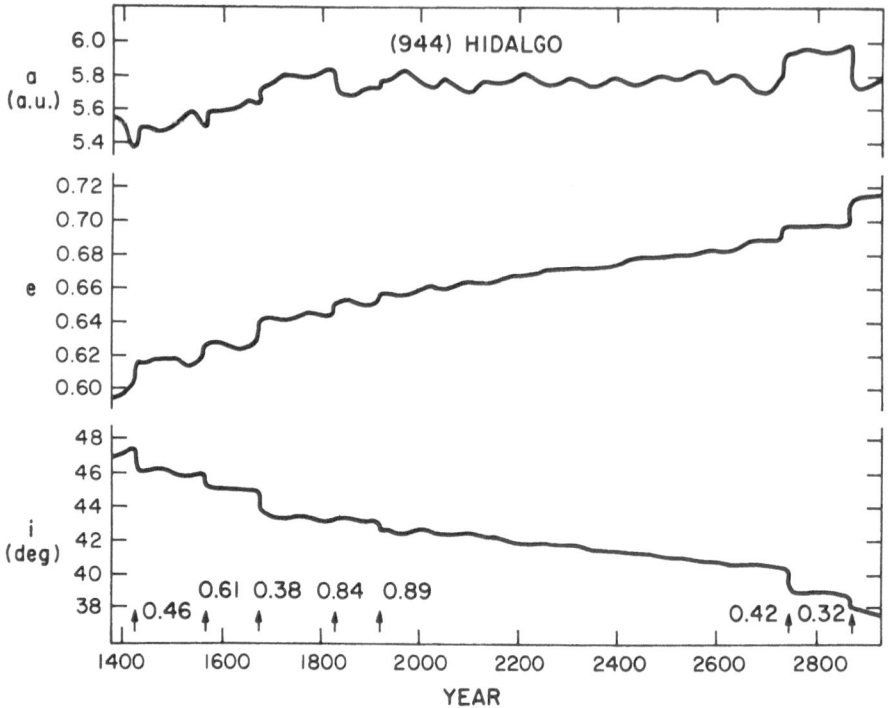

Fig. 1. The variations in the mean distance a, orbital eccentricity e, and orbital inclination i of Hidalgo during 1400–2900. The arrows at the foot indicate approaches to Jupiter, the least separations being stated in AU.

slight nongravitational forces, we made a solution for these effects by the method we have applied to a number of comets (Marsden, 1972). This caused the mean residual to be reduced to 1″36 (Marsden, 1970). However, as mentioned briefly in the note in press added at the end of the paper just cited, it seems more probable that the residuals are due to the error in the IAU value for the mass of Saturn, that of Bessel (1833). In 1924 Hidalgo passed only 3.9 AU from Saturn, about the closest possible. Further, because the mean motions of the two objects are very roughly in the ratio 2:1, there was another approach, though more moderate, of 5.3 AU in 1951 (and also one of

TABLE I

Dependence of the mean residual for Hidalgo on the mass of Saturn

Reciprocal mass of Saturn	Mean residual
3500.0	1″44
3499.0	1.26
3498.5	1.24
3498.0	1.26
3497.0	1.44

TABLE II
Residuals for 944 Hidalgo

UT	Δα cos δ	Δδ	UT	Δα cos δ	Δδ	UT	Δα cos δ	Δδ
1920 Oct. 31.78	0".0	+1".0	1934 Oct. 3.96	-0".9	-0".3	1937 Apr. 13.32	-0".1	+0".7
Nov. 19.77	+2.1	+1.2	4.97	+0.2	+1.6	14.36	+0.9	+0.4
19.82	-1.0	+1.1	7.94	-1.4	-0.2	May 12.24	+1.9	+0.8
19.85	-1.1	+0.1	8.88	-1.2	-0.7	13.28	+1.0	+0.9
Dec. 6.01	+2.0	-0.4	29.85	-1.7	-0.2	13.30	+2.7	+1.1
14.72	-1.1	-0.5	Nov. 7.20	-0.1	-0.9	June 10.20	+0.8	+0.6
1921 Jan. 15.83	+1.1	-1.4	7.21	+0.2	-1.4	10.23	+1.0	+0.7
31.79	-1.1	+1.0	12.09	-0.8	+0.3	1938 Apr. 27.34	-0.4	+0.8
Feb. 8.80	+0.5	-1.2	12.09	-0.3	+1.1	1948 Oct. 7.58	-2.8	+4.1
9.78	-0.9	+0.6	13.11	-0.5	+2.1	26.25	-0.9	0.0
12.80	+2.3	-2.2	15.30	-0.1	+0.3	26.27	-1.4	-0.1
Mar. 11.80	+0.7	+0.5	Dec. 7.12	-1.0	-0.9	Nov. 4.96	-1.4	-1.2
Nov. 9.18	+0.9	-1.5	1935 Mar. 2.09	+1.0	-1.6	4.97	-1.7	-1.2
26.13	+0.8	+0.5	Apr. 6.14	+0.7	-1.5	4.99	-0.8	-1.2
1922 Jan. 24.12	-0.1	-2.3	6.14	+0.4	-1.5	5.86	-2.8	+0.6
24.47	-0.5	-0.2	1936 Jan. 24.48	-0.1	-0.6	5.88	-2.7	-0.1
27.44	-1.0	+0.1	24.56	-0.1	-0.1	5.89	-3.1	+2.1
28.48	-0.5	-1.7	Mar. 28.30	-0.7	0.0	6.91	-0.1	+0.3
29.43	-1.4	+0.2	28.31	-0.1	-0.1	6.92	-1.9	-0.6
Feb. 5.01	+0.8	-1.0	29.33	-0.6	-0.2	6.96	-1.5	+1.2
26.42	-2.4	-0.1	Apr. 19.23	+0.4	+1.8	8.85	-0.5	+2.0
Mar. 3.43	+1.4	-1.4	26.30	+0.9	+1.2	1962 Oct. 2.42	+1.5	+0.1
4.43	-2.6	-1.2	26.31	+1.3	+1.3	2.43	+1.0	-0.5
9.46	+0.5	+2.1	27.26	+1.1	+1.5	27.26	+0.9	-0.4
22.24	-0.6	+1.2	27.27	+1.4	+1.2	27.26	+1.0	-0.6
22.26	-0.1	+1.3	May 22.17	+1.2	+0.4	Nov. 19.20	+2.3	-0.6
22.90	+0.3	-0.5	22.18	+1.7	+0.3	19.21	+2.1	-0.6
Apr. 1.01	+1.4	-1.0	23.18	+1.0	0.0	1964 Mar. 14.34	0.0	+2.2
15.14	-0.2	-1.0	23.18	+1.1	-0.2	20.33	+1.0	+0.6
15.16	-1.2	-2.7	24.18	+0.7	-0.1	20.36	+0.8	+0.4
22.20	+0.3	+0.5	24.19	+0.8	+0.2	Apr. 10.34	+1.5	+0.2
						10.37	+1.5	-0.2

4.1 AU in 1896, before Hidalgo was discovered). Additional orbit solutions have been made using several slightly larger values for the mass of Saturn. They indicate that the Hidalgo residuals can be substantially improved if the mass is increased by about 0.1 percent, and there is thus no need to solve for any nongravitational forces. The mean residuals corresponding to the various values of Saturn's mass are shown in Table I. The best fit comes from a reciprocal mass of 3498.5, and the individual residuals from this solution are listed in Table II.

In Table III we give a selection of determinations of the mass of Saturn – particularly the more recent determinations. During the nineteenth century the determinations from Saturn's satellites were generally more reliable than those from objects external to the system. More recently, the situation has been reversed, the satellite determinations seemingly more prone to systematic errors. The agreement among

TABLE III

Determinations of the mass of Saturn

Reciprocal mass (and mean error)	Reference	Object
3501.6 ± 1.2	Bessel (1833)	Titan
3502.2 ± 0.8	Hill (1895)	Jupiter, 1750–1888
3494.8 ± 1.1	Jeffreys (1954)	Satellites
3499.9 ± 1.2	Gaillot (1913)	Jupiter, 1750–1907
3497.6 ± 0.4	Hertz (1953)	Jupiter, 1884–1948
3499.7 ± 0.6	Clemence (1960)	Jupiter
3497.6 ± 0.2	Carr and Herget (1970)	P/Schwassmann-Wachmann 1, 1927–1965[a]
3498.7 ± 0.2	Klepczynski et al. (1970)	Jupiter, 1913–1968[b]
3498.5 ± 0.2[c]	Shapiro (1970)	All planets, 1750–1970
3498.5 ± 0.3[d]	This investigation	Hidalgo, 1920–1964

[a] Also fits observations in 1902.
[b] Also fits observations back to 1781.
[c] The true uncertainty is estimated at ±0.5.
[d] The mean error was determined by Herget (1972).

the four very recent determinations is certainly gratifying, and we could surmise that the slight disagreement of the P/Schwassmann-Wachmann 1 result is due, either to the influence of nongravitational forces or to systematic departures between center of mass and center of light. On the other hand, nongravitational forces *might* also be affecting Hidalgo, and there are obvious difficulties in measuring the position of Jupiter (and indeed all the major planets). All things considered, we tend to agree with Shapiro's (1970) suggestion that the true value of the reciprocal mass is contained in the range 3498.5 ± 0.5, although it would perhaps be worth while to make a further independent determination from a simultaneous study of the motions of several of the minor planets of aphelion distance greater than 4.0 AU.

References

Belyaev, N. A. and Chebotarev, G. A.: 1968, *Astron. Tsirk*. No. 480.
Bessel, F. W.: 1833, *Astron. Nachr*. **11**, 17.
Carr, H. J. and Herget, P.: 1970, private communication.
Clemence, G. M.: 1960, *Astron. J*. **65**, 21.
Gaillot, A.: 1913, *Ann. Obs. Paris Mem*. **31**, 105.
Herget, P.: 1972, this Symposium, p. 244.
Hertz, H. G.: 1953, *Astron. Pap. Washington* **15**, 215.
Hill, G. W.: 1895, *Astron. Pap. Washington* **7**, 17.
Jeffreys, H.: 1954, *Monthly Notices Roy. Astron. Soc*. **114**, 433.
Klepczynski, W. J., Seidelmann, P. K., and Duncombe, R. L.: 1970, *Bull. Am. Astron. Soc*. **2**, 247.
Marsden, B. G.: 1970, *Astron. J*. **75**, 206.
Marsden, B. G.: 1972, this Symposium, p. 135.
Shapiro, I. I.: 1970, private communication.

Discussion

B. Yu. Levin: How will the change in Saturn's mass influence the determination of the nongravitational forces on comets?

B. G. Marsden: In the best determined cases the figures should certainly not be changed by more than 1%.

43. ON THE DETERMINATION OF PLANETARY MASSES

P. HERGET

University of Cincinnati Observatory, Cincinnati, Ohio, U.S.A.

Abstract. A common method of determining the mass of a planet is to solve for the orbit of some other object several times with different trial values for the mass in question. It is pointed out that a parabola fitted to the values of the sums of the squares of the residuals may be used to obtain, not only the planetary mass, but also its probable error. Marsden's determination of the mass of Saturn from observations of Hidalgo is used as an example.

Undoubtedly the most preferable way to proceed in attempting to determine the mass of a planet from observations of a minor or some other major planet is to integrate simultaneously the trajectory and the variational equation, including partial differential coefficients with respect to this mass. On the other hand, it is a common practice to integrate only the trajectory with several different trial values of the mass in question, to make separate least squares solutions, often using two-body formulae for the partial differential coefficients, and finally to solve for the value of the mass from the minimum of the parabola which represents the sum of the squares of the residuals. Whenever this is done, the probable error of the resulting solution is equal to the probable error of unit weight multiplied by the square root of the latus rectum of this parabola.

The general form for the parabola is

$$(x - x_0)^2 = 4q(y - y_0),$$

or

$$A + Bx + Cx^2 = y, \tag{1}$$

where

$$A = \frac{x_0^2}{4q} + y_0$$

$$B = -\frac{2x_0}{4q}$$

$$C = \frac{1}{4q}.$$

To put x in dimensionless units and y in radians, let

$$x = \frac{M - M_0}{M_0}$$

$$y = \sum (v'')^2 (\text{arc } 1'')^2,$$

where M is the reciprocal mass and v'' is a residual in seconds of arc. Equation (1) is formed for each orbit solution, and the coefficients (although only C is required, since it is the reciprocal of the latus rectum of the parabola) are determined by least squares.

We illustrate this proposition with the data published by Marsden (1972, Table I) for determining the mass of Saturn from the observations of Hidalgo. The symmetry of the data points is not typical of most applications of this kind. Let $M_0 = 3498.5$, and since there are 94 observations the sums of the squares of the residuals can be constructed from the mean residuals. The equations of condition are thus

$$A + 4.29(10^{-4}B) + 18.383(10^{-8}C) = 194.92(\text{arc } 1'')^2$$
$$A + 1.43(10^{-4}B) + 2.043(10^{-8}C) = 149.23(\text{arc } 1'')^2$$
$$A = 144.53(\text{arc } 1'')^2$$
$$A - 1.43(10^{-4}B) + 2.043(10^{-8}C) = 149.23(\text{arc } 1'')^2$$
$$A - 4.29(10^{-4}B) + 18.383(10^{-8}C) = 194.92(\text{arc } 1'')^2.$$

The normal equations are (B being completely separable)

$$5A + 40.852(10^{-8}C) = 832.84(\text{arc } 1'')^2$$
$$40.852(10^{-8}A) + 684.217(10^{-16}C) = 7776.18(\text{arc } 1'')^2(10^{-8})$$

and elimination of A gives

$$C = (1.665 \times \text{arc } 1'' \times 10^4)^2 = \left(\frac{1}{12.4}\right)^2.$$

The probable error of x is thus

$$0.6745 \times 1''.24 \times \text{arc } 1'' \times 12.4 = 0.000050,$$

and the probable error of M is

$$0.000050 M_0 = 0.17.$$

Reference

Marsden, B. G.: 1972, this Symposium, p. 239.

44. THE INFLUENCE OF MINOR PLANETS ON THE MOTIONS
OF COMETS

K. A. SHTEJNS and I. E. ZAL'KALNE

Astronomical Observatory, Latvian State University, Riga, U.S.S.R.

Abstract. It is shown that in a sufficiently small interval there is a uniform distribution of the minimum distances between the orbit of a comet and the orbits of minor planets. It is also shown that there is a uniform distribution of the distance of the planet from the point of closest approach when the comet reaches that point. It is found that P/Daniel may collide with a microplanet of mass 4.05×10^8 g during 100 revolutions. It is possible for one of 60 short-period comets to meet during 100 revolutions a microplanet of mass 9.9×10^9 g and radius 8.6 m. The changes in reciprocal semimajor axis of a comet arising from close encounters with minor planets are also studied but found to be insignificant.

1. Approaches between Comets and Minor Planets

It is clear that very many minor planets remain to be discovered. The general opinion is that there exist minor planets with masses smaller than those of the comets. Such minor planets will be referred to as microplanets, and we shall extrapolate the statistical data of the known minor planets to microplanets with masses down to 10^8 g. We shall specifically consider the influence of minor planets on the motion of periodic comet Daniel, the orbital elements of which are $\Omega = 70°$, $i = 20°$, $\omega = 7°$, $\varphi = 35°$, $a = 3.6$ AU (Belyaev, 1966) and the assumed mass $m_k = 0.782 \times 10^{-19}$ solar masses. In this paper length is generally expressed in astronomical units, mass in solar masses and time in days. We have determined the minimum distances between the mean orbit of P/Daniel and the orbits of 1735 numbered minor planets (Chebotarev, 1968), the heliocentric velocity v_p of each planet, the relative velocity v'_k of the comet and the angle between the velocity vectors. We found that 13 planets have a minimum distance d of less than 0.1 AU, the actual distances being distributed almost uniformly over the interval (0.0, 0.1).

Let us demonstrate the uniformity of the distribution of the minimum distances d. In the vicinity of the intersection we may consider the trajectories of both minor planet and comet to be rectilinear. Figure 1 shows that the minimum distances between the comet orbit KK_1 and the planet orbits are equal to the distance from parallel lines to the plane Q_1, defined by OO_1 and KK_1. The microstructure in d will be between the plane Q_1 and the parallel plane Q_2 situated at a small distance. Since the number of orbits between Q_1 and Q_2 is proportional to the distance between these planes, the distribution is uniform. There will also be a uniform distribution if the path of the comet is through minor planet trajectories that can be expanded into several pencils of parallel trajectories.

The angle θ between the planet's velocity v_p and the comet's relative velocity v'_k is concentrated near 90°, this value being determined by the angle of intersection between the comet orbit and the nearly circular orbits of the typical minor planets.

Chebotarev et al. (eds.), *The Motion, Evolution of Orbits, and Origin of Comets*, 246–250. *All Rights Reserved.*
Copyright © 1972 by the IAU.

From the statistics we find that v_p is almost uniformly distributed over the interval
(0.008, 0.015) and v_k' is almost uniform in the range (0.003, 0.010). Because of the
small range of their variations we can replace θ, v_p, v_k' and also $\delta = v_k'/v_p$ by their
mean values. We can also demonstrate the uniform distribution of the distance h
from the point of closest approach which the planet has covered in unperturbed
uniform rectilinear motion (or must cover if $h < 0$) when the comet reaches that point.
Indeed, the time in which the minor planet is within this distance is proportional to
the distance. If the comet is at the point of closest approach of the orbits, the planet
must be somewhere on the circumference of its orbit, which for a radius of 2.90 AU
is 18.22 AU, and hence the probability $h/18.22$ corresponds to the length of h.

We need to know the distribution of the number of planets according to their
masses m_p and have therefore analysed in detail Kuiper's (1958) statistics of absolute
magnitudes. From observations of a number of minor planets on a plate he determined the so-called completeness factor, i.e., the possibility of finding all the planets

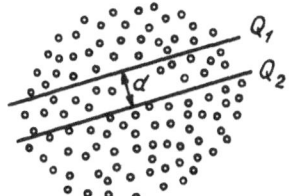 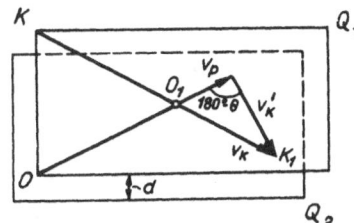

Fig. 1.   Demonstration of the uniform distribution of the minimum distances d between the orbit
of a comet and the orbits of minor planets.

on the plate and the dependence of the number of minor planets on their magnitudes.
On the basis of his data and the assumption that minor planets and microplanets
are spheres of density 3.5 g cm^{-3} we have found the relation between the number of
minor planets and their mass, namely, that the number N of planets in the interval
dm_p is

$$N\,dm_p = \frac{2.3 \times 10^{-14}}{m_p^{2.28}}\,dm_p. \tag{1}$$

Hence the density of the distribution is

$$p(N) = \frac{1.28}{m_p^{2.28}}\,\frac{(m_0 M_0)^{1.28}}{M_0^{1.28} - m_0^{1.28}}, \tag{2}$$

where m_0 and M_0 are the limiting values of the planet masses in the region considered.

2. Collisions

A collision between a minor planet and comet takes place if

$$\rho_p + \rho_k \geqslant \sqrt{A^2 + B^2} - A, \tag{3}$$

where ρ_p and ρ_k are the radii of the planet and the comet (both assumed spherical), and A and B are the semimajor and semiminor axes of the relative hyperbolic orbit of the planet and comet in their mutual sphere of action. If in uniform unperturbed rectilinear motion the minimum distance between the planet and comet is $\varDelta_m$, then (van Woerkom, 1948)

$$B^2 = \varDelta_m^2 = d^2 + h^2 \sin^2 \theta.$$

We also have

$$m_p = 2.5 \times 10^7 \rho_p^3, \tag{4}$$

and

$$A \simeq \frac{k^2 m_p}{v_k},$$

where k^2 is the gravitational constant. It follows from Equation (3) that

$$d^2 + h^2 \sin^2 \theta \leqslant (\rho_p + \rho_k)(\rho_p + \rho_k + 2A).$$

Taking into account the values of v_k' we have

$$A < 10^9 \rho_p^3.$$

The minor planet of largest mass has $\rho_p = 2 \times 10^{-6}$ AU; hence

$$\rho_p + 10^9 \rho_p^3 = \rho_p(1 + 10^9 \rho_p^2) < \rho_p(1 + 4 \times 10^{-3}) \approx \rho_p,$$

and the conditions under which collisions take place may be written as

$$d^2 + h^2 \sin^2 \theta \leqslant (\rho_p + \rho_k)^2, \tag{5}$$

i.e., the conditions existing for unperturbed motion.

From these conditions we can calculate the probable number of collisions of the comet with microplanets. The statistics show that 13 of the 1735 numbered minor planets have d in the range $(0.0, 0.1)$ and h in $(-9.11, +9.11)$. Since the d and h distributions are uniform and the values themselves independent, the number of microplanets with given d and h is proportional to the area of their variation regions in the d, h plane. Assuming $\rho_p = 0$, Equation (5) represents an ellipse with the semi-axes ρ_k and $\rho_k/\sin \theta$. The area of the ellipse is $\pi \rho_k^2/\sin \theta$, and the variation area of d and h for the above-mentioned 13 planets is $18.22 \times 0.1 = 1.822$. Hence, of these 13 planets the number that can collide with the comet is

$$\frac{\pi \rho_k^2 \times 13}{2 \sin \theta \times 1.822} \ll 1.$$

But the 13 minor planets correspond to 1735 discovered planets, and to each discovered planet there correspond

$$\frac{1}{1735} \int_{m_p}^{m_k} \frac{2.3 \times 10^{-14}}{m^{2.28}} \, dm$$

undiscovered ones. Hence, the total number of collisions per n revolutions of the comet will be

$$n \frac{\pi p_k^2 \times 13}{2 \sin \theta \times 1.822} \frac{1}{1735} \int_{m_p}^{m_k} \frac{2.3 \times 10^{-14}}{m^{2.28}} \, dm. \tag{6}$$

It follows that during 100 revolutions there is a good chance of a collision with a microplanet of mass 4.05×10^8 g. The lower limit of the integral is of great importance since the number of planets grows rapidly as the mass decreases, and it is therefore assumed in our calculations that there is a collision with a planet whose mass is equal to the lower limit. About 60 short-period comets of the Jupiter family have been discovered so far. This means that it is possible for one of them to meet a larger microplanet; and during 100 revolutions one of them may meet a microplanet of mass 9.9×10^9 g and radius 8.6 m. So far we have considered microplanets whose masses are greater than 10^8 g (or 5×10^{-26} solar masses). If we admit microplanets of mass 10^7 g, the number of collisions per revolution will become about 3.5 times greater, and a single comet may collide with an object of mass 1.2×10^7 g and radius 90 cm in a single revolution.

3. Perturbations in $1/a$

When estimating maximum perturbations (without collisions) it is possible to make use of H. A. Newton's formula

$$\Delta \frac{1}{a} = -\frac{4m_p}{S^2} \frac{h \sin^2 \theta}{d^2 + h^2 \sin^2 \theta}. \tag{7}$$

Let us determine the mean value $M[\Delta(1/a)]$ of the values of $\Delta(1/a)$ that arise when the minor planets come sufficiently close to the comet that d and h assume all possible values over the intervals $(0, d_0)$ and $(-h_0, h_0)$. No collisions will occur in an ellipse around the centre $d_0 = 0$, $h_0 = 0$, and we call this region Q_1. From symmetry it follows that

$$M[\Delta(1/a)] = 0.$$

For determining the dispersion we exclude the semicircle near the centre of the d, $h \sin \theta$ plane that has a radius smaller than that of the smallest planet, i.e., $p_0 = (d^2 + h^2 \sin^2 \theta)^{1/2} \approx 10^{-9}$. This will not cause an appreciable error since there are no collisions with minor planets more massive than the comet. In this case the dispersion is determined according to the formula

$$D\left(\Delta \frac{1}{a}\right) = \iiint_Q \left(\Delta \frac{1}{a}\right)^2 p(N) dd \, dh \, dm_p$$

$$= \frac{1.28(m_0 M_0)^{1.28}}{M_0^{1.28} - m_0^{1.28}} \frac{16}{S^2} \iint_{Q_1} \frac{h^2 \sin^4 \theta}{(d^2 + h^2 \sin^2 \theta)^2} \, dd \, dh \int_{m_0}^{M_0} \frac{m_p^2}{m_p^{2.28}} \, dm_p$$

$$\approx \frac{95(m_0 M_0)^{1.28}(M_0^{0.72} - m_0^{0.72})}{M_0^{1.28} - m_0^{1.28}} \iint_{Q_1} \frac{h^2 \sin^4 \theta}{(d^2 + h^2 \sin^2 \theta)^2} \, dd \, dh.$$

Introducing the polar coordinates

$$d = p \cos \varphi, \qquad h \sin \theta = p \sin \varphi,$$

we have

$$D\left(\Delta \frac{1}{a}\right) = 95 \frac{(m_0 M_0)^{1.28}(M_0^{0.72} - m_0^{0.72})}{M_0^{1.28} - m_0^{1.28}} \int\limits_{0 p_0}^{\pi p_1}\!\!\!\int \frac{\sin \theta \, p^2 \sin^2 \varphi}{p^4} \, p \, d\varphi \, dp.$$

Since $m_0 \ll M_0$ and $|\ln p_1| \gg |\ln p_0|$ we have

$$D\left(\Delta \frac{1}{a}\right) \approx 150 \sin \theta \ln p_1 M_0^{0.72} m_0^{1.28}.$$

It is obvious that in our case $D[\Delta(1/a)] < 10^{-28}$. When we modelled this process by the Monte Carlo method on an electronic computer we obtained $D[\Delta(1/a)] < 10^{-29}$.

Let us apply now Kolmogorov's inequality, which is that if mutually independent random values $\zeta_1, \zeta_2, \ldots, \zeta_n$ have finite dispersions, then the probability that

$$\left| \sum_{s=1}^{k} \zeta_s - M\zeta_s \right| < \varepsilon \qquad (k = 1, \ldots, m)$$

is not less than

$$1 - \frac{1}{\varepsilon^2} \sum_{k=1}^{m} D\zeta_k.$$

In our case $M\zeta_k = 0$, all the $D\zeta_k$ are equal, and m is the number of minor planets coming into the contour of radius p_1. The probability that the sum of the increments $\Delta(1/a)$ over 100 revolutions of the comet will not exceed 10^{-8} at any point, i.e.,

$$\left| \sum_{k=1}^{m} \zeta_k \right| < 10^{-8},$$

will thus not be less than

$$1 - 10^{16} \sum_{k=1}^{m} D\zeta_k. \tag{8}$$

From Equation (1), the number of minor planets m having d in the interval $(0.0, 0.1)$ is

$$m = \frac{13}{1735} \int\limits_{m_p}^{m_k} \frac{2.3 \times 10^{-14}}{m^{2.28}} \, dm \approx 10^8.$$

Then, substituting in Equation (8), we have

$$1 - 10^{16} \sum_{k=1}^{m} D\zeta_k > 1 - 10^{-4}.$$

References

Belyaev, N. A.: 1966, *Byull. Inst. Teor. Astron.* **10**, 696.
Chebotarev, G. A.: 1968, *Efemeridy Malykh Planet na 1969 God*, Nauka, Leningrad.
Kuiper, G. P., Fujita Y., Gehrels, T., Groeneveld I., Kent, J., Van Biesbroeck, G., and van Houten, C. J.: 1958, *Astrophys. J. Suppl.* **32**, 289.
van Woerkom, A.: 1948, *Bull. Astron. Inst. Neth.* **10**, 445.

PART IV

PHYSICAL PROCESSES IN COMETS

45. PHYSICAL PROCESSES IN COMETARY ATMOSPHERES

A. Z. DOLGINOV

A. F. Ioffe Institute of Physics and Technology, Leningrad, U.S.S.R.

Abstract. Formulae are obtained for the distribution of molecules in the cometary head, taking into account the conditions of hydrodynamic and free molecular flow in various regions around the nucleus. Experimental data are used to derive physical parameters near the nuclei of comets 1952 III, 1955 V, 1957 III, and 1960 II and the rate of decrease of mass. The possibility of chemical reactions in the region close to the nucleus is discussed. Gas condensation is shown to be a possible cause of dust formation under the conditions existing near the nucleus, and this process may be responsible for the major portion of the cometary dust component. The observed grouping of synchrones in the cometary tail can be explained on the assumption that the nuclear surface comprises two (or more) areas differing essentially in evaporation rate, the amount of matter ejected varying over the rotation period of the nucleus. Charged dust particles are shown to form, with electrons and ions, a common medium, i.e., dust plasma, which can be treated by the same methods used for ordinary plasma. Special investigations appear to be desirable when comets intersect meteor streams.

The nongravitational forces acting on a cometary nucleus depend upon the evaporation rate of the constituent matter and on possible ejections from the nucleus. For studying these phenomena, it is essential to know the characteristics of the cometary atmosphere, primarily those of its gaseous and dust components, which are directly dependent upon the processes taking place in the vicinity of the nucleus.

1. Neutral Gaseous Component of the Cometary Head

The distribution function of neutral molecules can be found (Dolginov and Gnedin, 1966; Dolginov *et al.*, 1970) from the kinetic equation

$$\frac{\partial N}{\partial t} + \mathbf{v}\frac{\partial N}{\partial \mathbf{r}} + \mathbf{g}\frac{\partial N}{\partial \mathbf{v}} = S - \frac{N}{\tau}, \tag{1}$$

where N is the number of molecules with velocity $\mathbf{v}$ at the point $\mathbf{r}$ at time t, τ is the lifetime of molecules prior to conversion to some other particles in the process of dissociation or ionization, $\mathbf{g}$ is the acceleration of the molecules due to light pressure and solar attraction, and S is the number of particles emitted by the source per unit time.

We are not in a position to consider directly the processes occurring in the region near the nucleus as we have no direct data on the constituent substances of the nucleus. None of the observed molecules, such as C_2, CN and OH, is original; rather, they are formed in the neighbourhood of the nucleus as a result of the decay of primary particles owing to the Sun's electromagnetic and corpuscular radiation or to chemical conversion processes. However, compared to the size of the head, the region where particle collisions are important and where formation of the observed molecules takes place is small (≤ 1000 km), and this justifies our viewing the neigh-

Chebotarev et al. (eds.), *The Motion, Evolution of Orbits, and Origin of Comets,* 253–259. *All Rights Reserved.*
Copyright © 1972 by the IAU.

bourhood of the nucleus as an effective point source of molecules. Suitable conditions are to be provided at the source boundaries to assure transition from continuous medium flow to free molecular flow. It has been shown by Hamel and Willis (1966) that molecular distribution at such a boundary is characterized by a common radial outflow velocity $\mathbf{u}$, and a certain longitudinal temperature T in a reference system moving at the velocity $\mathbf{u}$. The farther from the source, the closer the motion is to a radial trajectory, so that the transverse temperature (characterizing the spread in velocities normal to the radius) falls off rapidly to zero while the longitudinal temperature tends to a constant value. Granted the above assumptions, the problem is reduced to solving Equation (1) with the source function given by

$$S(\mathbf{r}, \mathbf{v}, t) = AG(t) \exp \left[-\alpha \left(\mathbf{v} - u\frac{\mathbf{r}}{r} \right)^2 \right] \frac{\delta(r)}{r^2} \delta \left(\frac{\mathbf{r}}{r} - \frac{\mathbf{v}}{v} \right)$$

$$A = \left(\frac{\alpha}{\pi} \right)^{3/2} \left\{ (1 + 2\alpha u^2)[1 + \varphi(\sqrt{\alpha}\, u)] + 2u\sqrt{\frac{\alpha}{\pi}} \exp\left(-\alpha u^2 \right) \right\}^{-1}. \qquad (2)$$

Here $\alpha = (m/2KT) = 1/v_T^2$ is the reciprocal of the square of the thermal velocity of the molecules v_T, m is the molecule mass, G is the source power, i.e., the total number of particles (of all velocities) emitted by the source per unit time, $\delta(r)$ and $\delta\left(\frac{\mathbf{r}}{\tau} - \frac{\mathbf{v}}{v}\right)$ are Dirac's one-dimensional and two-dimensional delta functions, respectively, and $\varphi(x)$ is the error function.

Equation (1) with the source function described by Equation (2) has been solved (Dolginov et al., 1970). Only some basic results will be mentioned here. The distribution function is very dependent upon the Mach number $M = u/v_T$. Within the regions close to the nucleus ($\alpha gr \leqslant 1$) but outside the regions of molecular collisions, we shall have, for $M \gg 1$,

$$N(\mathbf{r}, t) = \frac{\sqrt{\pi}\, MG}{\alpha r^2} \left[1 + \frac{\alpha g \mathbf{r}}{2\sqrt{\pi}\, M} \exp\left(-M^2 \right) \right], \qquad (3)$$

and for $M \ll 1$,

$$N(\mathbf{r}, t) = \frac{G}{2\alpha r^2} (1 + \sqrt{\pi}\, M)[1 + (1 - \sqrt{\pi}\, M)\alpha g \mathbf{r}]. \qquad (4)$$

It can be seen that at small distances from the nucleus the hydrodynamic outflow velocity leads to reduced anisotropy in the density distribution, and at $M \gg 1$ the distribution is practically isotropic. At large distances, where $\alpha gr \gg 1$, we shall have

$$N(\mathbf{r}, t) = \frac{G}{2r} \sqrt{\frac{\pi g}{2\alpha r}} \exp \left[-M^2 - \alpha gr(1 - \cos \psi) \right.$$
$$\left. - \frac{1}{\tau}\sqrt{\frac{2r}{g}} + 2M\sqrt{\alpha gr(1 - \cos \psi)} \right], \qquad (5)$$

where ψ is the angle between the radius vector of the molecule from the cometary nucleus and the prolongation of the radius vector of the comet from the Sun.

From Equation (5) it follows that when $M \ll 1$ the particle concentration at the periphery of the sunward portion of the cometary head falls off exponentially. This concentration distribution is due to light pressure, which forces the molecules back into the tail. In the case where $M \gg 1$, at $r < 2M^2/\alpha g$ the fall-off in concentration is much slower on the sunward side of the head than on the opposite side, while at $r > 2M^2/\alpha g$ the position is reversed. The fact is that at high hydrodynamic pressure the particles accumulate in the sunward part of the head, having no time to migrate towards the tail. On the side away from the Sun, the effects of hydrodynamic velocity and light pressure augment each other. It is therefore possible that the concentration will be higher on the sunward side than on the back side. This was in fact observed for periodic comet Encke and comet 1965 VIII (Ikeya-Seki). Normally, however, the distribution is drawn away from the Sun, which indicates a low M value.

The flux of light energy over a frequency range $\Delta\nu$ reaching a terrestrial observer from a sold angle $\Delta\Omega$ of the visible surface section of the coma and falling on a unit area normal to the line of vision is

$$I(t) = B_\nu(t) \int_0^\infty d\mathbf{r}_p \int_{\Delta\Omega} d\Omega \, N(\mathbf{r}_p - \mathbf{r}_N, t)$$

$$B_\nu(t) = c \int_{\Delta\nu} d\nu \, h\nu \, n_\nu \, \sigma_\nu(\vartheta, \varphi), \tag{6}$$

where n_ν is the number of quanta of the solar spectrum at frequency ν per unit frequency range and per unit volume in the observed region of the comet, $\sigma_\nu(\vartheta, \varphi)$ is the cross-section of the re-emission at frequency ν of the quantum by the molecule per unit solid angle in a direction with polar coordinates ϑ, φ from the incident radiation, c is the velocity of light, h is Planck's constant, $\mathbf{r}_N$ is the vector from the observer on the Earth to the cometary nucleus, $\mathbf{r}_p$ the vector to the point of the coma under observation and $\mathbf{r} = \mathbf{r}_p - \mathbf{r}_N$.

The explicit form of $I(t)$, which can be obtained by substituting Equations (3) to (5) into (6) and by integration, is given by Dolginov *et al.* (1971). Comparison of the isophotes observed for various comets shows fairly good agreement with theory. Values of parameters characteristic of the physical condition of the coma and obtained by comparing theory with experiment are shown in Table I for various comets.

TABLE I

Characteristics of the comas of various comets

Comet	r (AU)	Type of molecule	V (cm s^{-1})	T (deg)	g (cm s^{-2})	τ (s)
1955 V	0.93	C_2	10^5	1.7×10^3	0.395	1.35×10^5
1957 III	0.64	C_2	9.6×10^4	1.6×10^3	0.844	$\tau_0 = 3 \times 10^6$
						$\tau_1 = 3.4 \times 10^5$
1960 II	1.055	C_2	8.1×10^4	1.1×10^3	0.308	1.48×10^5
1952 III	1.2	CN	1.2×10^5	2.5×10^3	0.365	2×10^5
1957 III	0.64	CN	1.3×10^5	3.13×10^3	1.28	$\tau_0 = 1.8 \times 10^5$
						$\tau_1 = 1.26 \times 10^5$

It is noteworthy that in the case of comet 1957 III (Arend-Roland) two values, τ_0 and τ_1, are given for the lifetime of molecules in the solar radiation field. These have been calculated on the basis of isophotes observed on the sunward and rear positions, respectively. The difference can be explained on the assumption that the ultraviolet solar radiation responsible for dissociation and ionization is significantly absorbed by the coma molecules and dust particles. The results of processing the isophotes indicate that the lifetime of the molecules of this comet depends upon the angle ψ. To estimate the optical thickness of the coma, the lifetime of the C_2 molecule in comet 1957 III may be compared with that of the same molecule in, say, comet 1960 II (Burnham). If both of these comets were optically thin, the lifetime ratio would be equal to the ratio of the squares of the heliocentric distances, i.e., to 0.36 in this particular instance. The observed ratio equals 20. As the decrease in radiation intensity due to coma absorption is exponential, the ratio of these numbers gives $X_1 - X_2 = 4$, where X_1 and X_2 are the optical thicknesses of 1957 III and 1960 II, respectively. The ratio of C_2-molecule lifetimes for comets 1955 V and 1960 II is proportional to the ratio of the squares of the heliocentric distances, which is evidence that the comas of these comets are optically thin; i.e., $X_2 \ll 1$ and hence $X_1 \sim 4$.

It will be noted that it is not at all necessary for the optical thickness in the ultraviolet part of the spectrum to coincide with that of the visible region. Optical thickness may be used to obtain a lower estimate of the total rate G of molecules leaving the source. For comet 1957 III $G \simeq 10^{30}$ mol s^{-1}.

If Equation (6) is integrated over the total volume, an expression will be obtained for the absolute monochromatic brightness of the comet as a whole $I_v = B_v G\tau$. This expression is correct only for an optically thin comet, since τ is assumed to be constant throughout the head. Very few data are available on absolute monochromatic photometry of comets, and those available are usually inaccurate. Thus, G is determined very poorly, but for comets of medium brightness it appears to be of the order of 10^{29} to 10^{30} mol s^{-1}.

2. Dust Component of the Cometary Atmosphere

The usual explanation for the presence of dust in the cometary atmosphere is that the cometary nucleus is a conglomerate of various ices (such as CH_4 and NH_3) sublimating vigorously at relatively low temperatures and of contaminants in the form of dust particles of high melting-point. As the ices sublimate the dust particles are carried with the gas flows (Whipple, 1950).

An alternative mechanism (Dolginov, 1967) is that the dust particles form and grow due to gas condensation in the immediate proximity of the nucleus. As a matter of fact, some cometary gases, including such widespread gases as C_2, are present in the coma in an oversaturated state and must condense into dust particles if a sufficient number of 'seeds' is present and sufficient collisions occur to enable the seeds to grow to dust-particle size. In the 'dirty' conditions surrounding the nucleus, where there are plenty of ions, as well as microscopic solid particles contained in gas, there will always be condensation centres in sufficient number, considering that the condensing gases are not those which evaporate directly from the nucleus, but those formed as a result of dissociation or chemical reactions (e.g., C_2). The concentration of condensing gases can be defined by the expression

$$n(r) = \frac{Q}{vr^2}\left[1 - \exp\left(\frac{R-r}{v\tau}\right)\right], \tag{7}$$

where Q is the number of primary particles formed per unit time, v is the velocity of the primary particles, τ their lifetime, R is the radius of the nucleus, and r the distance from the centre of the nucleus to a given point. Particle growth is described by

$$dN_g/dt = \beta nvS[1 - \exp(-q\theta/T)], \quad N_g = 4\pi\rho^3/3\delta, \tag{8}$$

where δ is the number of atoms in a unit volume of dust-particle material, N_g is the total number of atoms in a dust particle, ρ the radius of the particle, $q = U/\Re$, where U is the heat of sublimation and $\Re$ the gas constant, $\theta = (T_H - T)T_H^{-1}$ is the degree of oversaturation of the vapour (T being the vapour temperature and T_H the vapour saturation temperature), β is the coefficient of adherence of vapour molecules to dust particles (assumed to be close to unity), and $S = 4\pi\rho^2$.

For carbon in the vicinity of the nucleus $T < 1000$ K, $T_H \simeq 2500$ K, $q \simeq 8 \times 10^4$ deg. The value of $\exp(-q\theta/T)$, determining the evaporation rate of the dust particles, is very small for the assumed values of θ, q, and T and can be omitted. Taking into account that $dr = v\,dt$, we may derive from Equation (8) an expression for the increase in dust-particle radius with increasing distance from the nucleus:

$$\frac{d\rho}{dr} = \frac{\beta Q}{vr^2}\left[1 - \exp\left(\frac{R-r}{v\tau}\right)\right], \tag{9}$$

the solution of which may be written

$$\rho = \beta\frac{Q\delta}{vR}\left[\frac{R}{v\tau}\ln\frac{v\tau}{R} - 0.577\frac{R}{v\tau} + \cdots\right] \quad R \ll v\tau$$

$$\rho = \beta\frac{Q\delta}{vR}\left[1 - \frac{v\tau}{R} + \cdots\right] \quad R \gg v\tau. \tag{10}$$

Further estimates are largely dependent upon the assumed lifetime τ_p of the primary molecules and density δ of the dust particles. Assuming $\delta = 1.2 \times 10^{-23}$ cm^{-3} (graphite), $\beta = 0.5$, $Q \simeq 10^{30}$ mol s^{-1}, $v = 10^5$ cm s^{-1} and $\tau_p = 10$ to 100 s, we obtain $\rho \simeq 10^{-4}$ to 10^{-5} cm, which agrees perfectly with the observed dust-particle sizes.

The quantity of dust particles in the cometary atmosphere may increase either (1) due to the fact that a nuclear layer has been exposed, allowing it to evaporate and yield a large quantity of parent molecules for condensing matter, or (2) as a result of an increased intensity of hard electromagnetic or corpuscular solar radiation, leading to a greater yield from the chemical reactions that produce condensing substances. Hard radiations may be a decisive influence on the progress of reactions, even though they provide but a small contribution to the total radiation flux from the Sun. This may lead to a correlation between flares in cometary brightness and solar activity.

A change in the chemical composition of the vapour towards increased quantity of condensing substances produces an increase in the brightness of the comet, even when the total amount of matter evaporating per unit time from unit surface area of the nucleus is constant. Increased brightness associated with condensation requires no additional energy; rather, it indicates liberation of energy. It is not necessary for outbursts in brightness to be associated with nuclear outbursts.

It should be emphasized that the conclusion regarding gas condensation into dust particles is unavoidable if the concentration of supersaturated vapours near the nucleus is 10^{12} to 10^{13} cm^{-3}, which is in agreement with the existing estimates.

3. Synchrones

To be able to estimate the nongravitational effect upon the comet's trajectory, it is essential to know the relationship between the intensity of particle emission and the characteristics of the surface layer of the nucleus, and also whether the emission is a continuous process or a series of outbursts. We can acquire information about this by studying a system of synchrone clouds of dust particles simultaneously leaving the region near the nucleus. The very presence of separated synchrones is evidence of the discrete nature of ejections. Normally, synchrones emerge in groups, the time intervals between successive synchrone ejections being in many instances roughly equal. Thus, the successive tail-synchrone ejections from comet 1910 I occurred on January 15.8, 16.4, 17.0, 17.6 and 17.9, i.e., on the average, every 0.6 day. A similar pattern is observed with other comets having a system of synchrones. The most natural explanation appears to be that the nuclear surface consists of two areas which differ radically in the amount of matter emitted per unit time. If such a nucleus is in rotation, the total amount of matter emitted is modulated by the rotational frequency, and this results in a group of almost equally spaced synchrones. Certainly, the periodicity cannot persist for a long time, as a given nuclear layer may be totally exhausted, or a similar layer may be laid bare in another region, and so on. Since the synchrone formation occurs at the same phase of the rotation, the ejection will affect the comet's motion as a single reactive impulse. It would be interesting to analyse the fluctuations in brightness along the syndynames, even in the absence of pronounced synchrones. Such analysis can give important information about changes in the rate of evaporation from the nucleus and hence about the composition and rotation of the nucleus.

4. Dust Plasma

We shall not consider here the complicated problem of ion motion in type I tails. When considering the dust component of the cometary atmosphere, however, it is necessary to take into account the fact that dust particles take up a charge that depends on the dynamical equilibrium of the processes of electron absorption from the surrounding plasma and electron ejection by photons and fast particles. A common medium, which may be called a dust plasma, composed of ions, electrons, and charged dust particles that may be regarded as a special kind of heavy ions, is formed at distances sufficiently removed from the nucleus for the neutral particles to become ionized. Dust plasma will have collective properties characteristic of ordinary plasma. In addition to the waves characteristic of a collection of electrons and atomic or molecular ions there will be waves associated with the presence of charged dust particles. In order to determine the specific parameters of dust plasma it is necessary to know the distributions of mass and charge among the dust particles, but information on both distributions is very uncertain. The estimates normally quoted for a dust particle in the interplanetary medium at a distance of 1 AU from the Sun lead to a positive charge of the order of 100 times that on an electron, in which case, a 10^{-5} cm particle will have a potential of 1.5 V. The Debye radius in the cometary dust plasma is determined by the molecular ion concentration, which is much larger than that of dust particles.

Various types of dust-plasma oscillations are possible.

Comets are so large that they can be expected frequently to cross meteor streams. It would be interesting to pre-compute the dates of such crossings and observe the possible effects. Specifically, it would be of interest to establish whether the anomalous behaviour described by Biermann and Lüst (1965) for the plasma tail of comet 1964 VI (Tomita-Gerber-Honda) was due to its intersecting a meteor stream. To explain this phenomenon by interaction with an interplanetary magnetic field would require that the strength of the field was much greater than actually observed.

References

Biermann, L. and Lüst, Rh.: 1965, *Mem. Soc. Roy. Sci. Liège Ser.* 5 **12**, 329.
Dolginov, A. Z.: 1967, *Astron. Zh.* **44**, 434.
Dolginov, A. Z. and Gnedin, Yu. N.: 1966, *Icarus* **5**, 64.
Dolginov, A. Z., Gnedin, Yu. N., and Novikov, G. G.: 1971, *Planetary Space Sci.* **19**, 143.
Hamel, B. B. and Willis, D. R.: 1966, *Phys. Fluids* **9**, 829.
Whipple, F. L.: 1950, *Astrophys. J.* **111**, 375.

Discussion

F. L. Whipple: Regarding condensation of dust near the comet, gas pressure should decrease with distance from the nucleus, and temperature should probably increase, contradicting the hypothesis. Also, dust and meteoroids must be of the same composition. Therefore iron, sodium, calcium, even hydrogen and many other elements must condense, not just carbon. Vaporization of these atoms in a comet nucleus seems unlikely.

A. Z. Dolginov: The absolute values of pressure and temperature, rather than their relative changes, are of importance for condensation. In principle, possible condensation processes for other elements may also be considered.

F. L. Whipple: I accept correction to my criticism if one allows possible growth of dust particles to dimensions of the order of microns.

G. Guigay: I should like to emphasize that synchrones are not always oriented in a radial direction; it was not the case for comet 1957 V (Mrkos), for example. Hence, the case you presented cannot be considered typical.

A. Z. Dolginov: The deflection of the synchrones from a radial direction is usually not large. I attempt here to explain only the large-scale properties of synchrones. The periodicity in the appearance of synchrones shows that the explanation is correct. To explain the deflection one must consider the development of the tail in more detail and take into account that the constituent matter of the synchrones is the dust plasma.

46. SOME REMARKS ON THE LIBERATION OF GASES FROM COMETARY NUCLEI

B. YU. LEVIN

O. Schmidt Institute of Physics of the Earth, Moscow, U.S.S.R.

Abstract. According to modern data water ice is the main constituent of cometary nuclei. The rate of its evaporation determines the rate of liberation of other constituents, including those that determine the photometric properties of comets. The estimates of the total mass loss per perihelion passage seem to give values about one order of magnitude smaller than those necessary to explain the nongravitational forces as due to the reactive force of material leaving the nuclei. Thus the conventional explanation of the nongravitational forces deserves further study from the point of view of the physical theory of comets.

1. Introduction

About 20 years ago Dubyago (1948) and Whipple (1950) ascribed the acceleration and deceleration of motion of some periodic comets (Encke, Biela, Brooks 2, and others) to the reactive force of material leaving the surface of a rotating cometary nucleus. Since then this explanation of the nature of nongravitational forces influencing the motions of some comets – both periodic and nonperiodic – has become widely accepted.

However, one difficulty exists: to explain the values of the nongravitational forces derived from observations, a mass loss of a few percent per perihelion passage is necessary (Dubyago, 1948; Whipple, 1950; Sekanina, 1969). Smaller values published by several authors are underestimates because the numerical value of the directional factor used in the calculations was 3 to 5 times too large. According to modern ideas on the evaporation of icy cometary nuclei, a mass loss of the order of 0.1 to 1% per perihelion passage is possible, but it seems impossible to explain larger mass losses.

Let us discuss this difficulty in detail.

2. Evaporation of Water Ice and the Photometric Properties of Comets

The discovery of the presence of atomic oxygen in cometary heads is important evidence which leads, with other evidence, to the conclusion that water ice (or snow) is the major constituent of cometary nuclei. For water ice, the heat of evaporation is large ($L = 12\,000$ cal mol^{-1} = 670 cal g^{-1}) and, as was shown by several authors (Biermann and Trefftz, 1964; Huebner, 1965; Delsemme, 1966; Rijves, 1966, 1970), its sublimation from an exposed nucleus unprotected by a nonvolatile insulating layer explains the production rate of H_2O molecules derived from the observed intensities of the O and OH lines. Smaller values of L derived earlier from the rate of variation of the so-called integral brightness with heliocentric distance, and based on the hypothesis that gases are desorbed from the nonvolatile layer, are fictitious for two reasons:

(1) For $L = 12\,000$ cal mol^{-1} the zone where comets are usually observed ($r \approx 1$ AU)

Chebotarev et al. (eds.), The Motion, Evolution of Orbits, and Origin of Comets, 260–264. All Rights Reserved.

corresponds to transition from intense sublimation (evaporation) at small heliocentric distances r to slow sublimation at large r. At sufficiently small r almost all the solar heat absorbed by the nucleus is used for sublimation of ice, and here the direct production of gas molecules from the nucleus as well as the liberation of nonvolatile dust particles varies according to the $1/r^2$ law. According to the experiments by Delsemme and Wenger (1970), grains of water ice must be stripped from the surface of the nucleus by the outstreaming gas. These grains evaporate when moving through the coma and thus increase the total gas production. It is probable that the ejection of grains also follows the $1/r^2$ law, and consequently the total gas production Q follows this law at small r.

On the other hand, when the comet is far from the Sun almost all the solar heat is re-emitted in the form of thermal radiation, and here Q closely follows the $\exp(-L/RT)$ law, where the effective temperature T is determined by radiative equilibrium ($T = T_0\sqrt{r}$).

In the transition zone, both sublimation and reradiation play an important role in the heat balance (Delsemme, 1966). In this zone inappropriate application of the desorption formula leads to fictitious numerical values of L much smaller than 12 000 cal mol^{-1}.

(2) The analysis of photometric observations of comets by means of the desorption formula was based on the proportionality between Q and the true integral brightness of the head I (Levin, 1943, 1947, 1948). However, observations usually give, not the total integral brightness I, but only the integral brightness I' of the central part of the head. If the limiting isophote has a constant brightness B, then $I' \sim Q^2/r^4$. But for many comets observed at $r < 1$ AU the decrease in r (and increase in Q) is accompanied by a decrease in the elongation of the comet from the Sun and, consequently, by an increase in the brightness of the sky background. Therefore, B increases, and this alters the dependence of I' on Q and r. For most comets it is thus impossible to derive the correct dependence of Q on r from the observed 'integral' brightness of the head.

Formerly (Levin, 1966), I regarded the strong dependence of integral brightness on heliocentric distance r – much stronger than the $1/r^2$ law – as an argument in favour of desorption of gases from a nonvolatile layer and against the sublimation of exposed ices. This was because at that time I thought the low values of L obtained from the photometric observations were real. If one assumes that cometary nuclei are composed of volatile ices with $L \approx 3000$–5000 cal mol^{-1} and unprotected by a nonvolatile layer, then at heliocentric distances of the order of 1 AU the sublimation would be intense and would follow the $1/r^2$ law. But this is not the case. At the present time the photometric behaviour of bright comets can be explained as due to the decisive role of the less volatile H_2O ice in the composition of nuclei. The rate of sublimation of the water ice determines the rate of liberation of all the other components, both more volatile and less.* For faint comets the observed integral brightness deviates substantially

---

* As some components of cometary nuclei are more volatile than water ice, and as nonvolatile components and grains of water ice are ejected without evaporation (Huebner and Weigert, 1966; Delsemme and Wenger, 1970; Delsemme and Miller, 1970), the effective heat of evaporation (per gram of ejected material) is somewhat smaller than 670 cal g^{-1}.

from the true brightness, and this has an additional influence on their photometric behaviour.

The smaller brightness of periodic comet Encke after perihelion, which I regarded as further evidence in favour of the desorption hypothesis (Levin, 1943, 1966), must now be explained in some other way, perhaps as a systematic error due to difference in observing conditions (brightness of sky background, etc.).*

Thus I withdraw my previous objections against the evaporation of exposed ices and agree that the total rate of gas production in comets is determined mainly by the rate of evaporation of the water ice. Therefore, the so-called Levin formula cannot be used to compute the latent heat of gas production in comets.

3. Evaporation of Exposed Ices and the Necessary Mass Loss

In the case of evaporation of exposed ices, although many details remain unclear or even unknown, it is possible to evaluate the thickness of the surface layer lost during one perihelion passage, as was calculated by Huebner (1967). For a typical comet with its perihelion inside the orbit of the Earth the thickness of this layer is a few metres, and it reaches a few tens of metres for sungrazing comets (Huebner, 1967). Since the relative mass loss is given by

$$\frac{\Delta M}{M} = \frac{4\pi R^2 \Delta R}{\frac{4}{3}\pi R^3} = \frac{3\Delta R}{R},$$

we obtain, for a nucleus with $R=3$ km and $\Delta R=3$ m, $\Delta M/M \sim 0.3\%$.

For smaller nuclei the mass loss can reach 1%, while for larger nuclei it is of the order of 0.1%. These numerical values for the mass loss are in the same range as those calculated by Dubyago (1948), Whipple (1950), and Sekanina (1969) to explain the nongravitational forces. However, all these authors used in their calculations a value of about 0.5 for the directional factor. This value seems somewhat high even for the radial component of the reactive force produced by evaporation from the sunlit hemisphere of a nonrotating nucleus. But it is several times too high when applied to the transverse component connected with the rotation of the nucleus. For a directional factor of about 0.1 Sekanina (1969) obtained for three comets mass loss rates of 0.1 to 0.5%, but for three others he obtained rates of 1% or more. The case of P/Schwassmann-Wachmann 2 is especially strange because of its large perihelion distance q (2.1 AU), while ΔR should decrease with increasing q.†

Thus there is a discrepancy of one order of magnitude between the value of the mass loss necessary to explain the nongravitational forces and that accounted for by the modern physical theory of comets.

---

* For nonperiodic comets (1915 II Mellish, 1965 VIII Ikeya-Seki) the similar effect can be explained by the development of a radial gradient in the composition of the outermost layers of the nucleus.
† In the zone where the $1/r^2$ law is applicable ΔR decreases proportionally to $q^{-1/2}$, if the role of eccentricity is neglected (Huebner, 1967), while at larger r it decreases more rapidly (Levin, 1947, p. 208).

4. Position of Maximum Ejection of Material

As cometary nuclei are composed of snow with low thermal conductivity, the surface temperature of a nucleus must quickly adjust to changes in the incident solar radiation caused by the nuclear rotation. At small r this is achieved by the use of solar heat for evaporation of ice, while at larger r both evaporation and reradiation play an important role. The maximum surface temperature must be close to the subsolar point.

Observations of comets seem to confirm these expectations. Drawings by visual observers of central parts of the coma, as well as photographs on which these parts are not overexposed, show emission fans directed towards the Sun. This can be seen on several figures in the *Atlas of Cometary Forms* by J. Rahe, B. Donn, and K. Wurm, on

Fig. 1.   Comet 1932 V
(Peltier-Whipple)
on 1932 August 24–25.

Fig. 2.   Comet 1932 V
(Peltier-Whipple)
on 1932 August 26–27.

Fig. 3.   Comet 1936 II
(Peltier)
on 1936 July 26–27.

some photographs of P/Encke (Roemer, 1961), and on unpublished photographs of comets 1932 V (Peltier-Whipple) and 1936 II (Peltier) taken for astrometric purposes with the 38-cm double astrograph at the Moscow Observatory (see Figures 1–3).

Although studies of the precise directions of these fans are lacking, it seems obvious that there are no large deviations from the sunward direction.

In most cases the behaviour of the nongravitational forces is in line with these data: the radial component is usually directed away from the Sun and is an order of magnitude larger than the transverse component (Marsden, 1969). Negative values of the radial component were recently found by Marsden (1970) for periodic comets Encke, d'Arrest, Wirtanen, Arend, and perhaps Pons-Winnecke, but the precision of these negative values is poor.

In all studies of nongravitational forces the possible displacement of the photometric centre of the asymmetric image of a comet from the true position of the nucleus was not taken into account. The probable importance of this effect was stressed many years ago by Mokhnach (1956), and the observations of the comet 1957 III (Arend-Roland) seem to confirm it (Debehogne, 1968). This effect must be taken into account in the study of such a delicate problem as nongravitational forces in comets.

5. Conclusion

At the present time there are difficulties concerning the amount of the mass loss from cometary nuclei. Resolution of them is indispensable for explaining the presence of nongravitational forces.

References

Biermann, L. and Trefftz, E.: 1964, *Z. Astrophys.* **59**, 1.
Debehogne, H.: 1968, *Acad. Roy. Belg. Bull. Cl. Sci.* **54**, 1040.
Delsemme, A. H.: 1966, *Mem. Soc. Roy. Sci. Liège Ser. 5* **12**, 69, 77.
Delsemme, A. H. and Miller, D. C.: 1970, *Planetary Space Sci.* **18**, 717.
Delsemme, A. H. and Wenger, A.: 1970, *Planetary Space Sci.* **18**, 709.
Dubyago, A. D.: 1948, *Astron. Zh.* **25**, 361.
Huebner, W. F.: 1965, *Z. Astrophys.* **63**, 22.
Huebner, W. F.: 1967, *Z. Astrophys.* **65**, 185.
Huebner, W. F. and Weigert, A.: 1966, *Z. Astrophys.* **64**, 185.
Levin, B. Yu.: 1943, *Dokl. Akad. Nauk SSSR* **38**, 82; *Astron. Zh.* **20** (4), 37.
Levin, B. Yu.: 1947, *Usp. Astron. Nauk* **3**, 191.
Levin, B. Yu.: 1948, *Astron. Zh.* **25**, 246.
Levin, B. Yu.: 1966, *Mem. Soc. Roy. Sci. Liège Ser. 5* **12**, 65.
Marsden, B. G.: 1969, *Astron. J.* **74**, 720.
Marsden, B. G.: 1970, *Astron. J.* **75**, 75.
Mokhnach, D. O.: 1956, *Byull. Inst. Teor. Astron.* **6**, 269.
Rijves, V. G.: 1966, *Komety i Meteory* No. 3, 3.
Rijves, V. G.: 1970, *Komety i Meteory* No. 17, 3.
Roemer, E.: 1961, *Publ. Astron. Soc. Pacific* **73**, 170.
Sekanina, Z.: 1969, *Astron. J.* **74**, 1223.
Whipple, F. L.: 1950, *Astrophys. J.* **111**, 375.

Discussion

A. Z. Dolginov: What is the basis for your statement that the nucleus consists mainly of water ice?

B. Yu. Levin: Estimates of the amount of gas evaporated from the nucleus, based on the hypothesis of water ice, are in agreement with observed intensities of atomic oxygen and hydroxyl in the head.

H. Alfvén: If the evaporation velocity is of the order of at most 100 m s^{-1}, the maximum value of ΔV necessary to account for the nongravitational effects cannot very much higher. Is this in agreement with the observed values of the nongravitational effects?

B. Yu. Levin: I think that we need an order of magnitude increase, either in the mass of matter ejected or in the velocity. The calculation allows for the relative velocity of the material leaving the nucleus, assumed to be 500 m s^{-1}.

D. O. Mokhnach: I want to say some words in support of nongravitational forces, and to refute my views of 15 years ago. Vsekhsvyatskij has shown that the secular decrease in the brightness of P/Encke is 2 to 2.5 magnitudes per 50 revolutions. Even if we assume that the decrease is only one magnitude, then the decrease in the evaporating surface of the nucleus would be as much as 60% and the corresponding mass loss would be about 25%. Most of the mass loss takes place near perihelion, and in fact, the mass loss in other parts of the orbit may be neglected. Hence, it was reasonable for Makover to assume that an abrupt acceleration occurs at the perihelion point. There are other effects associated with the variations in total brightness, such as the deformation of isophotes, and the deflection of the photometric centre; and systematic discrepancies between the mass centre and photometric centre can result only in an increase of the residuals of the orbit calculation. However, the mass loss, which is probably greater than I have calculated, yields nongravitational effects that may easily result in acceleration or deceleration of the comet's motion.

47. THE CHEMICAL COMPOSITION OF COMETARY NUCLEI

L. M. SHUL'MAN

Main Astronomical Observatory, Ukrainian Academy of Sciences, Kiev, U.S.S.R.

Abstract. The probable parent-molecules of radicals such as C_3 and N_2^+ are discussed, and it is concluded that cometary nuclei may contain complicated organic molecules, such as C_3H_4, CH_2N_2, and C_4H_2. It is suggested that these molecules are formed by radiation synthesis in solid phase. In a time interval of order 10^7 to 10^9 yr bombardment from cosmic rays would be expected to transform the chemical composition to a depth of 1 m. Solar cosmic rays do not penetrate as far, and as a result the surface layer of the nucleus can be enriched with unsaturated hydrocarbons. After a critical concentration of this explosive material is reached a further burst of solar cosmic rays can initiate an explosion and thus an outburst in the comet's brightness. This mechanism is the only one advanced to date that can explain the synchronism of the energy output over the whole nuclear surface.

1. Introduction

The present paper deals with the chemical composition of cometary ices, a subject about which we have only indirect information. It is well known that spectroscopic data give us information only about the radicals that have appreciable emissivity in the visible region of the spectrum. Many authors, beginning with Wurm, have treated this problem. As a result we have a long list of proposed parent-molecules (see Table I).

TABLE I

Parent-molecules

Observed radical	Proposed parent-molecules
CO^+, CO_2^+	CO_2
N_2^+	$N_2(?)$, $C_2N_2(?)$
C_2	C_2H_2, C_2H_4, $C_2H_6(?)$
C_3	$C_3H_8(?)$
OH	H_2O, H_2O_2
CH	CH_4
CH_2	CH_4
NH	NH_3
NH_2	NH_3, N_2H_4, CH_3NH_2
CN	C_2N_2, HCN

The choice of these molecules is based mainly on their simplicity. We ought to admit that simplicity is not a sufficient criterion for the selection and take into consideration other criteria too. Cometary ices must consist of sufficiently volatile substances. The molecules of the substances must contain the observed radicals, and the radicals must be weakly bonded in parent-molecules. The dissociation energy of an observed radical must exceed the escape energy of the radical from its parent-molecule. The parent-molecule structure must allow the formation of the observed radical in the easiest way, i.e., in the minimum number of elementary stages that have the maximum possible cross-section. We must reject any mechanisms for the formation of the cometary neutrals demanding three-body collisions in the inner coma.

2. Formation of the C_3 Radical

There are many substances that satisfy the above criteria. However, for several radicals the number of possible parent-molecules is very restricted, so we can predict the parent-molecules almost exactly. Radicals of this type include C_3. Each suggested parent-molecule must contain a group

$$C{=}C{=}C$$

or

with double bonds between carbon atoms. The molecules with

or

are not suitable owing to the relationship between the binding-energies of single (2.5 to 3.5 eV), double (4 to 6 eV) and triple (7.5 to 8 eV) bonds.

So we can assume that the simplest molecules able to produce C_3 are $H_2C:C:CH_2$, $HC:C:CH$ and $(CH_3)_2C:C:C(CH_3)_2$. In the last case C_3 escapes more easily because the binding energy of the CH_3 radical is ~ 3 eV, as opposed to ~ 4 eV for H.

Another mechanism, proposed by Jackson and Donn (1966), involves a disproportionation reaction following photodissociation of $HC:C.C:CH$, namely

$$C_4H_2 + h\nu \longrightarrow C_3H + CH \tag{1}$$

$$C_3H + R \longrightarrow C_3 + RH, \tag{2}$$

where R is an arbitrary radical. This supposition is based on the experimental results on the photolysis of diacetylene and on their interpretation given by Calloman and Ramsay (1957). Calvert and Pitts (1965) have found this explanation doubtful because a single bond in the C_4H_2 molecule is more likely to be broken up than a triple one:

$$C_4H_2 + h\nu \longrightarrow C_2H + C_2H. \tag{3}$$

Calvert and Pitts explain the origin of C_3 radicals in laboratory photolysis of C_4H_2 by another reaction of disproportionation:

$$C_2H + C_2H \longrightarrow C_3H + H. \tag{4}$$

The process (2) follows this reaction.

One can see an essential difference between the processes (1)–(2) and (3)–(4). The second one will take place only when the frequency of C_2H—C_2H collisions is sufficiently high. The number density of C_2H radicals must exceed 10^9 cm^{-3}. In addition, the number density of C_4H_2 must be much less than that of C_2H to prevent polymerization according to the scheme:

$$C_2H + H-C\equiv C-C\equiv C-H \longrightarrow H-C\equiv C-C\equiv C-\underset{\underset{H}{|}}{C}-\underset{\underset{H}{|}}{C}-. \tag{5}$$

It is interesting to note that the process (5) results in the formation of organic grains, and it may therefore be proposed as one of the possible mechanisms for the origin of dust in the atmospheres of comets.

A third possible way of forming the C_3 radical is fusion, e.g.

$$C_2 + C \longrightarrow C_3 + h\nu. \tag{6}$$

Besides difficulty with the probability of this process, we have here direct disagreement with observations. Indeed, according to Swings and Haser (1956), the C_3 radical is observed at greater heliocentric distances than C_2.

3. Formation of N_2^+ Ions

Another puzzle of similar character is the origin of the N_2^+ ions. The simplest supposition is that molecular nitrogen exists in the nuclei. But if this were so, nitrogen should have been evaporated at great heliocentric distances, owing to the low temperature and specific heat of evaporation. Another obvious difficulty is that emission of neutral nitrogen has not been observed, and if N_2^+ ions appear as the result of the ionization of N_2, the first negative system of the latter should certainly be present.

So we have evidence that N_2^+ is formed in some other way. If there is no free N_2 in cometary nuclei, we have to find parent-molecules that would give N_2^+ after dissociation, preferably with a charge exchange. The simplest molecules available are

$$\underset{\diagdown}{H}$$
$$N\overset{+}{=}N\overset{-}{=}N$$

and

$$\underset{\diagdown}{H}$$
$$\underset{\diagup}{\underset{H}{}}C\overset{+}{=}N\overset{-}{=}N$$

or

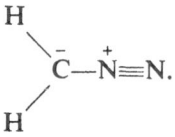

All these molecules have the groups $=\overset{+}{N}=\overset{-}{N}$ or $-\overset{+}{N}\equiv N$. The following processes may be proposed as mechanisms for the formation of N_2^+ ions:

$$HN_3 + H^+ \longrightarrow HN + N_2^+ + H \tag{7}$$

$$CH_2N_2 + H^+ \longrightarrow CH_3 + N_2^+ \tag{8}$$

$$\longrightarrow CH_2 + H + N_2^+ \tag{9}$$

or

$$HN_3 + h\nu \longrightarrow NH + N_2^+ + e \tag{10}$$

$$CH_2N_2 + h\nu \longrightarrow CH_2 + N_2^+ + e. \tag{11}$$

There are no experimental data on these processes and any laboratory experiments would be of significant interest.

4. Origin of Complex Parent-Molecules

In summary, we conclude that complicated molecules such as C_3H_4, CH_2N_2, C_4H_2 and others are to be expected in cometary nuclei. But how could they have been formed? It is very likely that the mechanism is radiation synthesis in solid phase induced by cosmic rays. Indeed, a cometary nucleus is exposed to cosmic rays for a long time before it approaches the Sun. The total intensity of galactic cosmic rays $I = 8 \times 10^8$ eV cm^{-2} s^{-1} sr^{-1}. The mean energy of a cosmic proton is $E = 2 \times 10^9$ eV. This corresponds to a penetration depth of

$$l = \frac{60 \text{ to } 80}{\delta} \text{ g cm}^{-2}, \tag{12}$$

where δ (g cm^{-3}) is the average density of the surface layer of the nucleus. Cosmic rays thus penetrate into the nucleus to a depth about 1 m.

A molecule in the surface layer absorbs the power

$$\frac{dD}{dt} = \frac{\pi I m}{l\delta} \text{ eV s}^{-1}, \tag{13}$$

and typical values of the radiation synthesis output are in the range

$$G \sim 0.01 \text{ to } 1 \text{ reaction eV}^{-1}. \tag{14}$$

This means that every molecule in the layer will be hit by a cosmic proton (and therefore able to take part in chemical reactions) during a time interval of

$$T = \frac{l\delta}{G\pi I m} = 2 \times 10^7 \text{ to } 2 \times 10^9 \text{ yr.} \tag{15}$$

This estimate shows that the chemical composition of ices in the surface layer of the nucleus may be entirely transformed during a time considerably shorter than the age of the solar system.

Unfortunately, there are no experimental data on the processes mentioned here. But although nobody has exposed the assumed cometary ices (especially the mixtures) to hard radiation, some laboratory results are of interest to the physics of comets. For example, it has been shown (Davis and Libby, 1964) that CH_4 exposed to the radiation of Co^{60} below a temperature of 77 K was transformed into viscous oil. The oil was a mixture of complicated molecules of the C_mH_n type, where the value of m approached 20. It would be very desirable if similar experiments were made with frozen mixtures.

5. Effects of Solar Cosmic Rays

A cometary nucleus is affected, not only by galactic cosmic rays, but also by solar ones. Solar cosmic rays (SCR) are softer and do not penetrate as far as those from the Galaxy. SCR are irregular in time because they are generated during active processes on the Sun.

The power spectrum of SCR is often approximated by the expression:

$$D(E) = D_0 E^\gamma \text{ proton cm}^{-2} \text{ s}^{-1} \text{ sr}^{-1} \text{ MeV}^{-1}, \tag{16}$$

where E (MeV) is the energy of a proton. The spectral index γ approaches values of 4 to 6 during solar flares, and $\gamma = 3$ when flares are absent. The free path l of a proton depends on its energy as

$$l = E^{7/4}/B\delta, \tag{17}$$

where B is a constant depending on the chemical nature of the absorber. For lead and air B is equal to 220 and 417, respectively. We shall accept the value $B = 300$.

It is easy to show that the power P absorbed at a depth h is given by the relation

$$P = \frac{2\pi(\gamma + 1)\Gamma(\tfrac{8}{7})\Gamma[(4\gamma - 3)/7]}{(\gamma - \tfrac{3}{4})\Gamma[(4\gamma + 11)/7]} \frac{D_0}{B^{4(\gamma - 2)/7}(h\delta)^{(4\gamma - 1)/7}} \text{ MeV g}^{-1}\text{ s}^{-1}. \tag{18}$$

For example (Brown and D'Arcy, 1959), the solar flare on 1959 July 14–15 had parameters $D_0 = 6 \times 10^{12}$; $\gamma = 4.5$. It follows that the surface layer (as deep as 1 mm) would receive ~ 0.5 eV per molecule in a time of the order 10^5 s. This energy is sufficient to destroy intermolecular bonds in cometary ices. The above-mentioned flare was extremely intense, and similar flares are seldom observed. Weaker bursts of SCR are more common and involve the radiation synthesis and accumulation of unsaturated hydrocarbons and other explosives in the surface layer of the cometary nucleus. This accumulation will build up until the critical concentration is reached. The next flare can initiate the inevitable explosion and hence an outburst in the brightness of the comet.

It is worth mentioning that SCR synchronize the output of energy over the whole nuclear surface, and thus the envelope thrown off as a result of the surface explosion will have spherical symmetry. All hypotheses on this subject proposed until now fail to explain fully the synchronization of the explosion.

It is clear that at small heliocentric distances intensive evaporation prevents the storage of explosives on the surface of a nucleus. But this mechanism gives us a key to understanding the outbursts in brightness and expanding shells observed for comets at large heliocentric distances.

References

Brown, R. R. and D'Arcy, R. G.: 1959, *Phys. Rev. Letters* **3**, 390.
Calloman, J. H. and Ramsay, D. A.: 1957, *Can. J. Phys.* **35**, 129.
Calvert, J. G. and Pitts, J. N.: 1965, *Photochemistry*, New York and London.
Davis, D. R. and Libby, W. F.: 1964, *Science* **144**, 991.
Jackson, W. M. and Donn, B. D.: 1966, *Mem. Soc. Roy. Sci. Liège Ser. 5* **12**, 133.
Swings, P. and Haser, L.: 1956, *Atlas of Representative Cometary Spectra*, Liège.

48. THE EVOLUTION OF COMETARY NUCLEI

L. M. SHUL'MAN

Main Astronomical Observatory, Ukrainian Academy of Sciences, Kiev, U.S.S.R.

Abstract. The evolution of a cometary nucleus depends on the size distribution of the dust content. If all the dust particles are small enough, they will all be thrown off into the coma; the secular decrease in the comet's brightness is due simply to the decrease in the nuclear radius, and the comet eventually disintegrates. On the other hand, if larger grains are present, they will start to form spots of a low-emissivity mineral envelope on the surface of the nucleus; in this case, the decrease in brightness is due to the increase in the area of the screened part of the surface, and the reduction in the nuclear radius is comparatively small. It is shown that after about 90% of the surface has been screened, the rate of screening increases considerably, as smaller particles become trapped between the larger ones; and there will be a corresponding acceleration in the brightness decrease. These nuclei are eventually covered with a solid mineral envelope, and they are transformed into asteroids.

1. Introduction

The icy model for a cometary nucleus is now generally accepted. However, some of the conceptions based on this model deserve criticism. The first one is the supposition that a so-called old comet has a nucleus covered by a continuous envelope of dust and stones. The second is the idea that the mineral envelope can be suddenly thrown off when the internal gas pressure reaches a certain level.

It will be shown in the present paper that comets that do not have exposed ice on their nuclear surfaces would be observed, not as comets, but as asteroids. A new modification of the icy model will be proposed: an icy surface that either becomes slowly covered by spots of a mineral envelope (spotted model) or remains homogeneous, depending on the size of the grains impregnated in the ice.

The merging of the separate islands of nonvolatile matter into one continuous envelope results in the 'death' of the comet, just as the total disintegration of a dustless nucleus does.

2. Effect of Grain Size

We assume that in its initial state the surface layer of a nucleus is a homogeneous mixture of ices, grains, and stones. The granulometric composition of the mineral bodies is given by the distribution function

$$F(a) = \begin{cases} n_d \dfrac{S a_{max}^s a_{min}^s}{a_{max}^s - a_{min}^s} \dfrac{1}{a^{s+1}}; & a_{min} < a < a_{max} \\ 0; & a > a_{max}, \, a < a_{min}, \end{cases} \tag{1}$$

where a is the radius of a grain and n_d the number density of grains. From meteoric data we adopt $s = 3$. A grain is in equilibrium at the nuclear surface if its radius satisfies the relationship

Chebotarev et al. (eds.), The Motion, Evolution of Orbits, and Origin of Comets, 271–276. All Rights Reserved.
Copyright © 1972 by the IAU.

$$a_{eq} = \frac{9}{32\pi} \frac{C_D mnv^2}{G\rho_d\rho_n R}, \tag{2}$$

where m, n and v are the molecular mass, number density, and velocity of the gas, respectively, ρ_d and ρ_n are the mean mass densities of a grain and the nucleus, R is the radius of the nucleus, G is the gravitational constant, and C_D is a drag coefficient given by the relation

$$C_D = \frac{2}{3S}\sqrt{\frac{\pi T_d}{T}} + \frac{2S^2 + 1}{S^3\sqrt{\pi}} e^{-S^2} + \frac{4S^4 + 4S^2 - 1}{2S^4} \frac{2}{\sqrt{\pi}} \int_0^S e^{-x^2}\,dx, \tag{3}$$

where T_d and T are the temperatures of the grain and gas, and

$$S = \frac{v}{\sqrt{(2kT)/m}} \approx 0.35 \tag{4}$$

(k being the Boltzmann constant) is a dimensionless parameter. Under typical conditions $C_D \approx 10$. All the grains with radii $a > a_{eq}$ will be stored at the surface and will form the nonvolatile envelope.

Three cases are possible. If

$$a_{eq} > a_{max}, \tag{5}$$

then all the dust particles can be thrown off into the atmosphere. Thus the nucleus evolves in a uniform manner, the composition of the surface layer remaining homogeneous. In this case, the radius decreases, and the evolution finishes with the total disintegration of the nucleus. Dustless comets, such as P/Encke (Whipple and Douglas-Hamilton, 1966), have the same uniform evolution. A second possibility is

$$a_{min} < a_{eq} < a_{max}, \tag{6}$$

in which case the grains with radii $a_{eq} < a < a_{max}$ will be stored on the surface, and the surface layer becomes nonuniform. This is the spotted model. The dusty crust grows until it covers the whole surface and the comet turns into an asteroid. The third case, $a_{eq} < a_{min}$, is similar to the second. It differs only by absence of dust in the cometary atmosphere, except for the dust formed by condensation of carbon (Dolginov, 1967).

3. Secular Decrease in Brightness

The second of the above-mentioned cases is the most general one. To treat it in detail we shall use a semiempirical approach, taking for the gas flux the expression

$$nv = n_1 v_1/r^n, \tag{7}$$

which corresponds to the empirical law of brightness variation

$$E = E_1/r^n, \tag{8}$$

r being the heliocentric distance, and where quantities with subscript 1 refer to $r = 1$ AU. Since v changes slowly one can obtain

$(\rho_n \sim \rho_d \sim 1 \text{ g cm}^{-3}, \quad v_1 \sim 10^4 \text{ cm s}^{-1}, \quad C_D \sim 10, \quad G \sim 6.7 \times 10^{-8},$

$n_1 \sim 10^{12} \text{ cm}^{-3}, \quad m \sim 5 \times 10^{-23} \text{ g})$

$$a_{eq} \text{ [cm]} = \frac{0.7}{R\text{[km]}r^n \text{ [AU]}}. \tag{9}$$

Below we shall need the following averaged quantities:

$$\langle a^2 \rangle = \frac{1}{n_d} \int_{a_{min}}^{a_{max}} a^2 F(a) \, da \sim 3a_{min}^2 \tag{10}$$

$$\langle a^3 \rangle = \frac{1}{n_d} \int_{a_{min}}^{a_{max}} a^3 F(a) \, da \sim 3a_{min}^3 \ln \frac{a_{max}}{a_{min}} \tag{11}$$

$$\langle a_s^2 \rangle = \frac{1}{n_d} \int_{a_{eq}}^{a_{max}} a^2 F(a) \, da \sim \frac{3a_{min}^3}{a_{eq}} \left(1 - \frac{a_{eq}}{a_{max}}\right). \tag{12}$$

Instead of the number density n_d we can use the fraction by volume f_v of the dust. Obviously,

$$f_v = n_d \pi a_{min}^3 \ln \frac{a_{max}}{a_{min}}. \tag{13}$$

If ξ is the fraction of the surface screened by the crust, then the rate at which the radius decreases is defined by the relation

$$\frac{dR}{dt} = \frac{(1 - \xi)mn_1 v_1}{(1 - f_v)\rho_n r^n}, \tag{14}$$

and thus the rate of screening is

$$\frac{d\xi}{dt} = \pi \langle a_s^2 \rangle n_d \frac{4\pi R^2(dR/dt)}{4\pi R^2} = \frac{3f_v(1 - \xi)mn_1 v_1[1 - (a_{eq}/a_{max})]}{4(1 - f_v)\rho_n a_{eq} r^n \ln (a_{max}/a_{min})}. \tag{15}$$

Substituting here the value of a_{eq} at perihelion and integrating over one revolution, we obtain the change in ξ per revolution from

$$\Delta \ln (1 - \xi) = -2.8 \times 10^5 \frac{R\text{[km]}\rho_n\text{[g cm}^{-3}]}{C_D v^{(p)}\text{[cm s}^{-1}]} \frac{f_v}{1 - f_v} \frac{1 - a_{eq}^{(p)}/a_{max}}{\ln (a_{max}/a_{min})}$$

$$\times \frac{q^{3/2}\text{[AU]}}{(1 + e)^{n - 1.5}} \int_0^{2\pi} (1 + e \cos \theta)^{n - 2} \, d\theta, \tag{16}$$

where $v^{(p)}$ and $a_{eq}^{(p)}$ are the values of gas velocity and equilibrium grain radius at perihelion, and q, e, and θ are the perihelion distance, orbital eccentricity, and true anomaly of the comet.

Finally, one can obtain an expression for the secular variation in the total absolute magnitude H_y:

$$H_y = H_y^{(0)} + 2.5\, Mv\, \varDelta \ln (1 - \xi), \tag{17}$$

where M is the modulus of common logarithms, and the decrease in absolute brightness changes linearly with the number of revolutions v. Such behaviour will take place until the particles with radii $a > a_{eq}$ touch one another at the surface of nucleus. It is easy to show that when the particles merge approximately $\pi/2\sqrt{3} \simeq 0.9$ of the total area will be screened.

At this moment the second layer of the crust will start forming. Of course, the smaller particles will be trapped in the pores between the particles stored in the first layer. This is equivalent to the rapid decrease of the quantity a_{eq} to the value

$$a'_{eq} = \left(\frac{2}{\sqrt{3}} - 1\right) a_{eq} \sim 0.15\, a_{eq}, \tag{18}$$

and the rate of brightness decrease will thus be multiplied by a factor of approximately 7. During the forming of the second layer of the crust the following relation holds:

$$H_y = H_y^{(0)} + 2.5\, Mv_1\, \varDelta \ln (1 - \xi) + 16.5\, M (v - v_1)\varDelta \ln (1 - \xi), \tag{19}$$

where v_1 is the number of revolutions before the formation of the first layer of the crust.

The straight lines in Figure 1 show the idealized decrease in brightness as the crust formation proceeds layer by layer. In fact, of course, the brightness variation will be smoothed out, as shown by the broken curve, for one should take into consideration that the crust formation is controlled by the actual distribution of the grains in the ice. We are dealing with a stochastic phenomenon. In reality, separate islands of crust

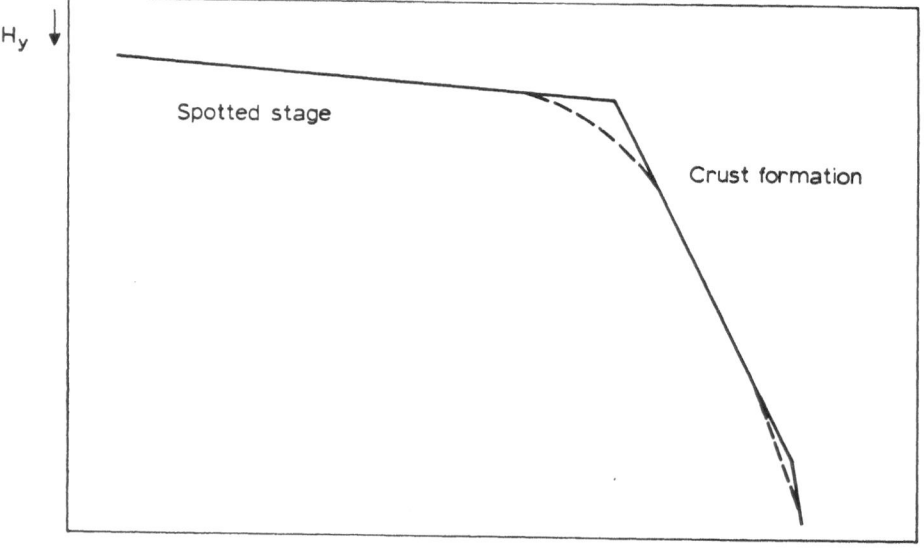

Fig. 1.   Decrease in a comet's absolute brightness with time.

are being formed. On some portions of the surface, only the first layer is being filled, while elsewhere the formation of the second layer has already begun. The picture resembles the thawing of dirty snow in spring.

4. Invisibility of Completely Crusted Comets

Now we shall show that a comet whose nucleus is shielded with a continuous crust would be invisible. The gas flow through the crust is controlled by the combined effect of sublimation, described by the isotherm

$$n_h k T_h = 10^{A - B/T_h}, \tag{20}$$

the energy balance at depth h

$$nv \frac{kB}{M} \approx K \frac{T - T_h}{h}, \tag{21}$$

and diffusion

$$nv \approx D \frac{n_h - n}{h}. \tag{22}$$

For rough estimates we do not need more precise forms for these equations. The subscript h refers to the values at the depth h (the thickness of the crust). A and B are the constants of the ice, and K and D are the coefficients of thermal conductivity and diffusion, respectively. The diffusion coefficient

$$D \approx \beta \bar{v} a_{min}, \tag{23}$$

where the dimensionless constant $\beta \leqslant 0.1$. Since $n \ll n_h$ and $\bar{v} \sim v \sim \sqrt{kT/m}$, Equation (22) can be reduced to the form

$$n \approx \beta n_h \frac{a_{min}}{h}. \tag{24}$$

From Equations (20), (21), and (22) we can obtain the iterative formula

$$T_h = B \left\{ A - \log_{10} \left[T_h (T - T_h) \right] - \log_{10} \left[\frac{KM}{\beta a_{min} B} \sqrt{\frac{m}{kT}} \right] \right\}^{-1}. \tag{25}$$

We shall adopt the values $\beta \approx 0.1$, $a_{min} \sim 10^{-6}$ cm, $A = 12$, $B = 1400$, $m \sim 5 \times 10^{-23}$ g. Following Whipple, we assume that the thermal conductivity is solely radiative, i.e., $K \sim 4 a_{min} \sigma \langle T^3 \rangle \sim 10^{-5}$. Thus we obtain $T_h \sim 120$ K; and for $h = 10$ cm, $nv \simeq 6 \times 10^8$ cm^{-2} s^{-1}. It is obvious that such a comet is much fainter than the comets that are usually observed. The dusty crust would make a comet initially of absolute magnitude 6 completely invisible.

Another property of comets with crusts is the thermal inertia of the shielding layer. The time scale of the thermal inertia is

$$t_i \sim (\rho c h^2 / K), \tag{26}$$

where ρ and c are the density and specific heat of the crust. Under the conditions assumed above $t_i \gtrsim 10^{11}$ s.

If the thermal conductivity coefficient is much higher than assumed here, the gas pressure at the icy surface can exceed the weight of the crust. But it would be wrong to conclude that the crust would then be thrown off. In fact, in this case we shall obtain a fluidized layer, possibly with channel structure. The fluidization does not prevent particles of the appropriate size from accumulating and the crust from growing. What happens is that the fluidization destroys contacts between grains, and thus the thermal conductivity decreases to its radiation value.

References

Dolginov, A. Z.: 1967, *Astron. Zh.* **44**, 434.
Whipple, F. L. and Douglas-Hamilton, D. H.: 1966, *Mem. Soc. Roy. Sci. Liège Ser. 5* **12**, 469.

49. ON THE SIZES OF COMETARY NUCLEI

V. P. KONOPLEVA and L. M. SHUL'MAN

Main Astronomical Observatory, Ukrainian Academy of Sciences, Kiev, U.S.S.R.

Abstract. Three methods have been developed for determining the sizes of cometary nuclei. They are all based on the spotted model for an icy nucleus. The first method follows from the relationship between the icy (unshielded) area and the total absolute brightness, although the quantity determined is actually a function of the radius and the shielding coefficient. The second method is based on analysis of the secular brightness decrease and the assumption that it is possible to establish when a new surface layer of dust starts to form; in practice, this method is found to be applicable only in the cases of P/Pons-Winnecke and P/Tuttle. The third method is also based on the secular brightness decrease, and it requires some assumption about the volume fraction of dust in the nucleus.

The total gas and dust production of an icy comet depends mainly on the size of the nucleus. There are few data on the radii of cometary nuclei, and these are inconsistent, as is clear from the estimates given in Table I for the radius of P/Halley at its 1910 return.

TABLE I

Estimates of the radius of the nucleus of P/Halley

R_{max} (km)	R_{min} (km)	R (km)	Author
—	—	0.75	Sekanina (1960)
—	—	1.0	Orlov (1945, 1960)
—	1.0	—	Dobrovol'skij (1953)
—	—	2.0	Sekanina (1962)
—	—	20	Vorontsov-Vel'yaminov (1946)
—	250	—	Dubyago (1950)
450	—	—	Bouška and Vanýsek (1967)
—	3.1–14.6	—	Whipple (1963)[a]

[a] This estimate refers to the 'original' P/Halley.

Many different methods have been used for the determination of nuclear radii. These have been based on the absolute brightness of the nucleus, the intensity of the continuous spectrum, dynamical properties, and so on. In addition to the references cited in Table I we mention here Vorontsov-Vel'yaminov (1945), Kostyakova (1966), Baldet (1951), Bobrovnikoff (1951), Hasegawa (1968), Houziaux (1959), Mianes *et al.* (1960), Richter (1958, 1963), Roemer (1966), and Whipple (1951). However, total absolute brightness has been used for this purpose for only a few comets (Dobrovol'-skij, 1952; Sekanina, 1960). Since absolute magnitudes are known for practically all comets, it is of interest to use this vast material to obtain estimates of the total emission area, and for periodic comets the radius of the nucleus as well.

It is assumed that the surface of a nucleus consists of exposed regions of ice, with spots of mineral crust (spotted model). Since the temperature of a nucleus and the gas

Chebotarev et al. (eds.), *The Motion, Evolution of Orbits, and Origin of Comets*, 277–282. *All Rights Reserved.*
Copyright © 1972 by the IAU.

density at the surface do not depend on the radius, there is a universal relationship between the area of emission of parent-molecules and the total absolute brightness. The universality of the relationship is violated only by the variety of chemical composition, and then only slightly.

The total number of C_2 molecules may be obtained in two ways:

(1) from the mass conservation law

$$N_{C_2} = 4\pi R^2 n_{01} v_{01} \tau_1 \alpha (1 - \xi), \tag{1}$$

where n_{01} and v_{01} are the density and velocity of gas at the nuclear surface, and τ_1 is the lifetime of C_2 molecules, these quantities being referred to unit heliocentric distance; R is the radius of the nucleus, α is a geometric factor, and ξ is the fraction of the surface screened by mineral crust (the shielding coefficient).

(2) from the total visual magnitude H. If f is the oscillator strength of the Swan system, then

$$f N_{C_2} = 10^{32.2 - 0.4H}. \tag{2}$$

From Equations (1) and (2) it follows that

$$R\sqrt{f(1 - \xi)} = \frac{10^{11.1 - 0.2H}}{\sqrt{4\pi\alpha n_{01} v_{01} \tau_1}} \text{ km}, \tag{3}$$

where the quantities in the denominator on the right-hand side are in c.g.s. units. This relation gives us the first method (henceforth referred to as method 1) for estimating the radius of a nucleus.

Assuming $f = 0.003$ (Golden, 1967) and $4\pi\alpha n_{01} v_{01} \tau_1 = 10^{22}$ mol cm^{-2}, we have calculated values of $R\sqrt{(1 - \xi)}$ for each apparition of all the periodic and nonperiodic comets observed up until 1968. With a few exceptions the values for the short-period comets decrease with time.

The question arises as to whether the brightness decrease is due to the shrinking of the nucleus (thawing) or the growth of the shielding coefficient ξ (screening). We have already discussed the evolution of nuclei (Shul'man, 1972) and have shown that the quantity $1 - \xi$ decreases with time by the law

$$\log (1 - \xi) = 0.53 \frac{R\delta_d}{C_D v_{op}} \frac{f_v}{1 - f_v} \frac{1 - a_{eq}/a_{max}}{\log a_{max}/a_{min}}$$

$$\times \left[\int_0^{2\pi} (1 + e \cos \theta)^{n-2} d\theta \right] \frac{q^{3/2}}{(1 + e)^{n - 1.5}} v_{tr}, \tag{4}$$

where $C_D \approx 10$ is the drag coefficient of the grain, f_v is the volume fraction of dust in the nucleus, e and q are the usual orbital elements, θ is the true anomaly, n the photometric parameter, v_{tr} the number of revolutions since the comet's origin, $\delta_d \approx 1$ is the mass density of the grain, a_{max} and a_{min} are the extreme radii of grains, a_{eq} is the radius of a grain in equilibrium at the nuclear surface, and $v_{op} \approx 10^4$ cm s^{-1} is the gas velocity at the surface at perihelion.

Equation (4) is valid for $\xi \leqslant 0.9$; otherwise the coefficient 0.53 must be replaced by 3.5, because after ξ has reached the value 0.9 a second layer of dusty crust starts to form, and the evolution proceeds more rapidly.

If observations exist of a comet both before and after the formation of the second layer, we have another method for determining the nuclear radius (method II). We obtain at two points v_1 and v_2 (measured according to the number of revolutions v) in the first part of the absolute brightness curve the function, following from Equation (3),

$$\psi(v) = \log\left(R\sqrt{1 - \xi}\right) = 0.1 - 0.2H \tag{5}$$

and form

$$\nabla_1 = \frac{\psi(v_2) - \psi(v_1)}{v_2 - v_1} = -0.2\frac{H(v_2) - H(v_1)}{v_2 - v_1}. \tag{6}$$

The situation is illustrated in Figure 1, the variation of ψ with v during this stage being denoted by the straight line AB. We do the same for two points v_1' and v_2' in the

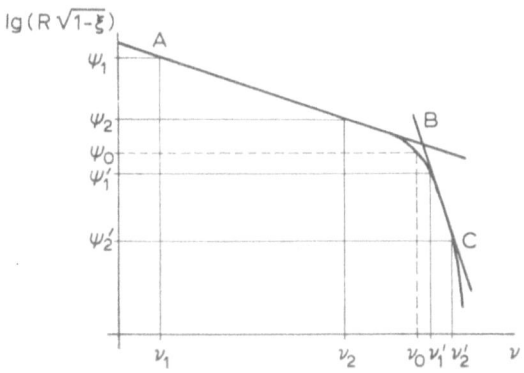

Fig. 1.   Variation of $\log\left(R\sqrt{(1 - \xi)}\right)$ with time.

second part of the brightness curve, this stage being denoted by the line BC. This gives us a possibility of testing the validity of the spotted model. If the model is correct, we must have that $\nabla_1(AB):\nabla_1(BC)\approx 0.15$. We can locate the point B (the number of revolutions from discovery then being denoted by v_0) and derive the radius of the nucleus from Equation (3), taking $\xi = 0.9$. Thus,

$$R = 10^{\psi(B) + 0.5}. \tag{7}$$

Equation (4) can be written in the form

$$\log(1 - \xi) = \nabla_2 R \frac{f_v}{1 - f_v} v_{tr}. \tag{8}$$

Assuming that $a_{eq}\ll a_{max}$ and $\log(a_{max}/a_{min})\approx 7$, we can easily calculate the ∇_2 coefficient for each comet. If screening is the main reason for the secular decrease of brightness, we can neglect the variation in radius. It follows from Equations (5), (6), and (8) that

$$\nabla_1 \simeq \tfrac{1}{2}\nabla_2 R \frac{f_v}{1-f_v},\qquad(9)$$

which enables us to calculate f_v. Finally, by evaluating Equation (8) at the point B we can obtain

$$\nu_{tr} = \left[\frac{1-f_v}{f_v R|\nabla_2|}\right],\qquad(10)$$

the square brackets indicating the integral part, and the true age of the comet follows by multiplying ν_{tr} by the revolution period. There is another possibility of checking the theory, for it is obvious that we must have $\nu_{tr}(B) \geqslant \nu_0$.

In practice we are generally not able to locate the point B, and method II turns out to be applicable only to the two short-period comets Pons-Winnecke and Tuttle: see Figure 2 and Table II, where ν_{ob} gives the total number of revolutions made by the comet since discovery.

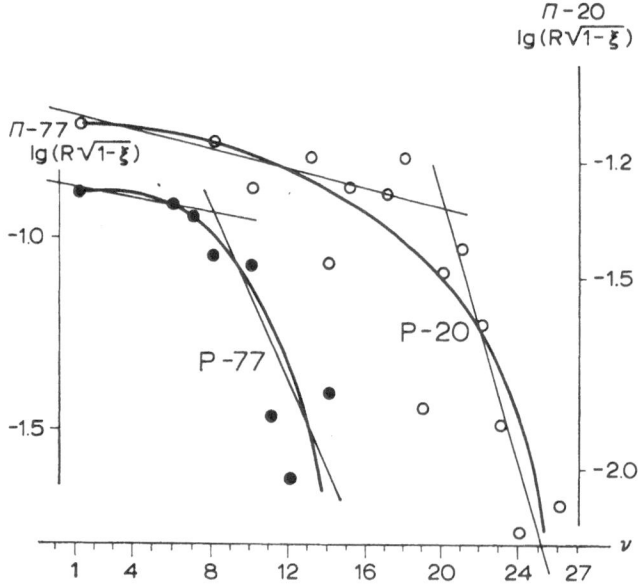

Fig. 2.   Variation of log $(R\sqrt{(1-\xi)})$ with time for P/Pons-Winnecke (P-20; points denoted by O) and P/Tuttle (P-77; point denoted by ●).

For a particular comet we have no real way of knowing whether it is in the process of forming its first layer (the AB region), its second layer (the BC region), or even some subsequent layer, although there is a high probability of observing it in the first stage. In general we obtain the product $R\sqrt{(1-\xi)}$ only, or using Equation (9), the product

TABLE II

Nuclear radii of P/Pons-Winnecke and P/Tuttle

Comet	v_0	f_v	R (km)	v_{ob}	v_{tr}	Origin
Pons-Winnecke	20	0.34	0.40	27	28	1814 ± 30
Tuttle	8	0.09	1.35	14	31	1554 ± 100

Earlier estimates for P/Pons-Winnecke are: 0.6–3.5 km (Roemer, 1966), 0.2 km (Baldet, 1951), 4–20 km (Bouška and Vanýsek, 1967; Richter, 1963), 82 km (Whipple, 1951).

$Rf_v/(1-f_v)$. In order to estimate the radius we must make some supposition about the dust content f_v (method III).

Values of ∇_1 have been obtained from the preliminary smoothed curves of absolute brightness for several periodic comets. A number of comets, e.g., Halley, Grigg-Skjellerup, Encke, and Brooks 2, have periodic variations in absolute brightness. Some comets, e.g., Grigg-Skjellerup, Perrine-Mrkos, and Holmes (Hasegawa, 1968), must have been discovered during outbursts, if the strong decrease in the absolute brightness after the first apparition is real; this question requires additional treatment, and in a number of cases we have excluded first apparition data from our analysis.

Estimates of the radii are given in Table III for most of the comets having $v_{ob} \geq 10$. Here, N is the actual number of apparitions, $v_i - v_k$ gives the revolution numbers of the

TABLE III

Derived nuclear radii for periodic comets with $v_{ob} \geq 10$

Comet	N	v_{ob}	$v_i - v_k$	$R_i\sqrt{(1-\xi)}$	$R_k\sqrt{(1-\xi)}$	$R(0.1)$	$\hat{R}(0.1)$	$R(0.3)$
Encke	48	56	11–56	0.46	0.08	11.37	0.54	2.95
			13–53	0.35	0.12	7.70	0.37	2.00
Grigg-Skjellerup	11	14	5–14	0.06	0.02	6.37	0.30	1.65
Tempel 2	14	19	1–19	0.30	0.11	1.46	0.07	0.38
			5–19	0.24	0.11	1.42	0.07	0.37
de Vico-Swift	3	21	1–21	0.57	0.02	4.13	0.20	1.07
Tempel 1	4	17	1– 3	0.41	0.16	8.32	0.40	2.16
			1–17	0.35	0.03	3.00	0.14	0.78
Tuttle-G.-K.	4	20	1–20	0.10	0.04	1.53	0.07	0.40
d'Arrest	11	18	4–18	0.31	0.07	3.11	0.15	0.81
Perrine-Mrkos	5	12	3–12	0.05	0.01	5.26	0.25	1.36
Kopff	9	10	1–10	0.42	0.07	3.98	0.19	1.03
Finlay	8	13	1–13	0.22	0.06	5.56	0.26	1.44
Biela	6	13	6–13	0.55	0.44	2.07	0.10	0.54
Wolf	11	12	1–11	1.05	0.03	4.50	0.21	1.17
Holmes	4	11	2–11	0.24	0.04	2.27	0.11	0.59
Borrelly	8	10	1–10	0.29	0.05	5.43	0.26	1.41
Brooks 2	10	11	1–11	0.66	0.02	5.36	0.26	1.30
Faye	15	17	4–17	0.91	0.06	4.23	0.20	1.10
Crommelin	6	19	9–19	0.83	0.10			
Halley	32	32	19–32	7.56	1.99	13.72	0.65	3.56

range over which the absolute brightness curve was smoothed, $R(0.1)$ and $R(0.3)$ are nuclear radii (in km) obtained by method III (the assumed value of f_v being given in the parentheses), and $\hat{R}(0.1)$ is the effective radius on the assumption that the comet is in its second stage of evolution.

The results obtained here can be used to forecast the absolute brightness of periodic comets at their next apparitions. For example, the absolute magnitude of P/Halley is expected to be about 5.6 in 1986.

References

Baldet, F.: 1951, *Astronomie* **65**, 289.
Bobrovnikoff, N. T.: 1951, in J. A. Hynek (ed.), *Astrophysics*, New York, Toronto and London, p. 302.
Bouška, J. and Vanýsek, V.: 1967, *Fyzika Komet*, Prague.
Dobrovol'skij, O. V.: 1952, *Byull. Stalinabad-Dushanbe Astron. Obs.* No. 6.
Dobrovol'skij, O. V.: 1953, *Byull. Stalinabad-Dushanbe Astron. Obs.* No. 8, 3.
Dubyago, A. D.: 1950, *Astron. Zh.* **27**, 5.
Golden, S. A.: 1967, *J. Quant. Spectr. Radiative Transfer* **7**, 225.
Hasegawa, I.: 1968, *Mem. Kyoto Univ. Ser. Phys. Astrophys. Geophys. Chem.* **32**, 37.
Houziaux, L.: 1959, *Acad. Roy. Belg. Bull. Cl. Sci.* **45**, 218.
Kostyakova, E. B.: 1966, *Problemy Kometnoj Fotometrii Inf. Byull.* No. 10, 65.
Mianes, P., Grudzinska, S., and Stawikowski, A.: 1960, *Ann. Astrophys.* **23**, 788.
Orlov, S. V.: 1945, *Golova Komety i Klassifikatsiya Kometnykh Form*, Leningrad.
Orlov, S. V.: 1960, *Priroda Komet*, Moscow.
Richter, N.: 1958, *Sterne* **34**, 11, 215.
Richter, N.: 1963, *The Nature of Comets*, Methuen, London.
Roemer, E.: 1966, *Mem. Acad. Roy. Sci. Liège Ser. 5* **12**, 23.
Sekanina, Z.: 1960, *Publ. Astron. Inst. Charles Univ. Ser. II* No. 28.
Sekanina, Z.: 1962, *Publ. Astron. Inst. Charles Univ. Ser. II* No. 36A.
Shul'man, L. M.: 1972, this Symposium, p. 271.
Vorontsov-Vel'yaminov, B. A.: 1945, *Astron. Zh.* **22**, 317.
Vorontsov-Vel'yaminov, B. A.: 1946, *Astrophys. J.* **104**, 226.
Whipple, F. L.: 1951, *Astrophys. J.* **113**, 464.
Whipple, F. L.: 1963, in *The Moon, Meteorites and Comets*, Vol. IV of the series: *The Solar System* (ed. by B. M. Middlehurst and G. P. Kuiper), University of Chicago Press, Chicago and London, p. 639.

50. SPLITTING AND SUDDEN OUTBURSTS OF COMETS AS INDICATORS OF NONGRAVITATIONAL EFFECTS

E. M. PITTICH

Astronomical Institute, Slovak Academy of Sciences, Bratislava, Czechoslovakia

Abstract. It is shown that sudden outbursts of comets, or even complete splitting of their nuclei into several parts, are rather frequent phenomena that need to be accounted for in the evolution of cometary orbits. The distribution of the points in which such events occur indicates that the tidal action of the Sun and Jupiter, and also solar radiation, are much more probable causes of these effects than are collisions with the asteroids. Splitting is found to be relatively more frequent for comets moving in hyperbolic orbits. The observed velocities of separation indicate that the disruption of the nucleus provides an effective mechanism for the ejection of the fragments from long-period into hyperbolic orbits, augmenting the loss by planetary perturbations. The outbursts can be responsible for smaller changes in the orbits, which after repetition might produce a quasi-secular variation.

1. Anomalous Changes in Comets

Anomalous changes – sudden outbursts of comets and complete splitting of their nuclei into several parts – are rather frequent phenomena in cometary evolution. Stefanik (1966) gives 13 well-documented cases of splitting of nuclei, and Vsekhsvyatskij (1966) gives more than 50 cases of observed sudden changes in the brightness of comets. During the last two centuries at least 40 comets appear to have suddenly increased in brightness shortly before their discovery, as was shown by Pittich (1969) on the basis of an analysis of the distribution of time intervals in which the comets remained undiscovered in spite of favourable observing conditions. The results have proven the reality of sudden changes in cometary brightness, even at considerable distances from the Sun. Only a part of the observed abrupt changes in brightness can be attributed to systematic errors in the estimates of the magnitudes.

These anomalies related to the evolution of comets prove that there exist external factors affecting comets. The following factors might be considered: the tidal action of the Sun and planets, especially Jupiter; collisions of comets with asteroids; and solar radiation. Their effect is spatially restricted, and anomalies in comets depend on the degree of intensity of these factors. The complete splitting of a comet nucleus can be considered as the extreme event.

2. Space Distribution of the Anomalies

It was Harwit (1968) who first considered the spatial distribution of the points in which the splitting of cometary nuclei occurred. The author (Pittich, 1971) gives similar data for nine further cases and for a number of cometary outbursts; the two figures presented here are taken from this paper. The space distribution of the splitting of 19 cometary nuclei is plotted in Figure 1; r is the distance from the Sun, d the distance from the ecliptic plane (full line), in astronomical units. Positive values of d

Chebotarev et al. (eds.), The Motion, Evolution of Orbits, and Origin of Comets, 283-286. All Rights Reserved.
Copyright © 1972 by the IAU.

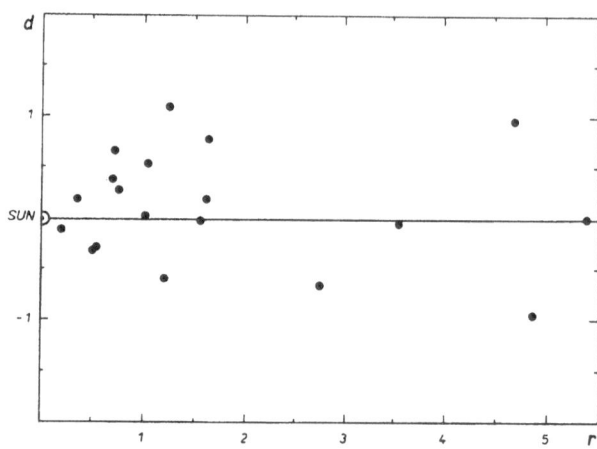

Fig. 1.   Positions of comets at the time of splitting.

indicate positions north of the ecliptic, negative values south of it. Figure 2, arranged
in the same manner as Figure 1, shows the positions of the outbursts, which occurred
shortly before (open circles) and after comet discovery (closed circles).

If selection effects of comet discovery are taken into account, the distribution of the
points in Figures 1 and 2 indicates that the tidal action of the Sun and Jupiter, and

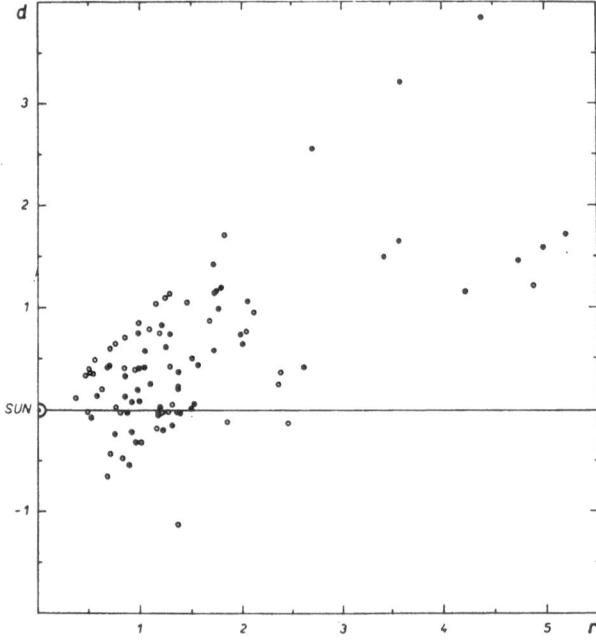

Fig. 2.   Positions of comets at the time of outbursts. Open circles: hypothetical outbursts preceding discovery. Closed circles: observed outbursts after discovery.

solar radiation, are much more probable causes of cometary splitting and outbursts than collisions with the asteroids. Explanation of the splitting and outbursts as due to collisions with asteroidal particles would require a concentration of the points within the main asteroid belt, i.e., roughly between $r=2.2$ and $r=3.2$ AU. Although this region is beyond the limits of observability of many comets, there appears to be a gap of avoidance rather than any concentration. On the contrary, a secondary maximum located outside the belt, nearer to the distance of Jupiter from the Sun, is indicated.

3. Indicators of Nongravitational Effects

Splitting is found to be relatively more frequent for comets moving in hyperbolic orbits than for the others. Approximately 6% of the known cometary orbits are hyperbolic, while among 19 comets with observed splitting of the nucleus, eight (i.e., 42%) belong to this category. This fact can be explained by the assumption that the orbit of the observed part of the nucleus has been affected by splitting.

This assumption is supported by the observed velocities of separation. They range from 2 to 40 m s^{-1}, with a characteristic value of about 15 m s^{-1} (Sekanina, 1966; Stefanik, 1966). These observed velocities lie within the range of the theoretical heliocentric velocity increments (Table I) necessary for the corresponding orbital changes of 'new' comets in Oort's sense ($a \sim 1000$ AU). The velocity excess between

TABLE I

Velocity increments

r (AU)	$V_{par} - V_{ell}$ (m s^{-1})			$V_{hyp} - V_{par}$ (m s^{-1})
	$a = 1000$	$a = 100$	$a = 10$	$a = -1000$
0.2	5	47	472	5
1.0	11	105	1066	11
2.0	15	149	1527	15
3.0	18	183	1898	18
4.0	21	212	2222	21
5.0	24	237	2522	24

parabolic and hyperbolic orbits, calculated for $a = -1000$ – i.e., for the extreme hyperbolic value observed (Porter, 1961; Marsden, 1966) – also agrees with the observed disruption velocities. As hyperbolic orbits of higher eccentricity have not been observed, it seems that hyperbolic orbits and the disruption of nuclei may have a close connection. The splitting of nuclei provides an effective mechanism, besides planetary perturbations, for the ejection of fragments from long-period into hyperbolic orbits. On the other hand, the splitting of a nucleus indicates the presence of nongravitational forces acting on the motions of comets.

For the compilation of Table I, using the formula

$$V^2 = k^2\left(\frac{2}{r} - \frac{1}{a}\right),$$

five values of the semimajor axis have been selected: three for elliptical orbits, one for parabolic, and one for hyperbolic. For elongated elliptical orbits the velocity increment required for ejecting from the solar system comets moving near the Sun is roughly proportional to a^{-1}, irrespective of the heliocentric distance r at which the acceleration occurs. On the other hand, the velocity increment necessary for such ejection varies with $r^{1/2}$. Therefore, the greatest changes in semimajor axis take place when the comet splits near the Sun. Table I shows that only 'new' comets can suffer this type of non-gravitational ejection from the solar system, as the necessary velocity increments are too large if a is of the order of 10 to 100 AU.

The splitting of a nucleus, however, is the extreme case of anomalous changes in comets. Greater or smaller outbursts, observed as flares in brightness, occur more frequently. In such events sufficiently strong nongravitational changes in the orbits can hardly be expected. In general, the outbursts can be held responsible only for smaller orbital changes, which after repetition might produce a quasi-secular variation.

The outbursts are about three times more frequent in comets in which an outburst has already been observed previously. This conclusion was obtained from 181 individual comets observed in 1925–1965. At least one outburst was recorded in 28 comets, and more than one outburst in 12 comets (Vsekhsvyatskij, 1966); i.e., the probability of the occurrence of an outburst is $p = 0.16$ in the former case and $p = 0.43$ in the latter case. Among 19 comets in which splitting of the nucleus occurred, there are nine cases in which outbursts have also been observed (Pittich, 1971), i.e., $p = 0.47$. This value is similar to the preceding one. Although the data on observed outbursts must be used with caution, in view of considerable errors in the magnitude data, which can simulate sudden changes in brightness, the statistical difference appears rather significant. A category of comets exists in which anomalous changes tend to repeat during their evolution.

It appears that the splitting and outbursts are very likely to be accompanied by nongravitational changes in cometary orbits. The occurrence of these anomalous phenomena can be used as suitable criteria for the selection of objects in studies of the nongravitational effects on cometary motion.

References

Harwit, M.: 1968, *Astrophys. J.* **151**, 789.
Marsden, B. G.: 1966, *Mem. Br. Astron. Assoc.* **40**, No. 2.
Pittich, E. M.: 1969, *Bull. Astron. Inst. Czech.* **20**, 251.
Pittich, E. M.: 1971, *Bull. Astron. Inst. Czech.* **22**, 143.
Porter, J. G.: 1961, *Mem. Br. Astron. Assoc.* **39**, No. 3.
Sekanina, Z.: 1966, *Bull. Astron. Inst. Czech.* **17**, 207.
Stefanik, R. P.: 1966, *Mem. Soc. Roy. Sci. Liège Ser.* 5 **12**, 29.
Vsekhsvyatskij, S. K.: 1966, *Problemy Kometnoj Fotometrii Inf. Byull.* No. 10, 3.

51. ON NONGRAVITATIONAL EFFECTS IN TWO CLASSES OF MODELS FOR COMETARY NUCLEI

O. V. DOBROVOL'SKIJ

Institute of Astrophysics, Dushanbe, U.S.S.R.

and

M. Z. MARKOVICH

Polytechnic Institute, Kalinin, U.S.S.R.

Abstract. Two types of cometary nuclei are considered: a homogeneous icy nucleus and an icy nucleus covered with a dispersive surface layer, 1 cm thick and having poor heat conductivity. The temperature of the evaporating ice surface of a nucleus rotating in a period of 6 hours about an axis perpendicular to the orbital plane was determined by numerical integration of the equation of heat conduction in the first case and of the Stefan problem in the second. For a nucleus of radius 10^5 cm and mass of 3.8×10^{15} g the reactive force was found to be about 10^7 dyn in either case, and the secular change in mean motion is of the order of the observed values. For the homogeneous nucleus the dependence of the reactive force on heliocentric distance is obtained.

In accordance with the ideas of Dubyago (1948) and Whipple (1950, 1951), an evaluation is made of the order of magnitude of the reactive force Φ_r and the corresponding nongravitational effect Δn in the mean motion of a comet for two types of icy models for cometary nuclei.

1. A Homogeneous Nucleus of H_2O Ice

In the case of a pure icy nucleus we take as the basis for our estimate of Φ_r Meshcherskij's formula for the effect $d\Phi_r$ on an element ds of the nuclear surface

$$d\Phi_r = \frac{dM}{dt} \mathbf{V}; \qquad d\mathbf{w} \approx \frac{1}{M}\frac{dM}{dt} \mathbf{V}, \tag{1}$$

where dM/dt is the rate of escape of mass from the surface element, $\mathbf{w}$ is the reactive acceleration, and $\mathbf{V}$ is the relative velocity of the escaping matter.

The nuclear temperature required for the determination of dM/dt is found from the solution of the heat-conduction equation

$$\frac{\partial T}{\partial t} = \frac{K}{\rho c}\frac{\partial^2 T}{\partial x^2}, \tag{2}$$

where $T(x, t)$ is the nuclear temperature, x the distance below the nuclear surface, K the coefficient of thermal conductivity, ρ the density, and c the specific heat of the nuclear material.

The boundary condition is written as follows:

$$q = -K(\text{grad } T)_{x=0} + \varepsilon\sigma T^4 + b, \tag{3}$$

where q is the heat received from the Sun, ε is the degree of blackness of the surface, σ the Stefan-Boltzmann constant, and b the heat expended vaporizing the ice. For a nucleus with its axis of rotation perpendicular to its orbit plane

$$
q = \begin{cases} \dfrac{(1-A)q_0}{r^2} \sin \omega t \cos \varphi & 0 \leqslant \omega t \leqslant \pi \\ 0 & \pi \leqslant \omega t \leqslant 2\pi, \end{cases} \tag{4}
$$

where q_0 is the solar constant, A the surface albedo, r the heliocentric distance, $\omega t = \lambda$ is the longitude of the surface element (counted from the terminator), and φ is the latitude of the surface element. The heat b spent on the evaporation of the ice is given by

$$
b = LN, \tag{5}
$$

where L is the heat of evaporation of a molecule, and N is the number of molecules evaporated from unit surface area per unit time:

$$
N = \frac{\alpha \mathscr{P}}{\sqrt{(2\pi m k T)}}, \tag{6}
$$

with α being the evaporation coefficient, $\mathscr{P}$ the vapour pressure of saturated steam at the temperature T, m the molecular mass, and k Boltzmann's constant. According to the experiments on the evaporation of ice in a vacuum, $\alpha = 1$ (Kajmakov and Sharkov, 1967). The initial condition takes the form

$$
T(x, 0) = f(x). \tag{7}
$$

The following values were adopted for the physical characteristics of H_2O ice: $\rho = 0.9$ g cm^{-3}, $\varepsilon = 0.64$, $A = 0.7$, $m = 3 \times 10^{-23}$ g and $L = 2 \times 10^{-20}$ cal; for K and c average values were taken for the range of temperatures 150–170 K from figures given by Shumskij (1955) and in *Termicheskie Konstanty Neorganicheskikh Veshchestv*, namely, $K = 4.42 \times 10^{-3}$ cal cm^{-1} s^{-1} deg^{-1}, and $c = 0.28$ cal g^{-1} deg^{-1}. The pressure $\mathscr{P}$ in dyn cm^{-2} can be found from the semiempirical relation

$$
\log \mathscr{P} = -2720/T + 13.75. \tag{8}
$$

Numerical integration of Equation (2) has been performed by the grid method (Markovich and Tulenkova, 1968) for a surface element at $\varphi = 60°$ and $r = 1$ AU, and with the initial condition $f(x) = 155$ K. The period of nuclear rotation P is taken as 6 h. The influence of the initial conditions, which are always to some extent arbitrary, can be partly or totally eliminated by computing the temperature of the nucleus for several successive rotations. Figure 1 shows the nuclear temperature computed for the second rotation. The lag of the maximum surface temperature T_0 compared to the maximum of solar radiation is about 3/8 h ($\frac{1}{16}P$).

By integrating Equation (1) over the whole nuclear surface we may determine the acceleration components S' outward along the radius vector and T' perpendicular to it (in the orbit plane); S' and T' are determined (approximately) from the equations

$$S' = \frac{3}{2\pi R_0 \rho} \Delta\varphi\Delta\lambda \sum_{\varphi=0}^{\pi/2} \sum_{\lambda=0}^{2\pi} Q(\lambda, \varphi)v(\lambda, \varphi) \cos^2 \varphi \sin \lambda$$

$$T' = \frac{3}{2\pi R_0 \rho} \Delta\varphi\Delta\lambda \sum_{\varphi=0}^{\pi/2} \sum_{\lambda=0}^{2\pi} Q(\lambda, \varphi)v(\lambda, \varphi) \cos^2 \varphi \cos \lambda, \qquad (9)$$

where R_0 is the radius of the nucleus, v is the average thermal velocity of molecules at the surface, and $Q=mN$ is the mass of matter escaping from unit surface area per unit time.

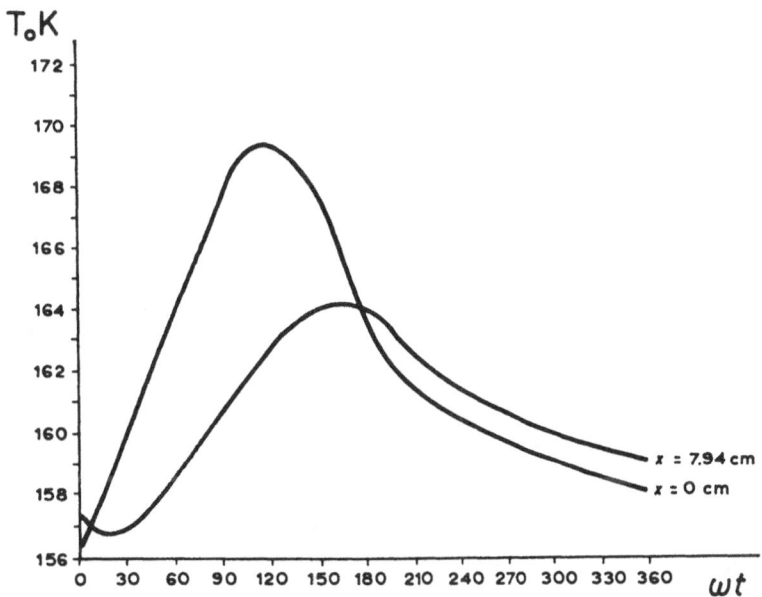

Fig. 1. Temperature variation T_0 on the surface and at a depth of 7.94 cm for a homogeneous icy nucleus of radius 1 km located at 1 AU from the Sun and rotating with a 6-h period.

For rough evaluation we took average values of $Q(\lambda, \varphi)$ and $v(\lambda, \varphi)$ for $\varphi=60°$ from the temperature curve in Figure 1. For $R_0=10^5$ cm and $\rho=0.9$ g cm^{-3}:

$$S' = 1.4 \times 10^{-8} \text{ cm s}^{-2}$$

$$T' = 8.5 \times 10^{-9} \text{ cm s}^{-2} \qquad (10)$$

$$\vartheta' = \arctan (T'/S') = 31°.$$

The change in the mean motion n for an orbit with semimajor axis a and eccentricity e is

$$\frac{dn}{dt} = -\frac{3e \sin \theta}{a\sqrt{(1 - e^2)}} S' - \frac{3\sqrt{(1 - e^2)}}{r}T', \qquad (11)$$

where θ is the true anomaly. Let us assume that S' changes symmetrically before and after perihelion. Then, in determining the average variation of n per revolution, we need take only T' into account; the action of S' in this case will be only to diminish the effective gravitational constant. According to Equation (10), and taking $e=0$ and $r=1$ AU, we obtain $n=2.6 \times 10^{-6}$ arc sec per day or 10^{-3} arc sec per day per revolution, in order of magnitude agreement with observational data (Dubyago, 1948; Whipple, 1950, 1951; Marsden, 1968; Richter, 1963; Sekanina, 1967, 1968; Sitarski, 1964). It can easily be shown that Δn will be of the same order for an elliptical orbit with semi-latus rectum 1 AU.

For the nuclear model discussed the dependence of w_r upon r can be obtained. According to Whipple (1950, 1951),

$$w_r = \zeta \eta \frac{q_0}{r^2} v_r \,(\text{eff}),$$

where ζ is a coefficient that depends on the properties of the nucleus (size, shape, etc.), and η is the efficiency of evaporation (which depends on r). For $r>1$ AU, $\eta = \eta_0 r^{-k}$, and according to the computations by Markovich, $k=1.5$. For $r<1$ AU, η fits very well the formula (Markovich and Tulenkova, 1968) $\eta = (1+r^{2.57})^{-1}$.

Adopting the approximation $T_0 = T_{00} r^{-\beta}$ ($\beta \approx 0.2$) for the temperature of the nuclear surface, we obtain $V_r\,(\text{eff}) \sim \sqrt{T_0} \sim r^{-\beta/2}$, whence

$$w_r \sim \frac{1}{r^{3.6}} \qquad\qquad (r > 1 \text{ AU})$$

$$w_r \sim \frac{1}{r^{2.1}(1 + r^{2.57})} \qquad (r < 1 \text{ AU}). \tag{12}$$

2. An Icy Nucleus with a Dusty Surface Layer

Let us now discuss a nuclear model in the form of an ice (H_2O) sphere of radius R_i, with a surface layer of thickness l consisting of finely dispersive and poorly heat-conducting powder.

For determining the temperature of the evaporating ice surface directly beneath the dispersive layer, we shall make use of the so-called Stefan problem (Lykov, 1952; Tikhonov and Samarskij, 1966). Mathematically, this problem is formulated as follows:

$$\frac{\partial T_1}{\partial t} = \frac{K_1}{\rho_1 c_1} \frac{\partial^2 T_1}{\partial x^2} \qquad (0 \leqslant x \leqslant l)$$

$$\frac{\partial T_2}{\partial t} = \frac{K_2}{\rho_2 c_2} \frac{\partial^2 T_2}{\partial x^2} \qquad (x \geqslant l) \tag{13}$$

$$T_1(x, 0) = f_1(x)$$

$$T_2(x, 0) = f_2(x) \tag{14}$$

$$T_1(l, t) = T_2(l, t) \tag{15}$$

$$q = -K_1(\text{grad } T_1)_{x=0} + \varepsilon \sigma T^4 \tag{16}$$

$$b = -K_1\left(\frac{\partial T_1}{\partial x}\right)_{x=l} + K_2\left(\frac{\partial T_2}{\partial x}\right)_{x=l} = -L\rho_2\frac{\partial R_t}{\partial t}. \tag{17}$$

The incident heat q is determined from Equation (4), and the subscripts 1 and 2 refer to the surface layer and to the ice, respectively.

The flow of gas through the boundary of a capillary opening, with molecular flow taking place in the case of comets, is determined with the aid of the formula

$$Q_m = \frac{2}{3}\frac{D}{l'}(p_2 - p_1)\bigg/\left(\frac{2}{\pi}\frac{\mu}{RT}\right), \tag{18}$$

where D is the diameter of the capillary, l' its length, μ the molecular weight of the gas, R is the universal gas constant, T the temperature of the gas, and $\Delta p = p_2 - p_1$ is the pressure difference at the ends of the capillary, p_1 being close to zero.

In calculating the flow of gas through porous materials, the corresponding effective values of l' and D are taken to be (Lejbenson, 1934, 1947)

$$l' = l\bigg/\left[\frac{6}{\pi}(1 - p')\right]$$

$$D = 0.52R_g, \tag{19}$$

where p' is the porosity, and R_g is the radius of the grains of dispersive material.

Heat spent for the sublimation of ice is determined from the relation

$$b = (\mathscr{P} - p_2)(2\pi m k T)^{-1/2}, \tag{20}$$

with $\mathscr{P}$ the pressure of saturated steam, determined from Equation (8), and p_2 the actual gas pressure in the pores of the surface layer directly adjoining the internal icy nucleus. Hence

$$p_2 = \mathscr{P} - \frac{b}{L}(2\pi m k T)^{1/2}, \tag{21}$$

where b is found from Equation (17).

The problem is solved by numerical integration for a surface element on the equator of the nucleus ($\varphi = 0$, $r = 1$ AU), the period P being taken as 6 h and the rotation axis assumed perpendicular to the plane of orbit. The formulae by Markovich and Tulenkova (1968) are used for the purpose.

The temperature of the evaporating ice surface (layer $x = l$) is determined by interpolation from the temperatures of four layers equidistant from the surface, two in the ice region and two in the surface layer.

The components of the reactive force are calculated from the relations:

$$S'' = \frac{3p'}{2\pi R_0 \rho}\Delta\varphi\Delta\lambda \sum_{\varphi=0}^{\pi/2}\sum_{\lambda=0}^{2\pi} Q_m(\lambda, \varphi)V_r(\lambda, \varphi)\cos^2\varphi\sin\lambda$$

$$\tag{22}$$

$$T'' = \frac{3p'}{2\pi R_0 \rho}\Delta\varphi\Delta\lambda \sum_{\varphi=0}^{\pi/2}\sum_{\lambda=0}^{2\pi} Q_m(\lambda, \varphi)V_r(\lambda, \varphi)\cos^2\varphi\cos\lambda,$$

where p' is the area of the capillary openings as a fraction of the total nuclear surface, and it is taken to be equal to the porosity of the surface layer material. To simplify the calculations $Q_m(\lambda, \varphi)$ is replaced, as earlier, by the average value $\frac{1}{2}Q_m(\lambda, 0)$.

The following physical characteristics of the surface layer of the nucleus have been adopted: $A=0.07$, $l=1$ cm, $R_g=5\times 10^{-4}$ cm, $p'=0.80$, $\rho_1=0.8$ g cm^{-3}, $c_1=0.2$ cal g^{-1} deg^{-1}, $K_1=3\times 10^{-6}$ cal cm^{-1} s^{-1} deg^{-1}. Heat exchange by radiation can be neglected in this case, for according to Reprintseva and Fedorovich (1968), the influence of temperature on heat radiation is important only in systems consisting of particles exceeding 1 mm in size; starting at 10 °C the role of radiation in such a system increases, and it reaches quite a significant amount at 100 °C. For the ice we take $c_2=0.36$ cal g^{-1} deg^{-1} and $K_2=4.7\times 10^{-3}$ cal cm^{-1} s^{-1} deg^{-1} as average values for the temperature range 150–200 K, the other characteristics being the same as in Section 1.

Figure 2 shows examples of temperature curves calculated for this model. The

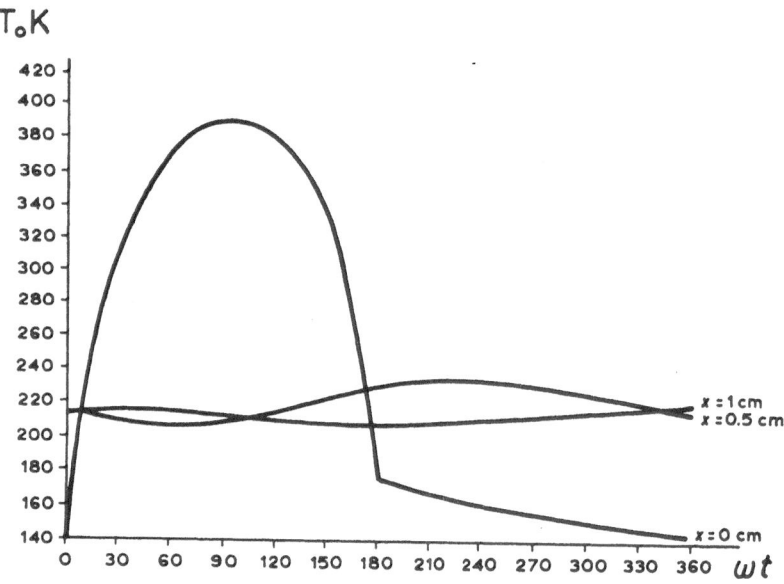

Fig. 2.   Temperature variation T_0 on the surface and at depths of 0.5 and 1.0 cm for an icy nucleus covered by a 1 cm dust layer.

temperature T_0 of the nuclear surface ($x=0$) on the illuminated side essentially follows the variation in solar radiation. The temperature of the evaporating ice surface does not depend on T_0: in our case it shows small variations which are not synchronous with the rotation of the nucleus. It is thus possible that the average value of the reactive acceleration will be close to zero.

For the particular case shown in Figure 2, with $R_0=10^5$ cm and $\rho=0.9$ g cm^{-3}, it follows for Equation (22) that

$$S'' = 7 \times 10^{-9} \text{ cm s}^{-2}$$

$$T'' = 9 \times 10^{-9} \text{ cm s}^{-2} \tag{23}$$

$$\vartheta'' = 52°.$$

Hence, for a circular orbit with $r = 1$ AU, from Equation (11) we have again $\Delta n = 10^{-3}$ arc sec per day per revolution.

Thus the icy model, which is widely accepted and agrees well with photometric data, automatically leads to nongravitational effects in the motions of comets, and they are largest near perihelion. It is also quite possible that nuclei coated with dust would produce nongravitational effects.

References

Dubyago, A. D.: 1948, *Astron. Zh.* **20**, 361.

Kajmakov, E. A. and Sharkov, V. I.: 1967, *Komety i Meteory* No. 15, 16.

Lejbenson, L. S.: 1934: *Neftepromyslovaya Mekhanika*, Gorgeolnefteizdat, Moscow, Groznyj, Leningrad and Novosibirsk, part 2.

Lejbenson, L. S.: 1947, *Dvizhenie Prirodnykh Zidkostej i Gazov v Poristoj Srede*, Gostekhizdat, Moscow-Leningrad.

Lykov, A. V.: 1952, *Teoriya Teploprovodnosti*, Gostekhizdat, Moscow.

Markovich, M. Z. and Tulenkova, L. N.: 1968, *Probl. Kosmich. Fiz.* No. 3, 25.

Marsden, B. G.: 1968, *Astron. J.* **73**, 367.

Reprintseva, S. M. and Fedorovich, N. V.: 1968, *Luchistyj Teploobmen v Dispersnykh Sredakh*, Nauka i Tekhnika, Minsk.

Richter, N. B.: 1963, *The Nature of Comets*, Methuen, London.

Sekanina, Z.: 1967, *Bull. Astron. Inst. Czech.* **18**, 15.

Sekanina, Z.: 1968, *Bull. Astron. Inst. Czech.* **19**, 210.

Sitarski, G.: 1964, *Acta Astron.* **14**, 323.

Shumskij, P. A.: 1955, *Osnovy Structurnogo Ledovedeniya*, AN SSSR, Moscow.

Tikhonov, A. N. and Samarskij, A. A.: 1966, *Uravneniya Matematicheskoj Fiziki*, Nauka, Moscow.

Whipple, F. L.: 1950, *Astrophys. J.* **111**, 375.

Whipple, F. L.: 1951, *Astrophys. J.* **113**, 464.

52. ROTATION EFFECTS IN THE NONGRAVITATIONAL PARAMETERS OF COMETS

Z. SEKANINA

Smithsonian Astrophysical Observatory, Cambridge, Mass., U.S.A.

Abstract. The effects of rotation of a cometary nucleus on the character of nongravitational parameters are discussed. It is suggested that the change of a nongravitational acceleration to deceleration (and vice versa) may be related to a precessional motion of the nucleus, or to secular variations in the lag angle coupled with severe orbital modifications.

1. Introduction

The results of Marsden's (1969, 1970) recent dynamical studies have conclusively shown that the inclusion of nongravitational terms in the equations of motion considerably improved the determinacy of the orbits of most short-period comets. It appeared, however, that the 'degree of obedience' to the nongravitational law applied by Marsden varied from comet to comet.

The components of the nongravitational acceleration were assumed to have the form:

$$Z_i = A_i f(r), \qquad i = 1, 2, 3,$$
$$f(r) = r^{-\alpha} \exp(-r^2/C), \tag{1}$$

where r is the solar distance, A_i are constants (nongravitational parameters), $i = 1$ for the radial component of the acceleration outward from the Sun, $i = 2$ for the transverse component in the orbit plane, perpendicular to the radius vector and in the direction of motion, $i = 3$ for the component normal to the orbit plane, and C and α are empirical constants.

In practice the dynamical disobedience of a comet appears in the form of systematic positional residuals within individual apparitions, and it is also reflected in a poorer quality of the nongravitational parameters. It should be pointed out that the deviations from Equations (1) are, on a secular scale, considerably smaller than the deviations from purely gravitational fits even for the most disobedient comets.

2. Two Groups of Short-Period Comets

The data on the nongravitational parameters of almost 20 short-period comets determined by Marsden (1969, 1970, 1972), Marsden and Sekanina (1971), and Yeomans (1972a, 1972b) suggest that the comets tend to discriminate into two basic groups.

Group I includes comets with A_1 always positive, well determined, and A_2 very small compared to A_1. Members of this group are P/Forbes, P/Daniel, P/Faye, P/Schwassmann-Wachmann 2, P/Whipple, and possibly also P/Schaumasse. P/Biela and P/Pons-Winnecke were likely nineteenth-century members of this group. P/Honda-Mrkos-Pajdušáková, P/Giacobini-Zinner, and P/Borrelly also seem to belong to this group,

Chebotarev et al. (eds.), The Motion, Evolution of Orbits, and Origin of Comets, 294–300. All Rights Reserved.
Copyright © 1972 by the IAU.

but some of their properties are rather intermediate in character between Groups I and II.

Group II covers comets with A_1 often negative, badly determined, and with A_2 comparable or nearly comparable with A_1 in magnitude. Definite members are P/Encke, P/Tempel 2, P/d'Arrest, and P/Wirtanen. Possible members are P/Tempel-Swift, P/Arend, and P/Pons-Winnecke (twentieth century).

There is no difference between the two groups in the degree of the determinacy of A_2, which is mostly excellent, and in the determinacy of A_3, which is always poor. Group I can be identified with the obedient comets, the less populated Group II with the unruly ones.

3. A Model for the Rotation of Cometary Nuclei

It would be very easy to disregard the small systematic positional deviations from Equations (1) on the grounds that they may be due to a displacement between the center of light and the center of gravity. However, some of the unruly comets show systematic residuals over periods of time during which the geometrical conditions varied considerably, and there should also have been large variations in the relative positions of the optical center and genuine nucleus.

While not denying the existence of occasional effects of photometric displacements on the positional residuals, we do not believe that they can have a dominant influence. We suggest instead that the dynamical disobedience may be due to the approximate character of Equations (1). In the following we compare the empirical nongravitational terms with a simple model for a rotating nucleus.

A simple, yet rather general, model of the nongravitational effects can be developed for a spherical nucleus rotating at a constant angular speed about an axis fixed in space. The momentum of ejected material gives rise to an impulse on the nucleus whose direction can be expressed in terms of the obliquity ε of the rotation plane to the orbit plane, the longitude φ of the subsolar meridian at perihelion from the ascending node of the equator on the orbit, and the lag angle Λ of the meridian of effective mass ejection behind the subsolar meridian. We have

$$
\begin{aligned}
Z_1 &= Z_0 f(r) a_1 [1 - p_1 \cos 2(\varphi + v)], \\
Z_2 &= Z_0 f(r) a_2 [1 + p_2 \sin 2(\varphi + v)], \\
Z_3 &= Z_0 f(r) a_3 p_3 \cos (H + \varphi + v),
\end{aligned}
\tag{2}
$$

where v is the true anomaly, Z_0 the magnitude of the impulse (or acceleration),

$$
\begin{aligned}
a_1 &= (1 - h^2) \cos 2H, \\
a_2 &= (1 - h^2) \sin 2H, \\
a_1 p_1 &= a_2 p_2 = h^2 \geqslant 0, \\
a_3 p_3 &= 2h(1 - h^2)^{1/2} \geqslant 0,
\end{aligned}
\tag{3}
$$

and

$$
\begin{aligned}
h &= \sin (\Lambda/2) \sin \varepsilon \geqslant 0, \\
\tan H &= \tan (\Lambda/2) \cos \varepsilon.
\end{aligned}
\tag{4}
$$

We point out that Equations (2) depend, in contrast to the empirical Equations (1), on the true anomaly: each of the two components in the orbital plane consists of a constant term and a periodic term, whereas the out-of-plane component has a periodic term only.

Equations (2) can be reduced to Equations (1) only if the amplitudes p_i of all three periodic terms are zero; then $A_1 = Z_0 a_1$, $A_2 = Z_0 a_2$, $A_3 = 0$. This would take place if $\varepsilon = 0°$ or $180°$ (rotational axis perpendicular to the orbit plane), or if $\Lambda = 0°$ (no delay in ejection mechanism). If none of these conditions is at least approximately satisfied, the effect of the periodic terms may result in poor determinacy of the empirical coefficients A_i.

Thus, the amplitudes p_i could have a decisive influence on the 'degree of obedience' of a comet. The greater the p_i, the more unruly the comet is. We shall now establish the meanings of the amplitudes p_i in terms of ε and Λ (they are independent of φ). A plot of p_1 against p_2 with ε and Λ as parameters is represented in Figure 1. An extensive analysis, comparing Equations (1) and (2), will appear in full detail elsewhere. Here we point out that, as a result, it is found that the lag angle Λ basically determines the degree of accuracy with which A_i coefficients can be derived when the periodic terms are neglected. The lag angle is also dominant in controlling the magnitude of the

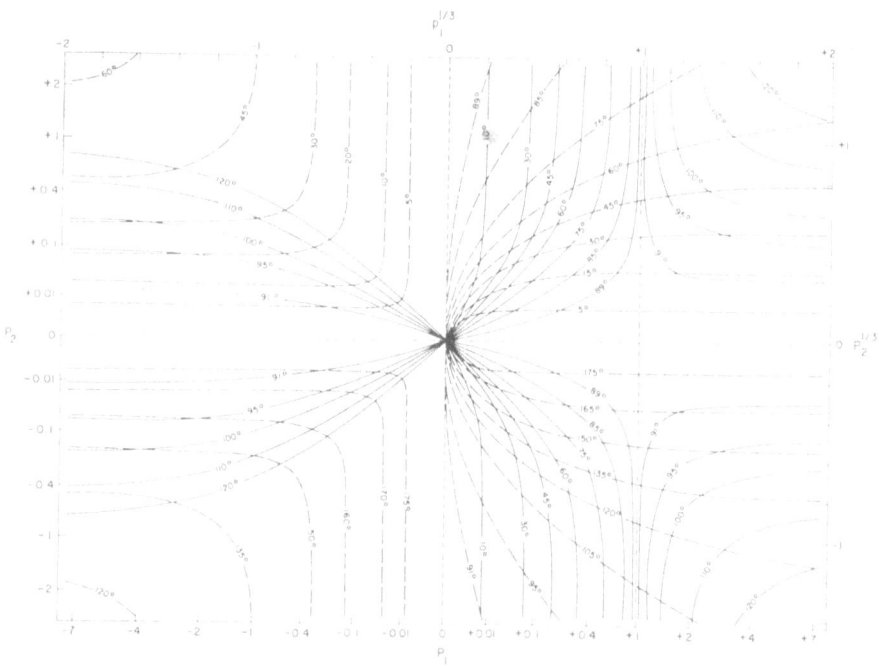

Fig. 1. Relative amplitudes p_1, p_2 of the periodic terms of the radial and transverse components, respectively, of the nongravitational acceleration in terms of the lag angle and obliquity. Solid curves are lines of constant lag angle; broken curves are lines of constant obliquity. Dotted lines run through the loci of singular solutions. The axes of abscissas and ordinates give, on a linear scale, the cube roots of p_1 and p_2, respectively.

transverse-to-radial component ratio A_2/A_1. The conclusions predicted from the analysis can be summarized thus:

(1) If the lag angle Λ is small, the radial component of the nongravitational acceleration is positive and well determined, and the A_2/A_1 ratio is small.

(2) If the lag angle is near to or in excess of 90°, the radial component can come out either positive or negative, and it is poorly determined. The transverse component A_2 may become comparable with A_1 in magnitude.

(3) The determinacy of A_2 is almost independent of the lag angle, and very good unless the obliquity ε approaches 90°.

(4) Positive A_2 does not necessarily refer to a direct sense of rotation, and vice versa. For obliquities near 90°, the sign of A_2 can be contrary to the expected one.

(5) The determinacy of A_3 is poor independently of ε and Λ.

We find that the properties of the empirical coefficients A_i of the two groups of short-period comets outlined in Section 2 are much like the expected properties of A_i for small and large lag angles, respectively. We therefore conclude that the two groups of comets differ from each other in the magnitude of the lag angle. Without bringing forward any argument here we point out that over a long time-scale the age of the comet (i.e., basically the number of revolutions about the Sun in an orbit with fairly small perihelion distance) is suspected to be the dominant influence on the lag angle.

4. Precession of the Cometary Nucleus. Periodic Comet Faye

A recent study of the motion of P/Faye between 1843 and 1970 (Marsden and Sekanina, 1971) has resulted in the surprising discovery that this comet had a secular acceleration until quite recently, but a secular deceleration since. The transverse-to-radial component ratio turned out to be smaller than for any other comet so far investigated and seems to be subject to variations with a period of about 140 yr (see Figure 2).

The variations can be interpreted as an effect of precession of the comet's nucleus. Two versions are suggested:

(1) Regular precession with the axis of precession perpendicular to the orbit plane. Then

$$\varphi = \varphi_0 + \frac{2\pi t}{P_{pr}},$$

$$\varepsilon = \text{const}, \tag{5}$$

where P_{pr} is the period of precession, equal to double the period of the variations in A_2/A_1. This model gives for P/Faye:

$$
\begin{aligned}
P_{pr} &= 37.8 \text{ revolution periods} \\
&= 280 \text{ yr}, \\
\varphi_0 &= 150° \text{ or } 330° \quad \text{(in 1843)}, \\
\varepsilon &= 88°.5, \\
\Lambda &= 27°.3.
\end{aligned} \tag{6}
$$

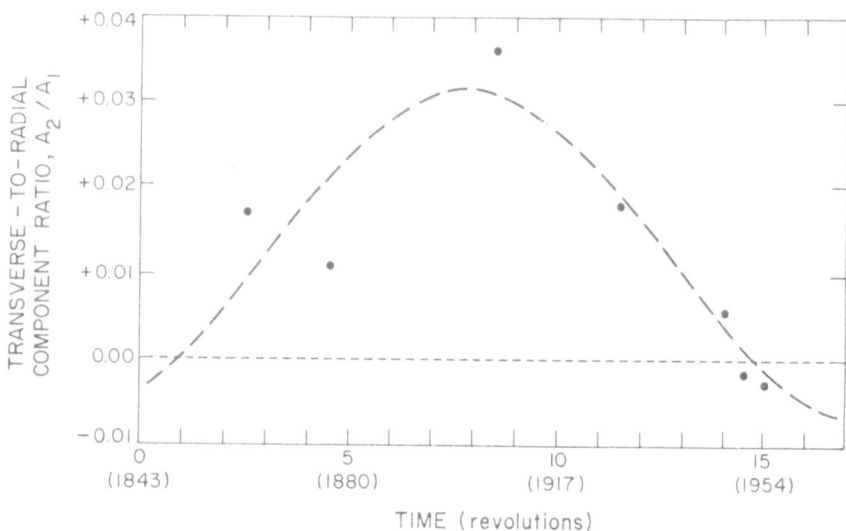

Fig. 2. Secular variations in the transverse-to-radial component ratio for P/Faye. The points are the empirical data from Marsden and Sekanina (1971). The dashed curve is the best sinusoidal fit; the dotted line is the line of A_2 sign change. Time is given in both the revolution-period units and years.

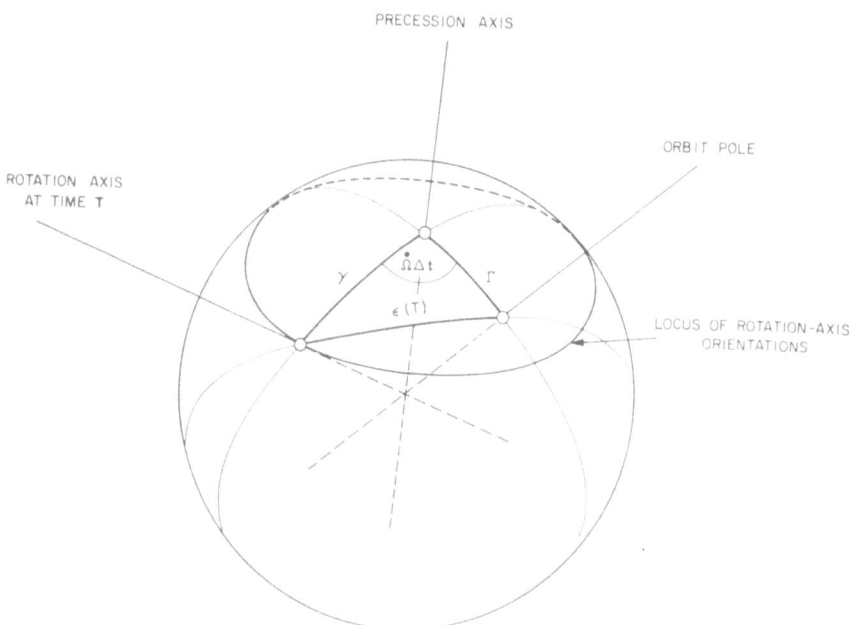

Fig. 3. Instantaneous obliquity $\varepsilon(T)$ in terms of the precession radius γ, precession axis deviation Γ and precession angular speed $\dot{\Omega} = 2\pi/P_{pr}$ for the variable-obliquity precession model. We denote $\Delta t = T - t_0$, where t_0 is the time of minimum ε.

(2) Variable-obliquity precession with the axis of precession tilted to the orbit plane. The obliquity varies with time according to (Figure 3)

$$\cos \varepsilon = \cos \gamma \cos \Gamma + \sin \gamma \sin \Gamma \cos \frac{2\pi(t - t_0)}{P_{pr}}, \tag{7}$$

where t_0 is the unknown epoch of minimum ε; $\varepsilon_{\min} = |\gamma - \Gamma|$; γ is the precession radius, and Γ is the deviation of the precession axis from the orbit pole. The precession period P_{pr} now equals the period of the variations in A_2/A_1. Applied to P/Faye this model gives:

$$P_{pr} = 18.9 \text{ revolution periods}$$
$$= 140 \text{ yr,}$$
$$t_0 = 1902, \tag{8}$$
$$\tan \gamma \tan \Gamma = 1.52,$$
$$90° < \gamma + \Gamma \le 102°,$$
$$\Lambda \ge 1°8.$$

The angles γ and Γ cannot be determined unequivocally.

5. Change in Sign of A_2 in the Presence of Severe Orbit Transformations: Periodic Comet Pons-Winnecke. A Final Remark

Marsden (1970) has found that P/Pons-Winnecke was secularly accelerated before 1875 [a result agreeing with that found by Oppolzer (1880)] but slightly decelerated after 1933. A detailed analysis of the motion of this comet suggests that the combined effect of nuclear rotation with a secular increase in the lag angle and of long-term orbit perturbations may be responsible for the change in the sign of A_2. The result is, however, rather uncertain numerically, because the comet is dynamically troublesome (regular close encounters with Jupiter). Apparently, the comet rotates in a retrograde sense ($95° < \varepsilon < 100°$).

The suspicion that P/Faye and P/Pons-Winnecke may have rotation axes almost in their orbit planes may be significant; from completely independent evidence we suggested earlier that this could also be the case with P/Halley and some other comets (Sekanina, 1967). Recently Gehrels et al. (1970) have concluded that Icarus has an obliquity of almost 90°. Uranus has an obliquity of nearly 90° too. The fact that there seem to be quite a few bodies of this sort in the solar system suggests that an effort should be made to discover whether there is a mechanism that could force the rotation axis of a body toward its orbit plane.

References

Gehrels, T., Roemer, E., Taylor, R. C., and Zellner, B. H.: 1970, *Astron. J.* **75**, 186.
Marsden, B. G.: 1969, *Astron. J.* **74**, 720.
Marsden, B. G.: 1970, *Astron. J.* **75**, 75.
Marsden, B. G.: 1972, this Symposium, p. 135.

Marsden, B. G. and Sekanina, Z.: 1971, *Astron. J.* **76**, 1135.
Oppolzer, T.: 1880, *Astron. Nachr.* **97**, 149.
Sekanina, Z.: 1967, *Bull. Astron. Inst. Czech.* **18**, 286.
Yeomans, D. K.: 1972a, this Symposium, p. 181.
Yeomans, D. K.: 1972b, this Symposium, p. 187.

53. A MODEL FOR THE NUCLEUS OF ENCKE'S COMET

Z. SEKANINA

Smithsonian Astrophysical Observatory, Cambridge, Mass., U.S.A.

Abstract. A study of the nongravitational deviations in the motion of P/Encke suggests that the nucleus of this comet can conceivably be approximated by a core-mantle model, composed of a porous solid core with embedded ices, surrounded by a thick envelope of free ices that gradually sublimates out.

1. Introduction

Recent investigations have brought more evidence on the total mass output from comets. Perhaps the most exciting and also most obvious are the recent observations of extensive hydrogen and hydroxyl atmospheres of comet 1969 IX (Tago-Sato-Kosaka) from OAO 2 and comet 1969i (Bennett) from OAO 2 and OGO 5 (Code *et al.*, 1970; Houck and Code, 1970; Bertaux and Blamont, 1970), which definitively confirm that the C_2 and CN bands, typical of the visual and photographic spectra, are produced by a very small fraction of the total output of material from the comets. In the light of this and other less direct evidence, Whipple's (1950) interpretation of the nongravitational effects in comets in terms of loss of volatile materials from the nucleus deserves full credit.

The enormous increase of information about the dynamics of comets from recent orbital calculations can also contribute to our understanding of the physical processes in, on and near the cometary nucleus, but an elaborate theory consistently interpreting the dynamical data is still missing.

2. Periodic Comet Encke

In order that we can study long-term variations in the mass output from comets, observations from as long a period of time as possible have to be considered and the results checked with the other evidence.

The most extensive information that can serve for this purpose is the list of data on the deviations of the motion of periodic comet Encke from the law of gravitation. The writer has compiled such a list for the period of almost two centuries (Sekanina, 1969a). Marsden's (1969, 1970) calculations on the motion of this comet suggest that at recent apparitions the radial component of the nongravitational acceleration may have been insignificant relative to the transverse component, or at any rate that it was not an order of magnitude larger than the transverse component. Both the analysis of the nongravitational terms in Marsden's equations of motion (Sekanina, 1972) and a theoretical study of the distribution in the gas-jet impulses over the nuclear surface of an Encke-type comet (Sekanina, 1970) support the idea that the effective direction of the nongravitational forces acting on such a comet is expected to be essentially perpendicular to the comet's radius vector.

Chebotarev et al. (eds.), The Motion, Evolution of Orbits, and Origin of Comets, 301–307. All Rights Reserved.
Copyright © 1972 by the IAU.

We shall measure the dynamical effect of the nongravitational acceleration in terms of the classical parameter κ, introduced by Encke (1829). It is defined by

$$\kappa = 600 \int_{t}^{t+P} \frac{d\mu}{dt}\, dt,$$ (1)

where μ is the daily mean motion of the comet. Expanding Whipple's (1950) ideas, the writer has shown that κ is generally proportional to the relative mass-loss rate of the comet per revolution (Sekanina, 1969b):

$$\kappa = C(\Delta M/M),$$ (2)

with the proportionality coefficient

$$C = -C_1 C_2 D(1 - e^2)^{1/2} \frac{\displaystyle\int_{-\pi}^{\pi} f(r)r\, dv}{\displaystyle\int_{-\pi}^{\pi} f(r)r^2\, dv}.$$ (3)

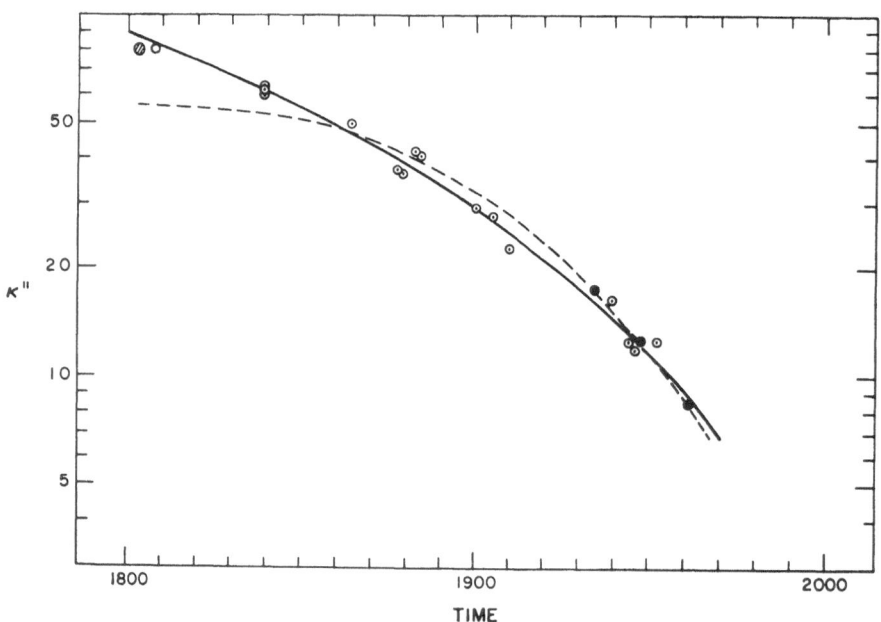

Fig. 1. Nongravitational parameter κ as a function of time for Encke's Comet. Circled points are κ values obtained in a classical way; solid circles are from Marsden's (1970) double exponential curve; the shaded circle is a provisional value, determined by the writer from Encke's 'pure elliptic revolution periods', covering the period 1786 to 1819; and the open circle is a value derived in the same way for the period 1795 to 1819. The full curve represents the best fit by means of Equation (10); the dashed curve is the Marsden double-exponential fit from 1927 to 1967 and extrapolated back.

Here C_1 is a positive constant, C_2 the transverse component in units of the total acceleration, D the effective velocity of ejection (i.e., integrated over the nuclear surface), e the orbital eccentricity, and $f(r)$ the law of variation of the nongravitational acceleration with solar distance r. In terms of the Whipple hypothesis the sign of C is just a matter of the sense of rotation. P/Encke has a secular acceleration, so that κ is positive. The variations in the nongravitational effects of this comet since the beginning of the nineteenth century are represented in Figure 1.

3. Models of Mass Output

We assume that the comet moved originally in an orbit that prevented it from being exposed to intense solar radiation. Whatever the internal structure of the nucleus, we assume that it was uniform at a certain time t_0, when the comet was gravitationally captured (presumably after a number of close approaches to Jupiter) and forced to move in a rather stable short-period orbit with perihelion distance comparable with its present value. The comet then started losing its volatile supplies at a significant rate.

We also assume that the nucleus of P/Encke is monolithic, spherical in shape, of radius R_0 at time t_0, and rotating rather fast. Assuming that specific, simple mechanisms govern the mass transfer in the nucleus and from its surface, we determine the relative rate of mass output per revolution $\Delta M_j / M_j$ as a function of the number of revolutions j since the time t_0.

For an icy nucleus with scattered dust impurities embedded in it, the law for the rate of mass loss is simply

$$\frac{\Delta M_j}{M_j} = \frac{3\alpha}{1 - \alpha j}, \tag{4}$$

where α is the linear rate of shrinkage of the nuclear radius in units of the original radius R_0 ($\alpha \ll 1$). Because of Equation (2), in this model the secular acceleration increases with time, in sharp contrast to what is observed for P/Encke.

Thus we shall consider instead a composite model, assumed to consist of a porous matrix of solid material, with ice filling the pores. The tensile strength of the nucleus is assumed to be sufficiently high to withstand the pressure from the diffusing volatile component, so that the nucleus of radius R_0 does not shrink with time. The mass distribution at t_0 in either of the two components is assumed to be uniform throughout the nucleus, the apparent ice density being ρ_0, the total apparent density ρ_c. During the first post-capture approach of the comet to the Sun, all the ice in a surface layer of thickness $\alpha_1 R_0$ is assumed to sublimate out. This loss is in part compensated by diffusion of volatiles from deeper layers to the surface. Before the comet comes to the Sun again, it is assumed that the uniform distribution of ice inside the nucleus is entirely restored at an apparent density of

$$\rho_1 = \rho_0(1 - \alpha_1)^3 < \rho_0. \tag{5}$$

During the second approach to the Sun, the surface layer of thickness $\alpha_2 R_0$ is depleted of ice, and so on. The output rate of this model follows the law

$$\frac{\Delta M_j}{M_j} = \frac{\dfrac{1 - (1 - \alpha_j)^3}{(1 - \alpha_j)^3} \displaystyle\prod_{k=1}^{j} (1 - \alpha_k)^3}{p_c/p_0 - \displaystyle\sum_{i=1}^{j} \frac{1 - (1 - \alpha_i)^3}{(1 - \alpha_i)^3} \prod_{k=1}^{i} (1 - \alpha_k)^3}. \tag{6}$$

If $\alpha_1 = \alpha_2 = \cdots = \alpha = \text{const}$, Equation (6) simplifies to

$$\frac{\Delta M_j}{M_j} = \frac{\exp \gamma - 1}{1 + [(p_c/p_0) - 1] \exp (\gamma j)}, \tag{7}$$

where

$$\gamma = \ln \frac{1}{(1 - \alpha)^3} > 0. \tag{8}$$

Equation (7) indicates that the nongravitational acceleration decreases with time, in qualitative agreement with what is observed for P/Encke.

Since the apparent ice density decreases with time, the thickness of the surface layer to be depleted each revolution may increase with time. Accepting the relation

$$\alpha_j \rho_{j-1}^{1/3} = \text{const}, \tag{9}$$

Equation (6) becomes

$$\frac{\Delta M_j}{M_j} = \frac{3\alpha(1 - \alpha j)^2}{p_c/p_0 - 1 + (1 - \alpha j)^3}, \tag{10}$$

where $\alpha = \alpha_1$, $p_c/p_0 \gg 1$. Equation (10) suggests that κ decreases with time if $p_c/p_0 \gtrsim 1.5$, but that it first increases and then decreases if $p_c/p_0 < 1.5$. Of course, only the former case is physically acceptable, and we again get qualitative agreement between the model and the observations.

The qualitative difference between the two versions of the composite model is that in the latter case the nucleus becomes entirely depleted of volatiles after completing $(1/\alpha)$ revolutions about the Sun since the time of capture (finite lifetime), whereas in the former version the nucleus has an infinite lifetime.

4. Failure of the Two Versions of the Composite Model. Core-Mantle Model. Conclusions

Applying the two versions of the composite model to P/Encke, we find that they fit the data almost equally well (mean residual in κ $\pm 2''10$ for the finite version, $\pm 2''19$ for the infinite), but the mass output rates differ by an order of magnitude (Table I), the ones resulting from the infinite version being much too high.

The finite version would be promising if it did not yield unacceptable values for the time of capture in terms of the density ratio p_c/p_0 (these two quantities cannot be separated from each other). It follows that the comet should have been captured

2600 yr ago if 99% of its mass were due to ice, 1000 yr ago if 88% were ice, and only 500 yr ago if 53% were ice. In fact, from the investigation by Whipple and Hamid (1952) of the Taurid meteor stream, it follows that P/Encke must have been captured at least 5000 yr ago; and from physical considerations we can estimate that ice, trapped in a porous but compact structure of the nonvolatile matrix of the composite nucleus, could not make up more than, say, 45% of the total mass.

TABLE I

Relative output rates of P/Encke computed from the composite model

Time	$\Delta M/M$ (percent per revolution)	
	finite version	infinite version
1800	0.34	6.4
1833	0.26	5.2
1867	0.18	3.8
1900	0.12	2.3
1933	0.069	1.3
1967	0.031	0.66

This discrepancy suggests that the composite model is quantitatively unacceptable and that ice must have played a much more important role in the history of Encke's Comet than the above model indicates. This has been the main reason for our suggesting a combination of the composite model (finite version) and the icy model into a core-mantle model.

The proposed core-mantle model has a solid, compact though porous core with embedded ices. The core is surrounded by a thick shell, or envelope, of free ices. It is physically reasonable to assume that the shell ices are contaminated by loosely distributed dust particles. After the comet has been captured and forced to orbit the Sun in a short-period path, the outer icy shell shrinks gradually owing to sublimation, as in the pure icy model. This process continues until the shell is completely sublimated out and the underlying solid core is exposed to solar rays. Free sublimation is then replaced by activated diffusion, and from this time on the model can be treated as a finite composite model. The mass output in the core phase is therefore given by Equation (10), where j is the number of revolutions from the time the core was dismantled. It can be shown that the mass-loss rate in the mantle phase is also described by an expression of the same form, provided that α is replaced by α_e, the rate of shrinkage of the envelope (in units of the initial nuclear radius R_0 per revolution), and that $\rho_c/\rho_0 - 1$ is replaced by $(R_c/R_0)^3(\rho_c/\rho_e - 1)$, where R_c is the radius of the core and ρ_e is the *apparent* density of the envelope. The process of dismantling of the core should be accompanied by a significant decrease in the output of volatiles due to the expected redistribution of the solar radiation effects on the new type of surface (with more energy needed for heating low-conductivity solid materials of the core). Quantitatively, the output deficit can be characterized by a coefficient Γ $(0 < \Gamma < 1)$, defined by

$$\lim_{j \to j_c+} (\Delta M_j)_{\text{core}} = (1 - \Gamma) \lim_{j \to j_c-} (\Delta M_j)_{\text{mantle}}, \tag{11}$$

where j_c is the time of core dismantling. In practice the dismantling process and the associated discontinuity in output occur rather smoothly. In physical terms this means that a layer of the nucleus was exposed, in which the tensile strength of the solid matrix, generally increasing from the surface toward the center, reaches a critical point; at this point the matrix is just able to withstand pressure from activated ices without losing any more substantial amounts of solid particles.

The application of the core-mantle model to P/Encke leads to the following suggestions about its evolution:

(1) The comet appears to be currently in the core phase. The averaged ejection velocity comes out 0.18 km s^{-1} (if $C_2 \simeq 1$, see Section 2), which suggests that for a thermal velocity of 0.6 km s^{-1} the anisotropy coefficient is as high as 0.3. The comet entered this phase about the year 1800 if the embedded ices made up originally 10% of the total mass of the core, about the year 1700 if they made up 20%, and about 1600 if 35%. The comet is expected to be entirely depleted of volatile substances in about 60 to 70 yr, or some 20 revolutions from now. This conclusion is in fairly good agreement with the photometric death date of the comet predicted by Whipple and Douglas-Hamilton (1966).

(2) Extrapolating the suggested evolution of the comet backward to its mantle phase, we find that a shell several meters in thickness would have been blown off during each revolution. We note that a sphere of pure water ice moving in the orbit of Encke's Comet would shrink at a rate of 2 to 3 m per revolution; contamination by highly absorbing meteoric particles would raise the rate to 4 to 5 m per revolution, while a sphere of unpacked water snow would diminish at a rate of 6 to 8 m per revolution. The corresponding brightness decrease with time would have been very slight, not more than 0.2 magnitude per century. The comet was probably not brighter than absolute magnitude 6 some 1000 yr ago.

(3) In contrast, the fading of the comet in its core phase is expected to be considerably greater, and progressively increasing. Assuming proportionality between the total gas output and the loss of constituents providing the C_2 bands that are dominant in the comet's visual spectrum, the expected luminosity deficit should currently amount to about 3 magnitudes per century, or 0.1 magnitude per revolution. This is in good agreement with the available photometric data (e.g., Sekanina, 1964; Whipple and Douglas-Hamilton, 1966).

(4) The estimate of the original dimensions of the nucleus of Encke's Comet depends much on the expected time of capture. If capture took place about 5000 yr ago (as indicated from association with the Taurids), the radius of the original nucleus would have been about 10 km; if it happened 20 000 yr ago, the radius would have been more like 25 to 30 km.

(5) Capture 5000 yr ago is likely because of its dynamical implications. The non-gravitational acceleration has produced a secular reduction in the comet's aphelion

distance. We can estimate that the aphelion distance amounted to about 4.7 AU 5000 yr ago, and that it would have been 5.2 AU 20 000 yr ago. In either case, the comet's aphelion might have been already inside Jupiter's orbit, but not until 5000 yr ago did the minimum distance between the two orbits become large enough to ensure the stability of that of the comet; before then, expulsion of the comet into a distant orbit would have been possible. Further, some 5000 yr ago the comet should have been near 3:1 resonance with Jupiter, and that might have helped to keep the comet out of the reach of Jupiter, although the comet had a large enough nongravitational acceleration eventually to escape from a possible libration. The Taurids are perhaps a product of the comet's early violent evolution in the innermost part of the solar system.

References

Bertaux, J.-L. and Blamont, J.: 1970, *Compt. Rend. Acad. Sci. Paris Ser. B* **270**, 1581.
Code, A. D., Houck, T. E., and Lillie, C. F.: 1970, *IAU Circ.* No. 2201.
Encke, J. F.: 1829, *Astron. Nachr.* **7**, 181.
Houck, T. E. and Code, A. D.: 1970, *Bull. Am. Astron. Soc.* **2**, 321.
Marsden, B. G.: 1969, *Astron. J.* **74**, 720.
Marsden, B. G.: 1970, *Astron. J.* **75**, 75.
Sekanina, Z.: 1964, *Bull. Astron. Inst. Czech.* **15**, 1.
Sekanina, Z.: 1969a, *Astron. J.* **74**, 944.
Sekanina, Z.: 1969b, *Astron. J.* **74**, 1223.
Sekanina, Z.: 1970, *Astron. Astrophys.* **7**, 109.
Sekanina, Z.: 1972, this Symposium, p. 294.
Whipple, F. L.: 1950, *Astrophys. J.* **111**, 375.
Whipple, F. L. and Douglas-Hamilton, D. H.: 1966, *Mem. Soc. Roy. Sci. Liège Ser. 5* **12**, 469.
Whipple, F. L. and Hamid, S. E.: 1952, *Bull. Roy. Obs. Helwan* No. 41, 30.

54. LABORATORY SIMULATION OF ICY COMETARY NUCLEI

E. A. KAJMAKOV and V. I. SHARKOV

A. F. Ioffe Institute of Physics and Technology, Leningrad, U.S.S.R.

Abstract. The problems of laboratory simulation of cometary phenomena are discussed. Results are given of experimental investigations of the thermodynamic parameters of H_2O ice and of frozen mixtures of electrolytes during sublimation under conditions of high vacuum and low temperature. At a distance of 1 AU from the Sun the equilibrium temperature of H_2O ice is shown to be about $-75\ °C$. Conditions are determined for the formation and disintegration of dust matrices during the sublimation of ices containing small dust particles. Dust particles are shown to leave the surface of the ice in the form of conglomerates of primary particles. The outflow velocities of the particles are found to be as high as several metres per second. The results obtained are discussed in relation to cometary phenomena.

The great strides made by man in gaining mastery over cosmic space have been responsible for the vigorous development of research in astrophysics in many directions. Because of their rapid variations, comets are of particular interest to astrophysicists. Since comets are sensitive to electrical and magnetic fields and to the solar wind they can be visualized as cosmic space probes. Observations of changes in their shapes and sizes, as well as in their spectra, permit us to study conditions prevailing in regions of space as yet inaccessible to man-made space probes (Beller, 1963).

According to the most widely accepted hypothesis, cometary nuclei consist of mixtures of various ices and particles of high melting point. Solar radiation causes sublimation of the ices, accompanied by ejection of the particles. There are many additional factors that greatly complicate the picture.

Most of the molecules, atoms and ions present in cometary atmospheres are formed as a result of ionization and dissociation of primary, or 'parent' molecules. Among these parent-molecules are H_2O, CO_2, NH_3, CH_4, $(CN)_2$ and several other organic and inorganic molecules of greater or less complexity (Donn, 1964). The best defined features in cometary spectra are bands corresponding to CN, C_2, OH and other molecules.

Many cometary phenomena await clarification. There are still a great many unsolved problems relating to the compactness and chemical composition of the nucleus, formation of the dust component in the coma, outbursts in brightness, etc. Individual hypotheses fail to explain all the observed facts.

Experimental research on comets is nowadays being conducted in many directions, using up-to-date techniques and modern instrumentation. We make particular reference to the use of telescopes in conjunction with sophisticated spectrographs and image converters, permitting radical reductions in exposure times; this method has already yielded important results (Rylov, 1967). Another new technique is the use of satellites and spacecraft for photography and spectroscopy of comets in the ultraviolet and infrared. Such studies ensure freedom from the influence of the Earth's atmosphere, enabling new data to be obtained and earlier results to be deciphered (Boggess, 1965).

.

Chebotarev et al. (eds.), The Motion, Evolution of Orbits, and Origin of Comets, 308–315. All Rights Reserved.

There is also the possibility of launching automatic probes towards comets. Plans currently under discussion in various countries envisage a flight around a comet, photography of the nucleus, sampling the atmosphere, and determination of the ionic composition (Haser and Lüst, 1966; Roberts, 1963).

Yet another direction of study is concerned with the simulation of cometary phenomena; this can be realized in laboratory conditions as well as out in space. Here we may include experiments in producing artificial clouds at high altitudes, observing the motions and changes in configuration of these clouds, and obtaining spectrograms. Experiments have been conducted with Na (Shklovskij and Kurt, 1959), NH_3 (Rosen and Bredohl, 1965), Ba and Sr (Föppl, 1965) and H_2O clouds, and many interesting results have been obtained. It will be noted, however, that such experiments are only indirectly associated with cometary problems, for they are conducted mainly for the study of the upper atmosphere. The molecules of most direct interest to cometary investigators have not yet been considered.

The possibility of creating an artificial cometary nucleus has more than once come under discussion, although it is not at all clear how or of what material it should be made. Some astronomers have suggested launching into space an icy rock (Swings, 1962), while others have proposed a ton of CO_2 or NH_3 (Shklovskij and Kurt, 1959). Launching such a large object into space would be a fairly involved and costly experiment, and we do not see that it would provide a great wealth of information. Considerable preparatory work would be necessary for the selection of suitable materials for both the gas and dust components, and at the same time, a great many technical problems would have to be solved.

Laboratory simulation of cometary phenomena has been conducted on a limited scale. The difficulty here is to provide conditions close to those of outer space. The solar wind and solar electromagnetic radiation interact with the gaseous component of a cometary atmosphere, causing ionization and photodissociation of the parent-molecules. Laboratory studies by Potter and del Duca (1964) have narrowed down the range of possible parent substances for the observed cometary radicals. Their experiments indicate that at a distance of 1 AU from the Sun simple compounds yielding C, CH and C_2 usually have life spans greater than 10 h, whereas in order to explain the observations the life spans should be an order of magnitude smaller. Thus it seems that pure ices (or mixtures of them) cannot be parent substances for the C-, C_2-, and CH-groups, and a search is indicated for other, less stable substances.

A very interesting experiment that bears on the problem of the origin of cosmic ices has been performed by Berger. This consisted of subjecting a frozen mixture of H_2O, CH_4, and CH_2 at a temperature of $-230\,°C$ to proton radiation from a cyclotron. Chemical analysis of the resultant products revealed traces of $CO(NH_2)_2$ (urea), CH_3COOH (acetic acid) and CH_3COCH_3 (acetone). Experiments such as this show that it is not at all impossible that fairly complex chemical compounds may be formed during the evolution of cometary nuclei.

Another example of laboratory simulation of cometary phenomena may be seen in the work of Danielsson and Kasai (1968). Their purpose was to study the ionization of molecules and the acceleration mechanisms in cometary atmospheres, as represented in the interaction of CO_2 ice and hydrogen plasma. As a result, they produced a 'comet tail' consisting mainly of heavy CO_2 ions and found that these are accelerated by plasma flow to speeds as high as 10^5 cm s^{-1}, which amounts to 10% of the plasma flow velocity. This type of experiment is of considerable scientific interest.

An extensive and complex programme of studies on cometary phenomena has been carried out for a number of years at the A. F. Ioffe Institute. The studies concerned the behaviour of various types of ice (including dustladen specimens) when they are subjected to the action of conductive and radiant heat and ultraviolet radiation under conditions of high vacuum and low temperature.

The initial experiments were on pure H_2O ice, of which many properties had been adequately investigated (Dorsey, 1940). An ice sample prepared from distilled water in a special cell was placed in a vacuum chamber (at 10^{-5} to 10^{-6} T) whose walls were cooled with liquid nitrogen. Embedded in the ice sample was a miniature electric heater for supplying energy to the ice and several copper-constantan thermocouples for measuring the temperature at various points. The cell was suspended on a spring balance, which served to determine changes in the weight of the ice sample due to sublimation in the vacuum. From these experiments the following relationships and parameters were successfully determined:

(1) The relationship between the 'equilibrium' temperature T of the ice and the specific power input W_{sp} supplied per unit area of the ice surface (see Figure 1). The equilibrium temperature is the temperature at which the energy input is consumed solely in molecular sublimation. The practical procedure for determining this is described by Kajmakov et al. (1972). If the water-ice albedo is taken to be 0.7, the curve may be used to determine the temperature of an H_2O cometary nucleus at various

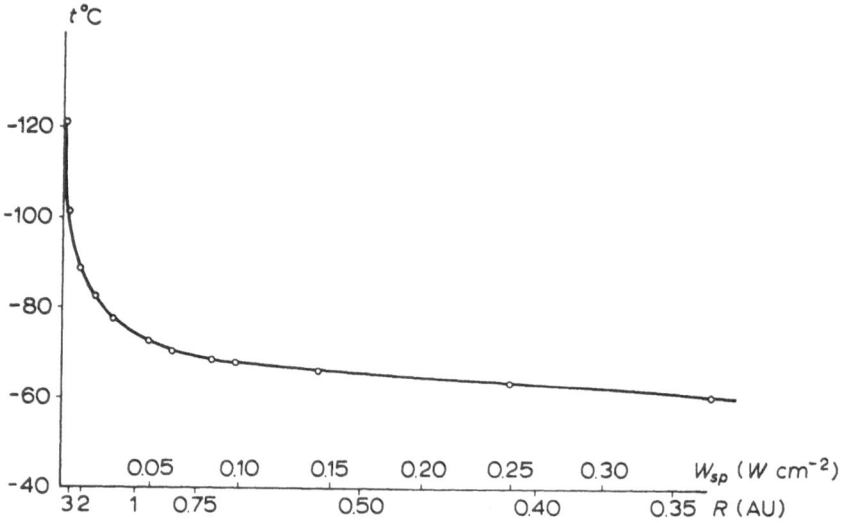

Fig. 1.   Temperature of H_2O ice T (°C) versus specific power input W_{sp} (W cm^{-2}). Corresponding heliocentric distances r (AU) are shown for an assumed albedo 0.7.

distances r from the Sun, provided that the comet's atmosphere does not have a shielding effect upon the nucleus. At $r=1$ AU the nuclear temperature would be $-75\,°C$.

(2) The ice consumption J at various temperatures. The solid curve in Figure 2 shows the experimental values, while the broken curve is computed on the basis of the kinetic theory of gases.

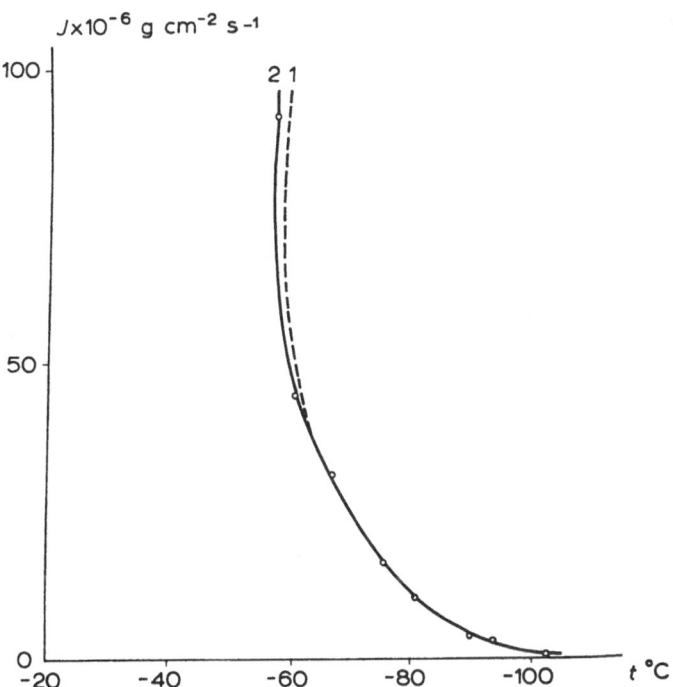

Fig. 2.   Consumption rate of H_2O ice J (g cm^{-2} s^{-1}) versus temperature T (°C). The broken curve (labelled 1) is theoretical, the solid curve (labelled 2) experimental.

(3) The heat of sublimation, i.e., the energy required for breaking the molecular bonds at the ice surface and the kinetic energy of the molecules. Within the temperature range -60 to $-90°C$ the sublimation energy was (670 ± 20) cal g^{-1}.

We have studied the sublimation of H_2O ice in a vacuum under the action of light beams. The light source used was a very high pressure mercury-quartz lamp whose energy distribution over the wavelength range 2300 to 10 000 Å is similar to that of sunlight. A light beam of known power was applied to the surface of the ice sample, and the ice consumption over the temperature range -55 to -60 °C was determined. For calculations involving the light power absorbed the ice albedo was again taken to be 0.7, and within the limits of error the experimental results based on light absorption coincided with those based on the application of heat energy to the ice.

In addition to experiments on pure water ice, we have conducted a series of experiments on frozen aqueous solutions of electrolytes. Studying the behaviour of such systems under conditions of high vacuum and low temperature is of interest both to the physics of solids and to the physics of comets. For instance, Delsemme and Swings (1952) are of the opinion that parent-molecules must be present in cometary nuclei, not as pure ices but as hydrates of the general form $B \cdot nH_2O$, where B is the appropriate parent-molecule and nH_2O is the number of water molecules with which it is combined.

By their outward behaviour *in vacuo* frozen solutions can be divided into two categories:

(1) those which form a porous salt matrix during sublimation, such as KCl or $CO(NH_2)_2$ solutions, and

(2) those which do not, such as NH_4OH and $(NH_4)_2CO_3$ solutions.

The first category is more common. Solutions of the LiCl type, which contain hydrated ions, are in a special category.

Initial results have been obtained on ice consumption and temperature as a function of energy input for solutions such as NH_4OH, $(NH_4)_2CO_3$, LiCl and KCl. Experiments on frozen solutions allow us to avoid many of the difficulties associated with ices of a different chemical nature, such as CO_2, C_2H_2, CH_4, etc. There is an extremely wide choice of substances, and it is possible to select combinations that would produce spectra similar to those observed for real comets affected by solar radiation.

It is known that sunlight is scattered by dust particles in cometary atmospheres and tails. The mechanism of dust-particle formation and acceleration, as well as other problems relating to the behaviour of the dust component of cometary atmospheres, are of particular interest, and we have therefore studied the behaviour *in vacuo* of ices containing dust components of various types (Kajmakov and Sharkov, 1969).

As before, these experiments were conducted in a cooled vacuum chamber. The ice sample containing dust contaminants was prepared as follows. A quantity of dust and a predetermined volume of distilled water were placed in a special container. After being thoroughly mixed and cooled down to $+1$ or $+2\ °C$, the mixture was quickly transferred to a nickel cell to be further cooled by liquid nitrogen. The cell and ice thus formed were suspended in the vacuum chamber over a glass plate. The measurements commenced when the pressure in the chamber dropped to 10^{-5} to 10^{-6} T. Either heat or light energy was applied to the ice, which assumed a certain steady-state temperature, related to the energy input. At this temperature sublimation of the ice occurred, accompanied either by the outflow of dust particles and conglomerates thereof, or by the formation of a dust matrix. Dust particles and conglomerates left traces on the glass plate, and this enabled us to determine the velocities and angular distribution of the outflowing particles and the difference in size between the initial particles and those which formed the traces. The angular distribution and velocities of the outflowing particles were determined by photometry of the traces. It was established that dust particles leave the ice surface almost perpendicularly (deviating from the normal by less than $15°$), and that the outflow angles are only slightly dependent on the nature, concentration and size. As the temperature of the ice increases, so does the number of particles with outflow angles substantially different from the normal. Figure 3 shows velocity versus ice temperature for various particles. The points labelled M on each curve correspond to the formation of dust matrices, conditions for the formation of which are as follows:

(1) The lower the density of the dust particles, the smaller their size and the higher the concentration and the probability of forming a matrix.

(2) The higher the temperature of the ice, the lower the probability of matrix formation, because as the temperature increases, so does the total impulse imparted to the dust particles by the sublimation products.

(3) The higher the albedo of the dust-laden ice, the lower the probability of dust-matrix formation. This conclusion was drawn from experiments made to ascertain the influence of a light beam (wavelength 3000 to 7000 Å) upon the surface of dust-laden ice.

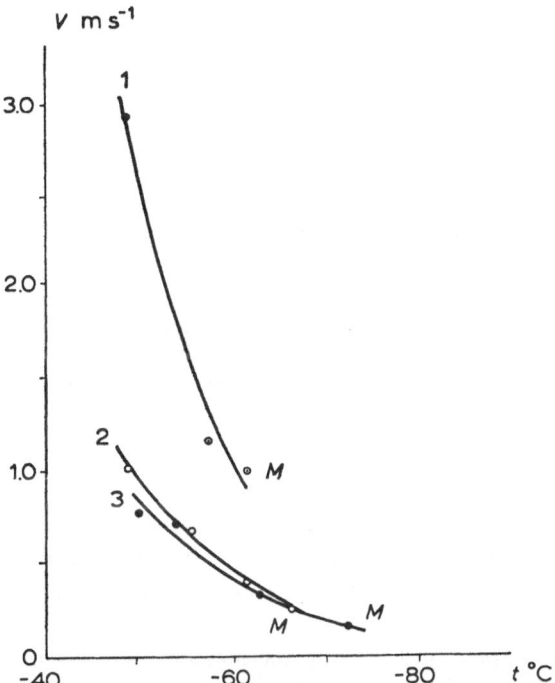

Fig. 3. Dust-particle velocity V (m s^{-1}) versus surface temperature of H_2O ice T (°C) for (1) Al_2O_3, particle size 2 μ, volume concentration 0.35%; (2) Al_2O_3, 2 μ, 1.05%; (3) SiO_2, 20 μ, 5%; M denotes the formation of a dust matrix.

If the traces left by the dust particles are examined under a microscope, the following pattern will be observed. At small distances from the icy surface the great majority of the initial particles produce conglomerates (Figure 4). Farther from the icy surface fewer form conglomerates and more separate particles are visible. The number of particles constituting conglomerates increases with increasing dust particle concentration and decreasing particle size and ice temperature. Particles 2 microns across and having a concentration by volume of 0.35% will form conglomerates with an average length of 5 to 7 mm and width 2 to 3 mm. The data obtained permit us to make the

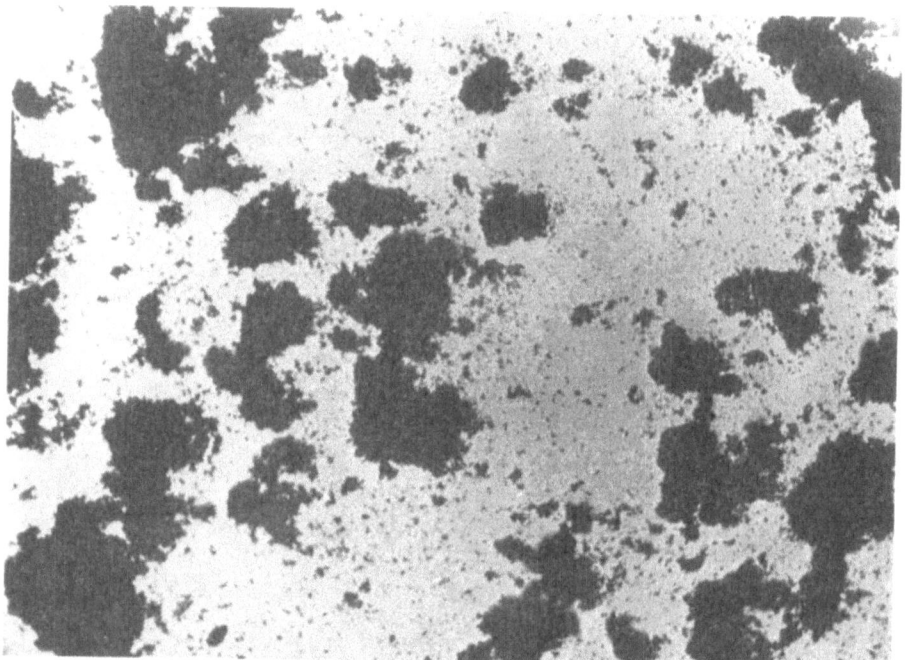

Fig. 4. Conglomerates of nickel particles. Primary particle size 2 μ, maximum conglomerate size 5 to 7 mm.

following assumptions and conclusions with regard to the physical processes taking place within comets:

(1) At considerable heliocentric distances a cometary nucleus containing a dust component will have a fairly low temperature, which will favour dust-matrix formation.

(2) As the comet approaches the Sun the temperature of a nucleus enclosed in a matrix will increase, and this may result in disintegration and shedding of the matrix. This process may take the form of an explosion, leading to a sharp increase in the amount of dust in the comet's head and an outburst in brightness.

(3) Once the matrix has been shed, the flow of volatiles from the icy surface increases. This causes the nucleus to be cooled down and reestablishes conditions for matrix formation. Thus the matrix may be formed and shed periodically as the comet approaches the Sun, the frequency of the cycles increasing with time.

(4) Most of the micro-sized dust leaves the cometary nucleus, not as separate particles, but in the form of conglomerates of varying size, shape and structure.

(5) The outflow of separate dust particles of submicron size from the ice surface is only possible at small heliocentric distances. The appearance of submicron dust at large distances from the Sun is presumably associated with some other mechanism, such as that suggested by Dolginov (1967).

(6) When conglomerate leaving the nucleus contains a volatile component, sublimation of this component may cause disintegration of the conglomerate, as well as its acceleration due to the exhaust thrust thus generated.

(7) Periodic meteor streams are remnants of comets. According to our views, meteors must be composed of conglomerates of particles. When such conglomerates enter the upper atmosphere they will be relatively bright objects, but due to their porous structure their reduced densities will be quite low.

[After the presentation of this paper a short film was shown, demonstrating some of the effects caused by sublimation of pure water ice and ice contaminated by dust particles.]

References

Beller, W.: 1963, *Missiles and Rockets* 12, No. 8, 22.
Boggess, A.: 1965, *Space Science*, London and Glasgow, p. 729.
Danielsson, L. R. and Kasai, G. H.: 1968, *J. Geophys. Res.* 73, 259.
Delsemme, A. H. and Swings, P.: 1952, *Ann. Astrophys.* 15, 1.
Dolginov, A. Z.: 1967, *Astron. Zh.* 44, 434.
Donn, B.: 1964, *Astron. J.* 69, 137.
Dorsey, N. E.: 1940, *Properties of Ordinary Water-Substance*, Reinhold, New York.
Föppl, H.: 1965, *Weltraumfahrt* 16, 139.
Haser, L. and Lüst, R.: 1966, *Raumfahrtforschung* 10, 196.
Kajmakov, E. A. and Sharkov, V. I.: 1969, *Trudy IV Zimnej Shkoly po Kosmofizike*, Apatity, part 2, 15.
Kajmakov, E. A., Sharkov, V. I., and Zhuravlev, S. S.: 1972, this Symposium, p. 316.
Potter, A. F. and del Duca, B.: 1964, *Icarus* 3, 103.
Roberts, J. G.: 1963, *Spaceflight* 5, 213.
Rosen, B. and Bredohl, H.: 1965, *Techn. Humn.* 62, No. 2, 5.
Rylov, V. S.: 1967, *Komety i Meteory* No. 15, 3.
Shklovskij, I. S. and Kurt, V. G.: 1959, *Iskusstv. Sputniki Zemli* No. 3, 66.
Swings, P.: 1962, *Space Age Astronomy*, Academic Press, New York and London, p. 370.

Discussion

J. M. Witkowski: What is the density of the matrix with respect to the density of the dust?

E. A. Kajmakov: Matrix density depends on dust concentration. Actually, if the concentration exceeds 1%, we can only guess that there has been ice sublimation. The dust structure in the matrix tends to reproduce the structure of the dust-laden ice. The density of the matrix may be one order of magnitude lower than that of dust-particle substance.

F. L. Whipple: I must congratulate you on your most successful and striking experiments. I deeply appreciate seeing in your film the effects I talked about 20 yr ago.

55. A NONGRAVITATIONAL EFFECT IN THE SIMULATION OF COMETARY PHENOMENA

E. A. KAJMAKOV, V. I. SHARKOV, and S. S. ZHURAVLEV

A. F. Ioffe Institute of Physics and Technology, Leningrad, U.S.S.R.

Abstract. Experimental results for the determination of the exhaust thrust developed by vacuum-sublimation products of H_2O ice and frozen aqueous electrolytes are described. The contribution of the exhaust thrust to nongravitational effects on comets is discussed.

1. Introduction

Comets are characterized by extremely violent processes (Dobrovol'skij, 1961). Modern studies of these processes involve the use of both the methods of classical optics and methods based on the latest advances in experimental techniques (Potter and del Duca, 1964; Danielsson and Kasai, 1968).

This paper deals with experiments for determining the exhaust thrust resulting from the sublimation of various types of ice in a laboratory simulation of cometary phenomena under conditions closely approximating those of outer space.

The particular interest in this problem is that when a comet passes near the Sun its orbital parameters tend to deviate from the values calculated purely from consideration of the gravitational attractions of the Sun and planets. For comets with small perihelion distances this deviation is the rule rather than the exception.

It is a reasonable assumption that such anomalies may be due to the comet's reaction to ejected mass. This hypothesis was first advanced by Bessel as long ago as 1836. It was further developed by Dubyago (1948), Whipple (1950), and Makover (1955, 1956).

Sekanina (1967) gave a theoretical treatment of the 'impulse' effect hypothesis, which assumes a local explosive process in a cometary nucleus of arbitrary shape, resulting in the ejection of fairly substantial amounts of material. The ejection manifests itself as an impulse noticeably affecting the motion of the nucleus.

Ejections from real comets have both solid-particle and gaseous components, and the complete phenomenon is rather complicated. Therefore, the first stage in our investigation of these processes was to study the pressure developed by ice sublimation products under conditions of high vacuum and low temperature that are similar to those in outer space.

2. Experimental Techniques

The experiment was carried out in the laboratory using a specially designed glass vacuum chamber cooled by liquid air and a torsion balance set up inside the chamber. The balance carried cells containing ice samples mounted so that the torque produced by the sublimating ice could be registered using a system of mirrors.

The cells were fitted with a miniature heating element and a copper-constantan thermocouple. The heater and thermocouple leads were thin (0.05 and 0.1 mm) wires, so that the balance torsion would be scarcely affected. The balance was calibrated by means of an air jet of known pressure. Its constant was found to be 4.4 divisions per milligram, and the calibration was checked from experiment to experiment.

The test samples of ice were prepared outside the vacuum chamber. The cells were placed inside and the evacuation began. After a while the bottom section of the chamber was cooled with liquid air. A certain delay in cooling occurred because the evacuation was accompanied by removal from the cell walls of frost that had settled there during the preparation of the ice samples and which produced undesirable effects upon the balance readings when sublimating.

Cooling of the chamber brought the temperature of the sample to -135 or -140 °C. The position of the balance index at this temperature was taken as the reference point, because the thrust was then far below the sensitivity range of the balance.

After cooling the samples and generating a sufficiently high vacuum (10^{-5} to 10^{-6} T) the heating elements were switched on and a certain amount of heat input was supplied to the ice; this was maintained at a constant level during the experimental run.

The power input W could be calculated from the formula

$$W = UI, \tag{1}$$

where U is the voltage and I the current, but this was not precisely the amount of power consumed in heating and sublimating the ice. As a matter of fact, the heat energy W applied to the test sample cell is expended in several ways: on the sublimation product (W_{subl}), residual gas molecules (W_{gas}), via the thermocouple and heater leads (W_{therm}) and by radiation (W_{rad}):

$$W = W_{subl} + W_{gas} + W_{therm} + W_{rad}. \tag{2}$$

The last three terms introduce an element of uncertainty into the evaluation of the sublimation energy. Calculation of the appropriate heat flows is a fairly complicated problem, and the following method was used instead (Kajmakov and Sharkov, 1969). The cells were filled with fine-grain silica sand, packed by means of a glass filter. The balance and cells were then cooled down and power supplied to the heaters, just as when the cells contained ice. The steady-state temperature of the cell (without ice) was measured as a function of the power input:

$$T_{s-s}^{cell} = f_1(W'), \tag{3}$$

where $W' = W'_{gas} + W'_{therm} + W'_{rad}$. Radiation losses are insignificant, and hence the difference between W_{rad} and W'_{rad} is negligible. It was also shown experimentally that W'_{gas} differs but slightly from W_{gas}. In experimental runs using ice samples, values of the steady-state temperature were obtained for various power input levels, giving

$$T_{s-s}^{cell+ice} = f_2(W). \tag{4}$$

Using Equations (3) and (4), it is possible to plot a graph of 'equilibrium' ice temperatures

$$T_{s-s}^{ice} = f_3(W) \tag{5}$$

for the case where all the energy input is spent entirely on molecular sublimation, i.e., on breaking the bonds between sublimating molecules and on their kinetic energy. By this method we were able to allow for the effect of specific conditions upon the experimental results.

The power-input calibration of the cells proved to be different for the two types of cell used. This caused further complications: during ice-sublimation runs it was found necessary to apply to the cells different amounts of power to maintain W_{subl} constant, or in thrust calculations to determine the contribution of each cell to the overall impulse recorded by the torsion balance. The method could be utilized if we made use of the relation between the flow of the sublimating molecules and W_{subl} (Kajmakov and Sharkov, 1969).

After establishing the equilibrium temperature for a given power input we measured the exhaust thrust. The time required for stabilizing the temperature, and hence the exhaust thrust, was mainly dependent upon the absolute temperature of the ice. If the power input were low, an experimental run could last as long as 10 to 12 h.

We have also conducted experiments using frozen aqueous electrolyte solutions (Kajmakov and Sharkov, 1972). Special attention was given to preparing these samples. High purity chemicals were used, and in order to avoid variations in concentration in different parts of the cell during freezing (salting out) the ice samples were prepared by freezing one layer at a time.

3. Measurements and Discussion

Using the techniques described above, we measured the exhaust thrust $\bar{p}$ developed by sublimating H_2O molecules from unit surface area for various specific power inputs to the ice. The thrust varied from (0.10 ± 0.01) to (4.0 ± 0.4) mg cm^{-2} over a power input range of 0.016 to 0.29 W cm^{-2}, and this corresponds to a temperature range of -75 to $-50\,°C$.

It will be seen from Figure 1 that the experimental relationship between thrust and temperature is generally in good agreement with theory. The theoretical thrust may be calculated from

$$\bar{p} = JV_x, \tag{6}$$

where J is the ice consumption and V_x is the mean molecular velocity in a given direction. The ice consumption is given by the familiar equation of the kinetic theory of gases:

$$J = \alpha p \sqrt{(M/2\pi RT)}, \tag{7}$$

where p is the saturated vapour pressure at the temperature T, M is the total molecular weight of the substance, R the gas constant, and α the Langmuir factor (allowing for the return and partial condensation of the sublimating molecules). The velocity V_x is given by

$$V_x = \sqrt{(kT/2\pi m)} \tag{8}$$

(Epifanov, 1965), where k is the Boltzmann constant and m the molecular weight. Since $R=kN$ and $M=mN$ (N being Avogadro's number), it follows that

$$\bar{p} = \alpha p/2\pi. \tag{9}$$

The factor α is assumed to be unity. This is justifiable, since at a pressure of the order 10^{-6} T and at low temperatures the free path of the sublimating molecules was twice the size of the sample.

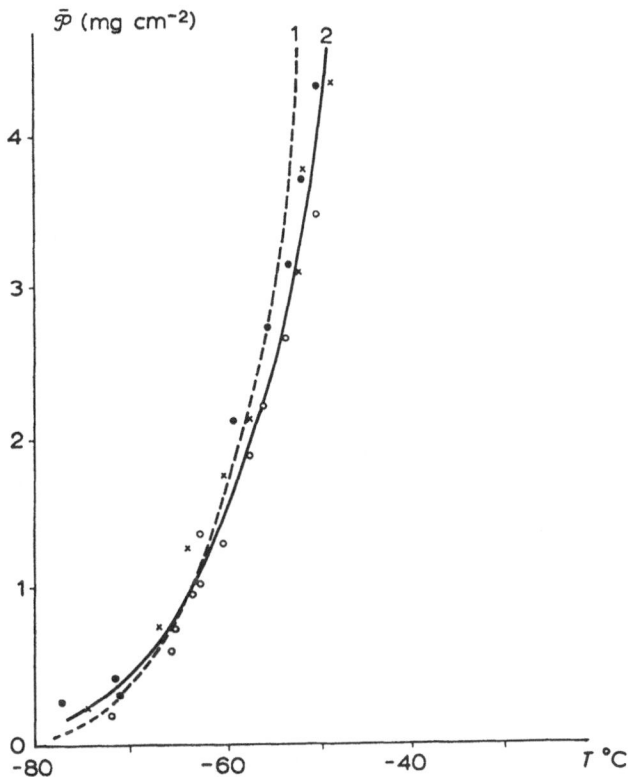

Fig. 1.   Exhaust thrust $\bar{p}$ (mg cm^{-2}) due to H_2O ice sublimation versus ice surface temperature T (°C). The broken curve (labelled 1) is theoretical, the solid curve (labelled 2) experimental.

In the upper part of the figure there is some deviation between the experimental and theoretical curves. This is because direct measurements of surface temperature involve considerable experimental difficulties, and the surface temperature was therefore estimated from temperature gradients measured for various power inputs.

In subsequent experiments the power and temperature ranges were expanded, and thrust relationships were obtained for temperatures -33 to -75 °C over a specific power input range of 0.016 to 0.46 W cm^{-2}. Above -50 °C (0.30 W cm^{-2}) the experimental curve deviates strongly from the theoretical one (see Figure 2) because of the 'greenhouse effect'. At high power inputs the quantity of molecules leaving the surface of the sublimating ice is so great that some of them collide and return to the surface. A blanket of H_2O vapour is formed over the surface; this hinders the sublimation products and causes a decrease in the thrust. As the temperature increases further there is a reduction in the rate of increase of thrust.

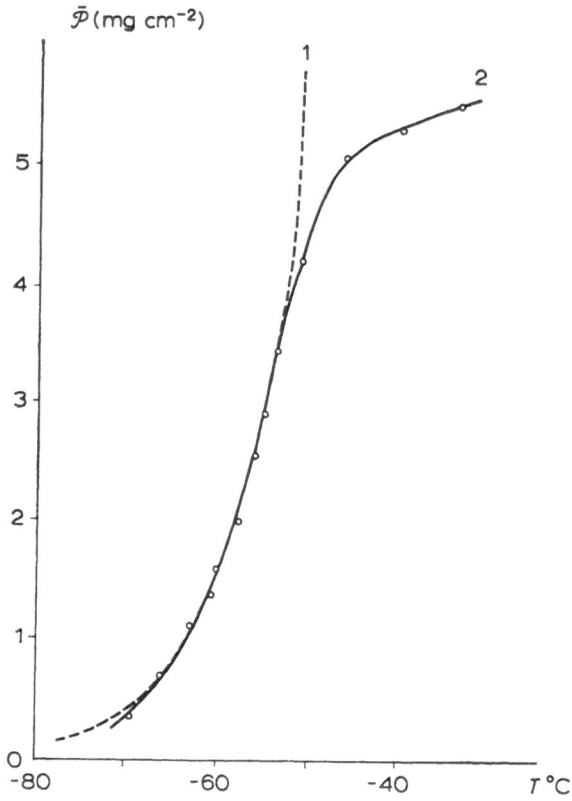

Fig. 2.   The greenhouse effect. See Figure 1.

There is a further difference between the curves at the lower end. This occurred at the termination of the experimental runs, the experimental curve being plotted as the temperature decreased. The difference may be explained as follows: when the ice surface drops down the cell, the cell walls begin to act as guides for the molecular outflow, and this leads to an increased thrust.

The theoretical curve was calculated on the assumption that there was an equally probable distribution of molecules at all outflow angles. Good correlation between experiment and theory bears out this assumption, showing that the sublimation products leave the surface in conformity with the cosine law. Earlier attempts at experimental confirmation of this, by other methods, had failed to give an unequivocal answer.

In ices prepared from 1.5N KCl solution, the temperature tended to rise as the water content decreased. This was due to the formation of a friable salt crust (matrix) over the ice surface. The temperature rise produces a certain 'pressure head' of sublimation products, and this in turn ensures a flow of sublimating molecules through the matrix pores, equivalent to the power input. Since the structure of the matrix varies with power input, the latter was maintained constant throughout the experiment.

As power is applied to the ice, there is a steep rise in thrust up to a certain level. Sublimation of H_2O molecules is then being effected through a very thin matrix layer that has formed in the process of evacuating and cooling the system, and the exhaust thrust tends therefore to reach the value for pure H_2O ice at the same power input. The shorter the time for evacuation and cooling, the closer the two values are. Subsequently, as the matrix increases in thickness, the thrust magnitude falls off slowly. It falls off more and more slowly until it practically reaches a plateau. The variation in thrust with time is illustrated in Figure 3 for two values of the power input.

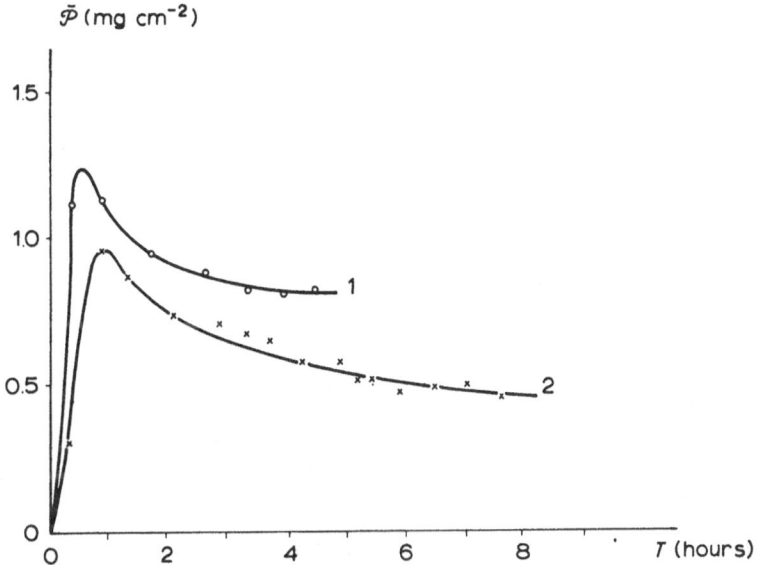

Fig. 3.   Exhaust thrust $\bar{p}$ (mg cm^{-2}) versus time t (hours) for frozen aqueous KCl solution at specific power inputs (1) 0.064 W cm^{-2}; (2) 0.058 W cm^{-2}.

Since the measurements conducted on the flow of sublimating material from KCl ice have shown them to be equivalent to flow from pure H_2O ice at identical power inputs, the fall-off in exhaust thrust with time and the difference generally from the thrust developed by H_2O ice can only be attributed to changes in the outflow velocity of the molecules.

If we assume that the plateau describes the thrust at a given power input, the estimated decrease in velocity due to the passage of the sublimating molecules through a dust-matrix layer 8 to 10 mm thick would be by a factor of 2 to 3.

Experiments were also conducted on 1.5N LiCl solution. Li^+ ions actively attract water to form a hydrate deposit (Konstantinov and Troshin, 1966), and it is interesting to compare the behaviour of matrix-forming substances having different types of ions.

Figure 4 shows that the thrust developed by sublimating molecules in frozen LiCl solution is lower than in the case of pure H_2O ice, although it is higher than for frozen KCl solution. In addition, it was found in the course of the experiments with LiCl

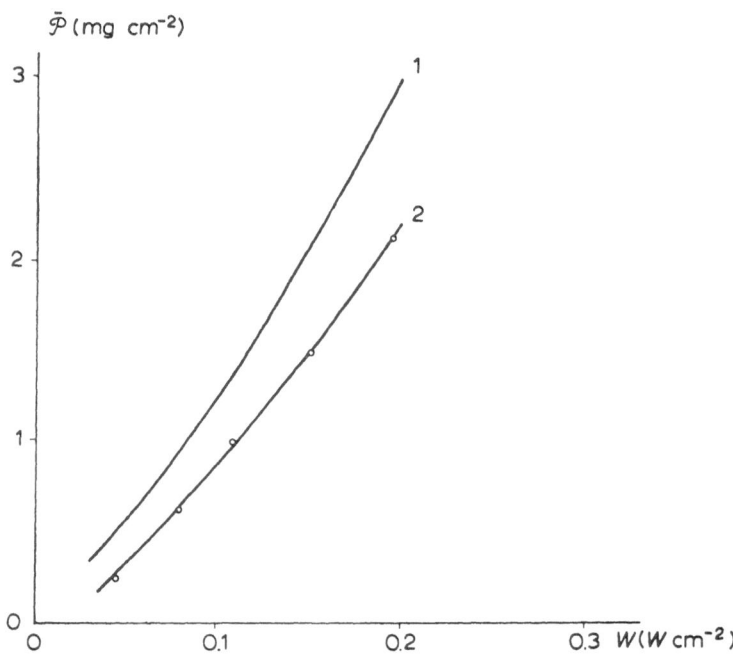

Fig. 4.   Exhaust thrust $\bar{p}$ (mg cm^{-2}) versus power input W (W cm^{-2}) for (1) H_2O ice; (2) frozen aqueous LiCl solution.

solution that the increase in thrust did not occur immediately upon application of power but after a lapse of about 2 h. It may be assumed that a surface layer of low permeability is formed on the ice, unlike the salt matrix of frozen solutions of the KCl type. This crust is so dense that the thrust developed is below the sensitivity limit of the balance. The crust is then thrown off by internal pressure and a rise in thrust follows.

It is possible that at lower power inputs the conditions for formation and growth of a matrix crust would be more favourable and the behaviour of LiCl ice similar to that of KCl ice.

4. Conclusions

The following conclusions can be drawn from the experiments and processing of the experimental data:

(1) Within the temperature range -75 to $-50\,°C$, the exhaust thrust varies from (0.10 ± 0.01) to (4.0 ± 0.4) mg cm^{-2}.

(2) These values for the exhaust thrust clearly indicate that ice sublimation products are discharged at all angles (according to the cosine law).

(3) At temperatures above $-50\,°C$, the exhaust thrust was observed to deviate sharply from theory owing to the greenhouse effect associated with the formation of a blanket of gas near the sublimating ice surface.

(4) A porous matrix on the sublimating ice surface leads to a reduction in the exhaust thrust as a result of a decrease in the velocities of the sublimating molecules.

References

Danielsson, L. R. and Kasai, G. H.: 1968, *J. Geophys. Res.* **73**, 259.
Dobrovol'skij, O. V.: 1961, *Trudy Stalinabad-Dushanbe Astron. Obs.* **8**.
Dubyago, A. D.: 1948, *Astron. Zh.* **25**, 361.
Epifanov, G. I.: 1965, *Fizika Tverdogo Tela*, Moscow.
Kajmakov, E. A. and Sharkov, V. I.: 1969, *Trudy IV Zimnej Shkoly po Kosmofisike*, Apatity, part 2, 15.
Kajmakov, E. A. and Sharkov, V. I.: 1972, this Symposium, p. 308.
Konstantinov, B. P. and Troshin, V. P.: 1966, *Izv. AN SSSR, Ser. Khim.* No. 11, 1907.
Makover, S. G.: 1955, *Trudy Inst. Teor. Astron.* **4**, 133.
Makover, S. G.: 1956, *Trudy Inst. Teor. Astron.* **6**, 67.
Potter, A. F. and del Duca, B.: 1964, *Icarus*, **3**, 103.
Sekanina, Z.: 1967, *Bull. Astron. Inst. Czech.* **18**, 15, 19, 286, 347.
Whipple, F. L.: 1950, *Astrophys. J.* **111**, 375.

PART V

ORIGIN AND EVOLUTION OF COMETS

A. ORBITAL STABILITY AND EVOLUTION

56. EJECTION OF BODIES FROM THE SOLAR SYSTEM IN THE COURSE OF THE ACCUMULATION OF THE GIANT PLANETS AND THE FORMATION OF THE COMETARY CLOUD

V. S. SAFRONOV

O. Schmidt Institute of Physics of the Earth, Moscow, U.S.S.R.

Abstract. The theory of planetary accumulation leads quite definitely to the conclusion that the formation of Oort's cometary cloud is the result of ejection of bodies to the outermost parts of the solar system due to encounters with the giant planets during their growth. Uranus and Neptune could have grown to their present dimensions only if the initial mass of solid material in their zones were substantially larger than that of these planets. The relative velocities of the bodies were increased through perturbations by the planetary embryos, and on reaching the escape velocity they would start to be ejected. Our concept of this process differs from that suggested by Öpik by the assumption that Jupiter and Saturn accreted hydrogen, not in solid but in gaseous state, and by the introduction of a more effective mechanism for the interaction with several embryos. In their final stages the embryos ejected amounts of mass an order of magnitude higher than the amounts accreted. Most of the mass was ejected into interstellar space by Jupiter, but the cometary cloud was created mainly by Neptune. The mass of the cloud is estimated to be about three times that of the Earth.

Study of the process of accumulation of the giant planets leads to the conclusion that in the final stage of their growth they should by their gravitational attractions eject a considerable number of solid bodies from the solar system. As the result of perturbations by the stars nearest to the Sun a small fraction of these bodies would remain at the periphery of the solar system, forming there the cometary cloud described in the classical work of Oort (1950, 1951).

For a correct evaluation of the amount of matter ejected it is necessary to discuss also other important factors that accompanied the process of accumulation of the planets: dissipation of gas from the solar system, and accretion of gas by Jupiter and Saturn. So far, there has been no single theory that takes account of all these factors simultaneously. It is possible, however, to divide the whole process into individual stages and to review at each of these only one or two of the most important factors. The results of such a review are briefly described below.

1. The Initial Stage in the Accumulation of the Giant Planets

Initially the planetary embryos grew on account of the solid bodies and particles that collided with them. The embryo masses were smaller than both m_a, at which value the accretion of gas begins, and m_e, when the ejection of bodies from the solar system begins. The process of embryo growth is quantitatively described by Safronov (1969). On the basis of this, the initial mass of the preplanetary cloud can be taken as equal to 0.05 to 0.06 $M_\odot$. Assuming that in the region of the giant planets about 1.5% of the mass of the cloud has condensed into solid particles, we find that the initial quantity of solid matter there was some 300 to 350 $M_\oplus$, i.e., about 6 or 7 times the quantity

Chebotarev et al. (eds.), The Motion, Evolution of Orbits, and Origin of Comets, 329–334. All Rights Reserved.
Copyright © 1972 by the IAU.

of solid matter in the giant planets. If there were less cloud material, Uranus and Neptune, which consist only of solid matter, could not have grown to their present size.

Gas took no part in the accumulation process but decreased somewhat the relative velocities of the small bodies and thus accelerated the growth of embryos. The effect was especially great when the bodies were small. At the same time gas was dissipated from the solar system. Apparently, the principal cause of the dissipation was a flow of fast particles emitted by the forming Sun during the time of its high electromagnetic activity (Schatzman, 1967). Towards the end of the initial stage the mass of gas in the zones of Jupiter and Saturn decreased by one order of magnitude. By this time the embryos reached mass m_a, at which point they began to absorb the remaining gas. The critical value m_a is expected to be between the masses of Mars and the Earth. The embryo of Jupiter grew to mass m_a in less than 10^8 yr. The Uranus and Neptune embryos grew much more slowly, and when they reached the mass m_a there was no gas left in their zones. These planets therefore had no gas accretion stage. Their initial stage terminated when their embryos grew to mass m_e, and they began to eject bodies from the solar system. For Neptune $m_e \approx 0.1\ M_\oplus$, and the time of growth in the first stage was over 10^9 yr.

2. Accretion of Gas by Jupiter and Saturn

According to the theory of stationary accretion of gas by gravitating bodies (Bondi and Hoyle, 1944; Bondi, 1952), the rate of increase in the mass m of a body as a consequence of absorption of gas that initially occupied an infinite space uniformly can be written

$$\frac{dm}{dt} = \frac{2\pi\alpha G^2 m^2 \rho_\infty}{(v^2 + c^2)^{3/2}} = Am^2\rho_\infty, \tag{1}$$

where ρ_∞ is the density of the gas far enough from the body that it is not perturbed by its gravitational field, v is the velocity of the gas with respect to the body, c is the thermal velocity of the molecules, and $1 < \alpha < 2$. Attempts to explain the formation of comets in the solar system by this mechanism (capture of interstellar matter by the Sun's gravitational field) are known to have failed, as have similar attempts to explain the formation of hot giant stars from main sequence stars. There are no reasons, however, for refusing to apply the accretion mechanism to the massive embryos of the giant planets.

Let m_g be the mass and U the volume of unexhausted gas in the whole zone of a planet. If the average density of this volume is taken in Equation (1), i.e. $\rho_\infty = m_g/U$, it can easily be found that the Jupiter embryo, with an initial mass equal to $M_\oplus$, would exhaust all the gas in its zone in 10^4 yr. In this case the characteristic accretion time for the remaining gas is about 4 orders less than the characteristic dissipation time.

However, the accretion of gas by the planetary embryos would be considerably hindered by the rotation of the preplanetary cloud. As an embryo exhausted the gas near its orbit the space would have been filled by gas coming from other heliocentric distances. Because of the low viscosity the radial displacement of the gas occurred with constant angular momentum (with respect to the Sun). This was accompanied by the development of an additional force δf tending to prevent the displacement δR in the heliocentric distance R, and (for unit gas mass) equal to

$$\delta f = -\frac{GM_\odot}{R^3}\,\delta R. \tag{2}$$

This force caused a considerable decrease in the density ρ of the gas in the direction of the embryo:

$$\frac{d\ln\rho}{dR} = -\frac{\mu}{\Re T}\,\delta f = -\frac{1}{h} \tag{3}$$

and

$$\frac{h}{R} \approx 10^{-3}\,\frac{R}{\delta R}. \tag{4}$$

Hence, the planet embryo could absorb gas only from a part of its zone. At the boundary the rates of accretion and dissipation are equal and

$$\delta R \approx 0.15R \approx 0.3\,\Delta R, \tag{5}$$

where ΔR is the half-width of the planet zone. Consequently, the planet embryo could absorb only about one-third of the gas remaining in the planet zone by the beginning of accretion.

The rate of gas dissipation can be found from the initial mass m_{g0} of gas in the planet zone. With m_{g0} equal to ten times the mass of Jupiter the dissipation time τ_d (i.e., the time corresponding to a decrease in density by a factor e) of gas in the Jupiter zone equals several tens of millions of years and turns out to be only a little less than the time of growth of the embryo to mass m_a.

3. Ejection of Bodies from the Solar System

The relative velocities of the bodies are determined by their gravitational perturbations during their close encounters, and they increased with the growth of the largest bodies. A comparison of the energy of random motions acquired in approaches and lost in collisions allows us to express (Safronov, 1969) the mean velocity v of the bodies in terms of the Keplerian circular velocity depending on mass m and radius r of the planet embryo (the largest body in the zone), namely,

$$v^2 = Gm/\theta r, \tag{6}$$

where θ is a dimensionless parameter of the order of several units. At some embryo mass m_0 the greatest velocities (somewhat greater than v) have reached the value $v_0 = 0.414\,v_c$, where $v_c = \sqrt{(GM_\odot/R)}$ is the Keplerian circular velocity of the embryo, and bodies whose vector $\mathbf{v}_0$ is directed along $\mathbf{v}_c$ escape from the solar system. Actually, the mean velocity v_E of ejected bodies is slightly greater than v_0, and the ejection occurs when $\mathbf{v}_E$ gets inside the cone of vertex angle 2φ and axis coinciding with the direction of $\mathbf{v}_c$ about the Sun:

$$v_E = v_c[\sqrt{(1 + \cos^2\varphi)} - \cos\varphi] = v_c u(\varphi). \tag{7}$$

The more effective the mechanism for the increase of the energy of the random motions of the bodies, the larger φ becomes and the greater the ratio λ of the mass of bodies ejected by the embryo to the mass of bodies that have coalesced with it.

We have found that in a revolving system of gravitating bodies the energy $v^2/2$ of the relative motion per unit mass increases as the result of encounters in unit time by the amount

$$\mathfrak{E}_1 = \beta'' v^2 / \tau_g, \tag{8}$$

where τ_g is the relaxation time during which the relative velocity vector $\mathbf{v}$ is deflected on the average by an angle $\pi/2$; the coefficient β'' is about 0.13. If the system includes a planet embryo, essentially more massive than other bodies and thus making the principal contribution to τ_g, Equation (8) would contain, instead of τ_g, the effective time of relaxation $\tau_{ge} \approx \tau_g'/2$, where τ_g' is the relaxation time dependent on all bodies but the embryo.

Let us introduce the following symbols: m for the mass of a planet embryo, m_b the mass of all other bodies in the planet zone, m_E the mass of bodies ejected from the zone beyond the boundaries of the solar system, $\mathfrak{E}_2$ the energy of random motions per unit mass per second lost in collisions, τ_s^* the average lifetime of a body prior to its collision with the embryo, τ_g^* the time for the deflection of $\mathbf{v}$ by the angle $\pi/2$ in encounters with the embryo only.

The equation for the conservation of energy of random motions of the bodies can be written

$$m_b \, dv^2 + (v_E^2 - v^2) \, dm_E + (v_m^2 - v^2) \, dm = 2m_b(\mathfrak{E}_1 - \mathfrak{E}_2) \, dt, \tag{9}$$

where v_m^2 is the mean square velocity of bodies colliding with the embryo m. Substituting for v and $\mathfrak{E}_1$ from Equations (6) and (8), then taking $dm/dt = m_b/\tau_s^*$ and retaining only the most important terms, we obtain

$$\lambda = \frac{dm_E}{dm} \approx \frac{2\beta'' v^2}{v_E^2 - v^2} \frac{\tau_s^*}{\tau_{ge}} \approx \frac{4\beta'' f}{\chi} \frac{v_e^2}{v_E^2 - v^2}, \tag{10}$$

where $v_e^2 = 2Gm/r = 2\theta v^2$, $\chi = \tau_g'/\tau_s^*$, and

$$f \approx 2 \ln \frac{D_m v^2}{Gm} - 1 \approx 4.$$

D_m, of the order of the radius of the sphere of action of the embryo, is the maximum approach distance that contributes to τ_g^*. With $v^2 \approx v_E^2/3$, $v_E \approx v_c/2$, $\beta'' \approx 0.13$, we have

$$\lambda \approx \frac{10}{\chi} \frac{v_e^2}{v_c^2}. \tag{11}$$

For Jupiter $\lambda \approx 200/\chi$. Saturn was the second principal perturbing body for the majority of the bodies of the Jovian zone. Since $\tau_g^* \propto 1/fm^2$, and with $D_m \propto r$, $f \approx 2 \ln cm^{-2/3}$,

$$\chi = \frac{\tau_g'}{\tau_g^*} \approx \frac{(m/m')^2}{1 + (4/3f) \ln (m/m')} \frac{R'^2}{R^2} \approx 20, \tag{12}$$

where primed quantities refer to Saturn. Consequently, for Jupiter $\lambda \approx 10$. The mass ejected by Jupiter was about a quarter of its own mass. Bodies with smaller velocities did not escape into Saturn's zone and for them χ was greater. Later, bodies from the Uranus and Neptune zones escaped into Jupiter's zone, and Jupiter effectively ejected them out of the solar system.

The value of $\lambda = \lambda'$ for Saturn for bodies moving rapidly through the zones of both Jupiter and Saturn can easily be found from Equation (10) on the assumption that $\tau'_{ge} = \tau_{ge}$ and $v = \text{const}$. Then $\lambda' \approx 0.6\lambda$. For slower bodies λ' is slightly less. In distinction to Jupiter and Saturn, which ejected bodies at practically their present mass, Uranus and Neptune ejected them during the whole course of their growth. Since $\lambda \propto v_e^2 \propto m^{2/3}$, the mean value of λ during this whole time is $\bar{\lambda} \approx \frac{3}{5}\lambda$, where λ refers to their present mass.

This evaluation agrees well with the above value for the whole mass of solid matter ejected from the preplanetary cloud, i.e., $\sim 300\ M_\oplus$, giving $\lambda \approx 6$. In Table I (column 4) are given the approximate values of the masses ejected by the giant planets on the assumption that $\lambda \propto (v_e/v_c)^2$ and allowing for the additional ejection of bodies arising from the later forming, more distant planets.

TABLE I

Masses ejected by the giant planets

Planet	$(v_e/v_c)^2$	m_p	m_E	$10^3(dm_c/dm_E)_{min}$	$(m_c)_{min}$	$10^3(dm_c/dm_E)$	m_c
Jupiter	21.4	9	100	1.2	0.12	2.3	0.2
Saturn	14.1	12	80	2.1	0.18	5.4	0.4
Uranus	10.4	14.6	50	4.2	0.2	13	0.6
Neptune	18.8	17.2	60	6.6	0.4	21	1.3

Total mass of the cometary cloud $0.9 M_\oplus < m < 2.5 M_\oplus$.

The mass of the solid material ejected to the outskirts of the solar system can also be estimated. The semimajor axis a of the orbit of a body moving with relative velocity v_E directed at angle φ to the circular velocity v_c at heliocentric distance R is determined by the expression

$$\frac{R}{a} = 1 - \frac{v_E^2}{v_c^2} - 2\frac{v_E}{v_c}\cos\varphi. \tag{13}$$

Hence

$$\frac{R}{a_1} - \frac{R}{a_2} = -2\frac{v_E}{v_c}(\cos\varphi_1 - \cos\varphi_2) \approx -2\frac{v_E}{v_c}\sin\varphi\,\Delta\varphi. \tag{14}$$

The lower limit for the mass of the cometary cloud can be found immediately by assuming the ratio of the mass dm_c ejected to the periphery of the solar system (in orbits with semimajor axes between a_1 and a_2) to the mass dm_E ejected into interstellar space to be equal to the ratio of corresponding solid angles between the cones φ_1 and φ_2 and inside the cone φ:

$$\left(\frac{dm_c}{dm_E}\right)_{min} = \frac{-\sin\varphi\,\Delta\varphi}{1 - \cos\varphi} = \frac{Rv_c}{2v_E(1 - \cos\varphi)}\left(\frac{1}{a_1} - \frac{1}{a_2}\right). \tag{15}$$

Such an assumption is valid for relatively close encounters when the vector **v** is deflected by more than 2φ. In this case **v** can appear at any point of the cone φ with equal probability.

For very elongated elliptical orbits a is about half the aphelion distance. Without committing much error it can be assumed that the bodies with orbital semimajor axes between $a_1 = 20\,000$ AU and $a_2 = 80\,000$ AU would undergo considerable stellar perturbations but still remain in circumsolar orbits. The values of $(dm_c/dm_E)_{min}$ and of the mass $(m_c)_{min}$ ejected to the outskirts of the solar system are calculated from Equation (15) for $\varphi = 35°$ and given in columns 5 and 6 of Table I. In this case the lower limit for the mass of the whole cometary cloud is found to be about one Earth mass; this is three times greater than the present value as calculated by Oort (1950, 1951) from the frequencies of the appearance of new comets.

It is known, however, that distant encounters are of great importance in the transport of energy. For small deflections of the vector **v** the ratio dm_c/dm_E would be significantly greater than that given by Equation (15). We shall not reproduce here the rather cumbersome expressions that take into account the contribution of distant encounters. An approximate estimate leads to the simple relation:

$$\frac{dm_c}{dm_E} \sim \frac{\Delta\varphi}{\varphi} \ln \frac{2\varphi}{\psi_1}, \tag{16}$$

where $\psi_1 \approx 2Gm/D_m v_E^2$ is the smallest deflection angle of **v**, corresponding to the largest value of D_m considered. The values dm_c/dm_E and m_c found from Equation (16), and assuming D_m equal to twice the radius of the sphere of action of each planet, are given in the last two columns of Table I. The result depends only slightly (logarithmically) on D_m. The mass of the cometary cloud is thus found to be about 3 Earth masses – in good agreement with the revised value of the present mass given by Öpik (1970). Jupiter has ejected into interstellar space more material than any other planet, although about half of the mass of the cometary cloud was supplied by Neptune. Because of the very slow growth of Neptune (and for that matter Uranus) most of the bodies ejected by it would by then have lost a significant portion of their volatiles.

References

Bondi, H. and Hoyle, F.: 1944, *Monthly Notices Roy. Astron. Soc.* **104**, 273.
Bondi, H.: 1952, *Monthly Notices Roy. Astron. Soc.* **112**, 195.
Oort, J. H.: 1950, *Bull. Astron. Inst. Neth.* **11**, 91.
Oort, J. H.: 1951, *Observatory* **71**, 129.
Öpik, E. J.: 1970, *Moon* **1**, 487.
Safronov, V. S.: 1969, *Ehvolyutsiya Doplanetnogo Oblaka i Obrazovanie Zemli i Planet*, Nauka, Moscow.
Schatzman, E.: 1967, *Ann. Astrophys.* **30**, 963.

57. ON THE STABILITY OF THE OORT CLOUD

E. M. NEZHINSKIJ

Institute for Theoretical Astronomy, Leningrad, U.S.S.R.

Abstract. An attempt has been made to estimate the rate of destruction of the Oort cloud by stars passing through it. Two different mechanisms of cloud dispersion have been investigated. Numerical estimates show that the cumulative dispersing mechanism plays the leading role. The lower limit for the half-life of the cloud is 1.1×10^9 yr.

1. Introduction

Recently there have appeared works (Sekanina, 1968; Vsekhsvyatskij, 1967; Zal'kalne, 1969) investigating the problem of destruction of the Oort cloud by stars passing through it. In these works it has been usual to discuss a model in which the total influence of the Galaxy is not taken into account; the cloud is considered to be a local formation whose properties and behaviour are fully determined by the Sun and passing stars.

New possibilities for research appear when the Oort cloud is regarded as the total collection of comets in the vicinity of the Sun within Hill's sphere, i.e., when the Jacobi integral is used for the system Sun-Galaxy centre-comet (Chebotarev, 1964, 1966; Antonov and Latyshev, 1972). With such an approach the stability of the cloud may be investigated without knowledge of the cometary orbits themselves. Trajectories of particles in the cloud are crucial for the investigation of the cloud's evolution, although (1) they are not known, and (2) in the general case, at least in the cloud's periphery, they differ significantly from conics (Chebotarev, 1964, 1966).

The need for statistical analysis of variations in cometary motion and of destruction of the cloud by the random passage of stars naturally evolves directly from the formulation of the problem. However, in stellar astronomy similar problems have been formulated and studied with sufficient profundity.

In this work an attempt is made to study the stability of the Oort cloud by the methods of stellar astronomy.

2. Model for the Oort Cloud

The following simplified model is discussed. The centre of the Galaxy and the Sun are taken to be two material points describing circular orbits about their common centre of mass (Ogorodnikov and Latyshev, 1969), at which point is located the origin of the rotating system of coordinates xyz. The xy plane is taken to be that of the motion of these two points, while the x-axis is directed along the line connecting them. The coordinates of the Sun and Galaxy centre are $x=d_1$, $y=z=0$ and $x=-d_2$, $y=z=0$, respectively (where $d_1 > 0$, $d_2 = 0$, since the mass of the Galaxy $m_2 \gg m_\odot$).

The Oort cloud is formed by particles of infinitely small mass located in the vicinity

Chebotarev et al. (eds.), The Motion, Evolution of Orbits, and Origin of Comets, 335–340. All Rights Reserved.
Copyright © 1972 by the IAU.

of the Sun within the Hill surface, and for which the Jacobi constant $C > C_0$ (where C_0 is the value of the Jacobi constant at the libration point L_1).

3. Mechanism for Sweeping Particles from the Oort Cloud

The passage of stars through the Oort cloud or near it causes variations in the Jacobi constants of the particles. Particles whose Jacobi constants C_1 become less than C_0 are unstable in Hill's sense (Moulton, 1914; Subbotin, 1968; Chebotarev, 1969) and therefore will not be regarded as belonging to the cloud, an assumption allowing us to evaluate the lower limit of the half-life of the Oort cloud.

If a particle belonged to the Oort cloud prior to the passage of a star and then left the cloud, the following inequality can be written:

$$C > C_0 > C_1; \tag{1}$$

whence

$$(C_0 - C) + (C - C_1) > 0, \tag{2}$$

where

$$C = 2\Omega - v^2,$$

$$C_1 = 2\Omega_1 - v_1^2, \tag{3}$$

$$\Omega = \tfrac{1}{2}n^2(x^2 + y^2) + k^2\left(\frac{m_\odot}{r_1} + \frac{m_2}{r_2}\right), \tag{4}$$

x, y, z being the coordinates of a particle prior to the star's passage; x_1, y_1, z_1 the coordinates of the particle after the star's passage; $r_1 = \sqrt{[(x-d_1)^2 + y^2 + z^2]}$, $r_2 = \sqrt{(x^2 + y^2 + z^2)}$, $\dot{v}^2 = \dot{x}^2 + \dot{y}^2 + \dot{z}^2$; k^2 is the gravitational constant, and n the angular velocity of the Sun. [In a more general case, if the Galaxy is considered to be more than a point mass (Antonov and Latyshev, 1971), the factor m_2/r_2 in Equation (4) is replaced by $\varphi(r_2, z)$.] Inserting Equations (3) and (4) into (2) we obtain

$$(C_0 - C) + 2(\Omega - \Omega_1) - (v^2 - v_1^2) > 0. \tag{5}$$

During the passage of a star through the cloud the coordinates of the particle change but little (since the velocity of the particle is several orders lower than that of the passing star). It can therefore be assumed that the perturbing star gives the particle a sudden impulse, changing only the latter's velocity:

$$\Omega - \Omega_1 = 0, \tag{6}$$

$$v_1 = v + \Delta v. \tag{7}$$

The inequality (5) can then be rewritten

$$(C_0 - C) + 2v\,\Delta v + (\Delta v)^2 > 0. \tag{8}$$

Obviously, the change in the Jacobi constant of the particle depends on the distance at which the star passes the particle. Let us therefore estimate the distance that would cause the latter's Jacobi constant C_1 to become less than C_0 and the particle to leave the cloud.

If $\Delta v \gg v$, the principal terms in inequality (8) are

$$(C_0 - C) + (\Delta v)^2 > 0. \tag{9}$$

Variation in the velocity of the particle (Δv) can be expressed (Ogorodnikov, 1958) in terms of the mass m and velocity V of the passing star, and the least distance p to it:

$$\Delta v = \frac{2k^2 m}{pV}. \tag{10}$$

Inserting this into the inequality (9) gives

$$p^2 < \left(\frac{2k^2 m}{V}\right)^2 (C - C_0)^{-1}. \tag{11}$$

We have thus obtained the radius of the tunnel all the particles of which are swept out by the passing star.

Let us now assume that a certain star consistently remains within the Oort cloud and moves with an average velocity V, sweeping particles out of a tunnel of average radius p. The density $v(t)$ is the same throughout the cloud but it changes with time. It should be noted that as soon as the star has passed the tunnel is again filled with particles.

Designating by $T = \text{const}$ and $M(t)$ the volume and mass, respectively, of the Oort cloud, we have

$$-\frac{dM}{dt} = \pi p^2 V v,$$

$$M = vT \tag{12}$$

$$-T\frac{dv}{dt} = \pi p^2 V v,$$

and integrating,

$$v = v_0 \exp\left(-\frac{\pi V p^2}{T} t\right).$$

If $qv = v_0 = v(0)$, then

$$t = \frac{T}{\pi V p^2} \ln q. \tag{13}$$

Equation (13) shows how long will elapse, under the above conditions, before the density of the cloud decreases by a factor of q.

If we designate by τ the average time spent by a star in crossing the Oort cloud, we may estimate, using formulae (11) and (13), the average number r of stars that would pass through the cloud before its density would decrease by a factor of q as the result of the sweeping out of particles. The result is

$$r = \frac{TV \ln q}{4\pi k^4 m^2 \tau \overline{(C - C_0)^{-1}}},$$ (14)

where $\overline{(C - C_0)^{-1}}$ is the average value of $(C - C_0)^{-1}$ for the particles of the Oort cloud.

4. Cumulative Mechanism for Dispersion of the Cloud

A passing star usually has a negligible effect on the velocity of most particles, and thus their Jacobi constants C usually remain larger than C_0. However, the cumulative effect of the passage of several stars can eject the particle from the Oort cloud.

It can be seen from the inequality (8) that in order for a considerable fraction of the particles to leave the cloud the mean square value of the increments in the particle velocities should be of the order

$$(\delta v)^2 = \bar{C} - C_0.$$ (15)

We now derive a formula for $(\delta v)^2$. (For more detail see Ogorodnikov, 1958.) For greater simplicity, let the distribution function $f(V)$ with respect to the relative velocities of stars have spherical symmetry, i.e., be independent of direction. Then in time t there will be on the average

$$2\pi pt V \, dp \, f(V) \, dV$$

passages of stars with velocities from V to $V+dV$ at distances from p to $p+dp$ from the particle. Assuming that the masses of all the passing stars are similar, we have

$$(\delta v)^2 = \int_{p_1}^{p_2} \int_0^\infty (\Delta v)^2 2\pi pt V \, dp \, f(V) \, dV$$

or, using Equation (10),

$$(\delta v)^2 = 8\pi k^4 m^2 t \int_0^\infty \frac{f(V) \, dV}{V} \int_{p_1}^{p_2} \frac{dp}{p},$$ (16)

where p_1 and p_2 are the minimum and maximum distances, respectively, to the passing star. Let us assume that $p_1 = p$ is the average radius of the tunnel given by inequality (11) and $p_2 = d$ is the average distance between stars. (If $p_2 > d$, multiple passages occur, i.e., several stars pass by a particle simultaneously at similar distances. It may therefore be supposed that such multiple passages do not produce any significant variations in the Jacobi constant.)

In so far as we are interested only in order of magnitude results, we may take

$$\int_0^\infty \frac{f(V)}{V} \, dV = \frac{\nu}{\bar{v}},$$

where $\hat{v}$ is the root mean square residual velocity of the stars. Then

$$(\delta v)^2 = 8\pi k^4 m^2 t v \frac{1}{\hat{v}} \ln \frac{p_2}{p_1}. \tag{17}$$

Using Equations (15) and (17) we find the cumulative half-life of the cloud to be

$$t = \frac{(\bar{C} - C_0)\hat{v}}{8\pi k^4 m^2 v \ln (p_2/p_1)}. \tag{18}$$

5. Numerical Estimates

We shall now estimate the half-life of the Oort cloud. For this purpose we take the following data (Ogorodnikov, 1958; Antonov and Latyshev, 1972): $d_2 = 10^4$ pc, $d = 2.15$ pc, $m = 10^{33}$ g, $m_\odot = 2 \times 10^{33}$ g, $m_2 = 2.6 \times 10^{44}$ g, $v = 0.1$ st pc^{-3}. The mean radius of the Oort cloud is 1 pc, $T = 3.35$ pc^3, $V = 20$ km s^{-1}, $\hat{v} = 20$ km s^{-1}, $\tau = 6.4 \times 10^4$ yr, $q = 2$.

Using Equation (14) we see that the density of the cloud will be decreased by half as the result of sweeping if 5.5×10^4 stars pass through the cloud. This will occur (Ogorodnikov, 1958) in 8.6×10^9 yr. On the other hand, from Equation (18) we find that the cumulative half-life of the Oort cloud is at least 1.1×10^9 yr. The cumulative effect is thus the more important mechanism, although more precise estimates will undoubtedly result in a much greater half-life.

Acknowledgments

I find it my pleasant duty to express my gratitude to G. A. Chebotarev for suggesting this problem and for the interest he has shown, and also to V. A. Antonov for his counsel and comments in the course of my work.

References

Antonov, V. A. and Latyshev, I. N.: 1972, this Symposium, p. 341.
Chebotarev, G. A.: 1964, Astron. Zh. 41, 983.
Chebotarev, G. A.: 1966, Astron. Zh. 43, 435.
Chebotarev, G. A.: 1969, Byull. Inst. Teor. Astron. 11, 625.
Moulton, F. R.: 1914, An Introduction to Celestial Mechanics, Macmillan, New York.
Ogorodnikov, K. F.: 1958, Dinamika Zvezdnykh Sistem, Moscow.
Ogorodnikov, K. F. and Latyshev, I. N.: 1969, Astron. Zh. 46, 1190.
Sekanina, Z.: 1968, Bull. Astron. Inst. Czech. 19, 223, 291.
Subbotin, M. F.: 1968, Vvedenie b Teoreticheskuyu Astronomiyu, Moscow.
Vsekhsvyatskij, S. K.: 1967, Priroda i Proiskhozhdenie Komet i Meteornogo Veshchestva, Prosveshchenie, Moscow.
Zal'kalne, I. E.: 1969, Astrometr. Astrofiz. No. 4.

Discussion

G. N. Duboshin: Your assumption that the passing stars do not influence the Sun would seem to be a very rough approximation.

E. M. Nezhinskij: If a large number of stars pass the Sun at considerable distances in various directions, they can be considered to have no net influence on the Sun's motion.

B. Yu. Levin: If each star produces a narrow tunnel, whereas total destruction stems from the passage of a large number of stars, it seems to me that Duboshin's objection is not important, because a star has a decisive effect only on the region very close to it.

58. DETERMINATION OF THE FORM OF THE OORT COMETARY CLOUD AS THE HILL SURFACE IN THE GALACTIC FIELD

V. A. ANTONOV and I. N. LATYSHEV

Astronomical Observatory, Leningrad University, Leningrad, U.S.S.R.

Abstract. An estimate is made of the maximum range over which the motion of a particle about the Sun and in the galactic field is stable in Hill's sense. An equation fitting the Hill surface is found, with allowance made for the fact that the galactic potential differs from that of a point mass. The volume enclosed by the surface is 3.35 cubic parsecs, the greatest distance from the Sun being 1.42 parsecs, this being along the line to the centre of the Galaxy.

It is well known that the motions of the planets of the solar system are not affected to any noticeable extent by stellar perturbations. On the other hand, it is not satisfactory to regard the solar system as isolated when one is studying the motion of a comet whose orbit has a semimajor axis of 2000 AU. What would happen if the orbit of the comet (or other particle) were extended even further? It is inevitable that the solar and galactic gravitational fields of force would become comparable in their effect on the particle. The orbit would lose all semblance to a Keplerian ellipse, and many of the conventional conceptions of celestial mechanics would be invalid. In such cases the classical methods of celestial mechanics must be complemented with other methods which have been used with various degrees of success in the solution of problems of stellar dynamics.

In this paper we consider a problem in which the particle would make many revolutions about the Sun at the greatest possible distance. The size of an appropriate surface was estimated by Chebotarev (1963, 1964, 1965, 1966), who proceeded from the assumption that all the mass of the Galaxy is concentrated at its centre. In actuality, the attracting material is distributed over an extended volume, and the potential of the real galactic field clearly differs from that of a point mass. On the average, a star can be expected to pass close enough to the Sun to have a significant influence once every 10^5 yr or so. Accordingly, a particle with a revolution period longer than this would respond to the smoothed gravitational field of several passing stars, rather than to their individual attractions. For such a particle the attracting matter is, as it were, spread over space in a continuous manner.

The galactic orbit of the Sun is taken to be almost exactly circular. Since the mass of the Sun is small compared to that of the Galaxy it is natural to expect that the dimensions of Hill's surface would be small compared to the distance R_0 from the Sun to the galactic centre.

We shall proceed now to the analytical derivation of Hill's surface. A rotating heliocentric system of coordinates is defined such that the x-axis is directed away from the galactic centre, the y-axis in the direction of galactic rotation and the z-axis towards the north galactic pole. The Jacobi integral in this system of coordinates is

Chebotarev et al. (eds.), The Motion, Evolution of Orbits, and Origin of Comets, 341–345. All Rights Reserved.
Copyright © 1972 by the IAU.

$$W + \frac{\dot{x}^2 + \dot{y}^2 + \dot{z}^2}{2} - \frac{\omega^2}{2}[(R_0 + x)^2 + y^2]$$

$$- u(x, y, z) - \frac{fM_\odot}{\sqrt{x^2 + y^2 + z^2}} = 0, \quad (1)$$

where x, y, z are the coordinates of the particle, ω is the angular velocity of the co-ordinate system (i.e., the angular velocity of the Sun in the Galaxy), f is the constant of gravitation, $M_\odot$ the solar mass, u the galactic potential and W the constant of integration.

If we take $\dot{x} = \dot{y} = \dot{z} = 0$, we obtain a family of zero velocity surfaces with parameter W:

$$W = \frac{\omega^2}{2}[(R_0 + x)^2 + y^2] + u(x, y, z) + \frac{fM_\odot}{\sqrt{x^2 + y^2 + z^2}}. \quad (2)$$

It is clear that a particle whose motion is governed by a Jacobi integral with a given value of W is not able to pass through the surface determined by Equation (2) and must therefore remain consistently on one side of the surface.

We expand the galactic potential u with respect to x, y, z, considering terms up to the second order only. This is simplified if we suppose that the Galaxy has a plane of symmetry ($z = 0$) and an axis of rotational symmetry. The distance R from the axis of symmetry is given by

$$R = \sqrt{(R_0 + x)^2 + y^2} \simeq R_0 + x + \frac{y^2}{2R_0}. \quad (3)$$

Taking into account the solar motion,

$$(\partial u/\partial R)_0 = -\omega^2 R_0, \quad (4)$$

it follows that

$$W' = \frac{1}{2}\left[\left(\frac{\partial^2 u}{\partial R^2}\right)_0 - \frac{1}{R_0}\left(\frac{\partial u}{\partial R}\right)_0\right]x^2 + \frac{1}{2}\left(\frac{\partial^2 u}{\partial z^2}\right)_0 z^2$$

$$+ \frac{fM_\odot}{\sqrt{x^2 + y^2 + z^2}}, \quad (5)$$

where W' is a constant and the subscript 0 indicates that the derivatives are to be taken at the origin. For convenience we write Equation (5) in the form

$$W' = \frac{1}{2}(\alpha x^2 + \gamma z^2) + \frac{fM_\odot}{\sqrt{x^2 + y^2 + z^2}}, \quad (6)$$

where α and γ may easily be expressed in terms of the local kinematic parameters of the Galaxy. Specifically, we make use of the well-known relationship between Oort's parameters A and B and the linear velocity of galactic rotation $\Theta = \omega R$:

$$A = \frac{1}{2}\left(\frac{\Theta}{R} - \frac{d\Theta}{dR}\right), \qquad B = -\frac{1}{2}\left(\frac{\Theta}{R} + \frac{d\Theta}{dR}\right). \quad (7)$$

Numerical values, according to Schmidt (1965), are

$$A = 15 \text{ km s}^{-1} \text{ kpc}^{-1}, \qquad B = -10 \text{ km s}^{-1} \text{ kpc}^{-1}.$$

Solar motion may be taken as a specific phenomenon of galactic rotation for $R = R_0$, the peculiar velocity of the Sun being small compared with Θ. Differentiating Equation (4) and comparing it with Equation (7), we obtain

$$\alpha = 4A(A - B); \tag{8}$$

and according to Agekyan *et al.* (1962), γ may be written

$$\gamma = -C^2, \tag{9}$$

where $C = 68 \text{ km s}^{-1} \text{ kpc}^{-1}$.

It is significant that the above procedure did not require any model for the Galaxy as a whole. Indeed, we have reduced everything to consideration of the observational parameters A, B, and C and have not had to make any assumptions about the structure of the Galaxy at great distances. For large values of W' Equation (6) gives almost spherical surfaces encircling the Sun. But our interest is in the critical Hill surface, i.e., a surface which ceases to be bounded as W' decreases further. A simple analysis shows that the critical value of W' is

$$W' = \tfrac{3}{2} f^{2/3} M_\odot^{2/3} \alpha^{1/3}. \tag{10}$$

For this value of W' two specific conical points, given by

$$y = z = 0, \qquad x = \pm \sqrt[3]{\frac{fM_\odot}{\alpha}}, \tag{11}$$

will appear on the surface. Apart from these points the surface is smooth and resembles a triaxial ellipsoid whose major axis coincides with the x-axis. Numerical values for the semiaxes are

$$x_{\max} = 1.42 \text{ pc} = 293 \times 10^3 \text{ AU}$$

$$y_{\max} = 0.92 \text{ pc} = 196 \times 10^3 \text{ AU}$$

$$z_{\max} = 0.74 \text{ pc} = 152 \times 10^3 \text{ AU}.$$

The triaxial form would be preserved in the case of a spherically symmetric gravitational field as well. Figure 1 shows a section by the plane $x0z$ of the surfaces for different values of W'.

The assumption that the closed Hill surface is small compared to galactic dimensions seems to be justified. It is interesting that this is the case for our Galaxy, for the situation would be very different in the inner parts of an elliptical galaxy. Then $\omega \approx$ const, the parameter A is small and the Hill surface is much larger, which indicates that a particle can return to its star after covering a distance several times the mean interstellar distance.

But in our Galaxy the volume of the critical Hill surface about the Sun is 28% of what it would be if all the stars were uniformly distributed. In absolute terms the volume

is 3.35 pc³. This estimate was obtained by means of an approximate numerical integration; despite a thorough investigation we have not been able to determine this
volume analytically.

The two critical points are merely libration points in the particle-Sun-Galaxy
system without the assumption of Newtonian gravitation. This modified three-body
problem, like the basic problem, contains five libration points (Duboshin, 1969), but
the other three points are far from the Sun and thus devoid of any physical meaning.

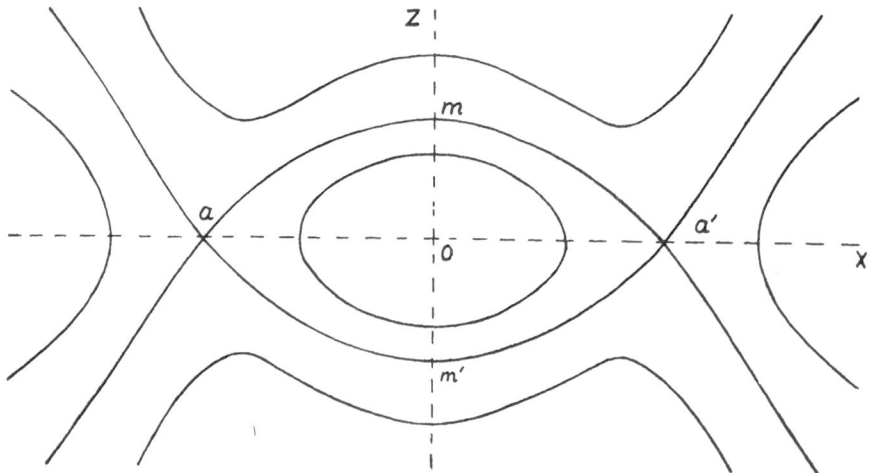

Fig. 1.   A section by the plane $x0z$ of the surfaces for different values of W'. The critical points
are a and a' and the critical surface $ama'm'$.

In conclusion, briefly we consider some further questions:

(1) Will a particle for which the surface given by Equation (6) is not closed be certain to escape from the Sun? The problem involves the existence of other integrals of
motion in addition to that of Jacobi. There is reason to believe that the meanderings
of the orbit are not limited; at least, something of this kind is observed in related
numerical experiments on galactic orbits of stars (Aarseth, 1966).

(2) What is the physical interpretation of the problem? Our approach is confined to
considering the geometry of possible orbits and does not involve any numerical
predictions concerning the density of the circumsolar cloud. Nor do we wish to state
the cosmogonical importance of particles moving in extremely extended orbits. It is
quite possible that our hypothetical particles bear no relation at all to comets that
could be seen from the Earth. On the other hand, statements concerning a circumsolar cloud at the distance of 1 to 2×10^5 AU can, in principle, be checked by analogy,
since similar particles from other stars should, even if very rarely, enter the solar
system and be detected. In the solar system, they will be easily distinguished by their
highly hyperbolic orbits. Further, we may expect to find, not only comet-like, but also
star-like satellites at these great distances from a particular star. Such extended

binary systems have recently begun to interest astronomers; and there is no *a priori* reason why the Sun should not have a dwarf, stellar satellite.

References

Aarseth, S. J.: 1966, *Nature* **212**, 57.

Agekyan, T. A., Vorontsov-Vel'yaminov, B. A., Gorbazkij, V. G., Deutsch, A. N., Krat, V. A., Melnikov, O. A., and Sobolev, V. V.: 1962, *Kurs Astrofiziki i Zvezdnoj Astronomii*, Fitmatgiz, Moscow, Vol. II, section 158.

Chebotarev, G. A.: 1963, *Astron. Zh.* **40**, 812.

Chebotarev, G. A.: 1964, *Astron. Zh.* **41**, 983.

Chebotarev, G. A.: 1965, *Analiticheskie i Chislennye Metody Nebesnoj Mekhaniki*, Nauka, Moscow and Leningrad (English translation: 1967, *Analytical and Numerical Methods of Celestial Mechanics*, Elsevier, New York), Chapter 6.

Chebotarev, G. A.: 1966, *Astron. Zh.* **43**, 435.

Duboshin, G. N.: 1969, *Astron. Zh.* **46**, 1279.

Schmidt, M.: 1965, in A. Blaauw and M. Schmidt (eds.), *Galactic Structure*, University of Chicago Press, Chicago and London, p. 513.

59. ON 'NEW' COMETS AND THE SIZE OF THE COMETARY CLOUD

G. T. YANOVITSKAYA

Main Astronomical Observatory, Ukrainian Academy of Sciences, Kiev, U.S.S.R.

Abstract. By means of the Oort-Schmidt definition of 'new' comets, Sekanina's catalogue of original and future orbits, and the 2σ and 3σ rules, we have compiled a list of 33 'new' comets. We draw some conclusions concerning the size of the cometary cloud.

Chebotarev et al. (eds.), The Motion, Evolution of Orbits, and Origin of Comets, 346. All Rights Reserved.
Copyright © 1972 by the IAU.

60. DIFFUSION OF COMETS FROM PARABOLIC INTO
NEARLY PARABOLIC ORBITS

K. A. SHTEJNS

Astronomical Observatory, Latvian State University, Riga, U.S.S.R.

Abstract. It is assumed that the accumulation of small, independent, random perturbations in the reciprocal semimajor axis of the orbit of a comet follows a normal distribution law whose standard deviation is a function of the inclination and perihelion distance and that for nongravitational forces it is a function of perihelion distances only; it is also assumed that secular accelerations do not change into decelerations, and vice versa. The standard deviations given by diffusion theory are in good agreement with the mean values of nongravitational impulses obtained from calculations from short overlapping arcs. The mean lifetime of a comet is found to be one hundred revolutions. To explain why many more near-parabolic comets are actually discovered than are theoretically expected the existence of comets of very short lifetimes must be accepted.

Considerable advances have been made in the study of separate stages of the evolution of cometary orbits and of the structure of cometary nuclei and atmospheres, but almost nothing has been elucidated concerning the actual origin of comets. This is because for some of the stages of cometary evolution we have only hypothetical conclusions and almost no observational data at all.

Let us consider the problem of the diffusion of comets (Oort, 1950). The existence of the cometary cloud seems indisputable to us. We understand the cometary cloud to be an aggregate of comets moving in nearly parabolic orbits. The only assumption made here is that we consider the number of long-period comets not to be restricted to the few hundred that have already been discovered. The number of comets regularly entering the inner part of the solar system is evidently much larger.

The orbits of these comets undergo diffusion, essentially in the following manner. Let us consider a number of comets with parabolic orbits and approaching the Sun for the first time. After passing through the inner solar system, approximately 50% will have elliptical orbits and the other 50% will leave the solar system. When the remaining comets pass by the Sun a second time, the probability of their leaving is much less, since the initial orbits are elliptical. For a given comet the process of accumulation of small perturbations is random, since the revolution period of the comet remains very much longer than that of Jupiter, and whenever the comet approaches the Sun it encounters an entirely different configuration of planets.

We assume that independent random accumulation (i.e., diffusion) takes place when the reciprocal semimajor axis $(1/a)$ is in the interval 0 to 0.025 AU^{-1}. We have demonstrated that the perihelion distances are practically invariable under diffusion (Shtejns and Kronkalne, 1964). In the present paper we consider that a normal distribution law holds and that its standard deviation σ is a function of the inclination i and perihelion distance. The effects of nongravitational impulses is taken into account in a similar way. It is assumed that the diffusion process takes place over a very long time and that a sufficient number of comets is involved. The disintegration of comets

Chebotarev et al. (eds.), The Motion, Evolution of Orbits, and Origin of Comets, 347–351. *All Rights Reserved.*
Copyright © 1972 by the IAU.

is taken into account by considering that after a particular number N of revolutions around the Sun they perish or become invisible. The number N is assumed to be a function of perihelion distance q.

If there were a sufficiently large number of known comets having the same perihelion distance and whose statistics $n = n(1/a, i, q)$ are not distorted by discovery conditions, we could find N in a purely empirical way. Knowing how the values of $1/a$ are distributed, i.e., knowing the standard deviation of the normal distribution σ for different hypothetical N, we can find $n = n(1/a, i)$. Choosing from all the theoretical $n(1/a, i)$ the one in best agreement with the empirical $n(1/a, i)$, we determine the maximum number of revolutions of the comet as the value of N with which the best theoretical distribution was found. But $n(1/a, i, q)$, even over the very small interval $(q, q + \Delta q)$, must be distorted by discovery conditions. Comets in the planetary zone diminish in brightness in the course of time, and consequently more new comets are discovered than is theoretically expected if discovery selection is ignored. New comets have $1/a \approx 0$. We feel that the most dependable results are obtained for $q \approx 1$ AU, when the observational conditions are the most favourable. Unfortunately, there are not enough known comets over a narrow interval ($\Delta q \approx 0.3$ AU), and if we have to use a longer one, we need to know $N = N(q)$, which is supposed to take the form

$$N = Aq^{1/2},$$

A being a constant. To avoid dependence of n on q we have determined the maximum number of revolutions according to the arithmetic mean value of $1/a$ and considered it independent of the discovery conditions. The data on $1/a$ have been taken from Porter (1961), although the original values have been replaced with those given by Sekanina (1966). Unlike our earlier study (Shtejns, 1964), parabolic and hyperbolic orbits are not considered here. As is well known, a parabolic orbit is usually determined from observations over a comparatively small arc, and it is not possible to detect a deviation from a parabola with sufficient reliability. Including parabolic orbits undoubtedly diminishes the mean value of $1/a$, since many of these comets must really have quite large positive values of $1/a$. Determination of the arithmetic mean of the original $1/a$ values also causes N to be underrated because these values have been determined mainly for orbits with negative values of $1/a$ near perihelion, the calculations having been made in order to prove Strömgren's idea of the nonexistence of hyperbolic original orbits.

For the determination of N we have taken comets with $0.9 < q < 1.9$ AU, the relevant data being summarized in Table I. It can be seen that there are 43 comets with direct motion and 27 with retrograde motion, and that this difference is due mainly to comets of large $1/a$. This phenomenon was discovered and explained by Oort. We have defined it as the first law of diffusion, namely: As a result of the diffusion of comets, orbits with the greater reciprocal semimajor axes have smaller inclinations; i.e., they concentrate towards the plane of Jupiter's orbit.

With the aid of the method of independent tests we have obtained theoretical values of $n = n(1/a, q, N)$. It should be noted that, in deriving the average value with respect to i, account has been taken of the fact that the normals to the orbital planes intersect

TABLE I

Comets for which $0.9 < q < 1.9$ AU

Interval of $1/a$ (AU^{-1})	Number of comets with	
	direct motion	retrograde motion
$0 < 1/a < 0.0015$	18	13
$0.0015 < 1/a < 0.015$	15	14
$0.015 < 1/a < 0.025$	10	0
Mean value	0.0069	0.0033
$1/a < 0$	4	3
Mean value	0.0063	0.0028

the celestial sphere uniformly. The results of these calculations are given in Table II, from which it may be seen that the arithmetic mean of $1/a$ is almost identical for $q = 1$ and $q = 2$. As for the observed comets, for $i < 90°$ there is a small decrease from $q = 1$ to $q = 2$. Therefore, the hypothesis $N = Aq^{1/2}$ is valid within this range. Comparing the data of Tables I and II, we come to the conclusion that $N = 100$ revolutions. This number should be considered as a minimum, because selection results in the brighter comets being discovered, and these have smaller values of $1/a$.

TABLE II

Dependence of $1/a$ on N and q

Maximum number of revolutions N		Mean value of $1/a$ (AU^{-1})	
		$q = 1$ AU	$q = 2$ AU
$i < 90°$	7	0.0014	0.0013
	60	0.0047	0.0050
	100	0.0063	0.0064
	180	0.0072	0.0072
$i > 90°$	7	0.0009	0.0009
	60	0.0033	0.0032
	100	0.0038	0.0040
	180	0.0061	0.0060

An estimate of mean cometary lifetime can be made over a definite interval $\Delta(1/a)$ according to the relative number of comets whose theoretical values for direct and retrograde motion are given in Table III. Comparing these data with the observational data of Table I, we see that considerably fewer comets with $1/a \approx 0$ and 0.02 are actually discovered than theoretically expected. According to Table III the most probable lifetime of a comet is also 100 revolutions.

The effects of nongravitational impulses are taken into account as follows. We consider that there is a normal distribution whose standard deviation σ is a function of q

only and that an acceleration does not change to a deceleration, or vice versa. Calculations for $N = 100$, $\sigma = 0.00006$ and $0.015 < 1/a < 0.025$ give five comets with direct and one with retrograde motion. For $N = 100$, $\sigma = 0.00015$, we have eight and four comets, respectively. Comparing this with the data of Tables I–III, we see that the relatively larger number of nearly parabolic orbits and orbits with $a = 90$ to 40 AU cannot be explained by the influence of nongravitational impulses on the nuclei of long-period comets. The value $\sigma = 0.00006$ is in good accord with the mean value of the nongravitational impulses obtained from calculations over short overlapping arcs. The effects of nongravitational impulses thus lead to smaller values for the maximum lifetimes of comets.

TABLE III

Relative number of comets with direct and retrograde motion

Interval of $1/a$ (AU^{-1})	$N(q=1\text{ AU})$:	Direct motion				Retrograde motion			
		7	60	100	180	7	60	100	180
$0 < 1/a < 0.0015$		28	12	10	8	22	10	8	6
$0.0015 < 1/a < 0.015$		15	30	30	31	5	17	19	19
$0.015 < 1/a < 0.025$		0	1	3	3	0	0	0	2

Let us check the second diffusion law, which states that orbits of large perihelion distance have on the average smaller semimajor axes. This law can be found from observations. The observational data are as follows: for direct motion, if $0.1 < q < 0.9$ AU, the mean value of $1/a$ is 0.0050 (0.0036 with $1/a < 0$), while for retrograde motion it is 0.0036 (0.0014 with $1/a < 0$). Comparing these data with those of Table I, we see that the mean value of $1/a$ increases with increasing q. For $q = 5.2$ AU we have no observational data. After 350 revolutions the theory yields 0.0100 for direct motion and 0.0054 for retrograde motion.

Appropriate conclusions can be made, not only from the mean value of $1/a$ but also from the relative number of comets in different intervals. The results are of interest for $q = 5.2$ AU, where capture takes place. Assuming a maximum number of 350 revolutions and if $0 < 1/a < 0.0015$, we have 5 direct and 5 retrograde comets; if $0.0015 < 1/a < 0.015$, we have 28 direct and 10 retrograde, and if $0.015 < 1/a < 0.025$, 21 direct and one retrograde. The maximum number of revolutions is chosen so as to demonstrate the reality of the third diffusion law, i.e., that there are more new comets with smaller perihelion distances. To explain why many more comets with nearly parabolic orbits are actually discovered than theoretically expected we can assume that some comets have very short lifetimes.

Acknowledgment

I thank I. E. Zal'kalne for statistical calculations and useful discussions.

References

Oort, J. H.: 1950, *Bull. Astron. Inst. Neth.* **11**, 91.
Porter, J. G.: 1961, *Mem. Br. Astron. Assoc.* **39**, No. 3.
Sekanina, Z.: 1966, *Publ. Astron. Inst. Charles Univ.* No. 48.
Shtejns, K. A.: 1961, *Astron. Zh.* **38**, 107.
Shtejns, K. A.: 1964, *Uch. Zap. Latv. Gos. Univ.* **68**, 39.
Shtejns, K. and Kronkalne, S.: 1964, *Acta Astron.* **14**, 311.

Discussion

F. L. Whipple: I am delighted to see this more thorough approach to the problem than I myself attempted a few years ago. We agree concerning the rather short lifetimes for many comets. The random distribution function of $\Delta(1/a)$ as shown by H. A. Newton and H. N. Russell is not quite symmetrical, the negative side having a long tail, so I wondered whether your perturbation function was completely symmetrical.

K. A. Shtejns: I assumed it to be symmetrical, but this does not play an important part in my investigation.

61. NEW ESTIMATES OF COMETARY DISINTEGRATION TIMES AND THE IMPLICATIONS FOR DIFFUSION THEORY

O. V. DOBROVOL'SKIJ

Institute of Astrophysics, Dushanbe, U.S.S.R.

Abstract. Numerical methods are used to establish the fraction of solar radiation utilized in the sublimation of ice on the surface of a cometary nucleus. An estimate can then be made of the lifetime of the nucleus. For a nucleus of radius 2×10^5 cm in an orbit of semilatus rectum 0.6 AU the lifetime is 300 revolutions. Comparison with diffusion theory reveals a significant deficiency of the theory in the case of comets with small reciprocal semimajor axes.

The rate of disintegration of comets and its relation to the statistics of cometary orbits is of great importance to the cosmogony of comets.

Since an icy model is the most probable one, that is what we shall discuss (Rijves, 1970). The lifetime τ of a comet can conveniently be determined in terms of the efficiency of the solar radiation η, defined as the ratio of the amount of solar radiation expended in disintegration to the total amount of solar energy incident to the nucleus. The mass dM lost by the nucleus during time dt will be

$$dM = -\frac{Q_0 r_0^2}{L r^2} s\, dt = -\frac{Q_0 r_0^2 s}{Lc} \eta\, dv, \tag{1}$$

where Q_0 is the solar constant, r_0 the astronomical unit, r the heliocentric distance, s the cross-section area of the nucleus, L the specific heat of disintegration, c the constant of the law of areas, and v the comet's true anomaly. If the nucleus is a sphere of density δ, the change in the nuclear radius R will be

$$dR = -\eta \frac{Q_0 r_0^2}{4cL\delta}\, dv.$$

In one revolution about the Sun the initial radius R_0 will decrease to

$$R = R_0 - \frac{2\pi Q_0 r_0^2}{4cL\delta}\, \bar{\eta},$$

$\bar{\eta}$ being the efficiency of the solar radiation averaged over the true anomaly.

We assume that after n revolutions R will diminish to zero; this number is then given by

$$n = \frac{2cLR_0\delta}{\pi Q_0 r_0^2 \bar{\eta}}.$$

Hence the total lifetime τ of the comet (in years) is

$$\tau = \frac{K}{\bar{\eta}} a^{3/2} p^{1/2}, \tag{2}$$

where a is the semimajor axis and p the semilatus rectum (both measured in AU), and K is the dimensionless quantity

$$K = \frac{2\delta L R_0 (\gamma M_\odot r_\odot)^{1/2}}{\pi Q_0 r_0^2}, \tag{3}$$

γ being the gravitational constant and $M_\odot$ and $r_\odot$ the mass and radius of the Sun.

Expressing r in AU, we may with reasonable precision write η in the form

$$\eta = \eta_0 r^{-\alpha}; \tag{4}$$

then

$$\bar{\eta} = \frac{\eta_0}{p^\alpha} J_\alpha, \tag{5}$$

where

$$J_\alpha = \frac{1}{\pi} \int_0^\pi (1 + e \cos v)^\alpha \, dv, \tag{6}$$

e being the orbital eccentricity; and obviously $J_0 = J_1 = 1$. For $\alpha \geq 2$ we have the recurrence formula

$$J_\alpha = \frac{2\alpha - 1}{\alpha} J_{\alpha-1} - \frac{\alpha - 1}{\alpha} (1 - e^2) J_{\alpha-2}. \tag{7}$$

Thus, within the range of validity of the approximation (4) we have

$$\tau = \frac{K}{\eta_0} \frac{a^{3/2} p^{\alpha + 1/2}}{1 + k_\alpha e^2}, \tag{8}$$

where k_α is given in Table I.

TABLE I

k_α as a function of α

α	k_α
0	0
1	0
2	1/2
3	5/6
4	13/12

Various determinations of α have been made for an H_2O nucleus. For instance, Markovich and Tulenkova (1968) and Dobrovol'skij (1966) found that when $r \geq 1$ AU, $\alpha = 3/2$ and $\eta_0 \approx 1/2$, while for smaller values of r, the coefficient η approaches 1 asymptotically. Thus, if the perihelion distance $q \geq 1$ AU, we shall have, approximately,

$$\tau = 2Ka^{3/2}p^2. \tag{9}$$

If $q < 1$, this may be regarded as the lower limit of τ, although when $p < 0.63$ AU the lower limit is more suitably given by

$$\tau_{\min} = Ka^{3/2}p^{1/2}. \tag{10}$$

Taking $L = 2.7 \times 10^{10}$ erg g^{-1}, we shall have, for a nucleus with $R_0 = 2 \times 10^5$ cm and $\delta = 1$ g cm^{-3},

$$K \approx 500; \tag{11}$$

so that, when $p = 0.6$ AU, $\tau \approx 300$ revolutions.

According to Shtejns (1961), diffusion theory gives the number of comets ν passing perihelion annually and having reciprocal semimajor axes $z = 1/a$ in the range $(z, z+1)$ as

$$\nu_z = A \exp\left(-gz^{1/2}\tau^{-1/2}\right), \tag{12}$$

where A and g are constants. Shtejns has discussed Equation (12) for $\tau = 7$ and 60 revolutions. Figures 1 and 2 show distributions with $\tau = 7$, 60, and 300 revolutions, the constant A having been selected in each case so as to satisfy the points given by Shtejns (1961).

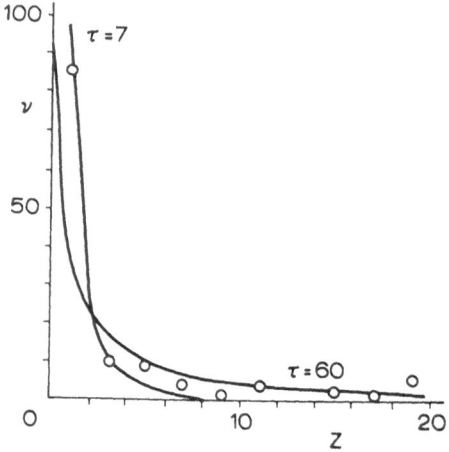

Fig. 1.   Distribution of the number of comets ν as a function of reciprocal semimajor axis z for lifetimes $\tau = 7$ and 60 revolutions. Observed points are denoted by circles.

Calculated curves for $\tau = 7$ and 60 revolutions coincide qualitatively with the points given by Shtejns; for $\tau = 7$ the fit is best for small z, while for $\tau = 60$ large z-values are represented more closely. For $\tau = 300$ revolutions a satisfactory description can be given only in the case of large z. Comets with small z do not fit the theoretical relationship at all: we could choose A so that the curve would pass through the first point of the observed distribution, but then it would not correspond to any of the others, and no reasonable variation of q would help eliminate the discrepancy.

We conclude that Equation (12) does not agree with the observations and should be replaced, or else the diffusion theory should be developed more rigorously, with account taken in particular of the dependence of τ on a in the original diffusion equation.

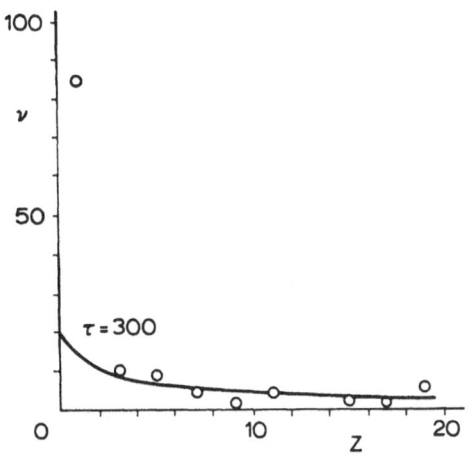

Fig. 2.   Distribution of ν as a function of z for $\tau = 300$ revolutions.

References

Dobrovol'skij, O. V.: 1966, *Komety*, Nauka, Moscow.
Markovich, M. Z. and Tulenkova, L. N.: 1968, *Probl. Kosmich. Fiz.* No. 3, 25.
Rijves, V. G.: 1970, *Komety i Meteory* No. 17, 3.
Shtejns, K. A.: 1961, *Astron. Zh.* **38**, 107.

62. COMETS AND PROBLEMS OF NUMERICAL CELESTIAL MECHANICS

S. K. VSEKHSVYATSKIJ

Astronomical Observatory, Kiev University, Kiev, U.S.S.R.

Abstract. A number of problems concerning the motions and orbital statistics of comets are suggested for numerical discussion.

Laplace (1805) was the first to consider cometary problems in a statistical manner. He established the probability for the appearance of comets moving in hyperbolic and almost parabolic orbits on the assumption that these comets came from interstellar space. It was subsequently demonstrated that this assumption was invalid. The lack of observed initial hyperbolic orbits serves as indisputable proof that comets originate in the solar system.

Statistical analysis of the directions of aphelion of the known comets leads us to the same conclusion. The pioneering work of Carrington and Svedstrup has been carried on with different material and by different methods by Fabry, Holetschek, Pickering, Eddington, Oppenheim, Bourgeois and Cox, Witkowski and Hurnik. In all these studies there was found to be no predominance of aphelia in the hemisphere including the apex of the Sun's way. However, there still appear in the literature the results of attempts to apply the idea of interstellar origin on account of the fact that the aphelion distribution hints somewhat at a connection with the structure of the Galaxy, or by calculating the directions of the apices of two comet streams (Witkowski, 1965). If one recalls that the observed comets have relative velocities of less than 0.1 km s^{-1} at the boundary of the Sun's sphere of action, and that the supposed comet streams could not bear any relation to interstellar regions – for which relative velocities would be expressed in kilometres and tens of kilometres per second – it becomes evident that there are no grounds for interpreting the results in such a manner. At relative velocities of less than 0.1 km s^{-1} (which is very small compared to the motion of the Sun in the Galaxy), cometary fragments (even those with periods of 10^9 yr) must have come in from distances not greater than a few tens of parsecs, and consequently they could not reflect the peculiarities of the structure of the Galaxy. There exist considerable deviations from a uniform distribution, but these can be assigned to the effects of perturbations by neighbouring stars (Vsekhsvyatskij, 1967). At present, calculation of the effects of individual stars on an individual comet is merely an academic matter, although it may not be impossible to make such calculations in the future.

We may make important conclusions from calculations of original and future orbits of comets. The results by Thraen, Fayet and Strömgren concerning the original elliptical character of the orbits of the majority of comets observed to have osculating hyperbolic orbits are confirmed in the recent summaries by I. Galibina and Z. Sekanina. This also very definitely shows that all the observed comets belong to the solar system, although it is true that we have to assume that beyond the bounds of the observed arc the motion of a comet is not influenced by internal or external forces other than gravitational attractions by the known planets.

The nongravitational effects are one of the most basic problems of cometary astronomy and cosmogony. Long ago, study of the motion of Encke's Comet, and later of the motions of other short-period comets, made it possible to determine cases of accelerations or decelerations unexplained by the influence of ordinary perturbations. I explained these effects 40 yr ago on the basis of the eruption theory as a consequence of relative youth and physical instability of a comet (Vsekhsvyatskij, 1931, 1933, 1950). An important contribution to the problem was made by F. L. Whipple and his disciples, in which the idea was advanced that the nucleus of a comet is basically composed of ice.

From the results by Whipple (1950, 1951), Hamid and Whipple (1953), Marsden (1968, 1969), and Sekanina (1968a, 1968b) concerning the nongravitational effects on cometary motion, as well as from the rich collection of numerical theories of the motions of numerous short-period comets, one can make two fundamental conclusions:

(1) nongravitational secular accelerations or decelerations in the motions of comets are appreciable in magnitude, and for comets that make close approaches to Jupiter they lead to considerable deviations from purely gravitational theory after only a small number of revolutions;

(2) nongravitational effects apparently decrease as a comet ages and are on the whole determined by explosive processes in the cometary nucleus. It follows that extrapolation of the motion of a comet over long intervals of time can have little meaning. This is especially true of calculations back into the past, when the nongravitational effects would have been larger. One has thus every reason to consider that the calculations on comets such as Lexell, Wolf, Brorsen, Oterma and others do not point to these comets leaving the Jupiter system before the time of their close approaches before discovery.

Therefore, to elucidate the real value of these extrapolations it would be desirable to recalculate the motions of these comets, varying the initial data (within the bounds of probable error and considering also the nongravitational effects) and comparing the results in order to determine the number of revolutions over which the extrapolations may be considered reliable. This is a particularly important problem. Unless one continues investigations similar to that by Leverrier (1857) on P/Lexell with due consideration of nongravitational effects as well, one cannot believe that the results are of real significance. The same is true of the calculations of several short-period comets over the interval 1660–2060 (Kazimirchak-Polonskaya, 1967), where extrapolation of the motion of P/Kearns-Kwee seems to result in a strongly hyperbolic orbit of a type never observed before. It follows from elementary considerations that the discovery of a sharply hyperbolic comet with perihelion distance less than 5 AU is many thousand times more probable than a hyperbolic orbit resulting from a close approach to Jupiter calculated for 1855. The fact that no such comet has ever been seen shows the unreliability of the extrapolated results.

In recent years detailed and valuable work on conventional orbital statistics has been published by Everhart (1967, 1968, 1969). Using much more numerical material, he has fully confirmed the correctness of the conclusions made earlier by H. A. Newton, those obtained in investigations in Kiev, as well as preliminary calculations by A. van Woerkom, Z. Sekanina and others on the accumulative effects of small perturbations. Everhart has shown that short-period comets (i.e., the newest and most rapidly disintegrating objects) could not be formed as the result of diffusion or capture from the system of long-period comets, confirming conclusions made long ago by R. A. Proctor, A. C. D. Crommelin and myself.

It is also important to continue statistical calculations on the eruptive theory. We have obtained remarkable agreement between the theoretical and observed distributions of elements of cometary orbits (Vsekhsvyatskij and Meshcheryakova-Babich, 1957), but it is of some interest to expand these calculations to all the giant planets, considering both the satellites and the planets themselves to be possible sources of the comets and varying the initial conditions.

Statistical analysis of the effects of stars passing through the hypothetical Oort comet cloud and study of the stability of motion near the boundaries of the Sun's sphere of action provides an independent means for interpreting the cometary problem. The radius of the Sun's sphere of action in the galactic field is some 70 000 AU.

Among other important numerical cometary problems we can mention the construction of theories of motion for 'new' (i.e., rapidly disintegrating) short-period comets which passed close to Jupiter shortly before discovery; if one assumes that the minimum separation was close to zero and then tries to obtain an orbit representing all the observations at the discovery apparition, new data may be obtained concerning the evolution of the nongravitational effects.

It is also desirable to compare the motions of comets in their preperihelion and postperihelion branches. This will not only clarify the nongravitational effects but will make it possible to study the effects of outbursts in cometary brightness, nuclear splitting, and so on, with a view to obtaining estimates of the masses of cometary nuclei. A thorough investigation of the motions of particularly interesting comets, coupled with consideration of their physical activity, can give new information about the influence of the solar wind, Poynting-Robertson effect and magnetic fields on the motions of comets and meteoroids.

References

Everhart, E.: 1967, *Astron. J.* **72**, 716, 1002.
Everhart, E.: 1968, *Astron. J.* **73**, 1039.
Everhart, E.: 1969, *Astron. J.* **74**, 735.
Hamid, S. E. and Whipple, F. L.: 1953, *Astron. J.* **58**, 100.
Kazimirchak-Polonskaya, E. I.: 1967, *Trudy Inst. Teor. Astron.* **12**; *Astron. Zh.* **44**, 439.
Laplace, P.: 1805, *Connais. Temps.* 1806.
Leverrier, U. J. J.: 1857, *Ann. Obs. Paris Mem.* **3**, 203.
Marsden, B. G.: 1968, *Astron. J.* **73**, 367.
Marsden, B. G.: 1969, *Astron. J.* **74**, 720.
Sekanina, Z.: 1968a, *Bull. Astron. Inst. Czech.* **19**, 351.

Sekanina, Z.: 1968b, *Probl. Kosmich. Fiz.* No. 3.
Vsekhsvyatskij, S. K.: 1931, *Astron. Nachr.* **243**, 286.
Vsekhsvyatskij, S. K.: 1933, *Astron. Zh.* **10**, 18.
Vsekhsvyatskij, S. K.: 1950, *Astron. Zh.* **27**, 23.
Vsekhsvyatskij, S. K.: 1967, *Priroda i Proiskhozhdenie Komet i Meteornogo Veshchestva*, Pros-veshchenie, Moscow, p. 150.
Vsekhsvyatskij, S. K. and Meshcheryakova-Babich, O. I.: 1957, *Astron. Zh.* **34**, 568.
Whipple, F. L.: 1950, *Astrophys. J.* **111**, 375.
Whipple, F. L.: 1951, *Astrophys. J.* **113**, 464.
Witkowski, J.: 1965, *Acta Astron.* **15**, 273.

63. THE EFFECT OF THE ELLIPTICITY OF JUPITER'S ORBIT ON THE CAPTURE OF COMETS TO SHORT-PERIOD ORBITS

E. EVERHART

Dept. of Physics, University of Denver, Denver, Colo., U.S.A.

Abstract. Single, random, close encounters of long-period comets with Jupiter are studied. In contrast to earlier work, Jupiter's orbit is taken to be elliptical, but this has no effect on the rate of capture into short-period orbits and neither does it influence the distribution of longitudes of perihelion.

Short-period comets are not distributed uniformly with regard to their longitudes of perihelion $\bar{\omega}$. Considerably more of these comets have perihelia on one side of the solar system than the other. The distribution is shown in Figure 1a for the 98 known short-period comets. The longitude of perihelion of Jupiter is marked by the arrow on the figure and, evidently, a preponderance of the short-period comets more or less line up their orbits with Jupiter's. Figure 1b shows that the same situation holds for a smaller group of 68 comets, these being all those of perihelion distance less than 2 AU and period less than 22 yr.

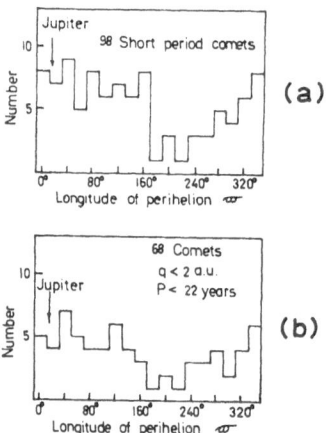

Fig. 1. Distribution of short-period comets with regard to longitude of perihelion $\bar{\omega}$.

The capture hypothesis is that short-period comets originate when long-period comets pass near Jupiter and lose heliocentric energy to Jupiter during the encounter. (It is possible in principle for other planets to have a similar effect in capturing a comet, but the effects of Jupiter vastly predominate.) The first quantitative discussion of the capture hypothesis was that by Newton (1893). More recently, the present author (Everhart, 1969) has tested this hypothesis extensively in a Monte-Carlo calculation, taking 10^9 hypothetical random parabolic comets interacting with the solar system, finding how many of these would be captured to orbits of short period, and obtaining

Chebotarev et al. (eds.), The Motion, Evolution of Orbits, and Origin of Comets, 360–363. All Rights Reserved.
Copyright © 1972 by the IAU.

the distributions to be expected among the captured comets in such quantities as period and inclination. This recent study, as well as that by Newton, assumed circular orbits for Jupiter and the other planets, a simplification that appeared to be reasonable for treating a rather complicated problem. However, because of symmetry a circular orbit for Jupiter could not possibly account for a nonuniform distribution of longitude of perihelion. Indeed, it was thought possible that taking Jupiter's orbit as elliptical would make changes in some of the other results as well.

Accordingly, the entire study of capture of comets by Jupiter is here repeated using an ellipse for Jupiter's orbit. The new calculation is somewhat more extensive and complicated than before, partly because of the transcendental equations that arise on finding the collision sites in the elliptical case, and partly because of the new requirement that Jupiter's mean anomaly must have a random value at the time each hypothetical comet comes to perihelion. The new result can be stated very simply: taking Jupiter's orbit as elliptical rather than circular makes no discernible difference in the results from the previous work. For example, the distribution of periods of the hypothetical short-period comets after capture has the same shape as before, and the scale factor is unchanged. The same result holds for the distribution of inclinations after capture and for all the other distributions plotted by Everhart (1969). The previous studies with a circular orbit for Jupiter gave distributions that did not agree with those of known short-period comets, and this lack of agreement is not changed in any way by taking Jupiter's orbit to be elliptical.

These results discussed so far have been averaged over Jupiter's orbit. One might inquire whether Jupiter captures a little more effectively near perihelion than near aphelion, or whether the opposite is true. Such an effect would show if the captured hypothetical comets were sorted according to where the encounter with Jupiter took place.

The results of such sorting are seen in Figure 2a, which shows the number distri-

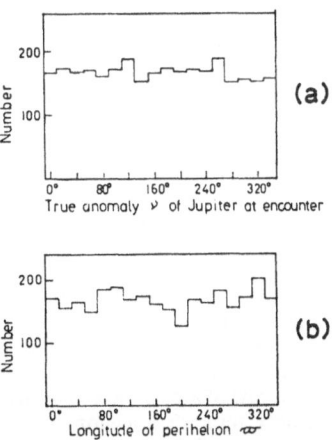

Fig. 2. Distribution of 3000 hypothetical comets captured to periods less than 22 yr with respect (a) to true anomaly v of Jupiter at the time of close encounter, and (b) to the longitude of perihelion $\tilde{\omega}$ of the captured hypothetical comets.

bution of hypothetical comets captured to periods less than 22 yr plotted versus the true anomaly of Jupiter at the time of the close encounter; and again in Figure 2b, which shows the number distribution plotted versus the longitudes of perihelion of the comets after being captured to periods of less than 22 yr. Both these histograms are uniform within statistical fluctuations. In the search for a possible effect many other distributions were examined: for example, those of inclination and period for comets encountering Jupiter in specified parts of its orbit. The same negative result is found in all cases. The eccentricity of Jupiter's orbit has no noticeable effect on the ability to capture comets to a short-period orbit in a single encounter.

Accepting this result then, we suggest three other possibilities that might account for the nonuniform distribution of $\tilde{\omega}$ seen in Figure 1.

(1) An observational selection effect. Although there has been no thorough study of observational selection of short-period comets, a somewhat similar asymmetry in a property of long-period comets can be so explained: it turns out that considerably more long-period comets have an argument of perihelion ω between 0 and π than between π and 2π. In a quantitative study (Everhart, 1967), this was shown to be an observational selection effect traceable to the fact that there have been more observers finding comets from northern observatories than from southern.

(2) The effect may be nonexistent. One may consider that the distribution of longitudes of perihelion is intrinsically uniform, but that statistical fluctuations in a small sample cause an apparent asymmetry. The well known chi-square test finds that there is a 5% chance of such an asymmetry arising statistically as in Figure 1a and a 25% chance as in Figure 1b. Unfortunately, one cannot repeat the plot with new data for 100 different short-period comets!

(3) The asymmetries seen in Figure 1 might conceivably be the result of Jupiter's orbital eccentricity if there were a small but cumulative effect in repeated encounters. This possibility has yet to be tested.

References

Everhart, E.: 1967, *Astron. J.* **72**, 1002, Fig. 5.
Everhart, E.: 1969, *Astron. J.* **74**, 735.
Newton, H. A.: 1893, *Mem. Natl. Acad. Sci. Washington* **6**, 7.

Discussion

E. I. Kazimirchak-Polonskaya: Which perturbations did you take into account?

E. Everhart: Those by the Sun when the comet was in Jupiter's sphere of influence and those by Jupiter when it was in the Sun's sphere of influence. The calculation used a hyperbolic encounter formula that agrees within a few percent with exact numerical calculations of such encounters.

L. Kresák: It appears surprising that you do not find any asymmetry in the expected distribution of perihelion longitudes. Because of the law of areas, there is both a greater probability of encounter and a slower mean relative motion near Jupiter's aphelion.

E. Everhart: The formula I derived for the change in energy during a close encounter depends on the impact parameter, the relative velocities, and the angle between certain vectors. It is not difficult to see that, to the first order, capture at Jupiter's perihelion and capture at aphelion should be about the same.

G. Sitarski: You showed earlier that the probability of capture of a comet by Jupiter is very small: for example, one observable periodic comet in some thousand years. Why do you now present new investigations based on the assumption that periodic comets originate through capture?

E. Everhart: I did this merely to see whether the eccentricity of Jupiter's orbit would make any difference. I did not expect to find more captures in the elliptical case than in the circular case, but I did expect to find some positive effects and was surprised not to see anything.

64. EVOLUTION OF SHORT-PERIOD COMETARY ORBITS DUE TO CLOSE APPROACHES TO JUPITER

O. HAVNES

*Astronomical Institute, Utrecht, The Netherlands**

Abstract. Evolution of short-period cometary orbits under the influence of close and moderately close approaches to Jupiter is studied. We have restricted the discussion to orbits in Jupiter's orbital plane and have neglected distant perturbations by Jupiter. The general evolution is a diffusion towards orbits of larger semimajor axes and a loss of comets, mainly due to ejection along hyperbolic orbits. We have found that the dynamical evolution of the orbits severely alters the assumed initial distribution of orbits during a time of some hundred years. We also tried to obtain an idea of the importance of close approaches as compared to the effect of moderate approaches to Jupiter for the evolution of the orbits. This was done by calculating the evolution of a group of comets twice, first considering the effect of perturbations up to a distance of 1.0 AU from Jupiter and then to a distance of 2.0 AU. The statistical results of the two calculations are in general agreement, indicating that close approaches to Jupiter mainly determine the evolution of the short-period comets.

1. Introduction

In any test of theories on the origin of comets, one of the major questions is whether or not the distribution of short-period comets can be satisfactorily reproduced. Numerical calculations on the orbits of known short-period comets over time intervals of 400 yr (Kazimirchak-Polonskaya, 1967) show that these orbits are generally not stable and that their evolution seems to be conditioned by close approaches to Jupiter. It is therefore of interest to examine to what degree a group of short-period comets is affected by such repeated close approaches to Jupiter and on what time scale this effect is of importance.

2. Method of Calculation and Results

We have attempted to find general features of the evolution of short-period comets by combining the evolution of many different orbits. Each cometary orbit is calculated by regarding the comet as part of a three-body problem consisting of the Sun, Jupiter, and the comet. We consider the planar case only. Kazimirchak-Polonskaya's (1967) work indicated that close approaches to Jupiter were most important for the evolution of the cometary orbits. We therefore neglected distant perturbations from Jupiter.

Three groups of comets were studied in which the perturbations by Jupiter on the comet were calculated when their mutual distance was less than a limit $\Delta = 1.0$ AU. Two groups of comets in direct orbits were considered, the first group having semimajor axes from 3.0 to 8.0 AU, the second group from 3.1 to 8.1 AU, both in steps of 1.0 AU. We also considered a group of comets in retrograde orbits, having semimajor axes from 3.1 to 8.1, also in steps of 1.0 AU.

---

* On leave from the Institute of Theoretical Astrophysics, University of Oslo, Blindern, Norway.

Chebotarev et al. (eds.), The Motion, Evolution of Orbits, and Origin of Comets, 364–369. All Rights Reserved.
Copyright © 1972 by the IAU.

The neglect of the more distant perturbations by Jupiter may introduce systematic errors in the statistical results. We therefore calculated the evolution of the first group of comets in direct orbits a second time where we included perturbations by Jupiter up to $\Delta = 2.0$ AU. The statistical results of the two calculations of different accuracy are in general agreement, indicating that the evolution of the short-period cometary orbits is mainly determined by close approaches to Jupiter.

The initial eccentricities e_0 which were considered are given in Table I as a function of initial semimajor axis a_0. For each pair of a_0 and e_0 we considered six different initial values of mean anomaly, varied in steps of 60° from 0° to 300°. Further details on the methods of numerical calculation and results for the cases when $\Delta = 1.0$ AU are given in Havnes (1970).

TABLE I

Minimum initial eccentricity e_s as a function of initial semimajor axis a_0

a_0 (AU)	3.0	4.0	5.0	6.0	7.0	8.0
Direct orbits	0.75	0.45	0.35	0.45	0.55	0.55
Retrograde orbits	0.75	0.25	0.15	0.25	0.35	0.35

The eccentricities which are considered are from $e_0 = 0.95$ to $e_0 = e_s$ in steps of 0.1. The values of e_s are chosen so that the orbits are such as could result from direct capture of parabolic comets (Havnes, 1970).

Each orbit is followed for a time interval of 1000 yr or until the comet is removed from the group, either by attaining a hyperbolic orbit or – in some cases – colliding with the Sun or Jupiter. The evolution of the first group of comets in direct orbits is shown in Figure 1, where the corresponding elements a and e are plotted at time $t = $ 100, 500, and 1000 yr. The comets having $e_0 = 0.95$ are omitted as they have short lifetimes (Shtejns, 1959). For the remaining comets the effects of limited lifetime have not been taken into account.

The loss of comets, in percentages of the initial numbers (the comets having $e_0 = $ 0.95 are not counted), are given in Table II for the different groups of comets. Figure 2 shows the loss of comets in more detail. We have given the percentage of comets removed from the system as a function of e_0 and a_0, where the figures are in terms of the number of comets in each group having the same e_0 and a_0. The loss seems to depend little on the initial eccentricity but generally decreases with initial semimajor

TABLE II

The percentage loss of comets in different groups

Group of comets	Δ (AU)	$t = 100$	$t = 500$	$t = 1000$ yr
Direct group number 1	1.0	2	9	12
Direct group number 1	2.0	1	7	14
Direct group number 2	1.0	1	8	15
Retrograde group	1.0	1	5	11

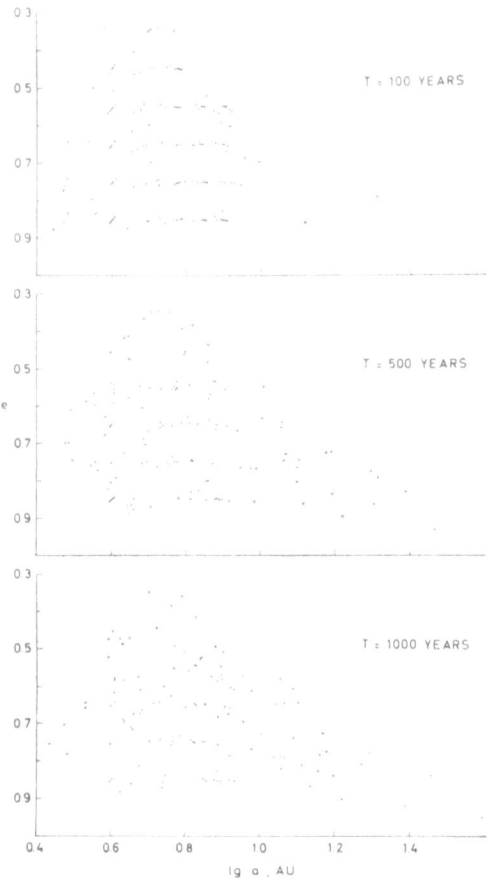

Fig. 1.   The evolution of a group of short-period comets (orbiting in direct heliocentric orbits)
due to close approaches to Jupiter is shown. The group of comets initially had semimajor axes in
the interval 3.0 to 8.0 in steps of 1.0 AU. The corresponding initial eccentricities were varied
from 0.85 in steps of 0.1 down to e_S as given in Table I. The distribution of corresponding elements
a, e is shown at times $t = 100$, 500, and 1000 yr.

axis. This last effect is to be expected as comets with the largest periods (semimajor
axes) generally have the smallest probability of passing close to Jupiter during a
certain time interval.

The general evolution of a group of short-period comets – according to our calculations – is a steady loss of comets and a diffusion towards orbits having larger semimajor axes. This diffusion is apparent from Figure 1 for the first group of comets in
direct orbits. In Table III we have demonstrated the diffusion for all the groups of
comets by calculating the average of the reciprocal semimajor axis $\langle 1/a \rangle$ at times
$t = 0$, 500, and 1000 yr.

From Figure 2 it appears that the loss of comets in the retrograde group is generally
less than for the corresponding direct groups. In Havnes (1970) it was stated that
'The rate of diffusion seems to be smaller for the retrograde comets than for the direct

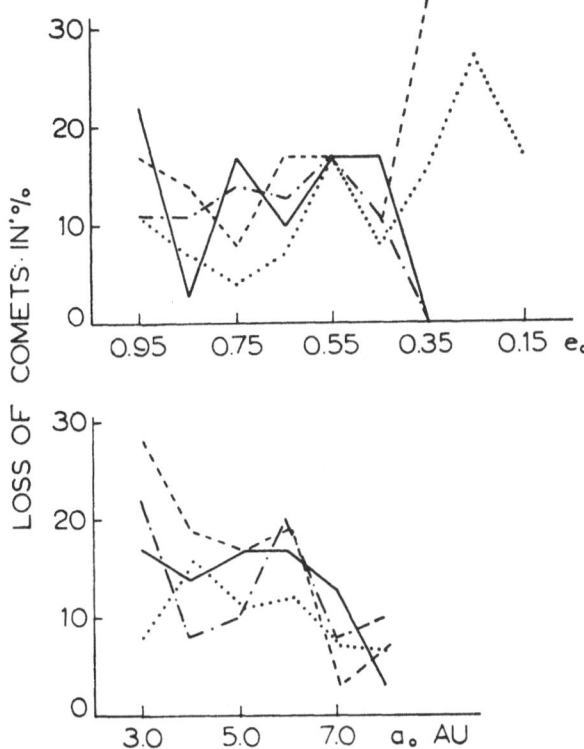

Fig. 2.   The loss of comets (due to ejection from the system or collision with the Sun or Jupiter) for the different groups of comets as a function of initial eccentricity e_0 (upper figure) or initial semimajor axis a_0 (lower figure). The line ——— gives the results for the first group in direct orbits (3.0 AU $\leq a_0 \leq$ 8.0 AU and $\Delta = 1.0$ AU) and —·—·— the results for the same group when $\Delta = 2.0$ AU. For the second group in direct orbits (3.1 AU$\leq a_0 \leq$ 8.1 AU and $\Delta = 1.0$ AU) the results are given by –––––, while · · · · · · gives the results for the retrograde group (3.1 AU $< a_0 \leq$ 8.1 AU and $\Delta = 1.0$ AU). The large spread in the results for direct orbits of $e_0 = 0.35$ is caused by the small number (6 orbits) that in each case had initial eccentricity $e_0 = 0.35$.

TABLE III

The average of the reciprocal semimajor axes $\langle 1/a \rangle$ for the different groups of comets

Group of comets	Δ (AU)	$t = 0$	$t = 500$	$t = 1000$ yr
Direct group number 1	1.0	0.193	0.167	0.150
Direct group number 1	2.0	0.193	0.172	0.143
Direct group number 2	1.0	0.189	0.158	0.141
Retrograde group	1.0	0.185	0.155	0.132

comets'. This statement resulted from a visual inspection of a plot for the retrograde comets similar to Figure 1. Here we find, according to Table III, that their rate of diffusion towards orbits of larger semimajor axes is somewhat larger than for the direct comets. However, the difference is small, and in view of this, we cannot explain the apparent nonexistence of retrograde comets in Jupiter's family as a result of a selection effect – differing between the direct and retrograde comets – in the evolution due to close approaches to Jupiter.

The effect of evolution on the distribution in semimajor axis of the first group of comets in direct orbits is shown in Figure 3. We have given the distribution for the two

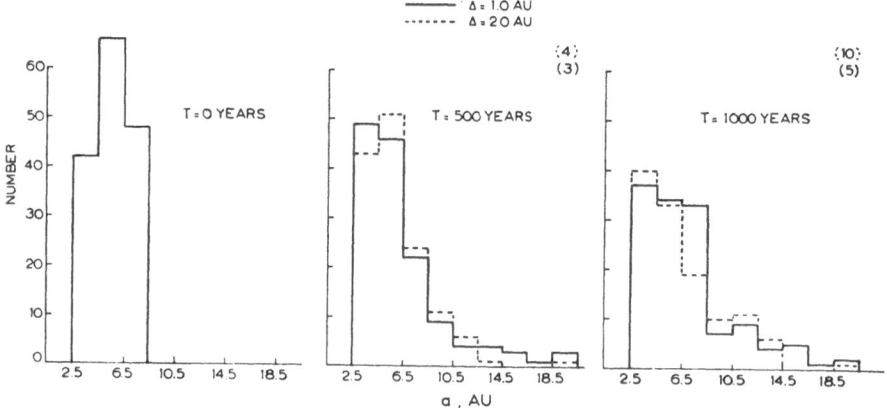

Fig. 3.   The distribution of semimajor axis for the first group of comets in direct orbits is shown at $t=0$, 500, and 1000 yr. The full and broken lines give the results by the calculations including perturbations out to distances of $\Delta = 1.0$ and 2.0 AU, respectively, from Jupiter. The distributions are shown up to $a = 20.5$ AU. The number of comets having $a > 20.5$ AU for the two groups are given in full parentheses and dotted parentheses, respectively.

calculations of different accuracy (perturbations by Jupiter on the comet for $\Delta = 1.0$ and 2.0 AU, respectively) at times $t=0$, 500, and 1000 yr. The distributions are in general agreement, indicating that the characteristics of the evolution of a group of short-period comets may be found by taking account of close approaches to Jupiter only.

3. Conclusion

Our calculations show that the distribution of a group of short-period comets changed considerably during a few hundred years. According to this, the evolution of the orbits due to close approaches to Jupiter should be taken into account when calculating theoretical distributions of such comets if their lifetimes are of this order. As we have considered the planar case only, the evolution of cometary orbits having $i \neq 0$ may be expected to be slower than for the case with $i=0$. However, as the inclinations of the short-period comets generally are small, calculations of the evolution of such a group should not give results differing considerably from our results.

The results obtained indicated that the evolution of the orbits is dominated by the effect of the close approaches to Jupiter. This conclusion was reached by comparing statistical results of two calculations of the evolution of a group of cometary orbits, taking into account perturbations by Jupiter on the comet out to mutual distances

$\Delta = 1.0$ and 2.0 AU, respectively. Even if the perturbations by Jupiter are taken into account at greater distances, we think that the close approaches will still determine the evolution of the short-period comets. Such approaches are fairly numerous, e.g., for the comets having $a_0 = 5.0$ AU and existing in bound orbits during the integration period of 1000 yr an average of 1.2 approaches to Jupiter closer than 1.0 AU occurred per 100 yr. On the other hand, for comets with large inclinations and thereby smaller probability of passing close to Jupiter, the perturbations by Jupiter at larger distances may be of importance.

References

Havnes, O.: 1970, *Icarus* **12**, 331.
Kazimirchak-Polonskaya, E. I.: 1967, *Astron. Zh.* **44**, 349.
Shtejns, K. A.: 1959, *Astron. Zh.* **36**, 512.

Discussion

B. G. Marsden: What method did you use for calculating the perturbations?

O. Havnes: In the outer parts I used Cowell's method in jovicentric coordinates, while if the distance became less than 0.2 AU I used the method of variation of the hyperbolic jovicentric elements.

65. A NEW ORBITAL CLASSIFICATION FOR PERIODIC COMETS

M. BIELICKI

Astronomical Observatory, Warsaw University, Warsaw, Poland

Abstract. A new classification for short-period comets is proposed – one based on minimum approach distances to the respective planets, rather than on aphelion distances.

Perturbations on the motion of a comet are of two general types:

(1) Slow variations during which time the comet does not pass close to a planet; and (2) the influences of close approaches to a planet.

The perturbations of the second type cause the cometary orbit to be drastically changed, and this can be considered the main factor in the evolution of the orbit over an interval of time comparable to the comet's lifetime. On the other hand, the present classification of periodic comets into planetary families according to their aphelion distances is quite artificial. Because of the inclinations of their orbits many cannot approach their supposed parent planets.

We therefore propose here a new definition, in which a comet is assigned to a particular family when the minimum distance between the orbits of the comet and corresponding planet is less than a certain value. The perturbations on a comet decrease very rapidly with increasing distance from a perturbing planet, the principal perturbing action during the close approach being of fairly short duration. If the above condition is fulfilled, we can consider that there exists a region, scarcely changing with time, where the cometary orbit is subject to large changes. The comet may repeatedly approach the planet in this region provided that

(1) The orbit of the comet remains elliptical after such an approach; and (2) no new region of approach is formed between the orbits of the comet and the same or some other planet.

In a particular interval of time a comet may develop one or more approach regions with one or more planets, or the comet may have no approach region. For the limiting distance for each planet we have adopted the radius of the sphere on which the perturbing acceleration due to the Sun is equal to the central acceleration in planetocentric motion (see Table I).

Considering the giant planets only the number of short-period comets in the various families now become: planetless, 21; Jupiter, 72; Saturn, 4; Saturn-Uranus, 1;

TABLE I

Limiting distances for planetary families (in AU)

Mercury	0.002	Mars	0.010	Uranus	0.67
Venus	0.010	Jupiter	0.50	Neptune	1.11
Earth	0.014	Saturn	0.62	Pluto	0.55

Uranus, 1. The Saturn family consists of comets Wild, Peters, Herschel-Rigollet and Grigg-Mellish. The Saturn-Uranus family consists of P/Tempel-Tuttle, while the Uranus family contains P/de Vico only. The Neptune family has no members. Many comets of the Jupiter family, such as d'Arrest, Kopff, Wolf, Borrelly, and Neujmin 2, were classified in the Jupiter family before; but some comets were classified in other families, as shown in Table II. It is interesting to note that the famous comets Encke and Halley both now belong to a planetless family.

TABLE II

Comet families according to the old and the new definitions

Comet	Old	New	Comet	Old	New
Encke	J	O	Olbers	N	O
Tempel 2	J	O	Herschel-Rigollet	>P	S
Kulin	J	O	Pons-Brooks	N	O
Giacobini	J	O	Brorsen-Metcalf	N	J
Arend	J	O	Westphal	N	J
Gale	S	J	Oterma	S	J
Neujmin 1	S-U	O	Lexell	>P	J
Stephan-Oterma	U	O	Halley	N	O

J = Jupiter, S = Saturn, U = Uranus, N = Neptune, P = Pluto, and O = planetless.

TABLE III

Effect of orbital changes arising from large perturbations

Comet	Time interval	Old	New
Schwassmann-Wachmann 2	1660–1730	S	J
	1730–2060	J	J
Whipple	1660–1775	S	J
	1775–2060	J	J
Neujmin 3	1660–1850	S	J
	1850–2060	J	2J
Gale	1660–1917	S	J + S
	1917–2060	S	J
Oterma	1660–1937	S	J
	1937–1963	J	J
	1963–2060	S	J
Brooks 2	1660–1886	S	J
	1886	U	J
	1886–2060	J	J
Lexell	1660–1767	J	2J
	1767–1779	J	J
	1779–2060	>P	J
Kearns-Kwee	<1855	hyperbola	J
	1855–1961	N	J
	1961–2060	J	J

As a result of orbital evolution a comet can change from one family to another, but this was more frequent with the older classification (see Table III).

Acknowledgments

I should like to thank G. Sitarski for the use of his computer programme for calculating minimum distances and K. Ziolkowski for his assistance.

66. THE MAJOR PLANETS AS POWERFUL TRANSFORMERS OF COMETARY ORBITS

E. I. KAZIMIRCHAK-POLONSKAYA

Institute for Theoretical Astronomy, Leningrad, U.S.S.R.

Abstract. The equations of motion of many of the comets of the Jupiter and Saturn families have been integrated for a period of 400 yr. Several examples illustrate the role of the major planets, especially Jupiter, in the evolution of these orbits: in such cases comets may be captured by Jupiter from Saturn's family or ejected from Jupiter's family into Saturn's, and so on; the possible escape of a comet along a hyperbolic orbit and the capture by Jupiter of a comet from a hyperbolic orbit are also demonstrated. Fictitious comets of the Uranus and Neptune families are also studied. It is concluded that a combination of diffusion theory and capture theory is in good agreement with the observations and makes it possible to consider both long- and short-period comets as part of a unified cometary complex.

1. Introduction

The research presented here is concerned with the evolution of cometary orbits inside the planetary system for a period of several centuries and with the determination of the roles of the giant planets (Jupiter, Saturn, Uranus, and Neptune) as powerful transformers in that evolution.

On account of the great amount of information available we are obliged to restrict ourselves here to discussing only a few important problems. They may be formulated as follows:

(1) Determination of the most interesting features of the great transformations and of the general evolution of cometary orbits.

(2) Elucidation of the influence of close approaches to Jupiter on the subsequent discovery of comets.

(3) Illustration of the agreement of deduction from classical capture theory with the results of our studies.

(4) Demonstration of the absence of contradictions between the observations and deductions from the capture theory.

(5) Presentation of the main stages in the evolution of cometary orbits by combining diffusion theory with capture theory.

First of all, to avoid possible misunderstandings that sometimes arise in discussions of our work (Vsekhsvyatskij, 1969, pp. 278–280; Sitarski, 1968), we shall elucidate the difference between 'numerical theories of cometary motion' and 'investigations of the evolution of cometary orbits'. The difference is very important as it concerns both the object and method of investigation.

In creating a theory of the motion of a comet the principal aim is to obtain the best conceivable agreement with observations over large intervals of time. Therefore, when the theory is being developed, all perceptible planetary perturbations and non-gravitational effects should be taken into account with high precision. The agreement has to be to seconds and even fractions of seconds of arc.

Chebotarev et al. (eds.), The Motion, Evolution of Orbits, and Origin of Comets, 373–397. All Rights Reserved.

It is different when we are studying the long-term evolution of cometary orbits: here the investigator is interested in the general process of orbital transformation and in the main factors affecting this process. It is sufficient to work to a precision of decimals of degree and even whole degrees.

When developing the theory of the motion of a comet the investigator is interested only in the period covered by the observations, in the course of which it is very rare for a comet to pass through Jupiter's sphere of action while the effects of passages through the spheres of action of the other planets cannot be detected at all. Very close approaches to Jupiter cause great difficulties, and when they do occur we must take special pains to avoid a gap in the theory during the time of close approach.

When studying the evolution of cometary orbits, however, the particular interest lies precisely in these approaches to the major planets, for one of the main purposes of these investigations is the establishment of the roles played by the major planets in the evolution of cometary orbits, and this role can be detected most obviously when a comet passes through the spheres of action of the planets. It thus becomes inevitable to extend the interval of time over which the study is made to a few centuries or even longer.

We have studied the evolution of the orbits of the short-period comets of different planetary families for an interval of 400 yr (1660–2060) and have ascertained that all the comets of Jupiter's family experience close or moderate approaches to Jupiter; in some cases there have been up to ten approaches during the 400-yr interval. The comets of Saturn's family approach both Jupiter and Saturn.

We did not discover any close approaches to Uranus and Neptune for the comets belonging to the families of these planets, and the period of 400 yr is too short for fully investigating whether such approaches really occur. Therefore, alongside the real comets we have introduced fictitious comets of the Uranus and Neptune families and have examined the transformations of their orbits in the spheres of action of those planets.

In studying the problem of the evolution of cometary orbits, we may apply either statistical methods – investigating fictitious objects on a large scale; or numerical methods – integrating the actual equations of motion of real and fictitious objects.

2. Real Comets in the Spheres of Action of Jupiter and Saturn

We are interested in modelling the true motions of comets as closely as possible. We have therefore conducted this investigation by our new method (Kazimirchak-Polonskaya, 1972), using our complex of double-precision programmes that minimize the error both in the initial data and in the integration. A heliocentric method was used throughout, and we took into account all perceptible planetary perturbations, and in the case of P/Wolf nongravitational effects too.

Starting from the very precise, continuous theory of the motion of P/Wolf for the whole period of observations, 1884–1967 (Kamieński, 1959; Kazimirchak-Polonskaya, 1972), we have integrated the motion of this comet back from 1884 to 1661. Initially, we allowed for the perturbations by Venus to Pluto and the nongravitational effects.

The nongravitational coefficients were varied according to the perihelion distance (Kamieński, 1961), which changed after each approach to Jupiter. We then repeated the calculation excluding the nongravitational effects. The resulting elements for an osculation epoch in 1661 are compared in Table I. It may be noted that the action of

TABLE I

Comparison of two integrations of P/Wolf over the interval 1884–1661, with and without nongravitational effects

	With nongravitational effects	Without nongravitational effects	Difference with-without
M	$0°54$	$359°99$	$+0°55$
ω	158.12	158.15	-0.03
Ω	218.62	218.48	$+0.14$
i	25.30	25.35	-0.05
e	0.4549	0.4538	$+0.0011$
a (AU)	4.000	4.001	-0.001
P (yr)	7.999	8.004	-0.005
q (AU)	2.180	2.185	-0.005

The elements are referred to the mean equinox 1950.0 and correspond to the epoch 1661 Aug. 20.5 Greenwich Mean Time.

the nongravitational forces does not affect the general course of the evolution of P/Wolf, even though in the 223-yr interval the comet made two close and two moderate approaches to Jupiter (see Table II).

TABLE II

Approaches of P/Wolf to Jupiter ($\Delta_{min} < 1.000$ AU)

T_{min}	Δ_{min} (AU)	T_{min}	Δ_{min} (AU)
1757 Jan. 15.4	0.465	1922 Sept. 27.1	0.125
1816 Jan. 4.0	0.898	2005 Aug. 14.2	0.542
1839 Nov. 30.3	0.250	2041 Mar. 10.1	0.600
1875 June 8.5	0.116		

In 1965–1966 N. A. Belyaev and the author completed the investigation of the orbital evolution of 45 short-period comets over 1660–2060. Belyaev studied the orbital evolution of ten comets of the Jupiter family, while the author worked on 35 comets belonging to the families of Jupiter, Saturn, Uranus, and Neptune. Up to the present time we have published the results for 22 typical comets, for which the best initial elements were available (Belyaev, 1966, 1967; Kazimirchak-Polonskaya, 1967a, 1967b, 1968). During the 400-yr interval those 22 comets made 156 approaches to Jupiter, 35 to Saturn, and 2 to Uranus; among them there were more than 40 passages through Jupiter's sphere of action and one passage through Saturn's.

Let us consider the main features of the orbital evolution of the most interesting short-period comets and the influence of Jupiter's perturbations on the discovery of these comets.

Figure 1 shows the orbits of the Earth, Jupiter, and five different averaged orbits of P/Wolf. We may note that this comet

(1) consistently remains in the Jupiter family, that planet securely maintaining the comet's aphelion (A) near its own orbit;

(2) revolves around the Sun along a 'pulsating ellipse' varying with an irregular periodicity depending on the character of the approaches to Jupiter; this feature was discovered by Kamieński (1959).

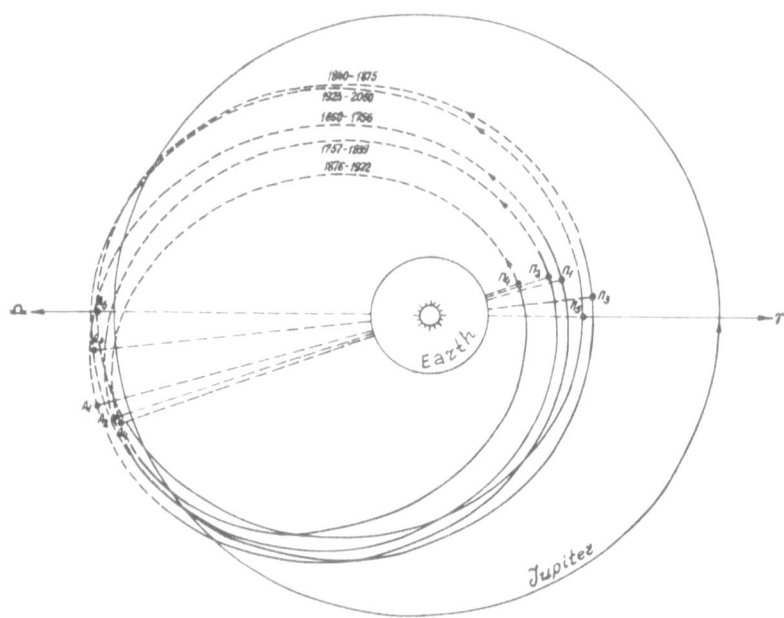

Fig. 1.   The orbit of P/Wolf.

The figure demonstrates graphically that for more than 200 yr the comet was inaccessible to observation on account of the large perihelion distance q. After passing through Jupiter's sphere of action in 1875, q was considerably reduced (from 2.4 to 1.6 AU), and consequently it became possible for the comet to pass near the Earth. In fact, on 1884 October 1, a month and a half before perihelion, the comet approached the Earth for the first time to a distance of 0.803 AU. The increased apparent brightness enabled the comet to be discovered (on 1884 September 17 in Heidelberg) as an object of the ninth magnitude.

An interesting, but different, type of orbital evolution is represented by P/Wolf-Harrington, which also remains a member of the Jupiter family for the 400-yr interval.

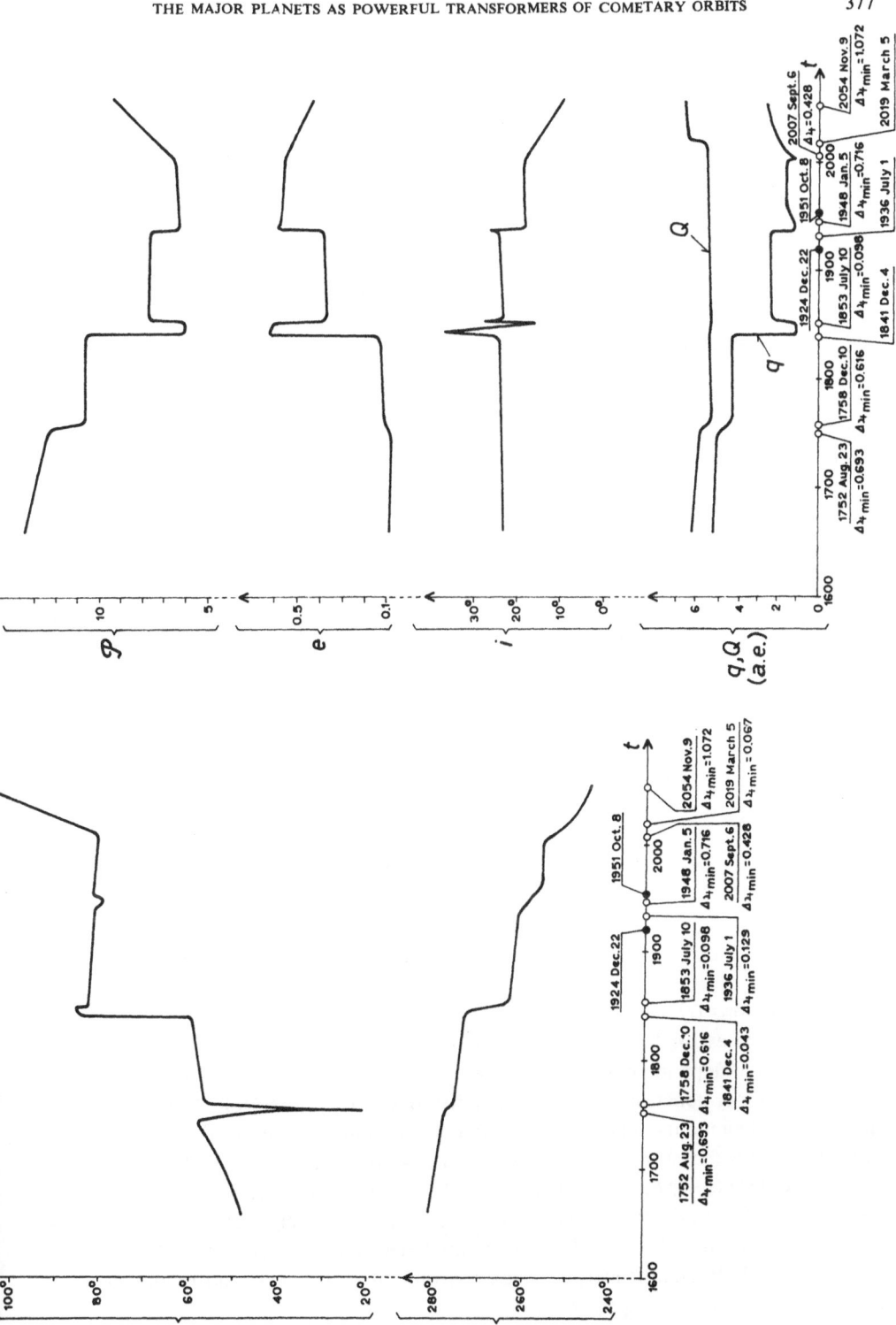

Fig. 2. Evolution of the orbital elements of P/Wolf-Harrington.

It makes nine approaches to Jupiter, four of them very deeply into its sphere of action (see Table III).

TABLE III

Approaches of P/Wolf-Harrington
to Jupiter

T_{min}	Δ_{min} (AU)
1841 Dec.	0.043
1853 July	0.098
1936 July	0.129
2019 March	0.067

Figure 2 demonstrates the changes in the orbital elements of this comet during 1660–2060, while Figure 3 illustrates the actual orbit over 1760–2000, taking into account

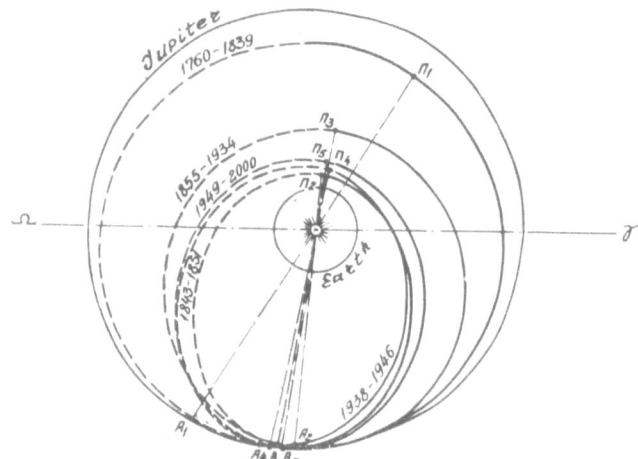

Fig. 3.   The orbit of P/Wolf-Harrington.

the perturbations by Venus to Uranus and starting from the elements by Wiśniewski (1960). The dates on the abscissa of Figure 2 show the nine approaches to Jupiter and the two independent discoveries of the comet (in 1924 and 1951). The more deeply the comet penetrates Jupiter's activity sphere, the more violent are the changes in the orbital elements. The 1841 encounter with Jupiter transformed the orbit from almost a circle (eccentricity $e = 0.10$ and period of revolution $P = 10.7$ yr) into an eccentric ellipse ($e = 0.60$, $P = 6.0$ yr) while the encounter in 1853 decreased the eccentricity again ($e = 0.37$, $P = 7.6$ yr). The comet was extremely faint in 1924 and not observed well enough for its orbit to be reliably determined; fortunately, the approach to Jupiter in 1936 significantly reduced q (from 2.4 to 1.6 AU) and made accidental rediscovery of the comet possible in 1951.

We shall discuss now another type of comet, one which in the past belonged to the Saturn family and was later captured by Jupiter into its family.

Figure 4 illustrates a series of orbits of P/Whipple, these being derived on the basis of Dinwoodie's (1960) elements, taking into account the perturbations by Venus to Neptune. From 1660 to 1770 this comet travelled in an orbit far removed from the

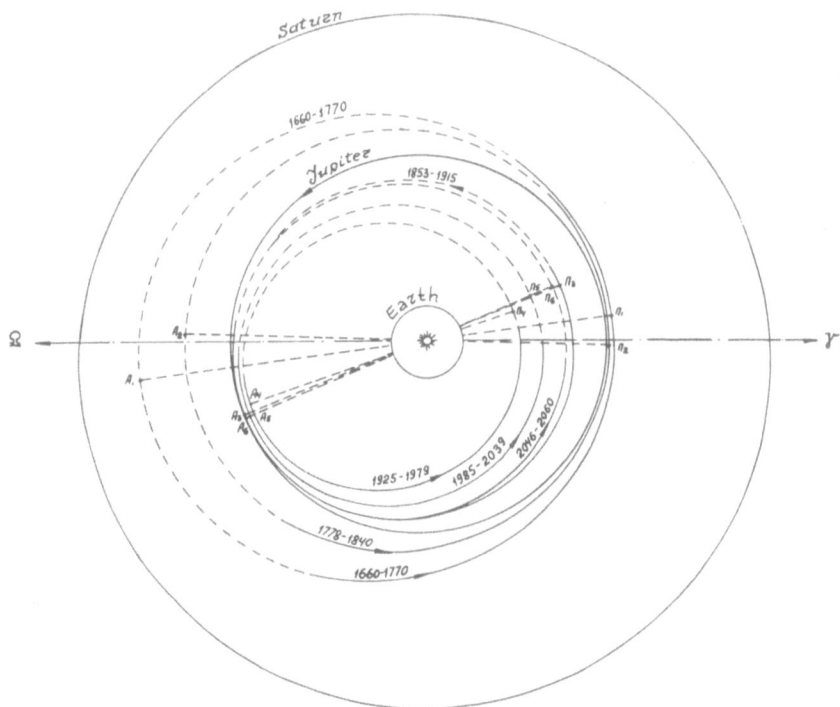

Fig. 4.   The orbit of P/Whipple.

Sun, and since its aphelion A_1 was located nearer the orbit of Saturn than that of Jupiter it was a member of the Saturn family. Its perihelion Π_1 was in the vicinity of Jupiter's orbit ($q = 5.2$ AU), allowing the possibility of close approaches to Jupiter. From 1770 to 1778 such an approach occurred (lasting 8 years on account of the almost circular form of the cometary orbit), and afterwards the comet moved on a rather different orbit. The second approach to Jupiter was even more characteristic, with planet and comet remaining within a mutual distance of less than 1 AU from 1840 to 1853. The comet's subsequent orbit was situated completely inside that of Jupiter. That planet had securely captured P/Whipple from the Saturn family. From 1915 to 1925 the comet was near Jupiter again, passing deeply into the sphere of action in 1922, and the resulting perturbations caused the perihelion distance to be decreased quite substantially, rendering discovery of the comet possible in 1933. Further encounters with Jupiter will take place during 1979–1985 and 2039–2046.

Capture from the Saturn family into the Jupiter family also occurred in the cases of P/Comas Solá (Belyaev, 1967) and P/Shajn-Schaldach. P/Schwassmann-Wachmann 1 was also captured in this manner, about the year 1730.

An exceptionally interesting type is represented by P/Oterma. We have investigated the evolution of this orbit, starting from the elements by Herget and Marsden (1961), taking into account the perturbations by Venus to Uranus. During the 400-yr interval there are four approaches to Jupiter (two of them very deeply into the sphere of action) and one to Saturn. The effects of the two closest encounters are of overriding importance, however, and Figure 5 shows only three averaged orbits for the comet. From 1660 to 1937 P/Oterma had an orbit with perihelion Π_1 slightly outside that of Jupiter

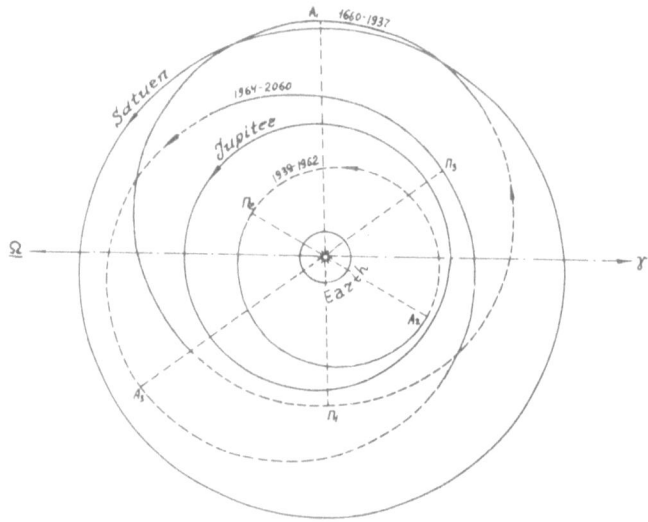

Fig. 5. The orbit of P/Oterma.

and aphelion A_1 beyond the orbit of Saturn; the revolution period exceeded 20 yr. At the end of 1936 the comet entered Jupiter's activity sphere and remained there for more than 2 yr, penetrating to $\Delta_{min} = 0.165$ AU. When the comet left the sphere its orbit was completely different: it was entirely between the orbits of Mars and Jupiter, with its aphelion A_2 on the inner side of Jupiter's orbit, and its perihelion distance down to 3.4 AU – resulting in the comet's discovery in 1943 already as a member of the Jupiter family. After three revolutions in its new 8-yr orbit it made an even closer approach to Jupiter in 1963 ($\Delta_{min} = 0.095$ AU) and was subsequently re-ejected into the Saturn family with its perihelion distance again up to 5.4 AU. Moderate approaches to Saturn (in 2008–2016) and to Jupiter (in 2023–2026) will cause further increases in the perihelion distance. During the complete 400-yr interval Jupiter causes the comet's line of apsides to advance 548° – more than one and a half revolutions.

Turning our attention now to the comets that remain in the Saturn family, we discuss briefly the orbits of P/Neujmin 3 and P/Gale. The former comet has nine approaches to Jupiter (that in 1850 well into its sphere of action) and six to Saturn, four of which have quite appreciable effects. On this comet the perturbations by Saturn are of the same order as those by Jupiter (e.g. several tens of degrees in ω and Ω). P/Gale is noteworthy in that during the 400 yr it experiences the greatest number of approaches: nine to Jupiter and eight to Saturn. In 1798 this comet passed through Saturn's sphere of activity (to $\Delta_{\min} = 0.17$ AU), although the actual circumstances of the approach did not produce any great changes in the orbital elements.

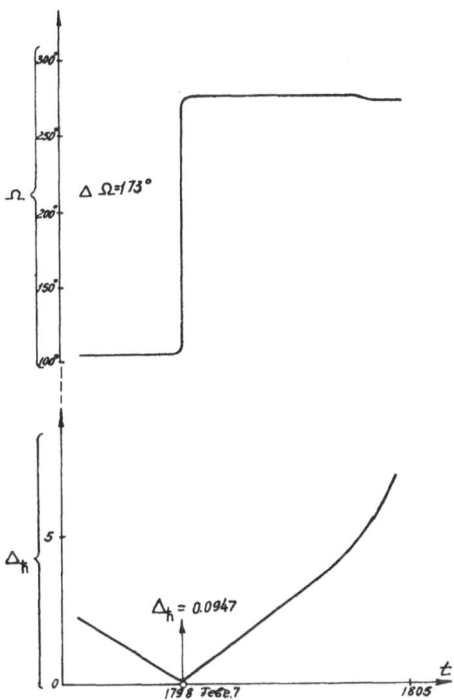

Fig. 6.   Perturbations in the longitude of the ascending node of a variant orbit of P/Gale in Saturn's sphere of action.

Since we were interested in the use of comets for revealing the quite unstudied perturbing effects of Saturn within its sphere of activity we varied the conditions of this approach of P/Gale and established that the perturbations by Saturn could be as large as those by Jupiter. Figure 6 illustrates the perturbations in the longitude of the ascending node of this fictitious orbit.

In the problem of the evolution of cometary orbits P/Neujmin 1 deserves particular attention. Although its aphelion is well beyond the orbit of Saturn, this comet belongs to the Saturn family. Our investigations have shown that between 1660 and 2060 there are no approaches to Jupiter at all, although there are six approaches to Saturn, those in 1892 and 1980 to minimum distances of 0.80 and 0.84 AU, respectively. The evolution of the orbit of this comet is thus determined to a considerable extent by the influence of Saturn.

Of much greater interest is P/Brooks 2, which in the past possessed an orbit resembling the present orbit of P/Neujmin 1. Proceeding from the theory of motion of this comet developed by Dubyago (1950, 1956) for the interval 1883–1960, we extended the calculations to the full 400-yr interval, taking into account the perturbations by Venus to Neptune but without considering the nongravitational effects on this comet. Figure 7 very strikingly illustrates Jupiter's capture of P/Brooks 2 from a situation between the Saturn and Uranus families. According to Dubyago this comet approached

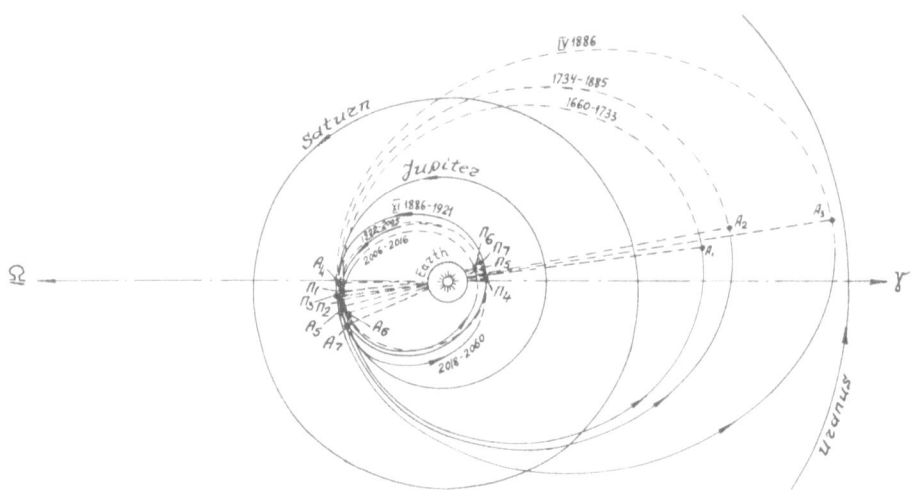

Fig. 7.   The orbit of P/Brooks 2.

Jupiter to within 0.00096 AU in 1886, and a really catastrophic transformation of the cometary orbit occurred: the ascending and descending nodes changed places, and the perihelion of the former orbit became the aphelion of the new; the perihelion distance decreased from 5.45 to 1.95 AU, and it was on account of this that the comet could be discovered in 1889. The revolution period, initially more than 40 yr, decreased to 7 yr. A further close approach to Jupiter occurred in 1922, and there will also be approaches in 2005–2006 and in 2017, but the comet nevertheless will remain in the Jupiter family at least until 2060.

The orbital evolution of P/Lexell is of exceptional interest. First discussed by Lexell (1777, 1778, 1781), the motion of this comet was subsequently treated in great detail – but with allowance only for the perturbations by Jupiter – by Leverrier (1844, 1857). Starting from Leverrier's (1844) elements for the 1770 apparition we have studied the evolution of the orbit of P/Lexell over 1660–2060, considering the perturbations by Venus to Pluto; see Figure 8. Until 1766 P/Lexell had a nearly circular orbit with $q=3$ AU, $P=10$ yr and was inaccessible to observation. Its close encounter with

Jupiter in 1767 ($\Delta_{min} = 0.020$ AU) transformed the orbit into an elongated ellipse with $q = 0.67$ AU, $P = 5.6$ yr, after which there was an exceptionally close approach to the Earth in 1770 (to within 0.015 AU), at which time the comet was discovered. After nine years it encountered Jupiter for the second time, on this occasion passing within 0.0015 AU of that planet. The minimum distance was rather similar to that of P/Brooks 2 in 1886, although the actual situations were appreciably different: P/Brooks 2 was near perihelion and approaching the Sun, while P/Lexell was near aphelion and receding; P/Brooks 2 was captured into Jupiter's family, while P/Lexell was ejected

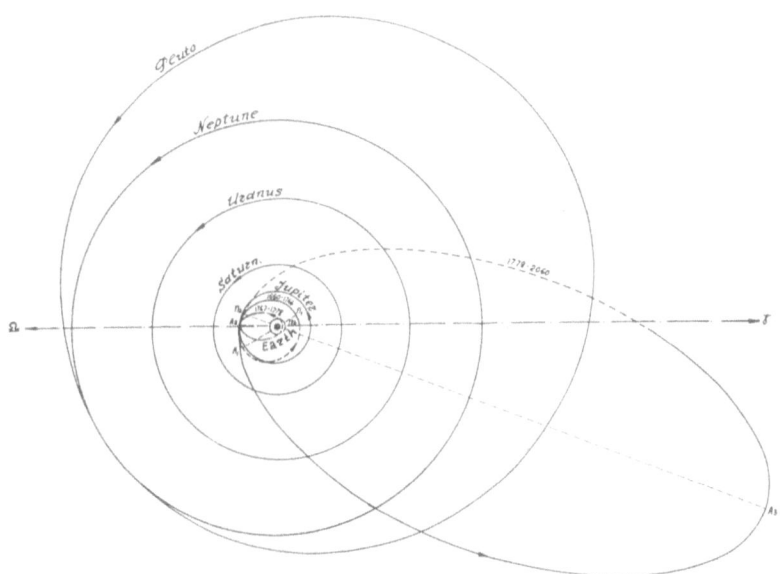

Fig. 8.   The orbit of P/Lexell.

from Jupiter's family into a transplutonian orbit with a revolution period of about 260 yr. The perihelion of this new orbit was close to Jupiter's orbit, not far from the aphelion of the former orbit, so P/Lexell again became unobservable.

We have compared our results on P/Lexell with those of Leverrier, Schulhof, and Callandreau. Following Leverrier's (1857) example, we varied the initial system of elements and derived several different possible evolutions of the orbit. As before, we confirmed Leverrier's results closely, finding that after the encounter with Jupiter in 1779 some of the solutions gave even more eccentric orbits, a few of them even being hyperbolic. Figure 9 shows one such ellipse and one hyperbola. All the solutions have their perihelia near Jupiter's orbit, confirming that the comet would definitely have been unobservable afterwards. Similar results were also obtained by Callandreau (1892, p. 49).

In connection with the case of P/Lexell the question naturally arises as to whether short-period comets exist that were captured in the past from interstellar space by Jupiter or some other major planet. If there is no real comet, is it possible to invent a fictitious comet with orbital elements similar to those of a real comet, and which would reveal the mechanism of such a capture?

The discovery of P/Kearns-Kwee in 1963 provided the answer to this interesting question. We investigated the orbital evolution of this comet in 1965, starting from the improved system of elements by Marsden (1964), based on observations over an interval of six months. The equations of motion were integrated forward to 2060 and back

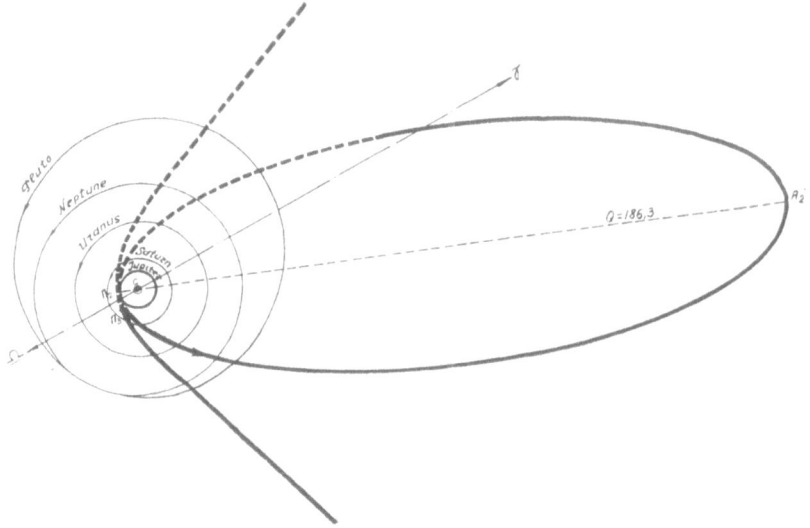

Fig. 9.   The possible escape of P/Lexell on a hyperbolic orbit.

to 1855, with allowance for the perturbations by Venus to Pluto. The results proved to be exceptionally interesting, showing for the first time the capture by Jupiter of a comet from a hyperbolic orbit. The original comet, moving in a hyperbola, passed within 0.042 AU of Jupiter in 1855 in the vicinity of its perihelion. The great perturbations by Jupiter transformed that orbit into an elongated ellipse whose aphelion (see Figure 10) was initially near the orbit of Neptune. Later, the aphelion A_3 moved to the boundary of the Neptune and Uranus families. In 1961 P/Kearns-Kwee approached Jupiter to within 0.032 AU, and that planet captured the comet into its own family, reducing q from 4.3 to 2.2 AU and enabling the comet to be discovered in 1963. There was simultaneously a rotation of the line of apsides by about 90°, and the revolution period decreased from 54 yr to 9 yr. With the help of later observations Marsden and Aksnes (1967) determined a more precise orbit and confirmed the enormous transformation of the comet's orbit in 1961. However, neither they nor Sitarski (1968) agreed with the capture of the comet from a hyperbolic orbit in 1855. This is naturally explained by differences in the initial elements for 1963 and by the increased discrepancies among the elements (especially the mean anomaly and mean motion) before the approach to Jupiter in 1961. However, no matter what the conclusions of other researchers may be, the nature of the problem is not changed, and the results obtained are still of significance for the problem of the evolution of cometary orbits. It is important that such a remarkable real object has been observed, and although the

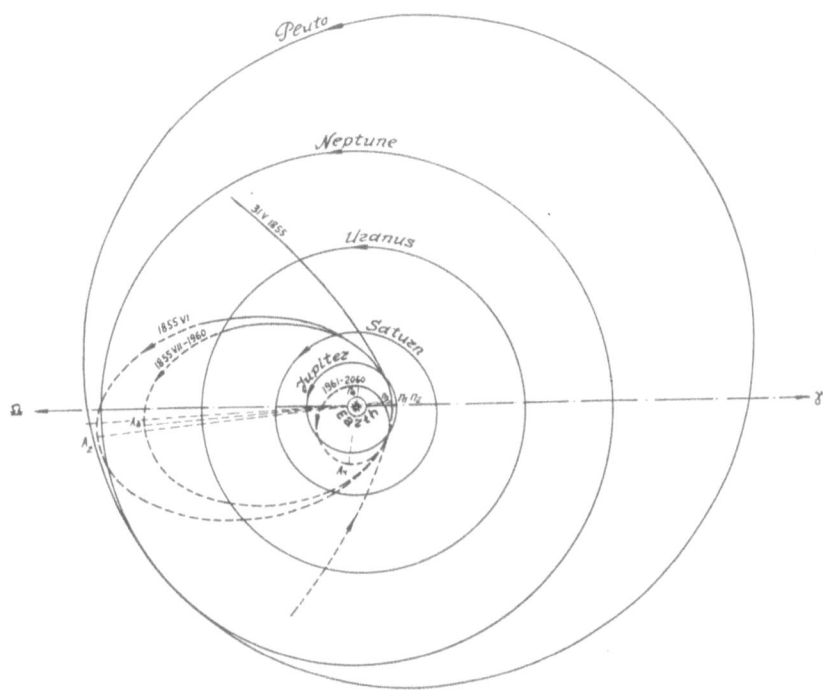

Fig. 10.   The orbit of P/Kearns-Kwee and the possible capture of this comet from a hyperbolic orbit.

starting orbit was not a definitive one we have been able to demonstrate the possibility, in principle, of Jupiter's capturing a comet from a hyperbolic orbit whose perihelion is near Jupiter's orbit. Our calculations have also shown a possible mechanism for this capture, first into an ellipse with its aphelion on the outskirts of the planetary system, and then into the Jupiter family.

3. Fictitious Comets in the Spheres of Action of Uranus and Neptune

We shall now discuss fictitious comets of the Uranus and Neptune families and possible transformations of their orbits in the spheres of action of those planets. Although we have studied many suitable fictitious objects we shall restrict ourselves here to only five: U-1 and U-2 in the Uranus family and N-1, N-2, and N-3 in the Neptune family.

Figure 11 illustrates the evolution of U-1 during the 10-yr interval spent by that

comet in the sphere of action of Uranus (1713–1723). The comet passed very deeply
into the sphere (the radius of which is 0.346 AU) and in 1719 attained its minimum
distance $\Delta_{min}=0.0104$ AU. The ascending node first regressed slowly, and then, in the
vicinity of Uranus, it changed its direction of motion and rapidly advanced 167°,
after which slow regression returned. The line of apsides turned through an angle of

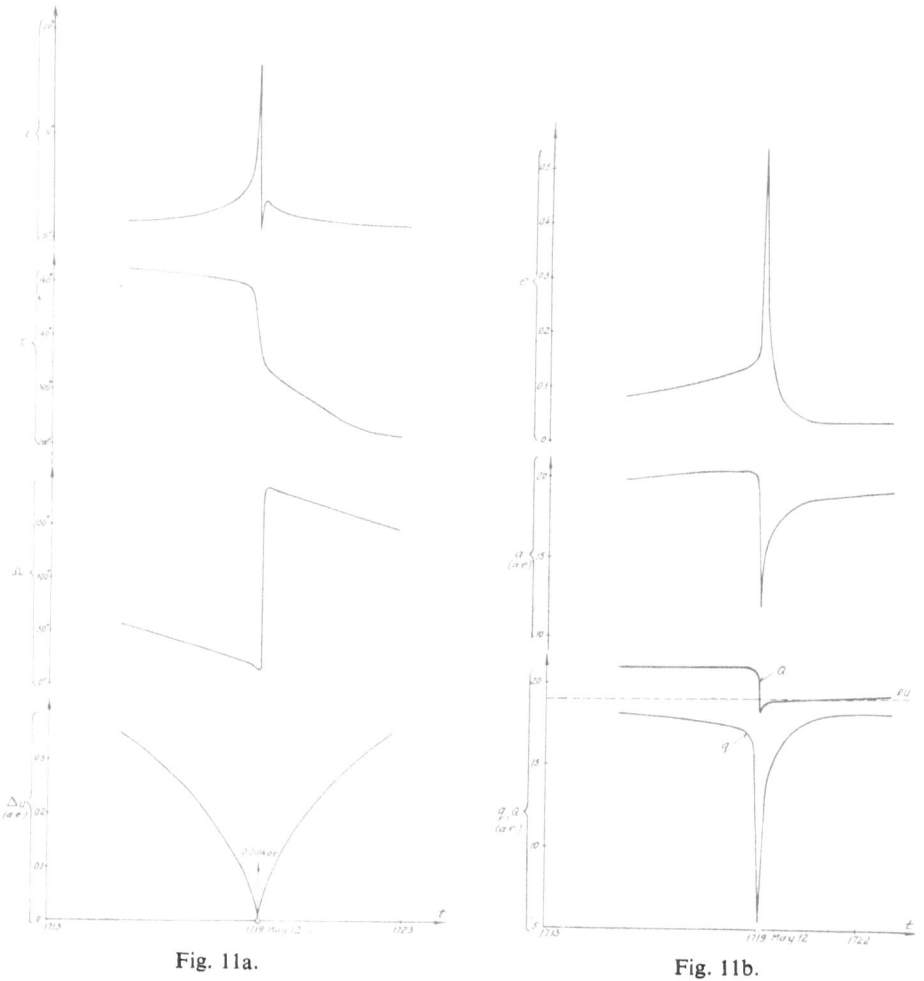

Fig. 11a.   Fig. 11b.

Fig. 11a–b.   Changes in the orbital elements of comet U-1 in Uranus' sphere of action.

320°; the orbital inclination initially increased very slowly from 1°.5 to 2°.5, then rose
rapidly to 16°.3, sharply decreased down to 0°.65, rose to 3°.6 and then slowly began to
decrease down to 1°.2, close to its initial value. On entering the sphere of activity the
comet's orbit was nearly circular (semimajor axis $a=19.2$ AU and eccentricity
$e=0.0614$) with its perihelion and aphelion just inside and just outside the orbit of

Uranus. As the comet penetrated Uranus' activity sphere its orbit changed violently into an elongated ellipse with very small perihelion distance and decreasing semimajor axis ($a_{min} = 11.9$ AU); the aphelion moved inward, crossing the orbit of Uranus. As the comet receded from Uranus its orbit changed more or less back to its initial form.

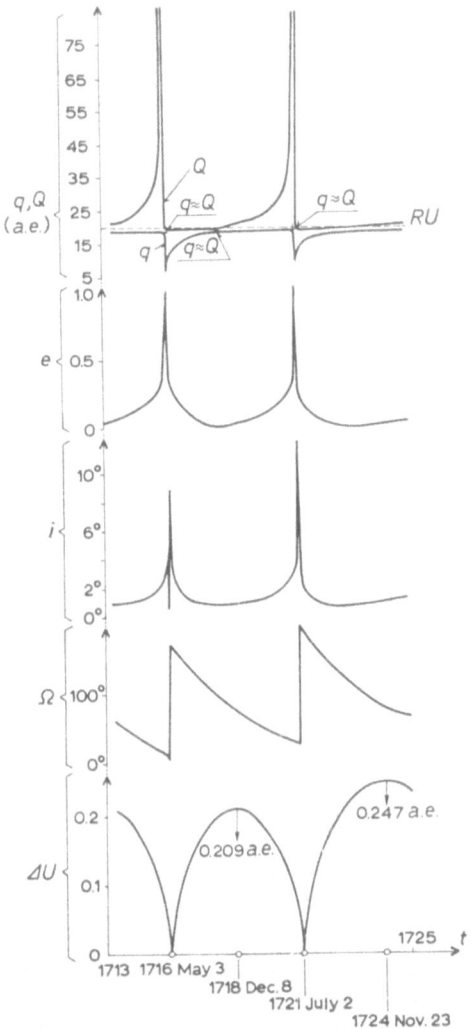

Fig. 12.   Changes in the orbital elements of comet U-2.

The orbit of comet U-2 evolves in quite a different manner during its encounter with Uranus (1713–1725) (see Figure 12). After reaching its minimum distance from Uranus in 1716 ($\Delta_{min} = 0.0031$ AU) the comet receded, but it did not leave the sphere of action. Its distance increased to a maximum ($\Delta_{max} = 0.209$ AU), after which there was a second minimum ($\Delta_{min} = 0.0015$ AU) and another maximum ($\Delta_{max} = 0.247$ AU).

During this time the cometary orbit changed twice from a low-eccentricity ellipse into a nearly parabolic orbit and back: on the first approach the eccentricity increased beyond 0.999, while on the second the orbit became hyperbolic. The variations in perihelion and aphelion distances (q, Q) are quite typical. When U-2 entered Uranus' sphere of action its perihelion was just inside the orbit of Uranus ($q=18.3$ AU). At the time of the first minimum approach q decreased suddenly to 7.30 AU; afterwards, the aphelion distance became approximately equal to the original q, and the perihelion distance slowly recovered to this value. Approximately the same process took place during the second minimum approach to Uranus.

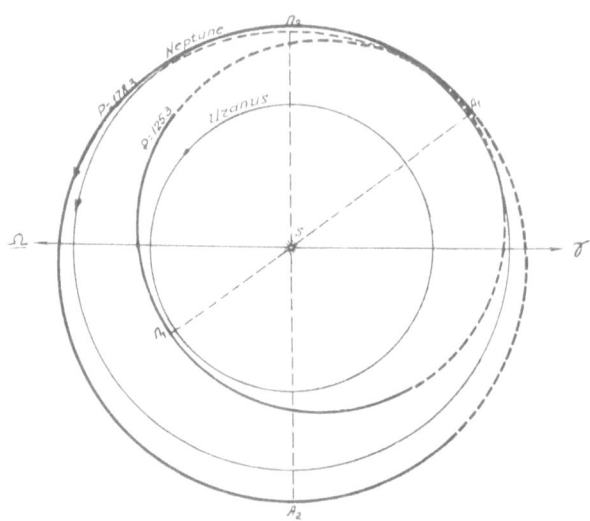

Fig. 13.   The orbit of comet N-1.

Comet N-1 initially had an elliptical orbit with $a=25.04$ AU, $e=0.199$, its perihelion Π_1 near the orbit of Uranus ($q=20.1$ AU) and aphelion A_1 near that of Neptune ($Q=30.0$ AU). It remained within Neptune's sphere of action (radius 0.580 AU) for about 13 yr (1708–1721), attaining $\Delta_{min}=0.140$ AU. Figure 13 shows that q increased by 9 AU, with the perihelion moving out to Neptune's orbit, and there was a slight increase in Q also. The line of apsides rotated through 125°7. On the other hand, the longitude of the ascending node (131°2) and inclination (1°77) were scarcely affected by the encounter.

Figures 14 and 15 illustrate the transformation of the orbit of comet N-2. Its close approach to Neptune in 1715 is unusual and interesting in that during an interval of little over a month there were three minimum approaches to Neptune ($\Delta_{min}=0.0012$ AU, followed by $\Delta_{max}=0.0063$ AU, then $\Delta_{min}=0.00093$ AU, $\Delta_{max}=0.0103$ AU, and finally $\Delta_{min}=0.0028$ AU). During each encounter the orbit of N-2 rapidly became retrograde and nearly parabolic and then direct and nearly parabolic. Figure 14 shows

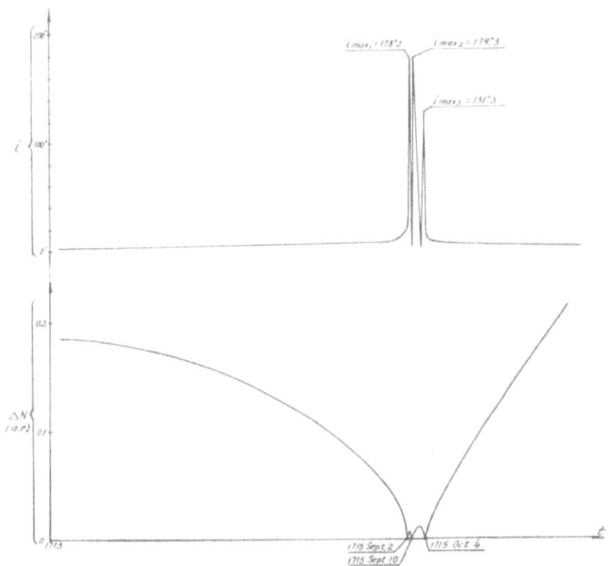

Fig. 14.   The three approaches of comet N-2 to Neptune and the changes in orbital inclination.

that the inclination had three distinct maxima ($i_{max_1} = 178°2$, $i_{max_2} = 179°3$, $i_{max_3} = 131°3$), corresponding to the three minimum approaches to Neptune. Figure 15 shows that after the comet's escape from Neptune's sphere of action the perihelion, initially

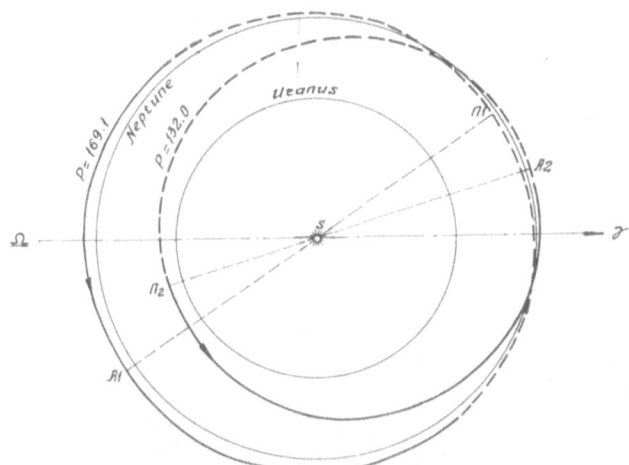

Fig. 15.   The orbit of comet N-2.

near the orbit of Neptune (Π_1), was moved to a point near the orbit of Uranus (Π_2). The aphelion remained slightly outside the orbit of Neptune, but the line of apsides advanced through 161°. The eccentricity increased from 0.029 to 0.170. The longitude of the ascending node advanced by 65°.

Comet N-3 occupies a unique place among the fictitious comets of Neptune's family. Figure 16 illustrates the variation in the distance of N-3 from Neptune during

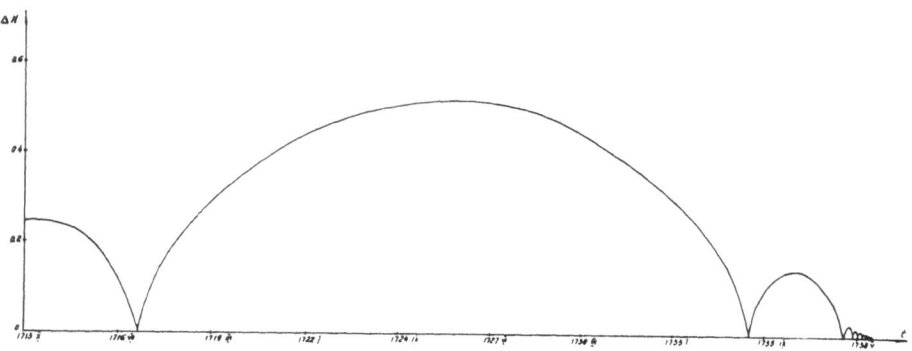

Fig. 16.   The multiple approaches of comet N-3 to Neptune.

1713–1738; Table IV gives Δ_{min} and Δ_{max} and the corresponding dates T_{min} and T_{max} for the same interval. After the first minimum approach the comet travelled almost to

TABLE IV

Minimum and maximum distances of comet N-3 from Neptune

T_{min}	T_{max}	Δ_{min} (AU)	Δ_{max} (AU)
1717 Feb. 8.9		0.00056	
1735 Mar. 14.9	1726 Mar. 27.5	0.00098	0.525
1738 Jan. 14.5	1736 Oct. 4.5	0.00092	0.148
1738 Apr. 25.9	1738 Mar. 8.8	0.00093	0.0303
1738 June 16.6	1738 May 19.2	0.00094	0.0193
1738 July 20.2	1738 July 3.4	0.00093	0.0143
1738 Aug. 12.2	1738 July 30.9	0.00092	0.0108
1738 Aug. 30.3	1738 Aug. 20.2	0.00092	0.0091
1738 Sept. 12.9	1738 Sept. 7.3	0.00092	0.0073
1738 Sept. 24.2	1738 Sept. 18.7	0.00092	0.0064
1738 Oct. 3.4	1738 Sept. 28.9	0.00093	0.0055
1738 Oct. 11.2	1738 Oct. 7.4	0.00092	0.0049
1738 Oct. 18.1	1738 Oct. 14.6	0.00092	0.0044
1738 Oct. 24.2	1738 Oct. 21.2	0.00094	0.0040
1738 Oct. 29.9	1738 Oct. 27.1	0.00090	0.0038
1738 Nov. 3.9	1738 Nov. 1.0	0.00092	0.0034
1738 Nov. 8.6	1738 Nov. 6.6	0.00091	0.0032
1738 Nov. 12.9	1738 Nov. 11.3	0.00099	0.0029
1738 Nov. 17.2	1738 Nov. 15.4	0.00089	0.0029
1738 Nov. 21.1	1738 Nov. 19.7	0.00091	0.0026
1738 Nov. 24.9	1738 Nov. 23.5	0.00088	0.0025

the periphery of Neptune's activity sphere before returning. Successive values of Δ_{min} stabilized quickly near 0.0009 AU, while the values of Δ_{max} decreased more slowly, the comet at the end of the integration run (at which point it seemed impractical to continue the calculation further, the step size having decreased to less than one hour) apparently having become a satellite of Neptune, the eccentricity of its neptunicentric orbit being about 0.5 and its revolution period less than four days.

Figures 17 and 18 illustrate the transformations in the various orbital elements (heliocentric) of N-3 during the first two minimum approaches to Jupiter. As happened with comet N-2, very eccentric retrograde orbits are produced for a short time, the inclination reaching maximum values $i_{max_1} = 174°3$ and $i_{max_2} = 172°6$. As the differences between Δ_{min} and Δ_{max} diminish the heliocentric orbit of N-3 asymptotically approaches that of Neptune itself.

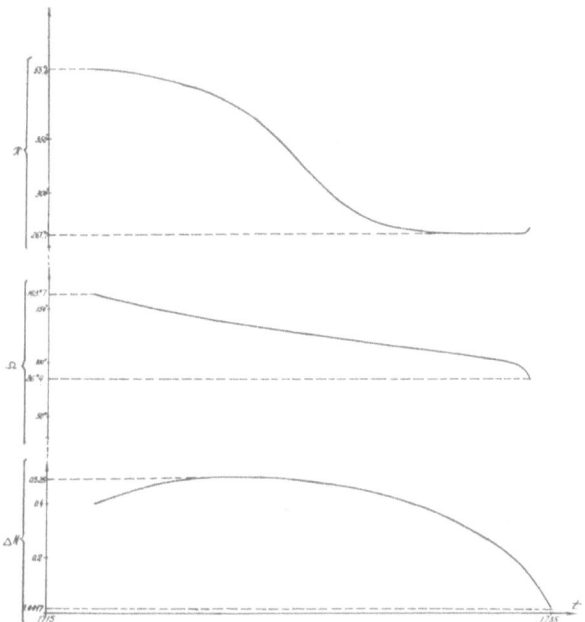

Fig. 17.   Perturbations in the longitudes of the ascending node and perihelion of comet N-3.

At the present stage of our work we have insufficient data for making any con-clusions or generalities about the future of comets of the N-3 type; however, there is reason to believe that Neptune's sphere of action may contain a great number of comets that remain there indefinitely, accompanying Neptune in its motion round the Sun. Similar comets may also exist in the activity spheres of Jupiter, Saturn and Uranus.

4. Discussion and Conclusions

From the data obtained for 45 real and several dozen fictitious short-period comets we can draw some conclusions and give answers to some of the questions raised at the beginning of this paper.

(1) The giant planets, Jupiter to Neptune, having large masses and extensive spheres of action, are powerful transformers of cometary orbits and appreciably affect their evolution; they change the spatial orientation of orbital planes and lines of apsides, vary the shapes and dimensions of the orbits, transfer comets from one planetary family to another and, in exceptional cases, remove them beyond the limits of the planetary system or capture them from hyperbolic orbits.

(2) Perturbations by the giant planets frequently create conditions favourable to the discovery of comets and on other occasions render comets inaccessible to observation.

(3) Scientific hypotheses on the origin of comets date back to Laplace and Lagrange. Laplace suggested that comets were of interstellar origin; Lagrange, on the other hand, supposed that they originated by eruption from the major planets, especially Jupiter. Each hypothesis has its adherents. Vsekhsvyatskij (1930, 1933, 1955, 1967, 1969, 1972) is a prominent supporter of Lagrange's ideas and has extensively developed the

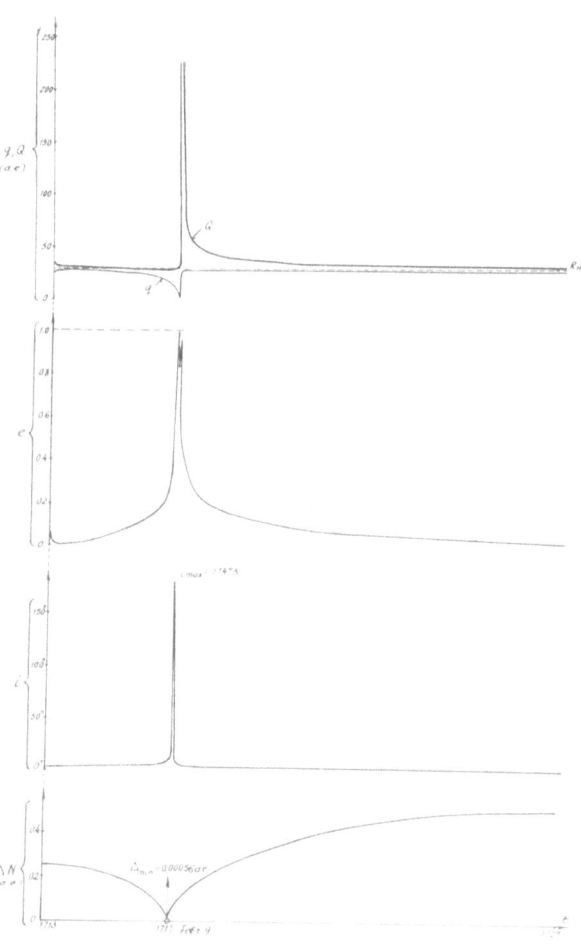

Fig. 18a.

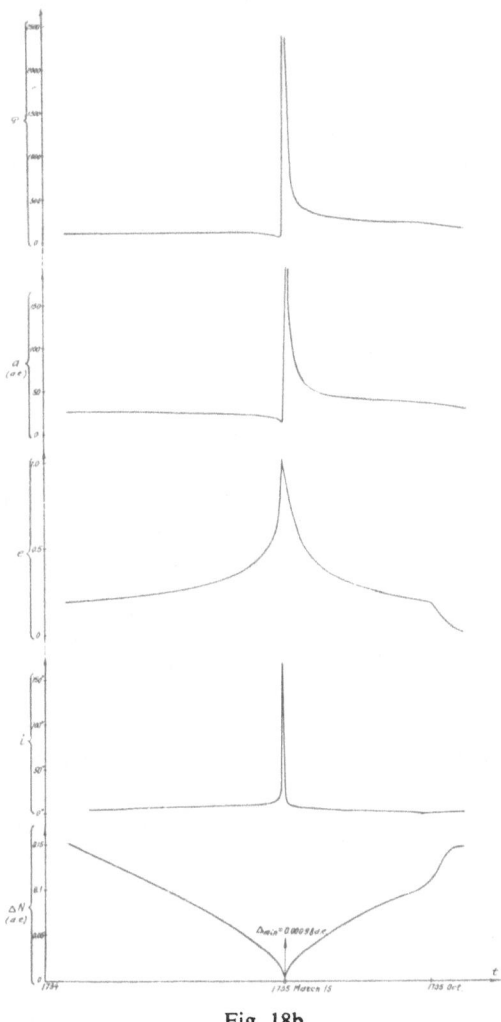

Fig. 18b.

Fig. 18a–b.   Changes in the orbital elements of comet N-3.

'eruption theory'. Shtejns (1960, 1961, 1962, 1964, 1972), Chebotarev (1972), as well
as Belyaev, the present author and others prefer the 'capture theory', and we shall
therefore discuss some of the conclusions of this theory in both its classical and its
modern form.

The capture hypothesis of Laplace was further developed by Tisserand (1889),
Schulhof (1891), Newton (1878, 1893), Callandreau (1892) and others. The classical
studies were concerned with the transfer of a comet from interstellar space into the
planetary system, i.e., transformation in Jupiter's activity sphere of an original parabolic
orbit into an elliptical orbit with direct motion and short revolution period. In the

initial studies the orbit of Jupiter was considered to be circular; in the sphere of Jupiter's action the jovicentric motion of the comet was taken to be Keplerian, perturbations by the Sun and other planets being ignored; outside that sphere perturbations (even those by Jupiter) were neglected. Under these idealized conditions Newton and Callandreau proved that, depending on the initial direction of motion, the orbit of the comet could be transformed by Jupiter into a short-period ellipse with direct motion, a retrograde ellipse, or a direct hyperbola. Everhart (1967, 1968, 1969, 1970, 1972) has continued these investigations more extensively and with greater precision. But he did not make any substantially new conclusions concerning the capture of comets by the major planets.

It is well known that all the short-period comets of Jupiter's family possess elliptical orbits with direct motion. Supporters of the eruption theory maintain that the absence of observable comets with strongly hyperbolic orbits and retrograde short-period elliptical orbits presents unsurmountable difficulties for the capture theory. However, Schulhof stated that Jupiter would convert comets into retrograde elliptical orbits only rarely, and then at great jovicentric velocities. Callandreau (1892, p. 49) concluded that comets leaving Jupiter's sphere of action on hyperbolic orbits cannot be observed, either because their perihelion distances are too large, or because the comets do not pass through the perihelion points of these hyperbolic orbits.

Our investigations are in full agreement with those conclusions. Ellipses with retrograde motion and small perihelion distances do appear, but only very rarely, and only for a short time while the comets (e.g., N-2 and N-3) are well inside the activity spheres of the giant planets, i.e., at very great planetocentric velocities, as Schulhof surmised.

And as for the comets that leave Jupiter's sphere of action along hyperbolic orbits, our studies of numerous variants of the close approach of P/Lexell to Jupiter in 1779 fully confirm Callandreau's conclusion that the perihelia of such orbits do not exist in the range of observability.

Finally, we shall sketch – from the standpoint of celestial mechanics – the general picture of the evolution of cometary orbits, shedding new light on the capture theory in conjunction with the theory of diffusion of nearly parabolic comets and consideration of stellar perturbations.

Oort (1950, 1951) advanced a hypothesis according to which an extended 'cloud of comets' surrounds the solar system at a distance of about 150 000 AU from the Sun. Fesenkov (1951) evaluated the diameter of the cloud as some tens of thousands of AU. Without discussing here the origin, stability, dimensions, and structure of this cloud, we shall regard the cloud as a collection of comets moving in nearly parabolic orbits and consider its formation the *first stage* in the evolution of cometary orbits. These problems are treated by Chebotarev (1964, 1966), Nezhinskij (1972), Antonov and Latyshev (1972), Yanovitskaya (1972), and others.

The *second stage* is determined by the stellar perturbations. This matter was first studied by Öpik (1932), then by Oort (1950) and Fesenkov (1951), and more recently by Shtejns (1962, 1964), Makover (1964), and Sekanina (1968). Shtejns established that stellar perturbations scarcely affect the semimajor axes of the cometary orbits, but they may very appreciably change the perihelion distances, even by hundreds of

AU. Stellar perturbations thus may either reduce the perihelion distances into the region of the inner planetary system or remove the comets entirely. Makover has confirmed these conclusions with concrete examples.

The *third stage* is the diffusion of comets, the very gradual accumulation of small planetary perturbations in the reciprocal semimajor axes of the near-parabolic cometary orbits. This process is very slow, continuing for millions of years. The total changes in the semimajor axes may become quite considerable. Van Woerkom (1948) was the founder of the theory of cometary diffusion. The theory was later developed by Oort (1950, 1951), Lyttleton (1953), Shtejns and Riekstyn'sh (1960), Shtejns (1961), Shtejns and Sture (1962), Shtejns and Kronkalne (1964, 1968), Kendall (1961), and Whipple (1962).

Shtejns includes in his investigations all the nearly parabolic comets and those long-period comets whose semimajor axes are not less than 40 AU. Using statistical methods he ascertains that diffusion produces a gradual reduction of the semimajor axes of a considerable fraction of the orbits of these comets without any change in perihelion distance. He also formulates the laws of cometary diffusion, taking into account the influence of an important astrophysical factor – the disintegration of cometary nuclei.

The first law of diffusion, discussed for the first time by Oort and then generalized by Shtejns, states that owing to diffusion comets having small semimajor axes also have small inclinations. The orbital planes of these comets gradually converge towards the plane of the ecliptic and consequently also to the mean plane of the orbits of the giant planets.

Shtejn's second law of diffusion states that cometary orbits having large perihelion distances also have small eccentricities.

The diffusion of comets thus reduces the orbital inclinations and leads to a gradual concentration of the perihelia of a multitude of invisible comets with nearly circular orbits on the periphery of the planetary system.

The *fourth stage* of the evolution is the capture of these invisible comets by the giant planets, and this has been the object of the investigations we have described here. Shtejns has studied by statistical methods most of the comets for which $a \geqslant 40$ AU; we have investigated by the numerical methods of celestial mechanics comets with $a \leqslant 50$ AU. Together, therefore, the two investigations encompass all comets, and for those satisfying 40 AU $\leqslant a \leqslant 50$ AU two independent methods have been applied. Diffusion, taking place over millions of years, has produced comets moving in low-inclination, direct, nearly circular orbits at the edge of the planetary system.

Prolonged approaches of these comets to the giant planets (especially Neptune, with its large sphere of action) result in great transformations of their orbits. Favourable conditions are created for throwing these comets into the inner part of the solar system, where they can be observed. There also exist comets with low-inclination hyperbolic orbits and very elongated elliptical orbits whose perihelia are situated near the orbits of each of the giant planets. Capture is possible only when they penetrate the activity sphere of a planet near perihelion. Although a single catastrophic capture by Jupiter may occur, it is more probable that the capture of comets is a very complicated process, taking place over several millennia as the comets are thrown from one of the giant planets to another. There is no doubt that Jupiter plays the dominant role.

Undoubtedly, a considerable fraction of the comets leave the solar system under the influence of stellar or planetary perturbations. This is confirmed by the researches of Galibina (1958, 1963, 1964), Shtejns (1962, 1964), Brady (1965), Sekanina (1966a, 1966b), and by our investigations on P/Lexell.

The results discussed here are necessarily of a preliminary character. We fully realize that after constructing sufficiently precise numerical theories for all the short-period comets, combining all apparitions, and after repeated investigations of their orbital evolution with the highest precision, we may arrive at substantial modifications to the general picture of evolution. However, there is reason to believe that our researches on the orbital evolution of the short-period comets and on the role of the giant planets in this evolution will be of some importance, for they reveal for the first time a real mechanism for possible transformations of cometary orbits and elucidate the probable connection between long-period and short-period comets. These studies, conducted by the precise methods of celestial mechanics, characterize an important stage in the general evolution of cometary orbits within the boundaries of the planetary system; without taking these results into account it is impossible to create any scientific theory on the origin of comets.

References

Antonov, V. A. and Latyshev, I. N.: 1972, this Symposium, p. 341.
Belyaev, N. A.: 1966, *Byull. Inst. Teor. Astron.* **10**, 696.
Belyaev, N. A.: 1967, *Astron. Zh.* **44**, 461.
Brady, J. L.: 1965, *Astron. J.* **70**, 279.
Callandreau, O.: 1892, *Ann. Obs. Paris Mem.* **20**, B1.
Chebotarev, G. A.: 1964, *Astron. Zh.* **41**, 983.
Chebotarev, G. A.: 1966, *Astron. Zh.* **43**, 435.
Chebotarev, G. A.: 1972, this Symposium, p. 1.
Dinwoodie, C.: 1960, *Quart. J. Roy. Astron. Soc.* **2**, 232.
Dubyago, A. D.: 1950, *Trudy Astron. Obs. Kazan* No. 31.
Dubyago, A. D.: 1956, *Byull. Astron. Obs. V.P. Engel'gardta* No. 32.
Everhart, E.: 1967, *Astron. J.* **72**, 716.
Everhart, E.: 1968, *Astron. J.* **73**, 1039.
Everhart, E.: 1969, *Astron. J*, **74**, 735.
Everhart, E.: 1970, *Astron. J.* **75**, 258.
Everhart, E.: 1972, this Symposium, p. 360.
Fesenkov, V. G.: 1951, *Astron. Zh.* **28**, 98.
Galibina, I. V.: 1958, *Byull. Inst. Teor. Astron.* **6**, 630.
Galibina, I. V.: 1963, *Byull. Inst. Teor. Astron.* **9**, 46.
Galibina, I. V.: 1964, *Byull. Inst. Teor. Astron.* **9**, 465.
Herget, P. and Marsden, B. G.: 1961, *Quart. J. Roy. Astron. Soc.* **2**, 158.
Kamieński, M.: 1959, *Acta Astron.* **9**, 53.
Kamieński, M.: 1961, *Acta Astron.* **11**, 33.
Kazimirchak-Polonskaya, E. I.: 1967a, *Trudy Inst. Teor. Astron.* **12**, 63, 86.
Kazimirchak-Polonskaya, E. I.: 1967b, *Astron. Zh.* **44**, 439.
Kazimirchak-Polonskaya, E. I.: 1968, *Astronomie* **82**, 217, 323, 432.
Kazimirchak-Polonskaya, E. I.: 1972, this Symposium, p. 95.
Kendall, D. G.: 1961, *Proc. Fourth Berkeley Symposium on Math. Stat. and Probab.* **3**, 99, 121.
Leverrier, U. J. J.: 1848, *Compt. Rend. Acad. Sci. Paris* **26**, 468.

Leverrier, U. J. J.: 1857, *Ann. Obs. Paris Mem.* **3**, 203.
Lexell, A. I.: 1777, *Acta Acad. Sci. Petropol.* **1**, 332; **2**, 328; *Mem. Acad. Roy. Sci. Pruss.* 638.
Lexell, A. I.: 1778, *Acta Acad. Sci. Petropol.* **1**, 317; *Acad. Sci. Petersb.* 1.
Lexell, A. I.: 1781, *Acta Acad. Sci. Petropol.* **2**, 351.
Lyttleton, R. A.: 1953, *The Comets and their Origin*, Cambridge.
Makover, S. G.: 1964, *Byull. Inst. Teor. Astron.* **9**, 525.
Marsden, B. G.: 1964, *IAU Circ.* No. 1857.
Marsden, B. G. and Aksnes, K.: 1967, *Astron. J.* **72**, 952.
Newton, H. A.: 1878, *Am. J. Sci. Arts*, New Haven, Conn. *Ser. 3* **16** (116), 165.
Newton, H. A.: 1893, *Mem. Natl. Acad. Sci. Washington* **6**, 7.
Nezhinskij, E. M.: 1972, this Symposium, p. 335.
Oort, J. H.: 1950, *Bull. Astron. Inst. Neth.* **11**, 91.
Oort, J. H.: 1951, *Observatory* **71**, 129.
Öpik, E.: 1932, *Proc. Am. Acad. Arts Sci.* **67**, 169; *Harvard Repr.* No. 79.
Schulhof, L.: 1891, *Bull. Astron.* **8**, 147, 191, 225.
Sekanina, Z.: 1966a, *Acta Univ. Carol. Math. Phys.* **2**, 3.
Sekanina, Z.: 1966b, *Bull. Astron. Inst. Czech.* **18**, 1.
Sekanina, Z.: 1968, *Bull. Astron. Inst. Czech.* **19**, 223, 291.
Shtejns, K. A.: 1960, *Uch. Zap. Latv. Gos. Univ.* **38**, 69.
Shtejns, K. A.: 1961, *Astron. Zh.* **38**, 107, 304.
Shtejns, K. A.: 1962, *Astron. Zh.* **39**, 915.
Shtejns, K. A.: 1964, *Uch. Zap. Latv. Gos. Univ.* **68**, 39.
Shtejns, K. A.: 1972, this Symposium, p. 347.
Shtejns, K. A. and Kronkalne, S.: 1964, *Acta Astron.* **14**, 311.
Shtejns, K. A. and Kronkalne, S.: 1968, *Izv. Akad. Nauk Latv. SSR* **9**, 59.
Shtejns, K. A. and Riekstyn'sh, E. I.: 1960, *Astron. Zh.* **37**, 1061.
Shtejns, K. A. and Sture, S. J.: 1962, *Astron. Zh.* **39**, 506.
Sitarski, G.: 1968, *Postępy Astron.* **16**, 159.
Tisserand, F.: 1889, *Bull. Astron.* **6**, 241, 289.
van Woerkom, A. J.: 1948, *Bull. Astron. Inst. Neth.* **10**, 445.
Vsekhsvyatskij, S. K.: 1930, *Monthly Notices Roy. Astron. Soc.* **90**, 706; *Astron. Nachr.* **240**, 273.
Vsekhsvyatskij, S. K.: 1933, *Astron. Zh.* **10**, 18.
Vsekhsvyatskij, S. K.: 1955, *Astron. Zh.* **32**, 432.
Vsekhsvyatskij, S. K.: 1967, *Priroda i Proiskhozhdenie Komet i Meteornogo Veshchestva*, Pros-veshchenie, Moscow.
Vsekhsvyatskij, S. K.: 1969, *Problemy Sovremennoj Kosmogonii*, Moscow, p. 240.
Vsekhsvyatskij, S. K.: 1972, this Symposium, p. 413.
Whipple, F. L.: 1962, *Astron. J.* **67**, 1.
Wiśniewski, W.: 1960, *IAU Circ.* No. 1729.
Yanovitskaya, G. T.: 1972, this Symposium, p. 346.

Discussion

S. K. Vsekhsvyatskij: Your statement on the unobservability of hyperbolic comets with perihelia on Jupiter's orbit is contradicted by the fact that other comets (P/Schwassmann-Wachmann 1 for instance) are observed far beyond a distance of 5 AU from the Sun.

E. I. Kazimirchak-Polonskaya: Comets more than 5 AU from the Sun are invariably faint and observable only because their orbits have been determined from observations made at smaller heliocentric distances, and we thus know where to find them. P/Schwassmann-Wachmann 1 was discovered only because of its unusual surges in brightness.

V. V. Fedynskij: May one consider that the 'pulsating ellipses', such as you noted for P/Wolf, is a typical mechanism in the evolution of the orbits of short-period comets?

E. I. Kazimirchak-Polonskaya: Yes. Analogous fluctuations were found by Belyaev in the case of P/Schwassmann-Wachmann 2 and by myself for P/Wolf-Harrington and other comets.

B. THEORIES OF COMETARY ORIGIN

67. THE ORIGIN OF COMETS

F. L. WHIPPLE

Smithsonian Astrophysical Observatory, Cambridge, Mass., U.S.A.

Abstract. The evolution of the solar system is surveyed, it being presumed that the Sun, Jupiter, and Saturn formed rather quickly and essentially with the composition of the original collapsing cloud of dust and gas. Just as the refractory material of the cloud is considered to have formed into planetesimals, from which the terrestrial planets collected, so is the icy material supposed to have produced comets, or cometesimals, from which Uranus and Neptune (and to some extent Saturn and Jupiter) were built up. The presence of a residual belt of comets beyond the orbit of Neptune is discussed, analysis of possible perturbative effects on P/Halley indicating that the total mass of such a belt at 50 AU from the Sun could not now exceed the mass of the Earth.

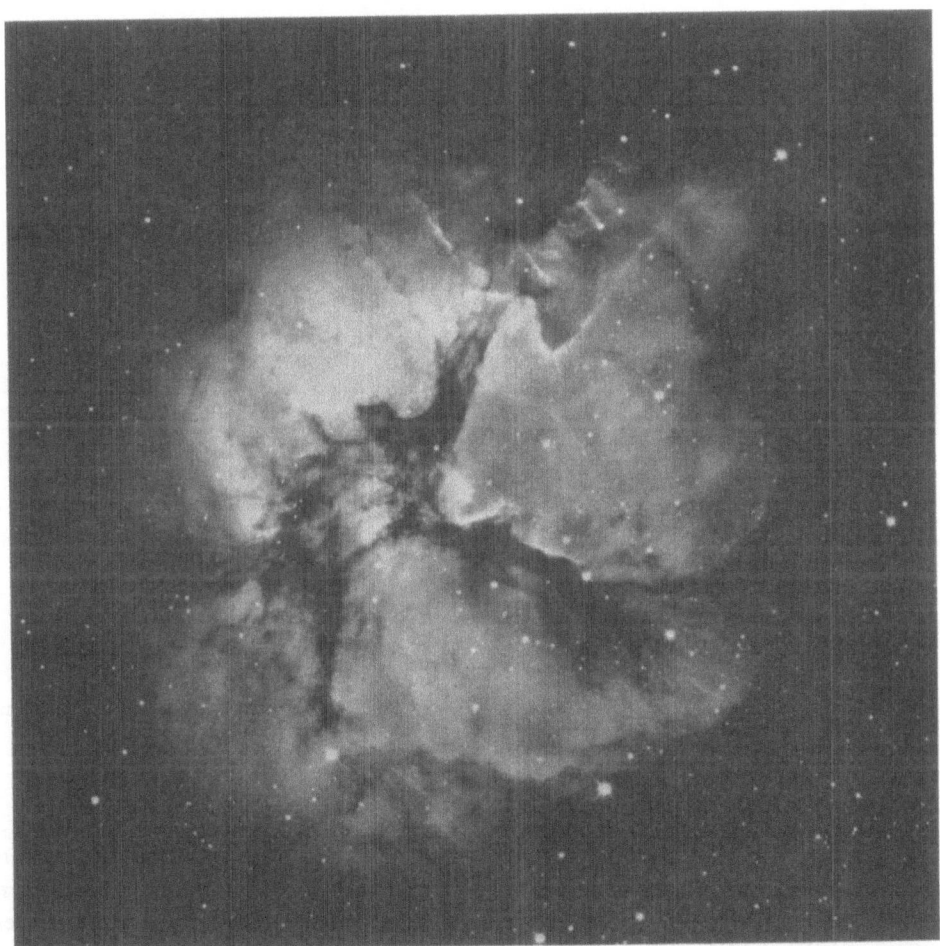

Fig. 1.   The Trifid Nebula (Lick Observatory).

Chebotarev et al. (eds.), The Motion, Evolution of Orbits, and Origin of Comets, 401–408. All Rights Reserved.
Copyright © 1972 by the IAU.

I shall deal only with a few aspects of the evolution of comets and of the solar system in order to point out a certain effect which may or may not be observable. I hope that you will look for it.

My basic assumption will be the usual one, namely that the solar system developed from a collapsing cloud of dust and gas. This is consistent with the earliest hypotheses of modern science – those of Kant and Laplace and with the concepts of Otto Schmidt.

I do not know whether the Trifid Nebula (Figure 1) represents the type of dust and gas cloud from which the solar system evolved, but it gives us some confidence that such condensations are not rare and isolated events.

The cloud is assumed to collapse because of instability, starting from a dimension of perhaps a large fraction of a parsec, and perhaps it would eventually look like the classical Laplacian discus (Figure 2). It is presumed to be a cool gas-dust nebula. One of the major problems, which we cannot discuss, is the rate of radiation loss against the energy gained by gravitational collapse. But presumably the gas in the outer regions cools after it collapses and takes the form of a Laplace-type discus.

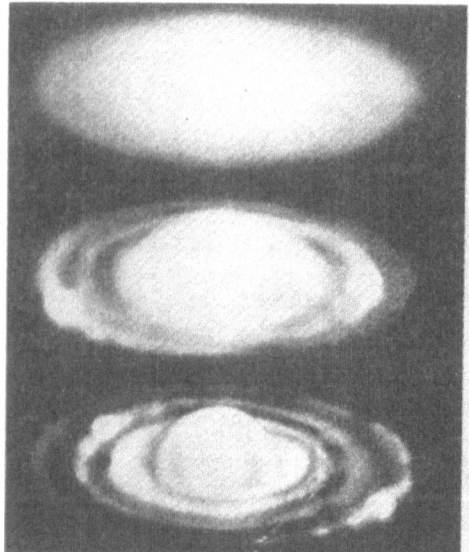
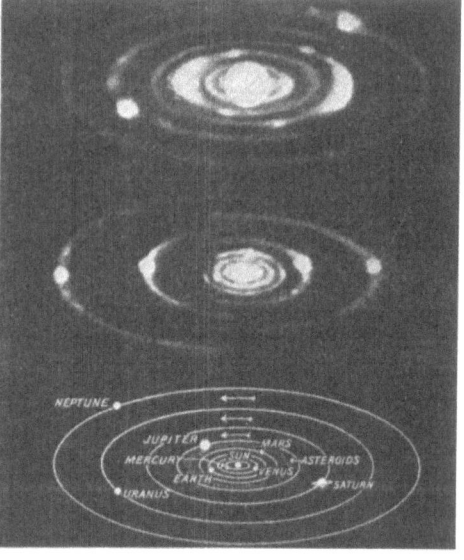

Fig. 2.   The Laplace nebula concept.   Fig. 3.   The Laplace solar-system formation concept.

Of course, we well know that Laplace's concept of rings being left behind (Figure 3) is not acceptable in terms of modern dynamics. We cannot get involved in details of these problems. Let us start with the concept that the solar system bodies were accumulated from the condensation of dust and gas, the Sun developing at the center.

Comet 1957 V Mrkos (Figure 4) typifies the comets, which I believe are aggregates collected from material frozen out of the original gas and dust cloud at large distances from the Sun.

Fig. 4. Comet 1957 V Mrkos (A. McClure).

The chemical elements and compounds observed in comets are listed in Table I. It is clear that most of the gases are simple compounds of carbon, nitrogen, oxygen, and hydrogen. But a comet close to the Sun gives evidence of other heavier atoms when the solar heat is great enough to vaporize them.

Harrison Brown was perhaps the first to point out clearly that the bodies in the solar system may be subdivided according to three natural divisions among the physical characteristics of the chemical elements: the *earthy* materials such as iron, silicon,

TABLE I

Comets, observed composition

Head: C_2, C_3, CH, CN, C^{12}, C^{13}
NH, NH_2, [O I] OH, H
Na, Si, Ca, Cr, Mn, Fe (Ni, Cu)
Tail: CH^+, CO^+, CO_2^+, N_2^+, OH^+ (CN)
and meteoritic dust

and magnesium that melt at rather high temperatures (see Table II); *icy* material which boils at rather low temperatures (and generally melts below 0 °C); and thirdly the *gases*, particularly hydrogen, helium, and the noble gases, that remain gaseous to extremely low temperatures. I question that at any time in the history of the solar system the temperature dropped low enough to freeze hydrogen and helium, because of the difficulty of radiating away the energy of collapse. Thus solid hydrogen or helium should not be among the icy materials that froze to form the comets.

TABLE II

Relative abundances of various atoms in the solar system[a]

Class of matter	Gaseous	Icy	Earthy
Elements and	H (1)	C (12)	Mg (24)
atomic weights	He (4)	N (14)	Si (28)
		O (16)	Fe (56)
			etc.
Sun	1.0	0.015	0.0025
Terrestrial planets			
and meteorites	Trace	Trace	1.0[b]
Jupiter	0.9	0.1	Trace
Saturn	0.7	0.3	Trace
Uranus, Neptune			
and comets	Trace	0.85	0.15

[a] By mass.
[b] Including oxygen.

    Now, if we look at the composition of bodies in the solar system (Table II), we see that, as presumed, the Sun is typical of the material with which we started. Among the terrestrial planets we find nearly 100% earthy materials and oxides with high melting points.

    Jupiter has almost solar composition, perhaps 90% gaseous material and 10% ices. So it must have collected initially from material very much like the Sun itself. It naturally follows that Jupiter must have formed rather quickly with almost the original composition of the gas-dust cloud. But the terrestrial planets, forming within Jupiter's orbit, must have collected, as suggested by many people, particularly Chamberlain and Moulton, from planetesimals made of the refractory materials – the earthy materials.

Next we note that Uranus, Neptune and the comets, insofar as we know, have almost exactly the same composition. Thus I maintain, as do Kuiper, Cameron, and others, that Uranus and Neptune were built up in the same fashion as were the terrestrial planets, except that the building blocks in the outer part of the system, where the temperature was colder, were comets, not planetesimals; we might say *cometesimals*.

I suggest that Saturn was formed almost simultaneously with Jupiter but, being farther out from the Sun, it collected a greater fraction of comets to increase its icy composition to perhaps 30% in mass.

Because of the relative abundances of the gases, ices, and earthy materials, the various classes of bodies in the solar system require approximately the same minimum quantity of original material (see Table III). In the case of the terrestrial planets one

TABLE III

Minimum original planetary masses

Objects	Present mass (Earth = 1)	Factor	Original material (Sun = 1)
Terrestrial	1.9	500	0.0028
Jupiter	317	10	0.0095
Saturn	95	30	0.0086
Uranus and Neptune	32	75	0.0072
Comets	1	900	0.0027
Minimum original mass	—	—	0.0308

must throw away all but perhaps 0.2% of the material in the form of gas and ices. Without going into detail, one needs a minimum of about one percent of the original solar mass to form respectively the terrestrial planets, Jupiter, Saturn, Uranus and Neptune and the comets separately. As you can see, I leave out of consideration a huge number of processes, all of which are controversial, interesting and important. But I go now to my main point, which may possibly be settled observationally.

In my view then, as the solar system developed, the comets formed beyond Saturn and collected as cometesimals into Uranus and Neptune. When Uranus and Neptune were formed, their masses perturbed the motions of the remaining comets in this region of the solar system. Many of the comets were thrown into the inner part of the solar system where some were captured by the planets (particularly Saturn) and the Sun, some were sublimated by solar heat as they are today, and some were thrown to infinity. Others were thrown to great distances in elongated orbits and formed the present-day comet cloud, whose stability was first discussed by Öpik and later by Oort, the great comet cloud extending to many thousands of astronomical units from the Earth. The remainder should have been formed very nearly in, and not have deviated far from, the plane of Jupiter or the other planets. They should still be moving in nearly circular orbits and should correspond at great distances, i.e., beyond Neptune, to the asteroids between Mars and Jupiter (see Figure 5). They would be comets, not asteroids, occupying this region of space. But they cannot be observed directly as comets, nor by total reflected light from the Sun, nor by obscuration of stars. So without sending space probes to that region of the space I think it is hopeless to observe such comets directly, in case they do exist.

But they should have a gravitational effect and their mass might amount to the order of the Earth's mass! If there is indeed a comet belt beyond Neptune, it would perturb the motions of the outer planets and may have produced the perturbations in the motion of Neptune that have been attributed to Pluto (Whipple, 1964). We know that

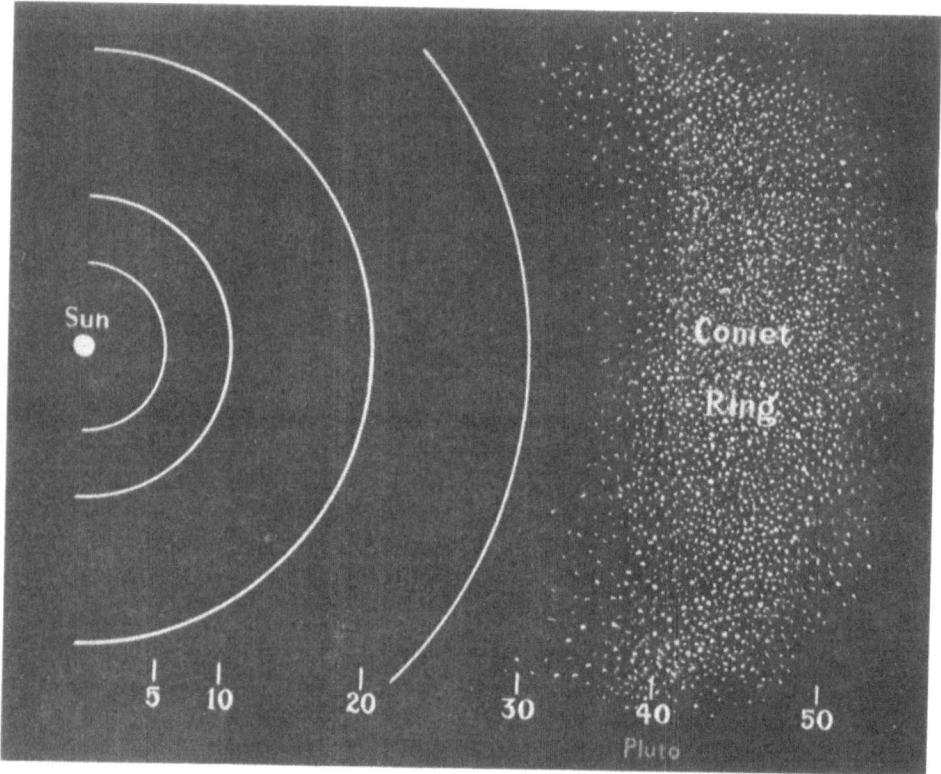

Fig. 5.   Region of possible comet belt.

Pluto could not have a mass of the order of one Earth's mass; it must be very much smaller and could not have appreciably perturbed Neptune.

Figure 6 shows the poles of the ecliptic, of the invariable plane, of Saturn's orbit, of Jupiter's orbit, and of Neptune's orbit. If the pole of the comet belt is somewhere in the general direction indicated in Figure 6, it could then produce the perturbations in the latitude of Neptune that were attributed to Pluto. And the mass required would be inversely proportional to the belt's inclination to Neptune's orbit plane. With an inclination of only a few degrees, a fraction of an Earth mass in the comet belt could have produced the perturbations attributed to Pluto.

Duncombe *et al.* (1968) have recently determined a new mass for Pluto, finding it to be much smaller than an Earth mass. I hope they will investigate to see whether there is evidence for a comet belt with a pole in the general region indicated in the figure.

So the question is not yet settled whether indeed a comet belt may have produced the perturbations in the motion of Neptune. There is, however, another method of searching for a comet belt; that is by its possible effect on the motions of long-period

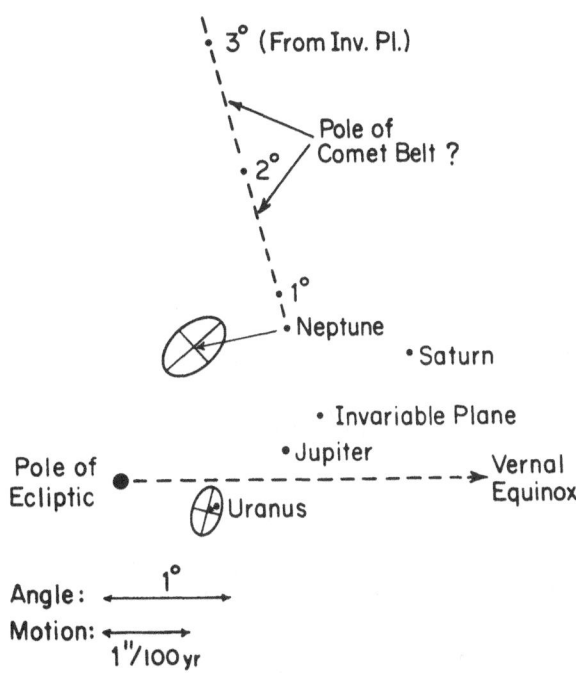

Fig. 6.   Pole of ecliptic and possible comet belt.

comets. Halley's Comet, for example (Figure 7), has its aphelion beyond Neptune and not far from Pluto's orbit. Consequently, it would be affected near aphelion by the gravitational attraction of a comet belt in that region of space. There are a few other comets that could be used, but Halley's is the most suitable. So Hamid, Marsden and I have looked into the last two apparitions of P/Halley to see whether there was any evidence for perturbations produced by an assumed comet belt beyond Neptune (Hamid *et al.*, 1968). We were not able to find direct effects in the motion of P/Halley to indicate the presence of a comet belt. But we could set upper limits to the total mass possible in that region of space. If there is the comet belt in a ring at solar distance 50 AU near the invariable plane, we can say with some confidence that its total mass does not exceed the mass of the Earth.

Because we still do not know whether such a comet belt exists, I hope that you will keep its possibility in mind. Perhaps you will find a more ingenious method for either discovering the existence of the comet belt or of proving that its mass is negligible.

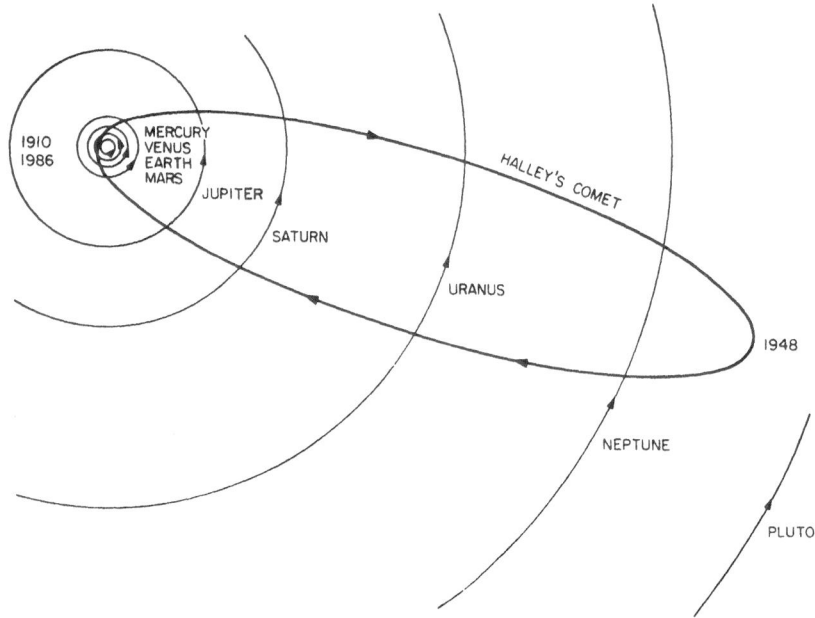

Fig. 7.   Orbit of Halley's comet.

References

Duncombe, R. L., Klepczynski, W. J., and Seidelmann, P. K.: 1968, *Astron. J.* **73**, 830.
Hamid, S. E., Marsden, B. G., and Whipple, F. L.: 1968, *Astron. J.* **73**, 727.
Whipple, F. L.: 1964, *Proc. Natl. Acad. Sci.* **51**, 711.

Discussion

S. K. Vsekhsvyatskij: How do you account for the large mean densities of Jupiter's satellites?

F. L. Whipple: I presume that Jupiter's satellites were formed in the same fashion as were most of the planets (except for Jupiter). The same properties exist in the system as in the solar system itself, namely that the denser bodies are near the primary and the less dense bodies farther away.

68. ON THE ORIGIN OF COMETS AND THEIR IMPORTANCE
FOR THE COSMOGONY OF THE SOLAR SYSTEM

V. G. FESENKOV

Committee on Meteorites, U.S.S.R. Academy of Sciences, Moscow, U.S.S.R.

Abstract. The processes in interstellar space, as well as the structure of nebulae – including the numerous small condensations around newly formed stars – suggest a relationship to the formation of the cometary cloud surrounding the Sun. Study of carbonaceous chondrites has revealed that in the earliest stage of the solar system's existence frequent periods of brief heating occurred, leading to the formation of very complex organic compounds and also of chondrules of crystalline structure. This was most probably a consequence of cometary collisions. The Tunguska phenomenon illustrates some consequences of a similar collision.

It is appropriate to consider that comets, as well as stars, are formed in interstellar space. This is, however, a very complicated magnetohydrodynamical problem. The magnetic field in our Galaxy is strong enough to confine cosmic rays and, in all probability, to determine the structure of the galactic spiral arms. Interstellar space, highly heterogeneous in composition, is constantly being affected by the galactic nucleus; shock waves from supernova explosions are propagated in it, sudden compressions of matter occur, there is interaction between its dust and gas constituents and a great amount of turbulence. All this results in the formation of heterogeneities of various dimensions: globules occurring only in the nebulae, the development of Herbig-Haro objects (star-like condensations) and stellar chains and associations that rapidly disperse into space. Such stellar chains can be observed, for instance, in the filamentary nebulae of Cygnus that formed at the boundary between dust clouds as the result of a supernova explosion about 70 000 yr ago.

Of great interest is the Orion nebula, which surrounds an unstable system of stars in the Orion Trapezium that developed some 23 000 yr ago. This nebula has been found to consist mainly of a large number of heterogeneities containing almost all the mass. Attempts are made to find, by analysis of spectroscopic data, the relative velocities of heterogeneities similar to comets. In that same nebula there were recently discovered, in infrared light, protostars at the beginning of collapse.

The chemical composition of comets fully corresponds to this concept of their origin. According to spectral data cometary heads contain molecular compounds such as C_2, C_3, CH, NH, CH_2, OH, CO_2, NH_2, to which correspond the parent molecules HCN, NH_3, CH_4, C_2H_2 and others; the composition of meteors suggests that silicates and various compounds involving heavy elements should also be included. The same compounds are found in interstellar space, this being compatible with the assumption of the development there of unstable bodies such as comets by the same processes connected with the formation of stars.

V. S. Safronov has expressed a similar viewpoint in developing his cosmogonical theory for the formation of the planets. He believes that the growth of the giant planets was accompanied by the gravitational ejection of part of their embryo material

Chebotarev et al. (eds.), The Motion, Evolution of Orbits, and Origin of Comets, 409–412. All Rights Reserved.

beyond the solar system, this process having led to the formation of an extended cometary cloud. He supposes that the total mass ejected constitutes from one-third to one-half of the mass of the giant planets (Safronov, 1969). This is equivalent to assuming that the planets developed through the successive combination of heterogeneities of cometary type and fairly complicated composition.

On the other hand, S. K. Vsekhsvyatskij has quite a different concept of the origin of comets – by cosmic vulcanism. His ideas are based in essence only on the supposition that in the inner regions of the solar system the short-period comets break up very rapidly, within a few centuries. Without going into details, we shall say that his concept cannot be accepted for a number of reasons. Firstly, it is quite impossible to produce by ejection from the planets the vast majority of comets in nearly parabolic orbits, which form a more or less uniform cloud in the vicinity of the Sun, extending – according to J. H. Oort and G. A. Chebotarev – for about 150 000 AU. Secondly, the highly unstable cometary matter is characteristic of the properties, not of planetary interiors, but of interstellar matter. Moreover, as Vsekhsvyatskij himself points out, a cosmic volcano calls for an unusually powerful source of energy, greatly exceeding that of nuclear reactions, and it could not possibly develop in bodies as small as planets – and especially their satellites.

Let us now dwell on some processes which occurred during the earliest stage of development of the solar system, prior to the formation of planets. Interesting data on the subject are supplied by a particular variety of meteorites – the carbonaceous chondrites. These objects retain an abundance of rare gases that corresponds to their cosmic abundance, and they are believed to be the oldest known matter. They are an amazing combination of organic hydrocarbon and nitrogen compounds and chondrules – small crystalline silicate balls. A detailed laboratory investigation of meteorites such as Orgueil, Cold Bokkeveld, and Murray has shown them even to contain such complex compounds as $C_6H_5[CH(CH_3)_2]$; $C_6H_4(C_2H_3)_3$, anthracene, phenanthrene and other compounds up to the bases of DNA. This once gave rise to the belief that these meteorites or their parent bodies could earlier have carried life. However, detailed experiments (by Studler, Hayatsu, Anders, and others) have shown that these complex compounds may be reproduced with great accuracy under conditions of thermodynamic equilibrium in Fisher-Tropsch reactions from a medium of typically cometary composition but much lower hydrogen content. Meteoric matter provides a catalyst, and rapid heating to about 800–1000 °C, followed by sudden cooling, is imperative. The formation of chondrules from the same material evidently requires much greater local heating, also followed by cooling.

Such bursts, which occurred during the preplanetary stage of the solar system, have been explained in various ways, e.g., by flares of the Sun itself before it joined the main sequence. In the opinion of H. C. Urey, mutual collisions of comparatively loose matter in a weak gravitational field generate a great deal of heat; many molten droplets are produced, and after rapid cooling they become chondrules. It is precisely this kind of process that would have occurred very frequently in the earliest stage of existence of a protoplanetary nebula that contained a large number of condensations of cometary type. On their mutual collision, accompanied by heating with subsequent expansion and rapid cooling, there would have resulted both complex organic compounds and, in similar quantities, chondrules. Together with other mineral matter, the chondrules subsequently formed more compact bodies – chondrites – and the latter gradually became part of much larger bodies. For a considerable length of time they experienced fairly complicated metamorphosis, accompanied by the formation of meteorites of different types. Such processes would most naturally occur in mutual collisions between comets – bodies of comparatively negligible masses, low density, and abundance of the simplest organic compounds with the inclusion of gases and meteoric matter, but naturally with a low hydrogen content.

The Tunguska meteorite, which was actually a small comet of minimum mass 10^{12} g, can be cited as an example illustrating what happens when a cometary nucleus is suddenly decelerated. The tail of the comet interacted with upper layers of the Earth's atmosphere and caused a short-lived anomalous luminosity of the night sky. When the nucleus of the Tunguska comet reached an altitude of approximately 10 km above the Earth's surface it produced a tremendous explosion, the shock waves from which felled the taiga forest for hundreds of kilometres around. We have shown that this corresponds to an average density of the nucleus equal to about 0.1 g cm^{-3}. It is known that material from the nucleus spread after the explosion over almost the whole Northern Hemisphere and considerably diminished the solar radiation received. A large number of silicate and magnetite balls, strongly resembling chondrules, fell on the ground in the vicinity of the epicentre, while along the whole trajectory of the bolide there was noted for the next few decades a considerable acceleration in the growth of vegetation; this was quite independent of the changes in the ecological conditions and probably connected with the precipitation of complex organic compounds.

We should point out that the above concept of the cometary origin of meteorites has also been suggested by other authors from other data. Yokoyama *et al.* (1968) have remarked that the asteroidal nature of meteorites does not conform well to their chemical and mineralogical composition; they emphasized, in particular, the absence in meteoritic matter of long-period isotopes formed by slow neutrons. Still earlier, Öpik (1966) discussed cometary nuclei as the principal source of meteorites. By comparing the orbits of comets, meteorites, and the asteroids of the Apollo group, Öpik concluded that all these bodies have a common origin, the comets being the initial bodies.

It thus seems most natural to assume that the planetary bodies formed from the heterogeneities of a protoplanetary nebula of cometary type. It was conventionally assumed that the protoplanetary nebula consisted of gas and dust of approximately the interstellar composition and that the dust particles gradually agglomerated into bodies of planetary size. This is, however, highly improbable. T. Gold believes that the supposed process of cementation of fine particles through their mutual collisions is the principal cosmogonical problem requiring solution. Urey also points out that there should be some forces, either electrostatic or ferromagnetic, to cement these particles together. Therefore he, and also Kuiper, prefer to believe that the protoplanetary nebula was divided in the course of fragmentation into a number of gas-dust masses or protoplanets. In these primary objects there were first formed small solid planets similar to the Moon in size, and they later combined into the present planets as the result of impact phenomena. The development of comogonical ideas increasingly leads to the conclusion that the comets – the heterogeneities contained in the proto-planetary nebula from the very beginning – were responsible for the complete evolution of the solar system and directly led to the formation of the planets.

References

Öpik, E.: 1966, *Adv. Astron. Astrophys.* **4**, 301.
Safronov, V. S.: 1969, *Ehvolyutsiya Doplanetnogo Oblaka i Obrazovanie Zemli i Planet.*, Nauka, Moscow, p. 186.
Yokoyama, Y., Mabuchi, H., and Labeyrie, J.: 1968, Symposium on *Origin and Distribution of the Elements*, Paris, p. 445.

69. THE ORIGIN AND EVOLUTION OF THE COMETS AND OTHER SMALL BODIES IN THE SOLAR SYSTEM

S. K. VSEKHSVYATSKIJ

Astronomical Observatory, Kiev University, Kiev, U.S.S.R.

Abstract. It has become evident that comets and other small bodies are indications of eruptive evolution processes occurring in many of the planetary bodies of the solar system. The total number of near-parabolic comets moving in the solar system is 10^{11} to 10^{12}, but as many as 10 to 15 percent of them are leaving the solar system with hyperbolic velocities. Taking into account also the number of short-period comets that degenerate into asteroids and meteor streams, we have estimated the total number of comets formed during the lifetime of the solar system as 10^{15} to 10^{16} (and total mass 10^{29} to 10^{31} g). The investigation of comets and other small bodies enables us to evaluate the scale of the processes of cosmic vulcanism and the tremendous internal energy of the planets, that energy being derived from the initial stellar nature of planetary material.

1. Introduction

Most of the modern cosmogonical hypotheses consider the numerous minor bodies of the solar system to be of little importance in shedding light on the history of the system. These minor bodies are generally regarded as relics of the primeval matter from which the Sun and planets condensed (Kuiper, 1951) or as objects that condensed in the outer regions of the primordial nebula at distances of 3–50 AU and were then thrown out to the periphery of the solar system (Oort, 1963). The hypotheses do not explain the forces that ejected these objects into parabolic orbits and the manner in which the orbits later became circular – to produce the hypothetical Oort cloud of comets.

This particularly speculative idea, which is not based on any analysis of observational data, is refuted by everything known nowadays about the structure and physical nature of asteroids and meteorites. It also seems to me that the significance of collisions has been overrated; consider how much more important volcanic and tectonic processes are, not only on the Earth, but also on the Moon (NASA, 1969). These results all speak in favour of the fact that the minor bodies have formed as the result of volcanic processes in the planetary bodies.

2. Data from Comets

Examination of the cometary data gives more obvious information (Vsekhsvyatskij, 1962, 1966, 1967). The existence of planetary families of short-period comets that cannot be explained either by 'capture' or by 'diffusion', the nature of cometary gases, the extensive supplies of ice in cometary nuclei (where the meteorite fragments are), and several other facts also prove that the comets are the youngest objects in the solar system.

It is important to state some principal conclusions:

Chebotarev et al. (eds.), The Motion, Evolution of Orbits, and Origin of Comets, 413–418. All Rights Reserved.
Copyright © 1972 by the IAU.

(1) Analysis of the 'Laplace problem', statistics of cometary perihelia, and the kinematics of the cometary system leave no doubt that all comets, and consequently the products of their disintegration, were created within the solar system, and, on the average, more recently than the planets.

(2) The existence of the families of short-period comets of Jupiter, Saturn, Uranus, and Neptune, and in particular the fact that Jupiter's comets were invariably in the vicinity of Jupiter not too long before discovery, demonstrate the recent formation of these comets by eruption in the planetary system. Jupiter's satellites are likely to be the immediate source of the youngest comets of the Jupiter family. The recent detailed investigations by Everhart (1969) have quite definitely confirmed that it is impossible to explain the observed distribution of orbits of the short-period comets on the supposition of gravitational capture.

(3) The catalogues by Sekanina (1966, 1968) list 35 comets the future orbits of which are known to be hyperbolic, so that these objects must therefore leave the solar system. Several of the objects have high absolute brightness and masses up to 10^{19} g. The amount of gaseous material alone lost by a comet of absolute magnitude $H_{10} = 0$ is 10^{12} to 10^{13} g per day, and the amount lost per revolution is typically $10^{15 - 0.4H_{10}}$ (Wurm, 1963). On the average one hyperbolic future orbit appears among every 4–6 nearly parabolic or long-period comets. This approximately corresponds to the results of van Woerkom (1948) on the accumulation of small perturbations. Taking into account the considerable number of comets with large perihelion distance q and all the intrinsically faint comets, we find that the annual loss from the system of comets cannot be less than 10^{18} g. During the whole period of existence of the solar system this corresponds to a loss of at least 10^{27} g. Comparison with the estimated total number of comets (10^{11} to 10^{12}) forces us to conclude that there must exist within the solar system sources for replenishing the cometary objects. Considering the number of comets that have left the solar system and the number of disintegrated short-period comets (which turned into asteroids, meteorites, and meteoroids), we estimate the total number of comets (with mean mass 10^{14} to 10^{15} g) created since the origin of the solar system as 10^{15} to 10^{16}. The total mass of material ejected from the surfaces of the planets is therefore 10^{29} to 10^{31} g.

(4) The existence of comets and other minor bodies provides the possibility of estimating the extent of cosmic vulcanism processes in the history of the solar system and the internal energy of the planets (some 10^{41} to 10^{43} erg) expended in the ejection of considerable quantities of planetary material into interplanetary and interstellar space.

These conclusions, made long before the space age began, led investigators to anticipate the high volcanic activity on the surfaces of the planets. Volcanic activity, not only on the giant planets, but also on Venus and Mars, has attracted attention, and the missions to the last-mentioned planets and the Moon have demonstrated the decisive role of volcanic processes in the evolution of planetary bodies. At the same time, recent investigations clearly confirm the existence of meteoritic masses in cometary nuclei and consequently that comets are fragments of the crusts and the frozen atmospheres of the planets.

3. Planetary Densities

It is clear that of all the physical characteristics of the planets, the one of greatest significance for studying the problem of the comets and other minor bodies is mean density. The physical and dynamical evolution of the planetary bodies took place as the result of loss of the lighter elements from the surface layers. This resulted in an increase in mean density. The absence nowadays of stellar abundances of hydrogen and helium in the terrestrial planets suggests that the amount of mass lost greatly exceeds the modern masses of these planets.

We may therefore use the data on mean density to establish the scale of the creation processes of comets, meteorites, and asteroids. The masses M, mean densities $\bar{\rho}_c$, and specific rotational energies $E_R = \frac{1}{2} I \omega^2 / M$ (where I is the moment of inertia about the axis of rotation and ω is the angular velocity of rotation) for various bodies are listed in Table I (Vsekhsvyatskij, 1971). It is apparent that there are two analogous

TABLE I

Masses, densities, and specific rotational energies

	M (grams)	$\bar{\rho}_c$ (g cm^{-3})	E_R (erg g^{-1})
Sun	2.0×10^{33}	1.41	9.6×10^8
Jupiter	1.9×10^{30}	1.39	2.6×10^{11}
Saturn	5.7×10^{29}	0.71	1.7×10^{11}
Uranus	8.7×10^{28}	1.6	2.5×10^{10}
Neptune	6.1×10^{28}	1.6	1.0×10^{10}
Earth	6.0×10^{27}	5.51	3.7×10^8
Pluto	5.5×10^{27}	10	9.5×10^8
Venus	4.9×10^{27}	5.3	5.1×10^3
Mars	6.4×10^{26}	4.0	9.4×10^7
Mercury	3.3×10^{26}	5.8	2.1×10^4
Ganymede	1.6×10^{26}	2.4	2.4×10^{6} [a]
Titan	1.4×10^{26}	2.3	2.4×10^{6} [a]
Triton	8.7×10^{25}	2.0	1.6×10^{6} [a]
Moon	7.4×10^{25}	3.35	3.6×10^3
Callisto	9.7×10^{25}	2.1	$10^6 - 10^{5}$ [a]
Io	7.2×10^{25}	4.0	$10^6 - 10^{5}$ [a]
Europa	4.7×10^{25}	3.8	$10^6 - 10^{5}$ [a]

[a] The period of revolution is adopted.

groups, the terrestrial planets and the Galilean satellites, each showing a decrease in mean density with increasing distance from the central body. This is regarded as an important indication of the eruptive evolution of the planets, proceeding most rapidly under the tidal influence of the central bodies.

The distribution of density in a planet is given by

$$\rho = \rho_0 (1 - \xi r^\lambda),$$

where ρ_0 is the central density, r is the fraction of the radius, varying from 0 at the centre to 1 at the surface, and ξ and λ are parameters. From this we may obtain the present mean density,

$$\bar{\rho}_c = \rho_0\left(1 - \frac{3\xi}{\lambda + 3}\, r^\lambda\right),$$

which enables us to estimate the initial radius r and establish the initial dimensions of the planet.

For the initial mean density of protoplanets we may take the present mean density of Jupiter or of the Sun, because their relative mass loss cannot be very significant. For the Earth we take $\rho_0 = 12$ g cm^{-3}, and it then follows that the initial radius was in the range 14 to 7.1 thousand kilometres, the loss of mass being anywhere between 2 to 3 and 0.2 times the present mass of the Earth. The loss of mass from Venus would be of the same order; much more mass could have been lost from Pluto – if the high mean density for this planet is correct.

The total amount of material ejected from all the planets since the origin of the solar system could exceed 10^{29} to 10^{30} g. This value is of the same order as the total mass of comets and other minor bodies created during the history of the solar system. This shows that inside the planets there must have been powerful energy sources far greater than anything derived from gravitational collapse or radioactive decay. It is quite natural to suppose that the energy supply was preserved from the initial stellar condition of the planetary material.

4. Minor Planets

The data from the system of asteroids agree well with our conclusions from the system of comets. Table II shows the distribution of the absolute magnitudes g of the 1746 permanently numbered objects. The first seven values, up to and including that for $g = 10$–11, probably characterize the real distribution of asteroidal sizes in the accessible

TABLE II

Distribution of absolute magnitudes of minor planets

g	4–5	5–6	6–7	7–8	8–9	9–10	10–11	11–12
	2	1	6	23	87	209	345	395
g	12–13	13–14	14–15	15–16	16–17	17–18	18–19	19–20
	332	230	95	13	6	1	0	1

region of space (up to 5.5 AU from the Sun), and from them we may obtain the following dependence of the number n and the mass M:

$$n = 0.000651 \times 2.94^g$$

$$\log M = 26.74 - 0.6\, g.$$

Figure 1 also shows the distribution, the above expression for n being shown by the broken line.

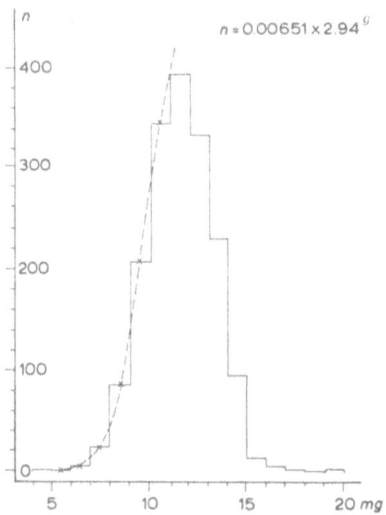

Fig. 1.   Distribution of absolute magnitudes g of the minor planets. The broken line shows the distribution extrapolated from the brighter objects.

The extrapolated number of minor planets of $g = 19.5$ is thus $n = 7.6 \times 10^6$, and their combined mass is $M = 10^{19} - 10^{20}$ g, from which it follows that the total mass of the asteroids up to $g = 19$-20 is of the order 10^{26} to 10^{27} g. Considering that the total volume of the planetary system and the volume of the visibility region are in the ratio $(40/5)^3 \simeq 500$, we deduce that the total mass of asteroidal material in the solar system is some 10^{28} to 10^{29} g, which is again approximately consistent with the cometary data.

5. Evidence from Meteoroids

According to Brandt and Hodge (1964) some 10^7 to 10^9 g of meteoroidal matter encounter the Earth daily and are pulverized in its atmosphere. Taking into account the Earth's attraction, this corresponds to a volume of approximately 10^{15} km^3. Supposing meteoroid density to be the same over the volume occupied by the system of planets (10^{30} km^3) the total mass of meteoroids must amount to 10^{23} g. Since these particles move along the cometary orbits, orbiting the Sun in decades or at most a few centuries, we may deduce that the total amount of finely dispersed substance produced during the existence of the solar system cannot be less than $10^{23} \times 10^{6 \text{ to } 7} = 10^{29}$ to 10^{30} g.

Among photographic and radio meteors a significant proportion have small perihelion distances ($q < 0.4$ AU) and large eccentricities; the majority have aphelion distances not exceeding 3 to 4 AU (Vsekhsvyatskij, 1967). More than a quarter of all the particles move in rather eccentric orbits relatively near the Sun. The Poynting-Robertson effect implies that the lifetimes of such particles are rather small and measured only in hundreds or at most thousands of years. Solar corpuscular radiation is still more effective at dispersing and sweeping out these particles. We conclude that the particles observed near the Sun arise in the inner region of the solar system. It seems impossible to explain their character by capture from long-period orbits or their creation as a result of the disintegration of comets.

It has already been suggested that objects with orbits of small perihelion distance and small semimajor axis (e.g., P/Encke, P/Wilson-Harrington, Icarus, the Apollo asteroids, etc.) could arise as a result of eruptive processes on Venus, the space missions having indicated temperatures there of the order 800 K, pressures of more than 100 atm, a large amount of dust in the upper atmosphere, and rapidly varying dark features above the cloud cover that appear to consist of clouds of volcanic ash.

The peculiarities of all the groups of minor bodies thus illustrate the rapid dynamical and physical evolution especially evident in the case of comets and meteoroids. Together with the results of analysis of meteorites they give evidence of the processes of ejection from the surfaces of satellites and planets (the Moon, the satellites of Jupiter and Saturn, Venus, Mars, and the Earth).

References

Brandt, J. and Hodge, P.: 1964, *Solar System Astrophysics*, New York and London.
Everhart, E.: 1969, *Astron. J.* **74**, 735.
Kuiper, G. P.: 1951, in J. A. Hynek (ed.), *Astrophysics*, McGraw-Hill, New York, Toronto and London, p. 400.
NASA: 1969, Apollo 11: Preliminary Science Report.
Oort, J. H.: 1963, in *The Moon, Meteorites and Comets*, Vol. IV of the series: *The Solar System* (ed. by B. M. Middlehurst and G. P. Kuiper), University of Chicago Press, Chicago and London, p. 665.
Sekanina, Z.: 1966, *Publ. Astron. Inst. Charles Univ.* No. 48.
Sekanina, Z.: 1968, *Publ. Astron. Inst. Charles Univ.* No. 56.
van Woerkom, A. J.: 1948, *Bull. Astron. Inst. Neth.* **10**, 445.
Vsekhsvyatskij, S. K.: 1962, *Publ. Astron. Soc. Pacific* **74**, 106.
Vsekhsvyatskij, S. K.: 1966, *Mem. Soc. Roy. Sci. Liège Sez. 5* **12**, 469.
Vsekhsvyatskij, S. K.: 1967, *Priroda i Proiskhozhdenie Komet i Meteornogo Veshchestva*, Prosveshchenie, Moscow.
Vsekhsvyatskij, S. K.: 1971, *Probl. Kosmich. Fiz.* No. 6.
Wurm, K.: 1963, in *The Moon, Meteorites and Comets*, Vol. IV of the series: *The Solar System* (ed. by B. M. Middlehurst and G. P. Kuiper), University of Chicago Press, Chicago and London, p. 573.

70. ON THE PROBLEM OF THE ORIGIN OF COMETS

J. M. WITKOWSKI

University Observatory, Poznań, Poland

Abstract. The various hypotheses on the origin of comets are surveyed, with particular reference to the phenomenon of comet-streaming, as revealed by statistical analysis of the directions of the perihelia of cometary orbits. The difficulties of reconciling comet-streaming with any of the hypotheses of cometary origin is pointed out. It is shown that in the case of the interstellar hypothesis these difficulties may be overcome by supposing that the velocities of incoming comets are decreased through the accretion of material from an Oort cloud of solar-system comets. An estimate is made of the total number of interstellar comets in the Galaxy.

Nowadays there exists a variety of hypotheses on the origin of comets. From a general point of view the hypotheses may be subdivided into two main groups: those presuming the interstellar origin of comets and those that suppose that comets originate in the solar system.

At the head of the first group is the Laplace hypothesis. Proceeding from the assumption of Kepler and Herschel about the availability of comets in interstellar space, Laplace determined the relative probabilities for comets with elliptical and hyperbolic orbits versus their velocities relative to the Sun. He assumed the radius of the Sun's sphere of action to be 100 000 AU and considered the relative velocities at which the comets enter that sphere all to be equally probable. He concluded that there would be a prevalence of elliptical and parabolic orbits, which was in agreement with the statistical data then available.

Laplace's statement of the problem itself has been repeatedly subjected to criticism. Schiaparelli and Niessl considered that proper account must be taken of the Sun's motion in space. Seeliger criticized the assumed velocity interval (zero to infinity). The investigations by Fabry, Moisseev, and Bobrovnikoff ascertained that the frequency of elliptical, parabolic, and hyperbolic orbits depends on the character of the velocity distribution function as well as on the Sun's motion (Hoffmeister, 1951, 1952). As a general result these investigations lead to the conclusion that the orbits of comets *cannot* provide an answer to the question of the origin of comets.

Not all the arguments put forward against the hypothesis of Laplace are fortunate. It was not without reason that Laplace considered only the velocities of comets relative to the Sun, neglecting the problem of the Sun's spatial motion. Not all authors have stated the problem of the character of the orbits in a sufficiently lucid way. The only possible orbits are ellipses and hyperbolas, since the statistical probability of circles and parabolas (as well as straight lines) is geometrically and dynamically infinitesimally small and need not be taken into consideration.

In the aforementioned hypotheses the *a priori* existence of ready-made comets in space is accepted. The hypothesis of Lyttleton (1948) differs from the preceding ones in that it considers the process of the passage of the Sun through a uniform interstellar cloud of dust (Bondi and Hoyle, 1944). The particles in that cloud are constrained by

Chebotarev et al. (eds.), The Motion, Evolution of Orbits, and Origin of Comets, 419–425. All Rights Reserved.
Copyright © 1972 by the IAU.

solar attraction to describe hyperbolas with the Sun as focus. These hyperbolas intersect along a line parallel to the relative velocity vector of the Sun and cloud and which passes through the Sun. Along this line accretion takes place, condensations being formed that according to Lyttleton represent newly formed comets. The hypothesis may be generalized (Gething, 1951; McCrea, 1953) by accepting a nonuniform structure for the cloud.

Lyttleton's hypothesis endeavours not only to explain the origin of comets, but also to give a physical model of cometary structure. While Lyttleton has in general terms succeeded in solving this problem, his hypothesis does not answer the questions raised by critics, and his model of cometary structure disagrees with observations (Richter, 1963).

A quite different way of solving the problem of the origin of comets is based upon the spatial distribution of the major axes of cometary orbits. This approach to the problem will be discussed later. Meanwhile let us turn to the hypotheses that presuppose that comets belong to the solar system. The first such hypothesis, advanced by Lagrange, was supplemented by Proctor and Tisserand, and at present is most extensively developed by Vsekhsvyatskij (1930, 1931, 1952). In accordance with this outlook comets are the products of volcanic eruptions upon the surfaces of the major planets or their satellites. However tempting this hypothesis may appear at first sight, it encounters considerable difficulties, the principal one being the great initial velocities of eruption needed (~ 67 km s^{-1} for Jupiter), velocities which are unacceptable from the viewpoint of the physics of these planets.

The hypothesis of Oort (1950, 1951) belongs to the same group. In general terms, it may be reduced to the following statements. The comets form part of the solar system, for otherwise the percentage of hyperbolic orbits would be considerably greater than is observed. Consequently, the comets accompany the Sun in its motion, i.e., they constitute a cometary cloud. Oort attributes the origin of that cloud to the disintegration of a planet that formerly moved around the Sun between the orbits of Mars and Jupiter; the fragments of this planet gave rise to the formation of both comets and minor planets. Perturbations by Jupiter and other planets caused some of the fragments to be diffused away, and this resulted in the formation of a large cloud of comets situated at present at heliocentric distances between 50 000 and 200 000 AU. Under the disturbing action of neighbouring stars the comets in this 'reservoir' may be either ejected from the solar system or directed back to the Sun, i.e., into the sphere of visibility of the terrestrial observer.

Oort's hypothesis is based to a great extent on the theoretical studies by van Woerkom (1948). However, as Oort remarks, further calculations are necessary. The principal statement in Oort's hypothesis is the denial of the existence of interstellar comets (i.e., the absence of 'original' hyperbolic orbits), as was supposedly proven by Strömgren (1914, 1916), Sinding (1948), and Galibina (1963). The next essential element of this hypothesis refers to the cometary reservoir on the outskirts of the solar system. Orlov (1939) does not believe the cloud could exist, and Lyttleton found in Oort's argument a substantial error in determining the position of the maximum concentration of the major axes of the cometary orbits.

The general survey of hypotheses considered here brings us to the following conclusions. None of the above-mentioned hypotheses solves the problem of the origin of comets. It should be readily apparent that statistical investigations of orbital forms cannot solve this problem. The existence of hyperbolic orbits suggests only that they have kept their initial character; on the other hand, the orbits of some interstellar comets could have been significantly modified at the edge of the solar system.

Oort's interpretation of the cometary cloud hypothesis, although interesting in itself, is too artificial. Lyttleton's hypothesis is generally more probable, conforming to our knowledge of interstellar space, but it requires more thorough elaboration.

We now turn our attention to a more comprehensive discussion of the problem of the distribution in space of the major axes of cometary orbits and the resulting deductions.

These investigations encounter difficulties because of the yet insufficient empirical data and their complex relationship with the observational conditions. Hoek discussed the matter more than a hundred years ago. Schiaparelli denied that there was any relationship in the distribution of the perihelia of cometary orbits with the galactic equator, whilst Holetschek admitted the existence of such a relationship but attributed it to observational conditions. According to Jantzen, the interpretation by Holetschek is not very convincing.

Svedstrup (1883) investigated the distribution of the perihelia of cometary orbits and pointed out (using a simplified ellipsoidal analysis) their connection with the galactic equator. His work received no response until Oppenheim (1922, 1924) again raised this problem. Oppenheim made use of the statistical methods of stellar astronomy, namely, ellipsoidal analysis. He examined separately the distribution of the perihelia of elliptical and parabolic cometary orbits as well as those of the minor planets. The ellipsoid for the minor planets has common features with the ellipsoid for elliptical cometary orbits, both ellipsoids being oriented in the plane of the ecliptic.

A quite different picture is obtained for parabolic cometary orbits. The ellipsoid is directed along the galactic equator, but the differences between the axes are small, from which it may be deduced that cometary material is distributed around the Sun symmetrically. Since the shortest axis is directed towards the vertices of star-streaming Oppenheim concluded that comets must be of interstellar origin. Under the disturbing action of the major planets the interstellar features of cometary orbits are smoothed out in the course of time; this is precisely what has happened in the case of the periodic comets, whose orbits have acquired planetary features, in particular the orientation of their orbital planes to the ecliptic.

About twenty years ago a series of investigations on the subject was initiated at the Poznań University Observatory. The calculations by Svedstrup and Oppenheim have been verified using more extensive statistical material relating to 451 long-period (parabolic) comets (Witkowski, 1953; Hurnik, 1959). It turned out that the results were not greatly influenced by the selection of material, for comets prior to 1700 and after 1900 gave practically identical ellipsoids for the perihelion distribution. This is at variance with the statements of Bourgeois and Cox (1934), who in their extensive studies attributed the results of Oppenheim to the influence of observational conditions.

It became necessary to check the conclusions by Bourgeois and Cox, all the more now that new studies of the distribution of cometary perihelia have appeared that speak in favour of Oppenheim's results (Tyror, 1957; Kresák, 1957). A critical analysis of the deductions by Bourgeois and Cox was performed at my suggestion by Hurnik (1959) on material embracing the aforementioned 451 comets, because Bourgeois and Cox derived 'Oppenheim's corrected ellipsoid' on the basis only of comets at $\alpha = \omega = 90°$, ω being the argument of perihelion of a cometary orbit and α the longitude of the Earth measured from the ascending node of the cometary orbit. Following the method of Bourgeois and Cox, Hurnik determined anew the 'selection factors' and carried out the calculations varying α and ω in 45°-steps between 0 and 360°. Hurnik's results demonstrated that the distribution of perihelia was independent of visibility conditions and that 'Oppenheim's corrected ellipsoid' displayed an orientation to the galactic equator. The selection factors thus gave a result contrary to that which had been expected by Bourgeois and Cox.

The character of Oppenheim's ellipsoid indicates that the longest axis is perpendicular to the plane of the Galaxy, while the shortest axis is directed to the vertices of star-streaming. It should now be considered fully confirmed that the distribution of the major axes of cometary orbits reflects the main features of the motions of the stars in space and supports the hypothesis of the interstellar origin of comets. A few suggestions can be made in favour of that hypothesis:

(1) The motion of interstellar dust particles and gas molecules in the neighbourhood of the Sun depends on the galactic gravitational field and thus should follow the motions of stars in that part of space. Consequently, the motions of the particles reflect the phenomenon of star-streaming. Under the influence of solar radiation pressure a small particle decreases its velocity component in the direction of the Sun. When this velocity component has been reduced to zero, the particle will become a member of the solar system. Let n be the ratio of light pressure to solar gravitational attraction, ρ the distance of the particle from the Sun, v the radial velocity component, T and ρ_0 the time and the distance of the particle from the Sun when $v=0$. We then obtain, for example,

if $n = 2$, $\rho = 10\,000$ AU, $v = 1$ km s^{-1}; then

$T = 50\,000$ yr, $\rho_0 = 1500$ AU;

if $n = 5$, $\rho = 40\,000$ AU, $v = 3$ km s^{-1}; then

$T = 65\,000$ yr, $\rho_0 = 720$ AU.

Over comparatively short intervals of time, condensation takes place in the gas-and-dust cloud in the Sun's vicinity; the probability of collisions between the particles is increased, leading to the formation of aggregations, and these would be too large to be affected by light pressure.

(2) In the opinion of many authors comet-like bodies exist in interstellar space in great quantities (Kepler, Russell, Lyttleton). According to Fehrenbach the total mass of interstellar meteorites inside a sphere of radius $r = 10^5$ AU amounts to 1/500 the mass of the Sun, or the mass of 10^{15} comets.

(3) These cosmic comets move along galactocentric orbits similar to those described by the stars.

(4) Assuming that the distance of the Sun from the centre of the Galaxy is 10^4 pc, that its velocity is 285 km s^{-1}, and that the mass of the Galaxy is 1.9×10^{11} M$_\odot$, we find, according to Laplace's definition, that the radius of the Sun's sphere of action is 60 000 AU. Comets entering this sphere of action from interstellar space may be captured by the Sun. It has now been established that capture is possible in the general problem of three bodies (Schmidt, 1951; Khil'mi, 1951; Sizova, 1952). Furthermore, it has long been known that under the influence of perturbations by Jupiter comets may be ejected from the planetary system; hence the reverse process, the capture of a cosmic comet by the solar system, is also possible. In accordance with the theory of probability most cases of capture occur at considerable distances from the Sun, so that comets moving along hyperbolic orbits will pass undetected at great distances from us. The only observable comets will be those captured in elliptical orbits of great eccentricity; comets with small orbital eccentricities will remain at great distances from the Sun and may in fact form the Oort cloud.

In summary the following conclusions may be inferred.

The main arguments against the interstellar origin of comets, namely the absence of hyperbolic cometary orbits (Strömgren) and the condition of zero relative velocity of the Sun and the comet (Schiaparelli) are not valid.

Studies of cometary statistics have detected the existence of comet-streaming similar to star-streaming. This phenomenon, as well as the determination of the solar apex from the distribution of the perihelia of cometary orbits (Tyror, Hurnik), are in favour of the hypothesis of interstellar origin.

As yet the mathematical aspects of the capture hypothesis have not been sufficiently developed. Specific initial conditions are required, and the process can hardly ensure a sufficient quantity of comets for the solar system.

Capture will be simplified if a decrease in the comet's heliocentric velocity takes place. Two possibilities should be taken into account: existence of a resisting medium or an increase in the mass of the comet. The first possibility should be rejected on the strength of the known data concerning interstellar space.

We may assume the existence of a spherical, uniform, gas-and-dust cloud surrounding the Sun at a distance of 100 000 AU and more. The particles in that cloud describe elliptical orbits of small eccentricity. For distances between 100 000 and 230 000 AU only direct motion ensures stability (Chebotarev, 1964). The existence of such a cloud is proven by, among other things, the process of 'renewal' of comets as they replenish their supplies of gas and dust near aphelion. An interstellar comet entering the sphere on a hyperbolic orbit will start to increase its mass. The capture cross-section of the comet is given by $D^2 = r_0^2 (1 + v_c^2/v_r^2)$, where r_0 is the radius of the comet, v_c the escape velocity, and v_r the relative velocity of the particle. During each astronomical unit of its passage through the gas-and-dust layer (density δ) the comet accretes mass $\Delta m_c = 1.5 \times 10^{13} \pi r_0^2 \delta$ g cm^{-3}. Assuming the thickness of the layer to be 1000 AU, $r_0 = 10^7$ cm, $\delta = 10^{-18}$ g cm^{-3}, $m_c = 10^{17}$ g, we find that the comet's relative mass accretion would be 10^{-5}, which is not sufficient for our purposes.

In order to decrease the comet's velocity it is necessary to assume that the cloud consists of a number of aggregations. These aggregations would accelerate the process of accretion and produce a comet consisting of a cosmic and a planetary component (Witkowski, 1968).

By analysing this hypothesis Piotrowski (1965) has shown that if the capture occurs at great distances from the Sun, it always leads

(1) to an increase in $1/a$;

(2) to a decrease in e for parabolic orbits;

(3) to a decrease in e for hyperbolic retrograde orbits and most often to an increase in e for direct, distinctly hyperbolic orbits; and generally;

(4) to an increase of the chance of discovery of interstellar comets with retrograde motion compared with comets moving along direct orbits.

Statements (3) and (4) are valid under the assumption of direct motions in the hypothetical cloud.

From a general point of view the modified hypothesis of capture may be brought into agreement with observational data. It ensures a constant supply of new comets, replacing the old ones which have disintegrated or have been thrown back into space.

Comets may be regarded as the visible links of a process of perpetual exchange of matter between the solar system and the rest of the universe. Comets transfer gas and dust from one region of our solar system to another, from the periphery to the central part of the system.

Finally I should like to mention an attempt at estimating the number of cosmic comets in the Galaxy. Our planetary system bears traces of an encounter with another planetary system several milliards of years ago. Pluto, and also the twin planets Uranus and Neptune, seem to speak in favour of such an encounter. Oort's cloud of comets itself may consist of debris following the close encounter of two planets from two different planetary systems. Accepting that one such encounter takes place once in four milliard years and that 10 interstellar comets enter our planetary system each year, we find that a sphere of radius 10^5 AU contains 5×10^{10} comets (the same order as that found by Oort for his cloud) and that the total number of cosmic comets in the Galaxy is about 5×10^{22}.

I have tried to give here arguments in favour of the interstellar origin of comets. Further work on the subject is necessary, and only the future will show which of the two hypotheses, the Laplacean or the Lagrangean, is correct.

References

Bondi, H. and Hoyle, F.: *Monthly Notices Roy. Astron. Soc.* **104**, 273.
Bourgeois, P. and Cox, J.-F.: 1934, *Bull. Astron.* **8**, 271; **9**, 349.
Chebotarev, G. A.: 1964, *Astron. Zh.* **41**, 983.
Galibina, I. V.: 1963, *Byull. Inst. Teor. Astron.* **9**, 46.
Gething, P. J. D.: 1951, *Monthly Notices Roy. Astron. Soc.* **111**, 468.
Hoffmeister, C.: 1951, *Sterne* **27**, 189.
Hoffmeister, C.: 1952, *Sterne* **28**, 229.
Hurnik, H.: 1959, *Acta Astron.* **9**, 207.
Khil'mi, G. F.: 1951, *Problema n Tel b Nebesnoj Mekhanike i Kosmogonii*, Moscow.

Kresák, L.: 1957, *Contr. Astron. Obs. Skelnaté Pleso* **2**, 19.
Lyttleton, R. A.: 1948, *Monthly Notices Roy. Astron. Soc.* **108**, 465.
McCrea, W. H.: 1953, in *La Physique des Comètes*, Liège, p. 337.
Oort, J. H.: 1950, *Bull. Astron. Inst. Neth.* **11**, 91.
Oort, J. H.: 1951, *Observatory* **71**, 129.
Oppenheim, S.: 1922, *Astron. Nachr.* **216**, 47.
Oppenheim, S.: 1924, in *Festschrift für H. v. Seeliger*, Berlin.
Orlov, S. V.: 1939, *Astron. Zh.* **16**, (1), 3.
Piotrowski, S. L.: 1965, *Acta Astron.* **15**, 281.
Richter, N. B.: 1963, *The Nature of Comets*, Methuen, London.
Schmidt, O.: 1951, *Izv. Akad. Nauk SSSR*.
Sinding, E.: 1948, *Publ. Mind. Medd. Kbh. Obs.* No. 146.
Sizova, O. A.: 1952, *Dokl. Akad. Nauk SSSR* **86**, 485.
Strömgren, E.: 1914, *Publ. Mind. Medd. Kbh. Obs.* No. 19.
Strömgren, E.: 1916, *Astron. Nachr.* **203**, 17.
Svedstrup, A.: 1883, *Astron. Nachr.* **107**, 113.
Tyror, J. G.: 1957, *Monthly Notices Roy. Astron. Soc.* **117**, 370.
van Woerkom, A. J. J.: 1948, *Bull. Astron. Inst. Neth.* **10**, 399.
Vsekhsvyatskij, S. K.: 1930, *Monthly Notices Roy. Astron. Soc.* **90**, 706.
Vsekhsvyatskij, S. K.: 1931, *Astron. Nachr.* **243**, 281.
Veskhsvyatskij, S. K.: 1952, *Astron. Zh.* **29**, 63.
Witkowski, J. M.: 1953, *Bull. Soc. Amis Sci. Lettres Poznań Ser. B* **12**, 205.
Witkowski, J. M.: 1968, *Observatory* **88**, 27.

PART VI

RELATIONSHIP WITH METEORS
AND MINOR PLANETS

A. ORBITAL EVOLUTION OF METEORS AND MINOR PLANETS

71. INVESTIGATION OF THE ORBITAL STABILITY OF MINOR PLANETS WITH COMETARY ECCENTRICITIES

G. A. CHEBOTAREV, N. A. BELYAEV and R. P. EREMENKO

Institute for Theoretical Astronomy, Leningrad, U.S.S.R.

Abstract. The evolution of the orbits of 19 asteroids of particular interest has been studied over the interval 1660–2060, perturbations by Venus to Pluto being taken into account. Information was obtained about the encounters with Venus, the Earth, and Mars. A few approaches of Hidalgo to Jupiter were noted. In distinction to the orbits of short-period comets, the orbits of the 19 asteroids are stable throughout the 400-yr interval.

1. The Group of Minor Planets with 'Cometary' Eccentricities

Among the known minor planets the group with so-called 'cometary' eccentricities is of major interest. The orbits of this group deserve to be thoroughly investigated in view of the possible genetic relationship between minor planets and comets, for these are the objects that might be 'former comets', i.e., cometary nuclei lacking comas, a problem that is being widely discussed nowadays (Gehrels *et al.*, 1970; Kresák, 1969; Marsden, 1970).

Another reason for studying this 'cometary' group of minor planets is that there may be close approaches to the Earth and other inner planets. Hermes, for instance, has approached the Earth to within 0.005 AU, and Apollo was only 0.00056 AU (85 000 km) from Venus in March 1858. The 19 minor planets studied in this work and their orbital elements are included in Table I in order of increasing mean daily motion μ. Most of the elements are taken from the 1967 edition of *Efemeridy Malykh Planet*. The orbits of the lost asteroids, indicated by asterisks, are based on observations at one opposition only and are therefore unreliable.

The structure of the belt of minor planets is characterized by the data shown in Table II (Chebotarev and Shmakova, 1970). Eight of the 19 planets under consideration here are in ring 2, one planet is in ring 3, and ten planets are outside the main belt, one of them being between the orbits of Jupiter and Saturn. The mean eccentricity *e* for the group is 0.50, as opposed to 0.15 for the whole system of minor planets.

Owing to the large eccentricities of their orbits, some of the asteroids pass inside the orbits of Mars, the Earth, and Venus and may have close encounters with these major planets; but the aphelion distances show that close approaches to Jupiter are impossible. The inclinations *i* are another distinct feature of the orbits of this group of minor planets, varying from $1°\!\!.5$ (Adonis) to $52°$ (Betulia); the mean value of $17°\!\!.4$ is almost twice the mean inclination ($9°\!\!.5$) for the majority of the minor planets.

Some of the short-period comets of Jupiter's family have orbits similar to those of minor planets with 'cometary' eccentricities; see Table III. Comparison of Tables I and III shows that the cometary orbits really are different, having large eccentricities and comparatively low inclinations. Of the 19 asteroids only three (Hidalgo, Adonis and Icarus), strictly speaking, have typical cometary orbits. It is interesting to note that

TABLE I

Orbital elements of minor planets with 'cometary' eccentricities

Number	Name	a (AU)	μ	e	ω	☊	π	i	q (AU)	Q (AU)	P (yr)	Epoch
944	Hidalgo	5.81	253″	0.66	57°.5	21°.0	78°.5	42°.5	2.00	9.64	14.04	1964 Mar. 6
	Jupiter	5.21	299	0.05	273.1	100.0	13.1	1.3	4.95	5.47	11.89	1950 Nov. 15
1508	1938 UO	2.77	770	0.42	92.4	14.8	107.3	28.7	1.61	3.93	4.61	1938 Oct. 29
1474	Beira	2.73	785	0.49	82.4	324.9	47.2	26.8	1.39	4.07	4.52	1941 Jan. 6
1134	Kepler	2.68	807	0.47	330.4	6.8	337.1	15.0	1.42	3.94	4.39	1953 Jan. 23
1036	Ganymed	2.66	819	0.54	131.1	216.3	347.4	26.3	1.22	4.10	4.33	1950 June 28
699	Hela	2.61	838	0.41	89.2	243.8	333.1	15.2	1.57	3.68	4.23	1955 Dec. 9
1009ᵃ	Sirene	2.62	838	0.46	184.6	229.6	54.2	15.8	1.41	3.83	4.23	1954 Jan. 18
719ᵃ	Albert	2.59	854	0.54	151.9	186.1	338.0	10.8	1.19	3.99	4.16	1911 Oct. 2
887	Alinda	2.52	886	0.54	348.1	111.0	99.1	9.0	1.16	3.88	4.00	1942 Jan. 31
1580	Betulia	2.19	1091	0.49	158.9	61.9	220.8	52.0	1.12	3.26	3.25	1963 May 31
ᵃ	Adonis	1.97	1284	0.78	39.5	352.5	32.0	1.5	0.43	3.51	2.76	1937 Nov. 6
1221	Amor	1.92	1331	0.44	25.5	171.2	196.7	11.9	1.08	2.76	2.67	1948 June 28
1627	Ivar	1.86	1394	0.40	167.0	132.9	299.9	8.4	1.12	2.60	2.54	1957 July 1
	Mars	1.52	1887	0.09	285.6	48.5	334.1	1.8	1.38	1.66	1.88	1900 Jan. 0
ᵃ	Apollo	1.49	1959	0.57	284.9	36.1	321.0	6.4	0.64	2.34	1.81	1932 Apr. 25
433	Eros	1.46	2015	0.22	178.1	304.0	122.1	10.8	1.13	1.78	1.76	1941 Jan. 6
1685	Toro	1.37	2218	0.42	126.5	274.0	40.5	9.4	0.79	1.94	1.60	1964 Aug. 23
ᵃ	Hermes	1.29	2421	0.47	90.7	35.4	126.1	4.7	0.68	1.90	1.47	1936 Feb. 25
1620	Geographos	1.24	2557	0.34	276.3	336.9	253.2	13.3	0.83	1.66	1.38	1961 Dec. 7
1566	Icarus	1.08	3172	0.83	30.9	87.7	118.6	23.0	0.19	1.97	1.12	1958 Sept. 24
	Earth	1.00	3548	0.02	—	—	101.1	—	0.98	1.02	1.00	1900 Jan. 0

ᵃ Lost objects.

TABLE II

The minor planet belt

	Interval of μ	$\Delta\mu$	Commensurability	N	a (AU)
Ring 1	610″– 740″	130	–	659	3.23–2.84
Gap	740 – 750	10	2:5	4	2.84–2.82
Ring 2	750 – 890	140	–	537	2.82–2.51
Gap	890 – 910	20	1:3	7	2.51–2.48
Ring 3	910 –1110	200	–	420	2.48–2.17
Total	610 –1110			1627	3.23–2.17

N gives the number of minor planets within the regions specified.

the orbit of Adonis resembles that of Encke's Comet. On the other hand, P/Wilson-Harrington has a typical asteroidal orbit similar to that of the planet 1627 Ivar.

The shapes and sizes of the asteroids play an important role in solving the problem of relationship between comets and minor planets. The latest investigations show Icarus to be nearly spherical (radius 0.54 km), while Eros and Geographos, for instance, are distinctly irregular in form (Eros being 35 by 16 by 7 km; Geographos 2.4 by 0.7 km).

2. Evolution of the Orbits of Minor Planets over 400 yr

The orbital evolution of the 19 asteroids is shown in Table IV, which gives the extremes in the orbital elements between 1660 and 2060. The equations of motion have been integrated on an electronic computer by Cowell's method with automatic variation of the integration step and allowance being made for perturbations by Venus to Pluto (Belyaev, 1967).

The principal result obtained is that the planetary orbits with cometary eccentricities are exceptionally stable compared with the orbits of typical short-period comets, the latter being characterized by close approaches to the major planets and drastic changes in the elements.

The greatest perturbations in longitude of perihelion are exhibited by 1009 Sirene ($\Delta\pi = 6°0$) and 944 Hidalgo ($\Delta\pi = -5°9$), while 1566 Icarus and Apollo ($\Delta\pi = \pm 0°2$) show the smallest. The planets of the Hilda and Thule groups are quite different (Chebotarev et al., 1970), the line of apsides of 1269 Rollandia moving through $\Delta\pi = 156°9$ and that of 279 Thule through $\Delta\pi = -335°3$ during the same 400-yr interval.

The perturbations in inclination are extremely small ($\Delta i = \pm 2°5$), except that for 1474 Beira $\Delta i = -8°8$. The eccentricity variations are also extremely small ($\Delta e_{max} = 0.05$). As a consequence of the minor variations in semimajor axis and eccentricity the perihelion and aphelion distances also change insignificantly, illustrating the long-term stability of the orbits. The maximum changes in perihelion distance are observed for 1627 Ivar ($\Delta q = 0.14$) and 944 Hidalgo ($\Delta q = 0.16$).

The motion of planet 887 Alinda is of special interest (Table V). The mean motion librates about the 1:3 commensurability with Jupiter ($\mu^* = 897″$). During the 400-yr

TABLE III

Short-period comets having period of revolution P less than 5.1 yr

Comet	μ	a (AU)	e	ω	☊	π	i	q (AU)	Q (AU)	P (yr)	T
Wilson-Harrington	1536.1	1.748	0.412	91°.9	278°.7	10°.6	2°.2	1.028	2.468	2.31	1949.78
Encke	1075.2	2.216	0.847	185.2	334.7	159.9	12.4	0.339	4.09	3.30	1961.10
Helfenzrieder	786.8	2.723	0.852	178.1	76.1	254.2	7.9	0.403	5.04	4.51	1766.32
Grigg-Skjellerup	724.1	2.864	0.704	356.3	215.4	211.7	17.6	0.855	4.88	4.90	1957.09
Blanpain	695.7	2.963	0.699	350.2	79.2	69.4	9.1	0.892	5.03	5.10	1819.89

TABLE IV

Variations in the elements of minor planets over 1660–2059

Planet	a (AU)	μ	e	ω	Ω	π	i	q (AU)	Q (AU)	P (yr)
699	2.61	840″	0.41	79°8	252°3	332°0	15°3	1.55	3.67	4.22
	2.61	839	0.41	93.5	240.3	337.7	15.3	1.55	3.67	4.23
719	2.59	853	0.55	140.5	194.2	334.7	9.4	1.17	4.01	4.16
	2.58	858	0.54	156.2	182.9	339.1	11.5	1.18	3.98	4.14
887	2.48	910	0.56	347.1	113.3	100.4	9.1	1.09	3.87	3.90
	2.47	916	0.58	350.6	109.5	100.0	9.5	1.05	3.89	3.87
944	5.67	263	0.63	58.6	24.1	82.7	44.9	2.12	9.22	13.49
	5.82	253	0.66	56.4	20.3	76.8	42.4	1.98	9.66	14.03
1009	2.62	836	0.46	175.7	233.6	49.3	15.7	1.42	3.82	4.24
	2.62	837	0.46	187.0	228.2	55.3	15.8	1.42	3.82	4.24
1036	2.66	817	0.56	124.4	223.5	347.9	24.5	1.18	4.15	4.34
	2.66	816	0.53	133.5	213.9	347.4	27.0	1.25	4.08	4.35
1134	2.68	809	0.48	320.0	13.4	333.4	14.1	1.40	3.96	4.39
	2.68	810	0.47	333.5	4.6	338.0	15.4	1.42	3.94	4.38
1221	1.93	1327	0.43	20.6	173.7	194.3	12.1	1.09	2.77	2.67
	1.92	1334	0.44	27.4	170.3	197.7	11.9	1.08	2.76	2.66
1474	2.74	784	0.48	79.6	330.9	50.5	35.5	1.41	4.07	4.53
	2.74	784	0.49	84.7	321.8	46.5	26.7	1.40	4.08	4.53
1508	2.77	771	0.42	88.4	21.0	109.4	28.6	1.60	3.94	4.60
	2.77	769	0.42	93.4	12.6	106.0	28.7	1.62	3.92	4.62
1566	1.08	3172	0.82	29.0	89.5	118.5	23.7	0.19	1.97	1.12
	1.08	3172	0.83	31.7	87.0	118.7	22.7	0.19	1.97	1.12
1580	2.19	1092	0.52	156.2	63.3	219.5	51.5	1.06	3.33	3.25
	2.19	1093	0.49	159.9	61.4	221.3	52.3	1.13	3.26	3.25
1620	1.24	2557	0.34	273.5	339.2	252.7	13.3	0.83	1.66	1.39
	1.24	2555	0.34	277.2	336.2	253.4	13.3	0.83	1.66	1.39
1627	1.86	1393	0.40	163.2	134.8	298.0	8.3	1.12	2.60	2.55
	1.86	1394	0.40	168.3	132.2	300.5	8.5	1.12	2.60	2.55
1685	1.37	2216	0.42	123.3	276.8	39.9	9.3	0.79	1.95	1.60
	1.37	2216	0.42	127.7	273.2	40.9	9.4	0.79	1.95	1.60
Apollo	1.53	1872	0.58	277.9	43.7	321.4	5.8	0.65	2.42	1.90
	1.49	1956	0.57	286.8	34.5	321.2	6.4	0.64	2.33	1.81
Hermes	1.30	2401	0.48	85.5	39.6	125.1	4.8	0.68	1.92	1.48
	1.29	2413	0.48	94.5	31.9	126.3	4.5	0.68	1.91	1.47
Adonis	1.97	1280	0.78	24.6	365.9	30.5	2.0	0.43	3.51	2.77
	1.97	1280	0.78	48.5	344.1	32.6	1.3	0.44	3.51	2.77

TABLE V

Orbital elements of minor planet 887 Alinda

Epoch	a (AU)	μ	e	ω	Ω	π	i	q (AU)	Q (AU)	P (yr)
1660 Feb. 8	2.48	910″	0.56	347°1	113°3	100°4	9°1	1.09	3.87	3.90
1700 Dec. 15	2.47	917	0.57	347.3	112.9	100.2	9.1	1.06	3.88	3.87
1800 Jan. 25	2.50	895	0.55	347.7	112.2	99.9	9.0	1.13	3.87	3.96
1900 Apr. 11	2.53	880	0.53	347.8	111.4	99.3	8.9	1.19	3.87	4.03
1942 Jan. 31	2.52	886	0.54	348.1	111.0	99.1	9.0	1.16	3.88	4.00
2000 Jan. 17	2.48	906	0.56	350.0	110.0	100.0	9.3	1.08	3.88	3.92
2059 Dec. 13	2.47	916	0.58	350.6	109.5	100.0	9.5	1.05	3.89	3.87

interval Alinda passes through the exact resonance twice. At the epochs 1660 and 1700 it is inside ring 3, and in 1800 it is inside the gap ($\mu = 895''$); between 1900 and 1942 Alinda is in ring 2, while around 2000 it enters the gap ($\mu = 906''$) once more, returning to ring 3 in 2059. Since the variations in eccentricity compensate for those in semimajor axis the aphelion distance remains practically constant, and there is no change in the line of apsides either.

Considering the character of close encounters (less than 0.10 AU) with the Earth, Mars, and Venus, the 19 planets may be divided into the following groups:

(1) No close approaches (433, 699, 799, 1009, 1036, 1474, 1508, 1580, 1627, 1685).
(2) Close approaches to the Earth only (887, 1221, 1620).
(3) Close approaches to Mars only (1134).
(4) Close approaches to Venus only (no planets).
(5) Close approaches to two planets (no planets).
(6) Close approaches to three planets (Adonis, Apollo, Hermes, 1566).

In addition there is 944 Hidalgo, which has four encounters with Jupiter (to less than 2 AU) between 1660 and 2060 (Table VI). These approaches did not appreciably change the orbit of Hidalgo.

TABLE VI

Encounters of 944 Hidalgo with Jupiter,
1660–2060

Date	Minimum distance (AU)
1673 Aug. 8	0.38
1756 Dec. 1	1.51
1827 July 3	0.84
1922 Oct. 13	0.90

References

Belyaev, N. A.: 1967, *Byull. Inst. Teor. Astron.* **10**, 696.
Chebotarev, G. A., Belyaev, N. A., and Eremenko, R. P.: 1970, *Byull. Inst. Teor. Astron.* **12**, 82.
Chebotarev, G. A. and Shmakova, M. Ya.: 1970, *Byull. Inst. Teor. Astron.* **12**, 104.
Gehrels, T., Roemer, E., Taylor, R. C., and Zellner, B. H.: 1970, *Astron. J.* **75**, 186.
Kresák, L.: 1969, *Bull. Astron. Inst. Czech.* **20**, 177.
Marsden, B. G.: 1970, *Astron. J.* **75**, 206.

72. EVOLUTION OF THE ORBITS OF SELECTED MINOR PLANETS DURING AN INTERVAL OF 1000 YEARS

M. A. DIRIKIS

Astronomical Observatory, Latvian State University, Riga, U.S.S.R.

Abstract. The long-term evolution of the orbits of 944 Hidalgo, 1036 Ganymed and 1134 Kepler is investigated.

The investigations of the evolution of cometary and asteroidal orbits performed by Kazimirchak-Polonskaya (1967, 1968), Belyaev and Chebotarev (1968), and Chebotarev *et al.* (1970) are very precise, but they are limited to a 400-yr time interval (1660–2060), because precise coordinates of the outer planets are available only for this interval (Eckert *et al.*, 1951).

For our studies we have made use of approximate coordinates of the major planets computed analytically using the principal perturbation terms. It thus becomes possible, in theory, to extend the studies of orbital evolution over unlimited time intervals. One must pay attention to two points, however, which restrict the possibility of obtaining plausible results:

(1) The accumulation of error after a large number (10 000 or more) of integration steps.

(2) The approximate theory used for the calculation of the coordinates of perturbing bodies is not sufficiently accurate if the time interval elapsed from the initial osculation epoch is very long (1000 yr or more).

Consequently we have generally limited our calculations to a time interval of about 1000 yr. We make use of a variation-of-elements method, since the changes in the elements are relatively small and we can use larger integration steps. Perturbations by the planets Venus to Saturn are taken into account.

So far we have studied the evolution of the orbits of the following asteroids: 944 Hidalgo, 1036 Ganymed, and 1134 Kepler. The orbits of these asteroids all have high eccentricities and are similar to the orbits of comets.

One of the most interesting of all asteroids, 944 Hidalgo has not only one of the highest orbital eccentricities (being exceeded in this respect only by 1566 Icarus), but also one of the highest inclinations (exceeded only by 1580 Betulia) and the greatest semimajor axis. Its evolution has therefore been investigated over an interval of more than 2000 yr (from 670 to 2900) (see Table I). This minor planet is exceptional in that it can come closer to Jupiter than the other planets can (see Table II). The closest encounters of 944 with Jupiter are as follows: 0.43 AU in 1211, 0.36 in 1673 and 0.50 in 2823. Marsden (1970) has found that all the minor planets avoid encounters with Jupiter except for 944 Hidalgo. The two planets having the next closest encounters are 279 Thule and 334 Chicago, which can approach to a distance of 1.1 AU from Jupiter. According to Marsden, 944 Hidalgo is a comet.

Chebotarev et al. (eds.), The Motion, Evolution of Orbits, and Origin of Comets, 437–439. *All Rights Reserved.*

M. A. DIRIKIS

TABLE I

Elements of 944 Hidalgo

Yr	ω	Ω	i	φ	μ	a (AU)
701	71°9	32°3	48°8	33°8	266″89	5.6120
903	69.5	30.6	47.9	34.8	260.76	5.6996
1104	64.4	29.0	47.9	35.9	272.72	5.5317
1300	61.6	27.0	46.6	37.1	265.68	5.6291
1510	59.9	25.3	45.4	38.6	266.83	5.6129
1713	59.3	23.3	43.3	40.0	254.30	5.7957
1907	56.8	21.8	43.0	40.6	255.91	5.7714
1964	57.6	21.0	42.5	41.0	252.75	5.8193
2101	55.8	20.1	42.4	41.6	260.17	5.7082
2309	55.1	18.1	41.6	42.2	253.27	5.8115
2502	54.1	16.4	41.1	42.9	258.19	5.7374
2709	53.4	14.2	40.5	43.5	258.21	5.7371
2903	51.4	12.2	39.2	44.6	258.68	5.7301

TABLE II

Minimal distances (AU)

Asteroid	Earth	Mars	Jupiter
944 Hidalgo	1.19	0.85	0.36
1036 Ganymed	0.33	0.07	1.90
1134 Kepler	0.43	0.08	1.52

The changes in the orbital elements of asteroids 1036 and 1134 are shown in Tables III and IV. In spite of the absence of close encounters with Jupiter we see that these changes have the same range as in the case of 944 Hidalgo.

TABLE III

Elements of 1036 Ganymed

Yr	ω	Ω	i	φ	μ	a (AU)
1351	115°8	232°8	22°8	35°0	816″62	2.6627
1451	118.3	229.8	23.3	34.6	815.64	2.6648
1551	121.0	227.0	23.8	34.3	816.05	2.6640
1651	124.4	223.5	24.5	33.9	816.26	2.6635
1750	126.7	221.0	25.1	33.4	815.85	2.6644
1850	128.7	218.7	25.6	33.1	816.43	2.6631
1950	131.1	216.3	26.3	32.9	818.60	2.6584
2050	133.4	214.0	27.0	32.2	817.26	2.6613
2150	135.1	212.1	27.5	31.5	815.44	2.6653
2250	136.9	210.3	28.1	31.1	815.88	2.6643
2350	139.4	208.2	28.8	30.6	817.36	2.6611
2450	141.0	206.6	29.3	30.0	816.38	2.6632

TABLE IV

Elements of 1134 Kepler

Yr	ω	Ω	i	φ	μ	a (AU)
1349	305°7	23°8	12°8	28°9	807″92	2.6818
1450	310.7	20.1	13.2	28.6	806.01	2.6860
1551	315.4	16.6	13.6	28.8	809.59	2.6781
1648	319.1	14.0	14.0	28.3	806.89	2.6841
1749	323.6	11.1	14.4	28.3	808.19	2.6812
1850	326.3	9.0	14.6	28.0	807.23	2.6833
1951	330.5	6.7	15.0	27.8	806.87	2.6841
2048	333.1	4.9	15.3	27.6	806.55	2.6848
2149	336.9	2.8	15.6	27.5	806.34	2.6853
2250	339.5	1.3	15.8	27.3	806.19	2.6856
2351	343.0	359.5	16.0	27.2	805.44	2.6873
2448	345.3	358.0	16.2	27.3	807.96	2.6817

In conclusion we can state that there is a significant difference between the motions of comets and asteroids:

(1) Close encounters with the major planets occur less frequently for asteroids than for comets.

(2) These encounters are never as close.

Consequently, the changes in the orbital elements of asteroids are never catastrophic. Small periodic (e.g., in a) and slow secular (e.g., in ω and Ω) variations prevail in the case of asteroids. The encounters cause the evolutionary changes in the elements to be accelerated, but in general this influence is of short duration.

References

Belyaev, N. A. and Chebotarev, G. A.: *Astron. Tsirk*. No. 480.

Chebotarev, G. A., Belyaev, N. A., and Eremenko, R. P.: 1970, *Byull. Inst. Teor. Astron.* **12**, 82.

Eckert, W. J., Brouwer, D., and Clemence, G. M.: 1951, *Astron. Pap. Washington* **12**.

Kazimirchak-Polonskaya, E. I.: 1967, *Trudy Inst. Teor. Astron.* **12**, 3.

Kazimirchak-Polonskaya, E. I.: 1968, *Astronomie* **82**, 323.

Marsden, B. G.: 1970, *Astron. J.* **75**, 206.

73. SECULAR PERTURBATIONS ON THE MINOR BODIES OF
THE SOLAR SYSTEM

I. V. GALIBINA

Institute for Theoretical Astronomy, Leningrad, U.S.S.R.

By means of the Gauss-Halphen-Goryachev method secular orbital variations have been studied for a period of 4000 yr (from −50 to +3950). The method has been applied to the periodic comets Halley, Brorsen-Metcalf, Pons-Brooks, Westphal, Olbers, Neujmin 1, and Encke. Investigations have demonstrated that the use of the method for most comets is not expedient as it does not allow for the possibility of approaches to the major planets and thus does not reflect the real evolution of the cometary orbits. Application of the method over the interval of 2000 years back from the epoch 1950.0 for the planets 279 Thule, 1162 Larissa, 1180 Rita, and 1202 Marina, as well as from the epoch 1850.0 for 1 Ceres has given adequate results and has displayed the stability of these orbits. Study of the secular perturbations on the Leonids over the interval of 4000 yr has confirmed the stability of that meteor stream. By means of the same method 14 minor meteor streams were investigated, and their orbits also proved to be stable. The availability of the various systems of osculating elements has permitted us to estimate for the first time the possibility of the encounter of those streams with the Earth over a 4000-yr period. For further details see Galibina (1970a, 1970b, 1971).

References

Galibina, I. V.: 1970a, *Byull. Inst. Teor. Astron.* **12**, 261.
Galibina, I. V.: 1970b, *Byull. Inst. Teor. Astron.* **12**, 470.
Galibina, I. V.: 1971, *Byull. Inst. Teor. Astron.* **12**, 870.

Discussion

B. G. Marsden: Two of the objects you have studied, P/Neujmin 1 and 279 Thule, are involved in librations about resonances with Jupiter. A purely secular treatment may thus yield results that are considerably in error. Is there a possibility that you could allow for these resonant effects in your calculations?

I. V. Galibina: I have not done so.

F. L. Whipple: What initial elements were adopted for P/Encke and which planetary perturbations were taken in account?

I. V. Galibina: The 1961 elements were adopted and the perturbations by Jupiter, Saturn, Uranus, and Neptune were considered.

V. V. Fedynskij: You have stressed the stability of meteor streams. The physical nature of meteoroids is not taken into account, however, and because of the Poynting-Robertson effect their lifetimes are much shorter than the interval over which their stability has been studied. This is especially relevant for small particles. I should like to suggest that in future the possibility of allowing for these nongravitational effects be investigated.

B. Yu. Levin: I do not agree that the lifetimes of the particles are so short. Consideration of the Poynting-Robertson effect will not appreciably influence the lifetime of a particle.

F. L. Whipple: The lifetime of a meteoroid is a few thousand years, comparable with the period covered by the investigation.

74. THE USE OF THE HALPHEN-GORYACHEV METHOD IN THE STUDY OF THE EVOLUTION OF THE ORBITS OF THE QUADRANTID AND δ AQUARID METEOR STREAMS

A. F. ZAUSAEV

Institute of Astrophysics, Dushanbe, U.S.S.R.

The Halphen-Goryachev method has been used for estimating the elements of the orbits of the Quadrantid and δ Aquarid meteor streams 4000 yr back from the epoch of osculation. Three meteor orbits were selected for each stream: the mean orbit, based on data by Shelton (1965) and by Jacchia and Whipple (1961), and also two variants. Two cases were discussed: in the first case only the perturbations by Jupiter were taken into account, and in the second those by Jupiter and Saturn. Designating the initial epoch of osculation as the year zero, we compared our results for the Quadrantids with those by Hamid and Youssef (1963) at the final epoch (the year -4000). The disagreements in the elements i, ω, and Ω were 3°, 8°, and 1°, respectively (in absolute value). In comparing the mean orbits of the two streams it was noticed that between the years -800 and -900, and also between -2300 and -2400, the eccentricities and inclinations to the ecliptic were practically identical. However, the minimum differences in Ω and ω were 174° and 153° between -800 and -900, and 15° and 30° between -2300 and -2400. Consequently, if only secular perturbations are taken into account we cannot obtain a positive answer to the question of connection between the Quadrantid and δ Aquarid streams during the 4000-yr interval. The orbital changes caused by the Poynting-Robertson effect will be negligible for the meteors discussed and can be disregarded.

References

Hamid, S. E. and Youssef, M. N.: 1963, *Smithsonian Contr. Astrophys.* **7**, 309.
Jacchia, L. G. and Whipple, F. L.: 1961, *Smithsonian Contr. Astrophys.* **4**, 97.
Shelton, J. W.: 1965, *Astron. J.* **70**, 337.

75. ON THE RATE OF EJECTION OF DUST BY LONG-PERIOD COMETS

V. N. LEBEDINETS

Institute for Experimental Meteorology, Obninsk, U.S.S.R.

Abstract. A model of the interplanetary dust medium that includes two subsystems, a spherical component and a flat one, has been proposed by the author as a result of analysis of radar meteor orbits. The disintegration of short-period comets is the main source of the flat dust cloud, while the disintegration of long-period comets is the main source of the spherical cloud. Assuming that dust particles fall into the Sun due to the Poynting-Robertson effect, one can estimate the intensity of ejection of dust from long-period comets necessary for the maintenance of the observed density of interplanetary dust. It has been found that the long-period comets should eject between 7×10^{14} and 2×10^{15} g of meteoric dust per year.

Analysis of the Kharkov catalogue of 12 500 orbits of meteors brighter than magnitude 7 (Lebedinets, 1968) has shown that the complex of small meteoric bodies producing radio meteors can be represented as a sum of two components: a 'spherical' component, comprising particles moving in orbits with random inclination i and more or less regular distributions of eccentricity e and perihelion distance q; and a 'flat' component, comprising particles moving in orbits with low inclinations ($i \leqslant 35°$), small semimajor axes ($a \lesssim 5$ AU), and fairly large eccentricities ($e \gtrsim 0.7$). The spherical component includes predominantly sporadic particles, while the flat component involves particles in streams. This shows that, on the average, the particles belonging to the spherical component appeared as independent celestial bodies before those of the flat component.

We have suggested that disintegration of long-period comets may be the principal source of particles in the spherical component (Lebedinets, 1968). The orbits of the particles are gradually transformed as a consequence of the Poynting-Robertson effect, and ultimately the particles fall into the Sun. As a result of the simultaneous action of these processes (disintegration of comets, transformation of particle orbits, and the particles falling into the Sun) there is a certain constant distribution of particle orbits. We have obtained formulae for this distribution with respect to a, e, and q; the distribution depends in essence only on the original distribution of injected particles by perihelion distance: $A(q_0)$.

By comparing the theoretical constant distribution with the observed distribution of orbits of the radio meteors, we have derived the function $A(q_0)$. Approximating it with a power function,

$$A(q_0) = A_0 q_0^n, \tag{1}$$

we find that the exponent n should be within the range $-1.5 \leqslant n \leqslant -1.0$.

Knowing $A(q_0)$, it is possible to evaluate the relative distribution of the spatial density of dust particles in the spherical component. Having calibrated this distribution with the results of measurements of particle densities near the orbit of the Earth,

Chebotarev et al. (eds.), *The Motion, Evolution of Orbits, and Origin of Comets*, 442–446. *All Rights Reserved.*
Copyright © 1972 by the IAU.

it is possible to evaluate the rate at which particles fall into the Sun and consequently the intensity of the process of dust ejection by the long-period comets.

Robertson (1937) obtained the following formulae for the orbital variations of an absolutely black, spherical particle due to radiative deceleration:

$$\frac{da}{dt} = -\frac{\alpha(2 + 3e^2)}{a(1 - e^2)^{3/2}}, \quad \frac{de}{dt} = \frac{5\alpha e}{2a^2(1 - e^2)^{1/2}}; \quad \alpha = \frac{3E_0 r_0^2}{4R\delta c^2}, \tag{2}$$

Here E_0 is the solar constant, r_0 the distance from the Earth to the Sun, R the radius of the particle, δ its density, and c is the velocity of light.

For the case when $e \neq 0$ Wyatt and Whipple (1950) obtained the relation

$$\frac{e^{4/5}}{q(1 + e)} = \frac{e_0^{4/5}}{q_0(1 + e_0)} = \frac{1}{C}, \tag{3}$$

where q_0 and e_0 are the values of q and e at some initial epoch, and C is a constant.

From Equations (2) and (3) we have

$$\frac{de}{dt} = -\frac{5\alpha(1 - e^2)^{3/2}}{2C^2 e^{3/5}}, \tag{4}$$

and from Equations (1) and (4) we may obtain the constant distribution of orbital eccentricities of particles injected into orbits with given q_0 and e_0 (≈ 1):

$$\left(\frac{dN}{de}\right)_{q_0} = -\frac{A(q_0)}{T(de/dt)}, \tag{5}$$

where T is the period of revolution. This distribution refers to particles passing through perihelion in unit time.

Using Kepler's third law and Equation (3) we may recast Equation (5) as

$$\left(\frac{dN}{de}\right)_{q_0} = \frac{2\sqrt{2}q_0^{1/2}A(q_0)}{5\alpha e^{3/5}}. \tag{6}$$

The partial density of particles at heliocentric distance r that have the particular q_0 is then

$$f(r, q_0) = \frac{1}{4\pi r^2} \int_{e_1}^{e_2} \frac{2}{dr/dt} \left(\frac{dN}{de}\right)_{q_0} de. \tag{7}$$

The radial component dr/dt of velocity depends on e, r, and q_0. The lower limit e_1 of integration is found from the condition $q_1 = r$ and Equation (3):

$$\frac{e_1^{4/5}}{1 - e_1} = \frac{r}{2q_0}. \tag{8}$$

When $q_0 \leq r$ the upper limit e_2 of integration is found from

$$e_2 = \frac{K - q_0}{K + q_0}, \tag{9}$$

K being the distance from the Sun to the boundary of the solar system; on the other hand, if $q_0 > r$, the upper limit is found with $q_2 = r$ and Equation (3):

$$\frac{e_2^{4/5}}{1 + e_2} = \frac{r}{2q_0}. \tag{10}$$

The radial component of the orbital velocity of the particle is found from the equations of elliptical motion:

$$r = a(1 - e \cos E), \qquad 2\pi t/T = E - e \sin E, \tag{11}$$

where E is the eccentric anomaly and t is measured from perihelion. It follows that

$$\frac{dr}{dt} = \frac{2\sqrt{2}\pi q_0^{1/2} e^{7/5}}{r(1 - e^2)^{1/2}} \sqrt{1 - \frac{1}{e^2}\left[1 - \frac{r(1 - e^2)}{2q_0 e^{4/5}}\right]^2}. \tag{12}$$

From Equations (7) and (12) we have

$$f(r, q_0) = \frac{A(q_0)}{10\pi^2 \alpha r} \int_{e_1}^{e_2} \frac{(1 - e^2)^{1/2} e^{-2}\, de}{\sqrt{1 - \frac{1}{e^2}\left[1 - \frac{r(1 - e^2)}{2q_0 e^{4/5}}\right]^2}}. \tag{13}$$

Integrating over q_0, we find the total spatial density of particles injected into orbits with any q_0:

$$f(r) = \frac{1}{10\pi^2 \alpha} \left\{ \frac{1}{r} \int_{0.01}^{r} A(q_0)\, dq_0 \int_{e_1}^{(K-q_0)/(K+q_0)} \frac{(1 - e^2)^{1/2} e^{-2}\, de}{\sqrt{1 - e^{-2}[1 - r(1 - e^2)/2q_0 e^{4/5}]^2}} \right.$$

$$\left. + \frac{1}{r} \int_{r}^{q_{02}} A(q_0)\, dq_0 \int_{e_1}^{e_2} \frac{(1 - e^2)^{1/2} e^{-2}\, de}{\sqrt{1 - e^{-2}[1 - r(1 - e^2)/2q_0 e^{4/5}]^2}} \right\} \tag{14}$$

For the particular power distribution of initial perihelion distances given by Equation (1) we obtain

$$f(r) = \frac{A}{10\pi^2 \alpha} \left\{ \frac{1}{r} \int_{0.01}^{r} q_0^n\, dq_0 \int_{e_1}^{(K-q_0)/(K+q_0)} \frac{(1 - e^2)^{1/2} e^{-2}\, de}{\sqrt{1 - e^{-2}[1 - r(1 - e^2)/2q_0 e^{4/5}]^2}} \right.$$

$$\left. + \frac{1}{r} \int_{r}^{q_{02}} q_0^n\, dq_0 \int_{e_1}^{e_2} \frac{(1 - e^2) e^{-2}\, de}{\sqrt{1 - e^{-2}[1 - r(1 - e^2)/2q_0 e^{4/5}]^2}} \right\}$$

$$= \frac{A_0 F(r, K, n)}{10\pi^2 \alpha}, \tag{15}$$

where the expression in braces is denoted by $F(r, K, n)$. In Equations (14) and (15) the lower limit of integration over q_0 is taken to be 0.01 AU, since at distances of less than this the equilibrium temperature of particles is higher than the temperature of evaporation of stone and iron meteorites; the upper limit is the maximum value q_{02}, which can be found from the condition that when $r > q_{02}$ the equilibrium temperature of a cometary nucleus is very low, and there is practically no ejection of dust.

From the radio observations of meteors we have found (Lebedinets, 1971) that in the vicinity of the Earth's orbit (i.e., at $r=1$ AU) one cubic kilometre contains on the average 10^{-5} particles with masses greater than $M=2\times 10^{-4}$ g, and of these 6×10^{-6} particles belong to the spherical component. Adopting a power distribution of particles through mass with exponent $S=2$, we find the integral density of the spherical component at $r=1$ AU to be

$$D_0 = 1.2 \times 10^{-9} \text{ g km}^{-3} = 4 \times 10^{15} \text{ g AU}^{-3}. \tag{16}$$

The differential density of particles with mass M is

$$f(1, M) = \left| \frac{dD(1, M)}{dM} \right| = \frac{D_0}{M^2}. \tag{17}$$

From Equations (15) and (17) we may find the unknown quantity A_0:

$$A_0 = \frac{10\pi^2 \alpha D_0}{M^2 F(1, K, n)}. \tag{18}$$

Substituting for the values of E_0, r_0 and c in Equation (2) we have

$$\alpha = \frac{3.55 \times 10^{-8}}{R\delta} = \frac{3.55 \times 10^{-8}}{(3\delta^2/4\pi)^{1/3} M^{1/3}} = \frac{C(\delta)}{M^{1/3}} \text{ AU}^2 \text{ yr}^{-1}, \tag{19}$$

where M is expressed in grams and δ in g cm^{-3}. With $\delta = 3.5$ g cm^{-3}, $C(\delta) = 2.48 \times 10^{-8}$. Hence

$$A_0 = \frac{10\pi^2 D_0 C(\delta)}{M^{7/3} F(1, K, n)}. \tag{20}$$

From Equations (1) and (20) we may obtain the total number N_Σ of particles which should be ejected by long-period comets in unit time, and their total mass M_Σ:

$$N_\Sigma(n, M_1, M_2) = \int_{0.01}^{5} q_0^n \, dq_0 \int_{M_1}^{M_2} A_0 \, dM$$

$$= \frac{15\pi^2 D_0 C(\delta)}{2F(1, K, n)} G(n) \left(\frac{1}{M_1^{4/3}} - \frac{1}{M_2^{4/3}} \right), \tag{21}$$

$$M_\Sigma(n, M_1, M_2) = \int_{0.01}^{5} q_0^n \, dq_0 \int_{M_1}^{M_2} A_0 M \, dM$$

$$= \frac{30\pi^2 D_0 C(\delta)}{F(1, K, n)} G(n) \left(\frac{1}{M_1^{1/3}} - \frac{1}{M_2^{1/3}} \right), \tag{22}$$

where

$$G(n) = \begin{cases} -\dfrac{1}{n+1} (5^{n+1} - 0.01^{n+1}) & n \neq -1 \\ 6.2 & n = -1. \end{cases} \tag{23}$$

Here M_1 and M_2 are the minimum and maximum particle masses.

Numerical integration of Equation (15) and the computation of the function $F(r, K, n)$ for different values of K and n were performed on a Minsk-22 computer. It was found that when $K > 1000$, $F(r, K, n)$ is practically independent of K. For the limiting values of $n = -1.0$ and -1.5 we have $F(1, \infty, -1.0) = 5.47$ and $F(1, \infty, -1.5) = 6.06$. With $M_1 = 10^{-13}$ g and $M_2 = 10^{13}$ g, $\delta = 3.5$ g cm^{-3}, Equations (22) and (23) give

$$\text{for } n = -1.0, \qquad M_\Sigma = 7 \times 10^{14} \text{ g yr}^{-1} \qquad\qquad (24)$$

$$\text{for } n = -1.5, \qquad M_\Sigma = 2 \times 10^{15} \text{ g yr}^{-1}.$$

It follows that the mass of dust ejected by all the long-period comets in one year is comparable to that of a single comet nucleus. During the lifetime of the solar system (5×10^9 yr) the total mass ejected would have been some 3×10^{23} to 10^{24} g, this constituting approximately one ten-thousandth part of the mass of the Earth.

References

Lebedinets, V. N.: 1968, in L. Kresák and P. M. Millman (eds.), 'Physics and Dynamics of Meteors', *IAU Symp.* **33**, 241.
Lebedinets, V. N.: 1971, *Space Res.* **11**, 307.
Robertson, H. P.: 1937, *Monthly Notices Roy. Astron. Soc.* **97**, 423.
Wyatt, S. P. and Whipple, F. L.: 1950, *Astrophys. J.* **111**, 134.

Discussion

F. L. Whipple: For masses greater than 10^{-5} g I find that erosion by smaller particles and collisions with comparable particles are more important than the Poynting-Robertson effect in eliminating the particles. For the zodiacal cloud I find that 10^7 g s^{-1}, or 3×10^{14} g yr^{-1} must be supplied by short-period comets to maintain the cloud. The spherical cloud does not appear to be very significant.

76. ÉVOLUTION SÉCULAIRE DES ORBITES DE PARTICULES MÉTÉORIQUES

J. DELCOURT

Centre National d'Etudes des Télécommunications, Issy-les-Moulineaux, France

Abstract. We study the secular evolution of the orbit described by a meteoric particle moving under the influence of planetary perturbations and the Poynting-Robertson effect. Using a simplified model, this study is made with only one disturbing planet. We present preliminary results on the long-term variation of the osculating elements of the orbit.

1. Introduction

Une liaison radioélectrique a permis d'observer les traînées ionisées créées par des particules météoriques dont la magnitude limite, estimée d'après la sensibilité de l'appareillage est de l'ordre de $M = 10$. L'étude présentée est faite conjointement avec le dépouillement de ces observations.

2. Hypothèses

(1) On ne considère qu'une seule planète perturbatrice, Jupiter. Dans cette étude préliminaire, son orbite est supposée circulaire et fixe.

(2) La particule météorique est supposée sphérique de diamètre $d \gg \lambda$, longueur d'onde du rayonnement incident; elle absorbe toute l'énergie incidente et la rayonne d'une manière isotrope. La force de Poynting-Robertson s'exprimera à l'aide d'un scalaire K.

(3) On néglige l'influence éventuelle du rayonnement corpusculaire du Soleil, qui pourrait être représenté par des équations analogues à celles de l'effet Poynting-Robertson.

(4) On néglige ici l'influence possible d'une charge électrique acquise par la particule.

3. Équations du Mouvement

Soit un trièdre d'inertie T', d'origine S; plan $x'Sy'$: orbite de Jupiter J. L'accélération totale agissant sur la particule P est

$$\mathbf{\Gamma}_T = -\frac{Gm_0}{r^3}\mathbf{r} + \frac{Kc}{r^3}\left(1 - \frac{\dot{r}}{c}\right)\mathbf{r} - \frac{K}{r^2}\mathbf{V} + Gm'\left(\frac{\mathbf{\rho}}{\rho^3} - \frac{\mathbf{r}'}{r'^3}\right)$$

$$= -\frac{\mu'}{r^3}\mathbf{r} + \mathbf{\Gamma}$$

$$\mathbf{\Gamma} = -K\frac{\dot{r}}{r^3}\mathbf{r} - \frac{K}{r^2}\mathbf{V} + \mathbf{\Gamma}_J$$

où m_0, m' sont les masses du Soleil S, de Jupiter J; $\mathbf{\Gamma} = \mathbf{SP}$, $\mathbf{r} = \mathbf{SJ}$, $\mathbf{\rho} = \mathbf{r}' - \mathbf{r}$; et $\mathbf{V}$ est la vitesse de P dans T'.

Chebotarev et al. (eds.), The Motion, Evolution of Orbits, and Origin of Comets, 447–453. All Rights Reserved.
Copyright © 1972 by the IAU.

On décompose $\mathbf{\Gamma}$ et $\mathbf{\Gamma}_J$ suivant les composantes radiale, perpendiculaire et binormale; respectivement (R^*, S^*, W^*) et (R, S, W). On obtient

$$R^* = -2K\frac{\dot{r}}{r^2} + R, \qquad S^* = -K\frac{\dot{v}}{r} + S, \qquad W^* = W,$$

où v est l'angle polaire de P sur son orbite.

4. Variations Séculaires des Éléments

On utilise la méthode de Gauss: on applique l'opérateur moyenne double en M et M' au système différentiel en a, e, i, Ω, ω, ε et on obtient

$$\bar{\dot{a}} = -K\frac{2 + 3e^2}{a(1 - e^2)^{3/2}} + \frac{2e}{2\pi n}\int_0^{2\pi}\bar{R}\sin E\,dE + \frac{2\sqrt{1 - e^2}}{n}\frac{1}{2\pi}\int_0^{2\pi}\bar{S}\,dE$$

$$\bar{\dot{e}} = -K\frac{5e}{2a^2(1 - e^2)^{1/2}} + \frac{1 - e^2}{2\pi na}\int_0^{2\pi}\bar{R}\sin E\,dE + \frac{(1 - e^2)^{3/2}}{2\pi nae}\int_0^{2\pi}\bar{S}\,dE$$

$$- \frac{\sqrt{1 - e^2}}{2\pi nae}\int_0^{2\pi}\bar{S}(1 - e\cos E)^2\,dE$$

$$\bar{\dot{i}} = \frac{\cos\omega}{2\pi na\sqrt{1 - e^2}}\int_0^{2\pi}\bar{W}(\cos E - e)(1 - e\cos E)\,dE$$

$$- \frac{\sin\omega}{2\pi na}\int_0^{2\pi}\bar{W}(1 - e\cos E)\sin E\,dE$$

$$\bar{\dot{\Omega}} = \frac{\sin\omega}{2\pi na\sqrt{1 - e^2}\sin i}\int_0^{2\pi}\bar{W}(\cos E - e)(1 - e\cos E)\,dE$$

$$+ \frac{\cos\omega}{2\pi na\sin i}\int_0^{2\pi}\bar{W}(1 - e\cos E)\sin E\,dE$$

$$\bar{\dot{\omega}} = -\frac{\sqrt{1 - e^2}}{2\pi nae}\int_0^{2\pi}\bar{R}(\cos E - e)\,dE + \frac{1 - e^2}{2\pi nae}\int_0^{2\pi}\bar{S}\sin E\,dE$$

$$+ \frac{1}{2\pi nae}\int_0^{2\pi}\bar{S}(1 - e\cos E)\sin E\,dE - \bar{\dot{\Omega}}\cos i$$

$$\bar{\dot{\varepsilon}} = -\frac{2}{2\pi na}\int_0^{2\pi}\bar{R}(1 - e\cos E)^2\,dE + \bar{\dot{\omega}}(1 - \sqrt{1 - e^2})$$

$$+ 2\bar{\dot{\Omega}}\sqrt{1 - e^2}\sin^2\frac{i}{2},$$

où

$$\bar{R} = \frac{1}{2\pi} \int\limits_0^{2\pi} R \, dM',$$

et également $\bar{S}$ et $\bar{W}$; $\tilde{\omega} = \Omega + \omega$, $\epsilon = \tilde{\omega} - n\tau$.

5. Méthode de Gauss-Halphen

Cette méthode est utilisée pour calculer les composantes $\bar{R}$, $\bar{S}$, $\bar{W}$ de $\bar{\Gamma}_J$. On en trouve un exposé succinct par Musen (1963).

De l'expression de $\mathbf{\Gamma}_J$, on tire aisément

$$\bar{\mathbf{\Gamma}}_J = \frac{Gm'}{2\pi} \int\limits_0^{2\pi} \frac{\boldsymbol{\rho}}{\rho^3} \, dM'.$$

Le lieu de $\boldsymbol{\rho}$ est un cône elliptique de sommet P. On démontre que

$$\bar{\mathbf{\Gamma}}_J = -\frac{Gm'}{\pi a'^2 z'} \, \mathbf{\Phi} \cdot \mathbf{r},$$

avec

$$\mathbf{\Phi} = \tfrac{1}{2} \int\limits_{E'} \frac{\boldsymbol{\rho}}{\rho^3} (\boldsymbol{\rho} \times d\boldsymbol{\rho}); \quad (E' : \text{orbite de Jupiter}).$$

$\mathbf{\Phi}$ ne dépendant pas de $|\boldsymbol{\rho}|$, E' peut être remplacée par une section quelconque du cône. Le problème est alors de déterminer les composantes de $\mathbf{\Phi}$ et de $\mathbf{r}$ dans un trièdre approprié. En particulier, en prenant un trièdre $Pxyz$ de sommet P, coïncidant avec les axes principaux du cône et en prenant pour courbe d'intégration la section par un plan $z = 1$, les 3 composantes diagonales non nulles de $\mathbf{\Phi}$ s'expriment, à l'aide des intégrales elliptiques complètes de Legendre, à partir des coordonnées de P dans T'. Mais la détermination des axes principaux d'un cône de sommet et de base connus demande la résolution d'une équation du 3e degré (méthode de Gauss-Hill). Halphen a montré que l'on peut éviter cette résolution. On prend pour contour d'intégration la biquadratique gauche, intersection du cône et de la sphère centrée sur P, de rayon unité (courbe unicursale de paramètre s) et on fait le changement de variable $s \to u$:

$$\sqrt{(s-A)(s-B)(s-C)} = -\tfrac{1}{2}p'u,$$

où A, B, C sont les grandeurs des axes principaux.

La méthode est résumée dans les points suivants (Halphen, 1886):

(1) les composantes de $\mathbf{\Phi}$ s'expriment en fonction de la période réelle ω de pu et de $\eta = \zeta\omega$ (où pu, ζu sont les fonctions elliptiques de Weierstrass).

(2) ω et η s'expriment en fonction des invariants elliptiques g_2 et g_3 de pu.

(3) le changement de variables $s \to u$ relie les fonctions symétriques des racines de pu, soit g_2 et g_3 et les fonctions symétriques des racines de l'équation en s du cône (trois invariants non nuls du cône).

(4) ces invariants du cône s'obtiennent eux-mêmes à partir d'une configuration donnée du cône, c'est-à-dire finalement, en fonction des coordonnées de P dans T'. Ainsi, on aboutit à

$$\overline{\Gamma}_J = -\Psi \cdot \mathbf{r},$$

où

$$\Psi = A\mu + B\nu,$$

et μ, ν sont des dyadiques symétriques dont les composantes sont des polynômes respectivement de degrés 6 et 2 suivant les coordonnées de P (dans T'). A et B sont des scalaires dont les expressions sont:

$$(1) \qquad B = \frac{4}{\pi} g_2^{-5/4} \left[KF(\tfrac{1}{12}, \tfrac{5}{12}; \tfrac{1}{2}; \xi) - (\operatorname{sgn} g_3) \frac{\pi}{4\sqrt{3}K} \xi^{1/2} F(\tfrac{7}{12}, \tfrac{11}{12}; \tfrac{3}{2}; \xi) \right]$$

$$(2) \qquad B = \frac{4}{\sqrt[4]{12}} g_2^{-5/4} \left(\frac{1+\sqrt{\xi}}{2} \right)^{-1/6} F\left(\tfrac{1}{6}, \tfrac{1}{6}; 1; -\frac{1-\sqrt{\xi}}{1+\sqrt{\xi}} \right), \qquad g_3 > 0$$

$$(1) \qquad A = +(\operatorname{sgn} g_3) \frac{40}{3\pi\sqrt{3}} K g_2^{-11/4} \xi^{1/2} F(\tfrac{13}{12}, \tfrac{17}{12}; \tfrac{3}{2}; \xi) - \frac{8}{K} g_2^{-11/4} F(\tfrac{7}{12}, \tfrac{11}{12}; \tfrac{3}{2}; \xi)$$

$$- \frac{154}{27K} g_2^{-11/4} \xi F(\tfrac{19}{12}, \tfrac{23}{12}; \tfrac{5}{2}; \xi)$$

$$(2) \qquad A = -\frac{10}{9} \sqrt[4]{12} g_2^{-11/4} \left(\frac{1+\sqrt{\xi}}{2} \right)^{-7/6} F\left(\tfrac{7}{6}, \tfrac{1}{6}; 2; -\frac{1-\sqrt{\xi}}{1+\sqrt{\xi}} \right), \qquad g_3 > 0$$

où $\xi = 27 g_3^2 g_2^{-3}$ est l'invariant de Klein, $K = K(1/\sqrt{2}) = \Gamma^2(\tfrac{1}{4})/4\sqrt{\pi}$, où $K(k)$ est l'intégrale elliptique complète de deuxième espèce, et $F(\alpha, \beta; \gamma; z)$ est la fonction hypergéométrique de Gauss.

Apparemment, Musen ne considère pas le cas où $g_3 < 0$. Les formules (2) qu'il utilise ne sont valables que pour $g_3 \geqslant 0$; d'ailleurs, elles proviennent des développements au voisinage de $\xi = 1$ des solutions de l'équation hypergéométrique de Bruns, et de ce fait, convergent très lentement au voisinage de $\xi = 0$, c'est-à-dire précisément lorsque g_3, passant par 0, change de signe. Nous avons montré qu'au voisinage du plan de Jupiter, g_3, considéré comme fonction du rayon polaire r de P, devient négatif pour $a'/\sqrt{2} < r < a'\sqrt{2}$ (a' étant le rayon de l'orbite de J).

6. Effet Poynting-Robertson

L'influence du seul effet Poynting-Robertson sur les orbites des très petits corps a été étudiée (Robertson, 1937; Wyatt et Whipple, 1950). Avec les hypothèses du (2) on montre que

$$K = 7.1 \times 10^{-8} \, d^{-1} \, \delta^{-1} \, \text{UA}^2 \, \text{an}^{-1},$$

δ étant la densité de la particule. Les calculs ont été effectués avec plusieurs valeurs de K; en particulier $K = 2 \times 10^{-6} \, \text{UA}^2 \, \text{an}^{-1}$ (Figures 1 à 5) correspond à une particule sphérique, de densité $\delta \approx 3$ et de diamètre $d \approx 100 \, \mu\text{m}$.

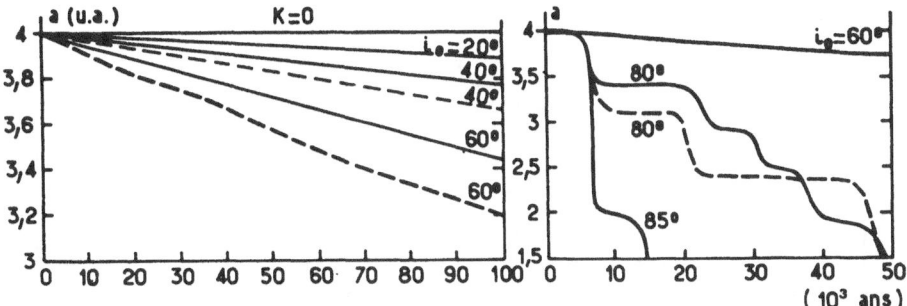

Fig. 1. Variation de a ($a_0 = 4$; $e_0 = 0,4$ et $0,6$; $\omega_0 = 90°$).

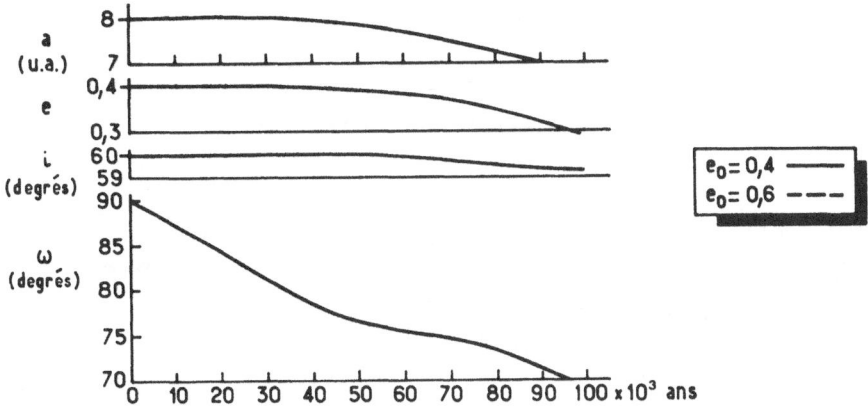

Fig. 2. Variation de a, e, i et ω ($a_0 = 8$; $e_0 = 0,4$; $i_0 = 60°$, $\omega_0 = 90°$).

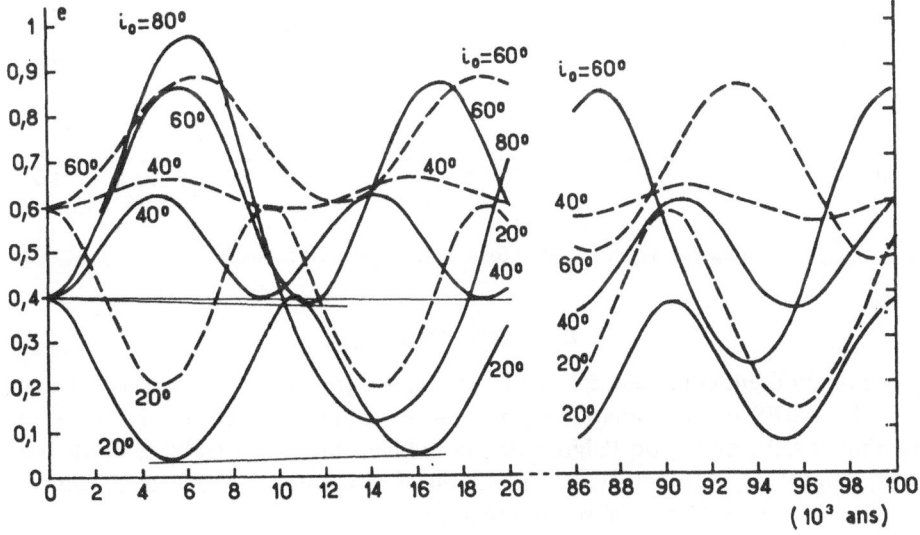

Fig. 3. Variation de e ($a_0 = 4$; $e_0 = 0,4$ et $0,6$; $\omega_0 = 90°$).

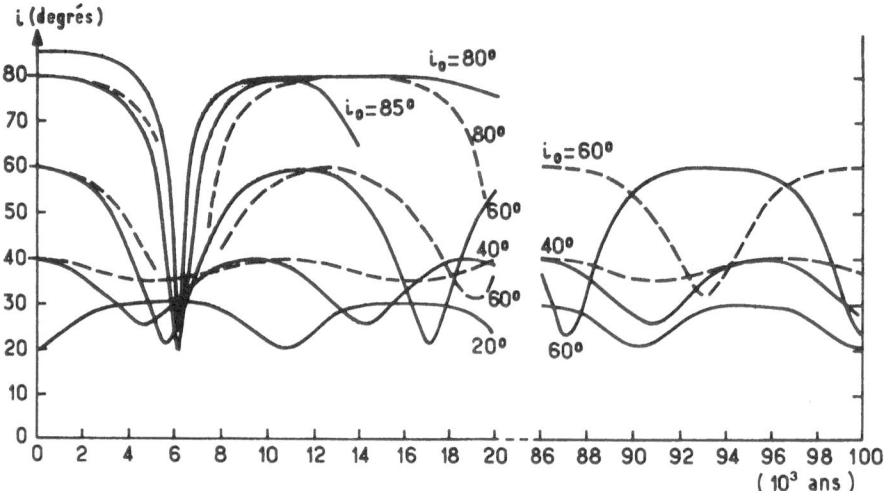

Fig. 4.   Variation de i ($a_0 = 4$; $e_0 = 0,4$ et $0,6$; $\omega_0 = 90°$).

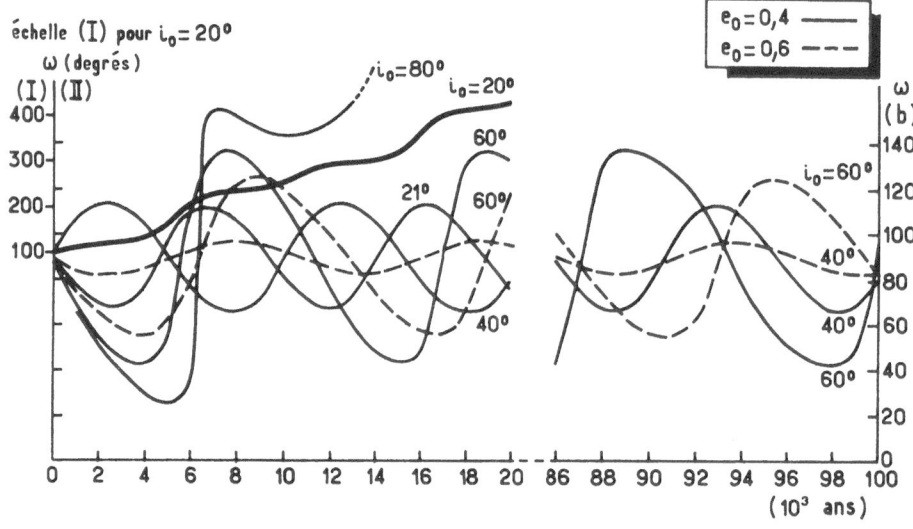

Fig. 5.   Variation de ω ($a_0 = 4$; $e_0 = 0,4$ et $0,6$; $\omega_0 = 90°$).

7. Intégration Numérique

Le système différentiel moyenné en M et M' est intégré numériquement (sur calcula-teur C II—10070) par une méthode à pas liés. On utilise un algorithme proposé par Hamming (1959), dérivé de l'algorithme de Milne, qui a une stabilité analogue et permet, en général, d'obtenir un ordre plus élevé de l'erreur de troncature, soit, dans le cas présent, $O(h^6)$, h étant le pas d'intégration.

Un test de proximité des deux orbites est appliqué en permanence; toute approche en deçà d'une certaine limite choisie (par example, 0.3 UA) est signalée et constitue dans le programme actuel, un point d'arrêt de la solution. Une étude plus complète sur ce sujet est en cours.

8. Résultats

Les calculs ont été effectués:
(1) dans le cas intérieur pour $a_0 = 3$; 4; $e_0 = 0,4$; 0,6; $i_0 = 20, 40, 60, 80, 85°$; $\omega_0 = 0, 45, 90°$.
(2) dans le cas extérieur pour $a_0 = 8$; 9; 10; $e_0 = 0,4$; 0,6; $i_0 = 20, 40, 60, 80°$; $\omega_0 = 90°$.

Des résultats partiels sont illustrés par les figures 1 à 5. Les variations calculées de $\Omega(t)$ ne sont pas montrées car le système étudié possède la symétrie de révolution (orbite de J supposée circulaire). On a déterminé la transition circulatoire-oscillatoire pour $\omega(t)$ qui, par example pour $a_0 = 4$, $e_0 = 0,4$ est comprise entre $i = 20°$ et $i = 21°$. Cette valeur de i est voisine de celle obtenue par Kozai (1962) qui élimine les termes à courte période par la méthode de von Zeipel. Un résultat à noter consiste en la diminution sensible du semi-grand axe a, pour les grandes inclinaisons; dans ce cas, l'excentricité varie beaucoup et la distance périhélie q étant très faible au voisinage de $e = 1$, les termes de Poynting-Robertson deviennent importants. Dans ces conditions, la durée moyenne de vie physique de la particule météorique peut être réduite fortement alors que, dans la présente étude de dynamique, on ne considère que la durée de vie orbitale. Les approches avec Jupiter qui se produisent pour certaines conditions initiales nécessitent, pour leur exploitation, une étude plus complète qui est en cours.

Remerciements

L'auteur remercie M. J. Kovalevsky, du Bureau des Longitudes, Paris, pour l'intérêt constant qu'il porte à cette étude, et tient à exprimer sa vive gratitude à ses collègues du C.N.E.T. Lannion, MM le Clerc et Leguen, pour leur contribution essentielle à la programmation numérique.

Références

Halphen, G. H.: 1886, *Traité des fonctions elliptiques*, Paris, Tome I.
Hamming, R. W.: 1959, *J. Assoc. Comput. Machinery* **6**, 37.
Kozai, Y.: 1962, *Astron. J.* **67**, 591.
Musen, P.: 1963, *Rev. Geophys.* **1**, 85.
Robertson, H. P.: 1937, *Monthly Notices Roy. Astron. Soc.* **97**, 423.
Wyatt, S. P. et Whipple, F. L.: 1950, *Astrophys. J.* **111**, 134.

Discussion

E. N. Kramer: In what way does consideration of the Poynting-Robertson effect influence the results?

J. Delcourt: For particles of magnitude 6–7 and for a period of 20 000 yr the Poynting-Robertson effect does not play an important part. But since we intend to increase the period of investigation the effect was taken into account in the equations. For smaller particles the effect would be of importance over the 20 000-yr period.

77. DEFORMATION OF A METEOR STREAM CAUSED BY AN APPROACH TO JUPITER

B. YU. LEVIN

O. Schmidt Institute of Physics of the Earth, U.S.S.R. Academy of Sciences, Moscow, U.S.S.R.

A. N. SIMONENKO

*Committee on Meteorites, U.S.S.R. Academy of Sciences,
Moscow, U.S.S.R.*

and

L. M. SHERBAUM

Astronomical Observatory, Kiev University, Kiev, U.S.S.R.

Abstract. The increase in the width of a meteor stream with time is usually regarded as due to differences in the secular variations of the slightly different orbits of individual meteoroids. However, the ordinary differential planetary perturbations acting on different parts of an elliptical ring of meteoroids have a more important effect. Model calculations are presented for a filament-like stream that makes close approaches to Jupiter. They confirm the major importance of deformations of the stream by direct planetary perturbations that act even on a filament-like stream of zero thickness. The disturbed part of the stream takes the form of a bend, dispersing along the initial orbit and producing a significant decrease in the spatial density of the meteoric particles.

The increase in the width of a meteor stream, leading in time to its dispersion and ultimate merging into the sporadic background, is usually regarded as due to differences in the secular perturbations on the slightly different orbits of individual meteoroids (Ahnert-Rohlfs, 1952). Some ten years ago one of the present authors (Levin, 1956) stressed the major importance of deformation of a stream arising from the differential planetary perturbations on different parts of the elliptical ring. This differential effect deforms even a filament-like stream of zero thickness. The maximum perturbations are experienced by the part of the stream that passes the point of closest approach to the orbit of a perturbing planet simultaneously with that planet. Model calculations on the effect have been made (Sherbaum, 1970) by Cowell's method (using an electronic computer) for three similar filament-like streams that approach Jupiter closely; the streams can be regarded as three filament-like parts of the same broad stream.

TABLE I

Parameters for the streams

	I	II	III
a	4 AU	4 AU	4 AU
e	0.5	0.5	0.5
q	2 AU	2 AU	2 AU
i	8°	5°	4°12′
ω	30°	52°59′	108°
Ω	30°	6°53′	311°43′
$\varDelta_{min}$	0.154 AU	0.327 AU	0.511 AU

Chebotarev et al. (eds.), The Motion, Evolution of Orbits, and Origin of Comets, 454–461. All Rights Reserved.
Copyright © 1972 by the IAU.

The orbits of the streams are of identical size and shape and have their perihelia in the same direction, but they differ in orientation and consequently in the distances Δ_{min} of closest approach to Jupiter (see Table I).

In Figure 1a the three orbits cannot be distinguished from one another. We have marked the part of the stream whose behaviour we have studied. Approximately in the middle of this part is situated the 'closest approach particle' (CAP) which later passes at the minimum distance from Jupiter. The distance from CAP to the ends of the

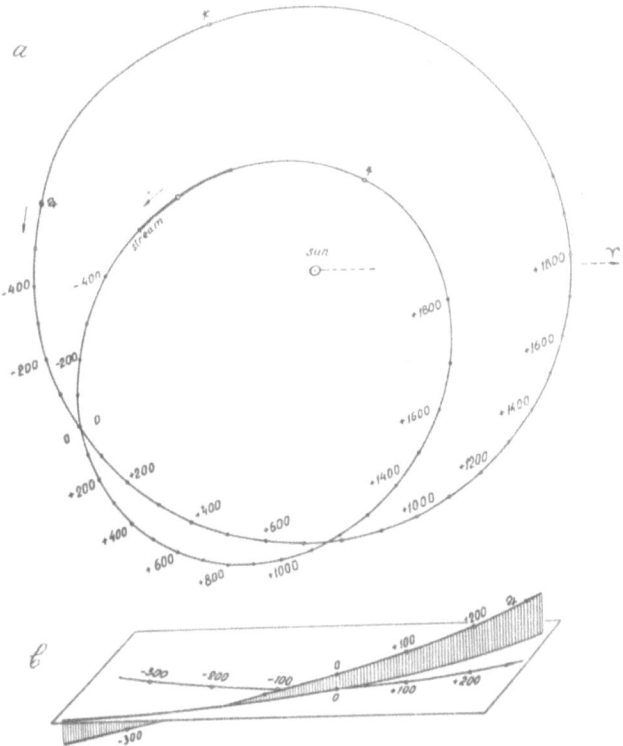

Fig. 1. (a) The orbits of Jupiter and the meteor stream. The positions of CAP (the 'closest approach particle') and Jupiter in their respective orbits are marked by dots at intervals of 100 days before ($-$) and after ($+$) the moment of closest approach (0). (b) the position in space of the orbits of Jupiter and stream I in the region of approach.

section of the stream studied is $25°$ in mean anomaly M_0 or 202 days in time. Figure 1b shows the position in space of the orbit of the first stream and that of Jupiter in the region of approach. Streams II and III pass below the first one.

Figure 2 shows for streams I and II the distances of the particles from Jupiter and the orbital elements of the particles at different moments of time before and after the closest approach to Jupiter. Because of the perturbations by Jupiter the streams are deformed, and the minimum distances from Jupiter are smaller than the Δ_{min} given above, namely 0.125 AU for stream I, 0.297 AU for stream II and 0.490 AU for stream III.

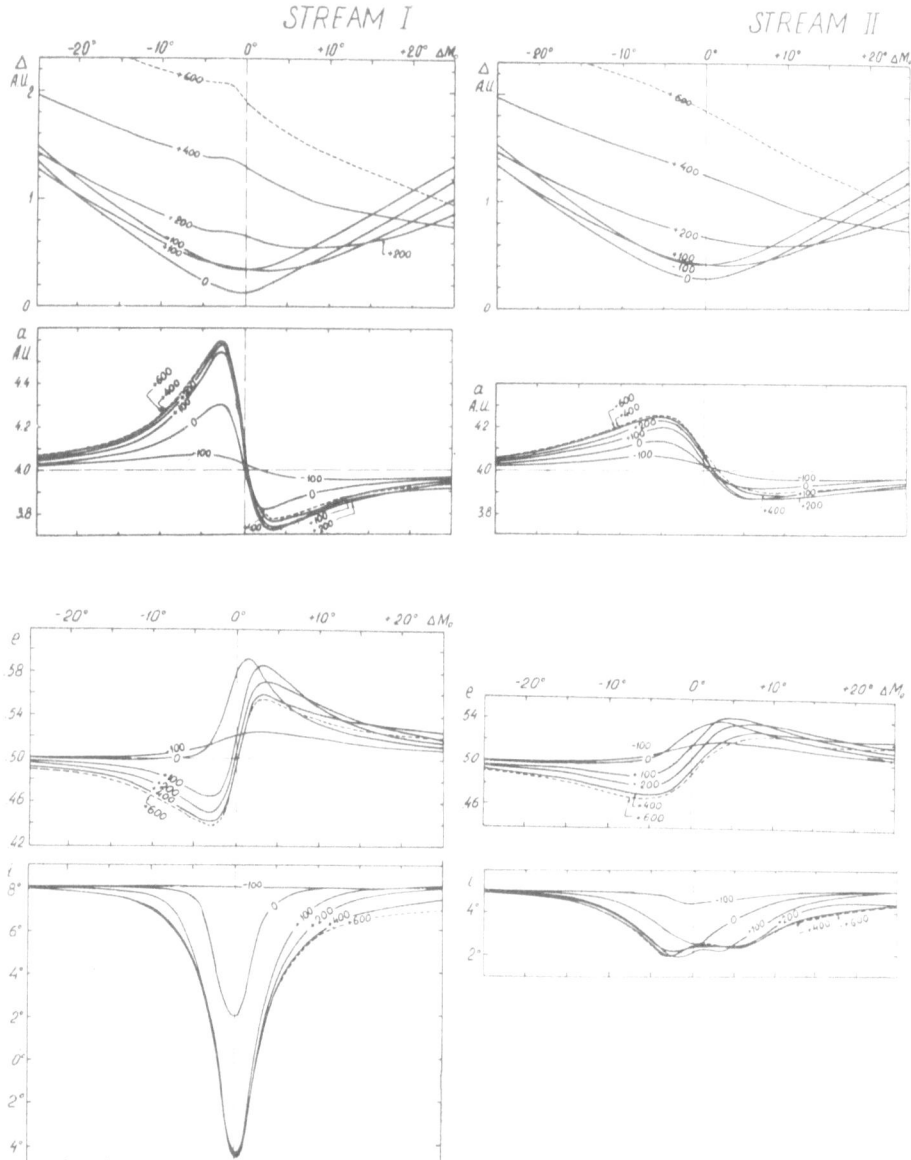

Fig. 2.   The distances of particles of streams I and II from Jupiter and the orbital elements of the
particles. The abscissa gives the distances of the particles from CAP in terms of
the initial mean anomaly.

For the particles which come closest to Jupiter, i.e., for those situated near CAP,
the changes in semimajor axis a are small. This is because during the closest approach
the velocity vectors of these particles are approximately perpendicular to the direction
to Jupiter. Jupiter's attraction changes mainly the directions of the velocity vectors and

not the magnitudes. For particles moving ahead of CAP a diminishes monotonically with time, while for particles moving behind CAP it increases monotonically. The maximum and minimum of a occur for particles situated 2 to 3° from CAP for stream I and at somewhat larger distances for streams II and III. In the examples calculated a increases more rapidly than it decreases. At $t = -100$ days the extremal changes in a for particles of stream I are smaller than 0.1 AU. At the moment of the closest approach of Jupiter to the perturbed stream ($t = 0$) these extremal changes of a reach 0.3 AU, and then during the 100 days after approach they rise to 0.54 AU.

The changes in eccentricity e are, roughly speaking, opposite in phase to the changes in a. There is a minimum of e for particles moving behind CAP and a maximum for those moving ahead. At $t = +400$ days the extremal values of e for stream I are 0.44 and 0.55. For particles near CAP the velocity vector turns first in one direction and then the other. Thus, before the closest approach e increases, and after the approach it returns almost to the original value of 0.5.

The changes in inclination i for the particles in stream I are substantial. At the time of closest approach the inclination of particles close to CAP changes at a rate of about 10′ per day. The total change in i within 600 days after closest approach reaches 12°.5; the value of i goes through zero, and there is thus an interchange of the ascending and descending nodes. This interchange of nodes is depicted here by passage below the line $i = 0$; the motions of the particles remain direct.

The orbital elements change rapidly during the period of closest approach (from -100^d to $+100^d$), but the variations subsequently become more gradual. At $t = 400$ or 500^d after closest approach the perturbations become much smaller than before, and the further changes in the elements are about 0.1 to 0.01 of those accumulated previously. By that time Jupiter is more than 0.7 to 0.8 AU from all parts of the stream.

The changes in a and e for particles in streams II and III during the period of their approach to Jupiter are similar but markedly smaller than for stream I. The changes in i are also small.

Before being perturbed by Jupiter the particles moved one after another along the same orbit. After the perturbations they move along different orbits. Although the orbital elements have already changed substantially by the time of closest approach the particles continue to be situated near the original orbit. But gradually a bend develops in the stream.

Figure 3 shows the portion of stream I that suffers the greatest perturbations. The bends represent the instantaneous locations of particles that are about to move along different trajectories. At the time of closest approach the most perturbed particles form a small bend directed towards Jupiter. They deviate from the initial orbit by less than 0.1 AU. At the top of the bend are the particles closest to CAP. The deformation increases with time and becomes more and more pronounced after Jupiter has moved away from the stream and only slightly influences its motion. The form of the bend changes. Particles which followed CAP have increased velocity: during the first stage of the development of the bend they outrun CAP, while the particles that moved ahead are decelerated and lag behind.

The deformations of streams II and III are similar to those described above for stream I, but they are smaller and develop slowly.

The further development of the bend in stream I is determined mainly by the changes occurring in orbits of individual particles during a short time around closest approach. For streams II and III the development of the bend after closest approach is somewhat more complicated because of the larger relative role of the previous and subsequent small perturbations by Jupiter.

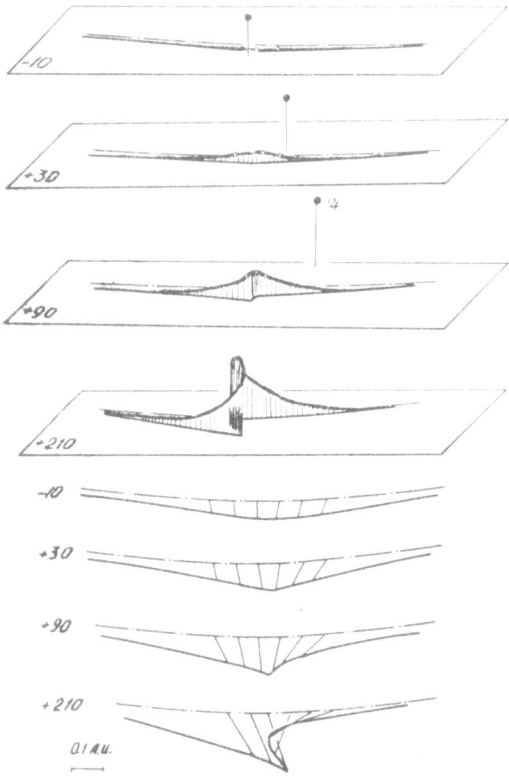

Fig. 3. The portion of stream I that suffers the greatest perturbations. The arc considered is 23° in terms of the initial mean anomaly. The four upper diagrams represent the locations of the bend with respect to the plane of the initial (unperturbed) orbit at various times, while the four lower ones are their projections on that plane.

Figure 4 shows the development of the bends of all three streams during one revolution, while Figure 5 illustrates subsequent spatial positions of the bend in the first stream relative to the plane of the initial orbit. The diagrams show the CAP and two points that were initially situated at $\pm 1°\!.2$ in terms of the initial mean anomaly. At the beginning these particles are situated close to one another. Under the perturbations by Jupiter they start to deviate together from the initial orbit, but in the course of time they drift apart because of the differences in their orbits.

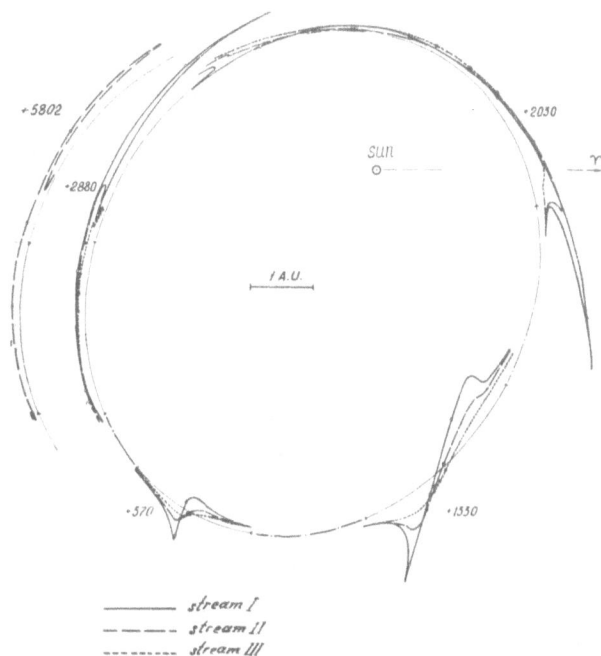

Fig. 4. Development of the bends in the streams during one revolution, projected on to the ecliptic plane. The corresponding unperturbed sections of the stream (50° in length in terms of the initial mean anomaly) are shown by a thin line.

The deviations of particles from their unperturbed positions are of the same order of magnitude both in the plane of their initial orbit and perpendicular to it. Differences in the orbits lead to rapid changes in the size and form of the bend. As already mentioned, the development of the bends in streams II and III proceeds in a similar way, but more slowly. In Figure 4 the bend in the stream II is shown after two revolutions ($t = +5802^d$).

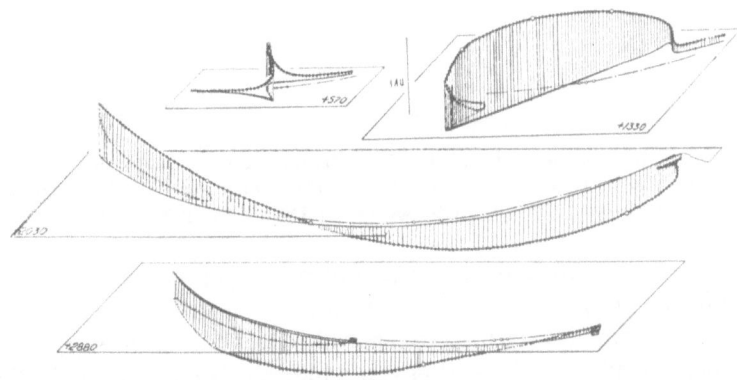

Fig. 5. Development of the bends in stream I relative to the plane of the initial orbit.

The linear density of particles in the bend rapidly diminishes. For stream I it decreases by more than an order of magnitude during one revolution. Due to the large dependence of the perturbations on the distance to Jupiter, there are corresponding increases in the distances between particles in the bends on neighbouring filamentary streams. Consequently, in the examples calculated, i.e., for a stream closely approaching Jupiter, the spatial density of particles in the perturbed section decreases after one revolution by 3 to 4 orders of magnitude, and it continues to decrease during subsequent revolutions. Near the point of closest approach, however, the simultaneous passage of different parts of the bend can lead to temporary increases in density.

It must be noted that the subsequent extensions of the bends already formed is accompanied by the production of 'younger' bends during the regular passage of Jupiter near the point of closest approach. After several revolutions the particles of stream I will thus have dispersed over a region up to 2 AU in width. This dispersion is much larger than that resulting from ejection of particles by a cometary nucleus (see Figure 6). Actually, Jupiter disperses not only the initial stream but also the bends

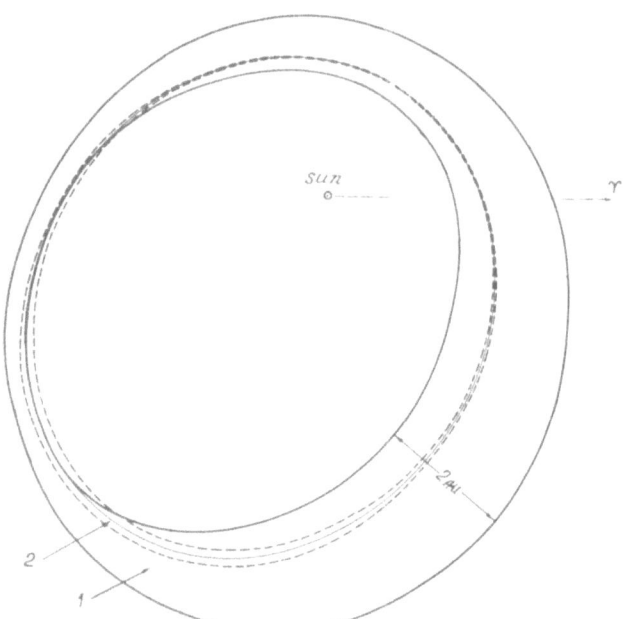

Fig. 6. Region 1 is that occupied by particles of stream I after a single approach to Jupiter. Region 2 is that occupied by particles ejected isotropically (at a velocity of 0.1 km s⁻¹) from a cometary nucleus near perihelion (the true ejection velocities are possibly smaller than this).

already formed, and therefore the dispersion of particles proceeds more rapidly and over a larger region of space.

Dispersion occurs rapidly even for a stream which does not have close approaches to Jupiter. Ordinary small perturbations also produce bends (see Figure 7). Because of the larger inclination ($i = 30°$) this stream does not approach Jupiter's orbit any closer than 0.74 AU. Nevertheless, the point of intersection illustrated describes complicated loops increasing in size with time. To each revolution there correspond 2–3 loops. During the third or fourth revolutions the loops are already about 10^6 km in extent, and by the tenth revolution they extend up to 20×10^6 km.

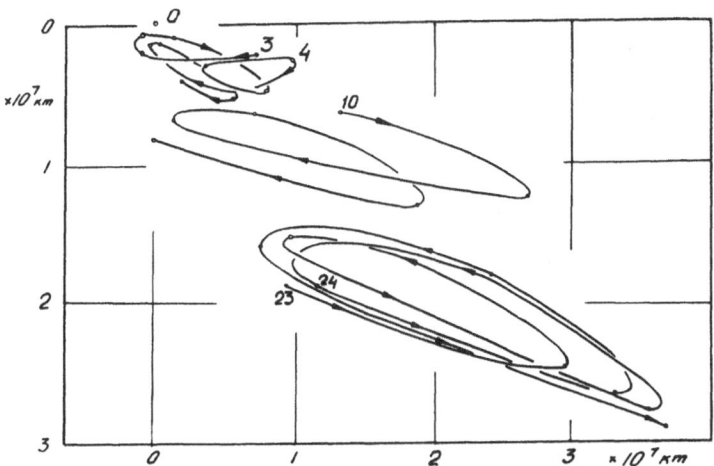

Fig. 7.   The displacement with time of the point of intersection of a filament-like stream ($a = 4.5$ AU, $e = 0.5$, $i = 30°$, $\omega = 100°$, $\Omega = 9°$) with a plane perpendicular to the stream at a point situated 90° from the initial perihelion.

References

Ahnert-Rohlfs, E.: 1952, *Veroeffentl. Sternw. Sonneberg* **2**, 38.

Levin, B. Yu.: 1956, *Fizicheskaya Teoriya Meteorov i Meteor noe Veshchestvo v Solnechnoj Sisteme*, Akad. Nauk SSSR, Moscow. (See also Levin, B. Yu.: 1961, *Physikalische Theorie der Meteore und die meteoritische Substanz im Sonnensystem*, Akademie-Verlag, Berlin.)

Sherbaum, L. M.: 1970, *Vestn. Kiev. Gos. Univ. Ser. Astron.* No. 12.

Discussion

Yu. V. Evdokimov: How many meteors were used in your calculations and why should they initially have the same orbit?

B. Yu. Levin: We considered 16 points located at equal intervals in mean anomaly. Meteors leaving the nucleus of a comet with almost zero velocity will initially have practically the same orbit but will be at different distances from the comet.

Yu. V. Evdokimov: In our calculations on the Draconids we used meteors which correspond very closely to the real ones. It seems to me that even if the meteors leave the comet with small velocities, they do not have the same orbit.

78. ORBITAL EVOLUTION OF THE α VIRGINID AND α CAPRICORNID METEOR STREAMS

E. I. KAZIMIRCHAK-POLONSKAYA and N. A. BELYAEV

Institute for Theoretical Astronomy, Leningrad, U.S.S.R.

and

A. K. TERENT'EVA

Astronomical Observatory, Kiev University, Kiev, U.S.S.R.

Abstract. The motions of two typical meteor streams of the Jupiter family, the α Virginids and the α Capricornids, were considered over the interval 1902–2003, perturbations by Venus to Neptune being taken into account. The substantial transformations of their orbits in Jupiter's sphere of action and its neighbourhood were studied. Conclusions are made concerning the role of large perturbations by Jupiter in dispersing meteor streams and altering the visibility conditions of meteor showers.

1. Introduction

It is well known that at the present time several hundred meteor streams exist in the solar system, the majority of them belonging to the Jupiter family. The purpose of this investigation is to study the evolution of the orbits of typical meteor streams of the Jupiter family and to examine the great transformations of their orbits under the influence of strong perturbations by Jupiter.

Two streams have been selected: the α Virginids and the α Capricornids. They are sufficiently intense, act for prolonged periods of time (more than 20 days), contain large bolides and have been under observation for not less than a hundred years.

Analysis of photographic and visual observations (Terent'eva, 1966) reveals that the α Virginid system is very dispersed, and that it is best represented by separate small streams acting for not less than two months. Among those small streams it is most advisable to select stream No. 48, whose aphelion is located in the immediate vicinity of Jupiter's orbit. For the investigation of the α Virginids we have used the orbit of meteor No. 7333, which is included in the catalogue of 413 exact orbits derived by Jacchia and Whipple (1961) from observations made in the U.S.A. during 1951–1954. The initial system of elements is as follows:

$$
\begin{aligned}
&\text{Epoch} = 1953 \text{ April } 15.28 \text{ UT} \\
&M_0 = 351°035 \qquad\qquad\quad a = 2.787 \text{ AU} \\
&\left.\begin{aligned}
\omega &= 289.0 \\
\Omega &= 25.0 \\
i &= 2.3 \\
\pi &= 314.0
\end{aligned}\right\} 1950.0 \quad
\begin{aligned}
e &= 0.861 \\
q &= 0.388 \text{ AU} \\
Q &= 5.186 \text{ AU} \\
P &= 4.653 \text{ yr.}
\end{aligned} \qquad (1)\\
&n = 0°211835
\end{aligned}
$$

Wright *et al.* (1956) have made a special study of the α Capricornid stream, investigating the radiants and orbits of 12 double-station and 50 single-station photographs

Chebotarev et al. (eds.), *The Motion, Evolution of Orbits, and Origin of Comets*, 462–471. *All Rights Reserved.*
Copyright © 1972 by the IAU.

obtained by the Harvard Observatory during 1899–1953. They concluded either that the α Capricornids are a single, continuous stream from July 16 to August 22, or that they consist of double (or multiple) independent streams occurring during July 17–August 1 and August 1–22, respectively (see Figure 1). The authors consider the second suggestion more probable, and they also believe that the stream may be associated with P/Honda-Mrkos-Pajdušáková. For the purpose of our investigation

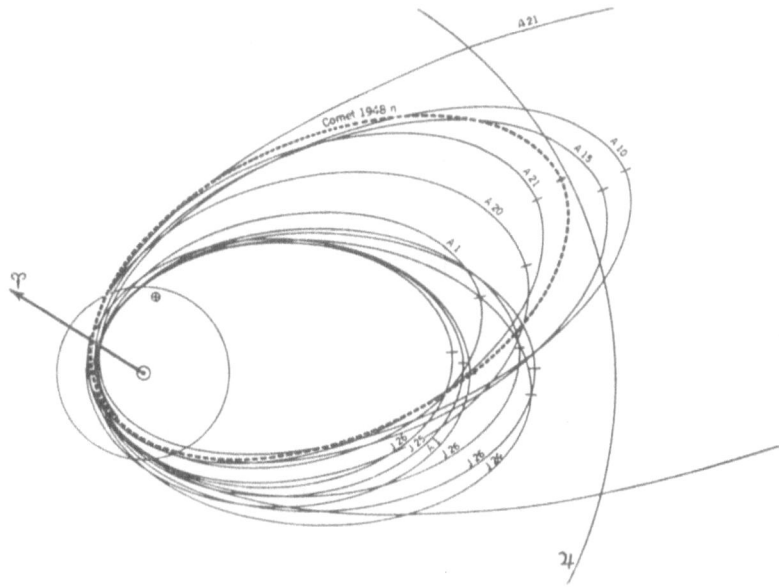

Fig. 1.   Orbits of α Capricornids and Comet 1948n P/Honda-Mrkos-Pajdušáková
(from Wright *et al.*, 1956).

the orbit of meteor No. 3567 has been selected. It belongs to the August α Capricornids, and in Figure 1 its orbit is designated as A 15. The initial system of elements is:

$$Epoch = 1952 \text{ August } 15.3216 \text{ UT}$$

$$M_0 = 352°046$$ ⎫
$$\omega = 272.1$$
$$\Omega = 142.34$$ ⎬ 1950.0
$$i = 0.4$$
$$\pi = 54.4$$ ⎭
$$n = 0°180141$$

$$a = 3.105 \text{ AU}$$
$$e = 0.827$$
$$q = 0.538 \text{ AU}$$ (2)
$$Q = 5.673 \text{ AU}$$
$$P = 5.471 \text{ yr.}$$

In our studies we have for the first time considered the three-dimensional form of a meteor stream. This was accomplished by making suitable variations in some of the initial elements (M_0, a, i). We studied the motion of ten α Virginid particles and five α Capricornid particles.

The integration of the differential equations of motion of those particles has been performed on the electronic computer BESM-4, using both the double-precision

TABLE I

Approaches of the particles of the α Virginid meteor stream to Jupiter

Particle	1908		1931		1955		1967		1979		1991		2002/2003	
	T_{min}	Δ_{min} (AU)	T_{min}	Δ_{min} (AU)	T_{min}	Δ_{min} (AU)	T_{min}	Δ_{min} (AU)	T_{min}	Δ_{min} (AU)	T_{min}	Δ_{min} (AU)	T_{min}	Δ_{min} (AU)
I	–	–	Oct. 13	0.497	Sept. 8	0.298	–	–	–	–	Mar. 28	0.316	Dec. 5	0.395
II	–	–	Oct. 11	0.507	Aug. 23	0.175	–	–	May 22	0.145	–	–	–	–
III	–	–	Oct. 16	0.489	Aug. 27	0.178	–	–	June 16	0.131	–	–	Jan. 22	0.373
IV	–	–	Oct. 12	0.536	Oct. 2	0.384	–	–	–	–	June 7	0.592	–	–
V	–	–	Oct. 11	0.489	Aug. 22	0.125	–	–	Apr. 6	0.360	–	–	–	–
VI	Apr. 23	0.509	Dec. 3	0.073	Aug. 14	0.062	Dec. 20	0.934	–	–	–	–	May 31	0.204
VII	–	–	Oct. 15	0.253	Aug. 21	0.112	–	–	Apr. 11	0.869	–	–	–	–
VIII	–	–	Oct. 12	0.606	Aug. 26	0.140	–	–	–	–	–	–	–	–
IX	–	–	Oct. 11	0.509	Aug. 26	0.177	–	–	May 21	0.146	–	–	–	–
X	–	–	Oct. 12	0.520	Aug. 28	0.206	–	–	June 8	0.160	–	–	Jan. 5	0.496

programme by Kazimirchak-Polonskaya (1972) and the single-precision Cowell programme by Belyaev (1972). Perturbations by Venus to Neptune were taken into account, the integration being carried over an interval 50 yr before and 50 yr after the epoch of osculation and using step-sizes from 0.3125 to 20.0 days.

2. The α Virginids

In the case of the α Virginids it was ascertained that the main particle I with the initial orbit (1) and the particles II, III, ..., X with variant orbits make a succession of approaches to Jupiter. Table I lists the approaches for which $\Delta_{min} < 1.0$ AU. It is evident that in 1931 and 1955 all ten particles approached Jupiter closely. The values of Δ_{min} and T_{min} varied quite significantly, however, and the evolution of the various orbits before 1931 and after 1955 was quite different.

Table II illustrates the orbital evolution of particle I in the form of orbits averaged over four intervals of time, each being separated by close approaches to Jupiter. It is evident that

(1) the orbital elements that determine the shape and size (a, e) as well as the direction of the line of apsides (π), are exceptionally stable during the 100-yr interval, in spite of close approaches to Jupiter;

(2) By contrast, the elements determining the spatial orientation of the orbit (Ω, i) undergo appreciable changes. The ascending node, for instance, regresses through 86°.4, and this shifts the time of appearance of the shower by about three months.

TABLE II

Orbital evolution of particle I of the α Virginid meteor stream

	Ω	π	i	e	a (AU)	P (yr)	q (AU)	Q (AU)
1903–1931	55°.8	313°.3	2°.4	0.87	2.74	4.54	0.35	5.1
1932–1954	25.0	314.0	2.3	0.86	2.78	4.65	0.39	5.2
1955–1990	341.6	314.6	4.6	0.86	2.71	4.46	0.38	5.0
1991–2003	329.4	315.3	7.3	0.85	2.74	4.53	0.41	5.1

Analysis of the results for the α Virginids leads to the following conclusions:

(1) The particles may be divided into two groups according to the character of their approaches to Jupiter. Particles IV and X, whose approaches to that planet are of the same type as those of particle I, exhibit orbital evolutions of a similar character.

(2) On the other hand, the orbits of particles II, III, V, VII, VIII, IX, and especially VI (which penetrated very deeply into Jupiter's sphere of action) are characterized by stability only of the semimajor axis, period and aphelion distance, while all the other elements are subject to rather substantial changes. The regression of the line of nodes is enormous.

Figures 2 and 3 illustrate the evolution of the orientation elements Ω, π, i, eccentricity and perihelion distance for particles III and VI, the differences in evolution rate for which are particularly pronounced. The evolution of the orbital inclination of particle III is of exceptional interest. From the beginning of 1902 to the middle of 1955 the inclination remains invariable at about 2°3. On 1955 June 22 the particle

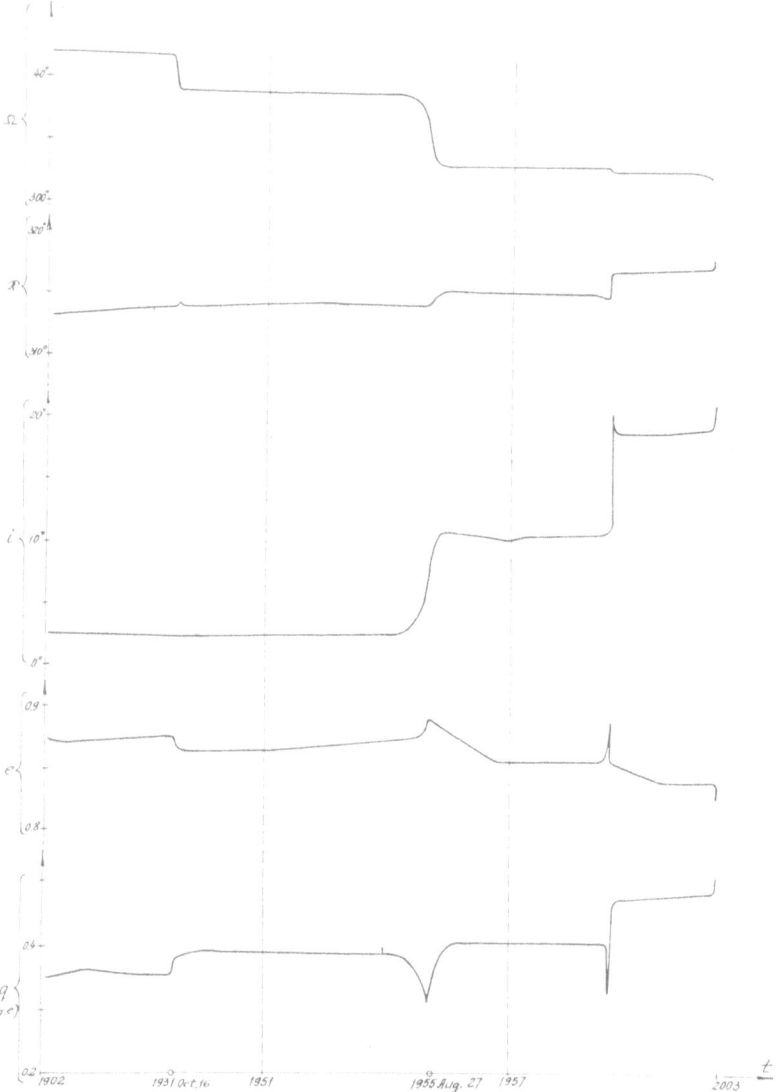

Fig. 2.   Evolution of the orbital elements of particle III of the α Virginid stream. Between the vertical broken lines the time scale is expanded.

enters Jupiter's sphere of action, attains its least distance from Jupiter ($\Delta_{min}=0.178$ AU) on August 27 and leaves the sphere on October 26. During this 4-month period i rises rapidly from 2°3 to 10°2, and then maintains the latter value for 24 yr. At the second and even closer approach to Jupiter (to $\Delta_{min}=0.131$ AU on 1979 June 16) it increases even more sharply, from 10°.2 to 19°.5 (it remains in Jupiter's sphere of action from April 15 to August 15 of the same year); at the beginning of 1980 it drops slightly, to 18°.4, and at the approach to Jupiter in 2003 ($\Delta_{\min}=0.373$ AU) it continues to increase up to 20°.3.

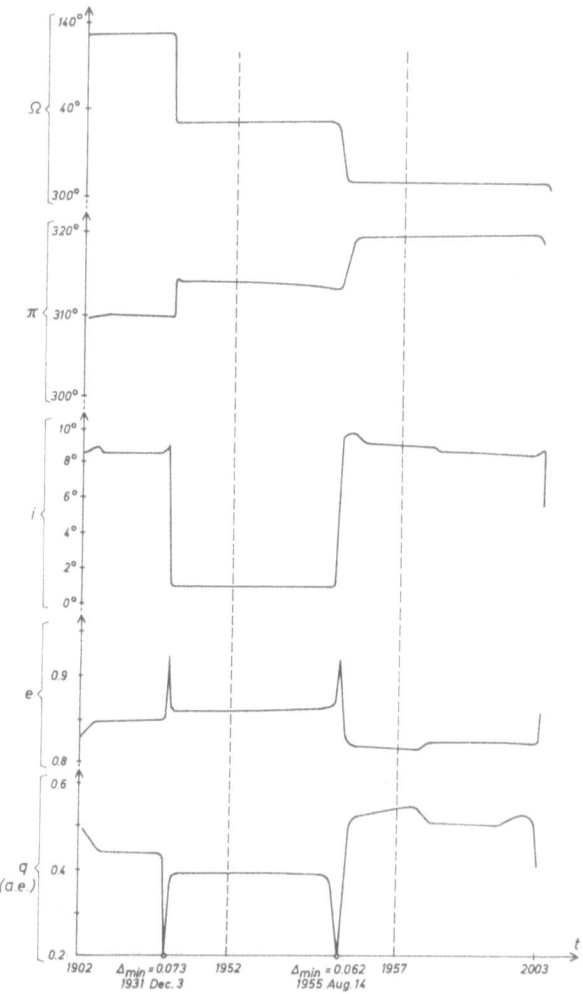

Fig. 3.   Evolution of the orbital elements of particle VI of the α Virginid stream. Between the vertical broken lines the time scale is expanded.

The longitude of the ascending node of the orbit of particle VI behaves in a similar way. It is nearly constant (127°.4 to 126°.6) from 1902 to 1931. On 1931 September 28 particle VI enters Jupiter's sphere of action, has an exceptionally small perijove distance ($\Delta_{\min}=0.0726$ AU on December 3) and leaves the sphere on 1932 February 7. During this interval Ω regresses from 126°.6 to 25°.3. It then remains invariable until the end of 1954, and during the second, even deeper passage through Jupiter's activity sphere (1955 June 9 to October 23, $\Delta_{\min} = 0.0623$ AU on August 14) changes from $25°.0$ to $316°.6$. The stability of Ω is then maintained until the 2003 approach. During the period under study (1902–2003) the total change in Ω is $175°.4$ (from $127°.4$ to $312°.0$).

The most significant aspect in the evolution of the α Virginids is shown in Figure 4, where the variations in Ω are illustrated for all ten particles from 1954 March 15 to 1957 January 2. Initially, all the particles had orbits with ascending nodes at longitudes

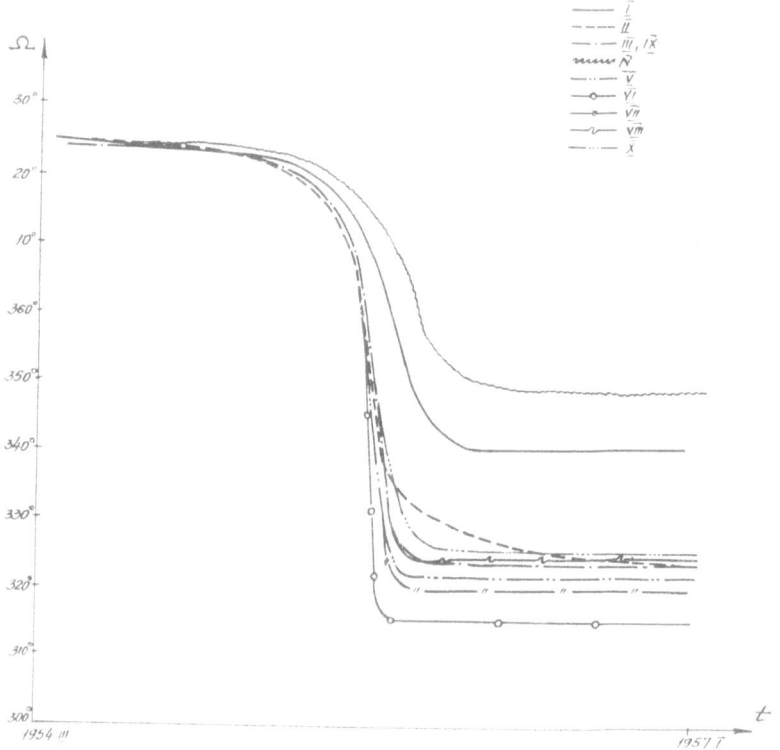

Fig. 4.   Perturbations in the longitudes of the ascending node of α Virginid members 1954–1957.

near 25°. Owing to the different conditions of the approaches to Jupiter in 1955, the dispersion in the lines of nodes of the particles increased to 35°. It is evident that this dispersion will increase even more significantly as the different particles continue to make further approaches to Jupiter, not only under different conditions but at different times (see Table I).

3. The α Capricornids

Table III summarizes the circumstances of the approach of the main particle of the α Capricornid stream to Jupiter. Table IV and Figure 5 illustrate the orbital transformations in the form of three averaged orbits.

TABLE III

Approaches of the main particle of the α Capricornid
meteor stream to Jupiter

Date	Δ (AU)	
1911 Aug. 24	0.950	
1912 Jan. 6	0.335	Enters sphere of action
Mar. 14	0.0323	Δ_{min}
May 19	0.326	Leaves sphere of action
Oct. 21	1.043	
1970 June 15	0.954	
Nov. 12	0.311	Enters sphere of action
1971 Jan. 29	0.110	Δ_{min}
Apr. 5	0.309	Leaves sphere of action
Oct. 2	1.031	

TABLE IV

Orbital evolution of the main particle of the α Capricornid meteor stream

	Ω	π	i	e	a (AU)	P (yr)	q (AU)	Q (AU)
1902–1911	66°	61°	33°2	0.73	3.14	5.6	0.85	5.4
1912–1970	142	54	0.4	0.83	3.10	5.5	0.54	5.7
1971–2003	228	49	8.2	0.81	3.05	5.3	0.57	5.5

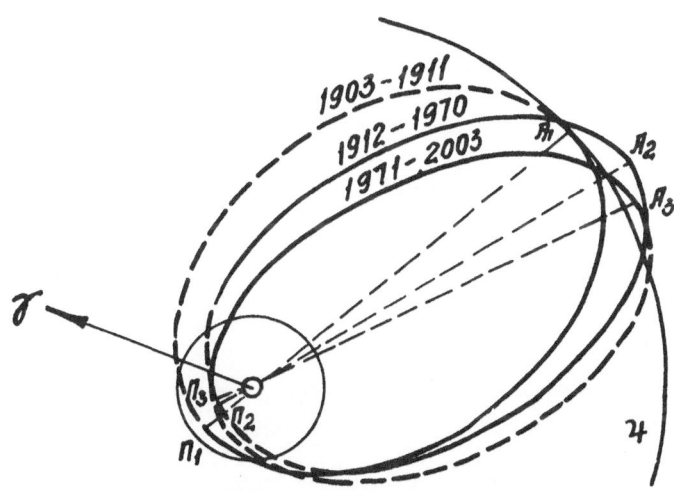

Fig. 5. The orbit of the main particle of the α Capricornid stream.

The evolution of this meteor stream is quite different from that of the α Virginid
stream. During the interval 1902–2003 the ascending node advances by 162° and the
line of apsides regresses through 12°; the inclination varies considerably, first rising

from 33°2 to 45°8, then falling to only 0°4 and subsequently rising to 8°2. The perihelion distance q undergoes appreciable changes during the first approach to Jupiter,
starting at 0.85 AU, diminishing to a minimum value of 0.23 AU, and then increasing
to 0.54 AU, where it remains, with minor variations, to the end of the period under
investigation; the eccentricity also undergoes appreciable perturbations during the
first approach to Jupiter. However, as with the α Virginids, the semimajor axis, period
and the aphelion distance of the α Capricornids are quite stable. Figure 6 shows the
detailed evolution of the orbital elements during the very close approach to Jupiter
in 1912.

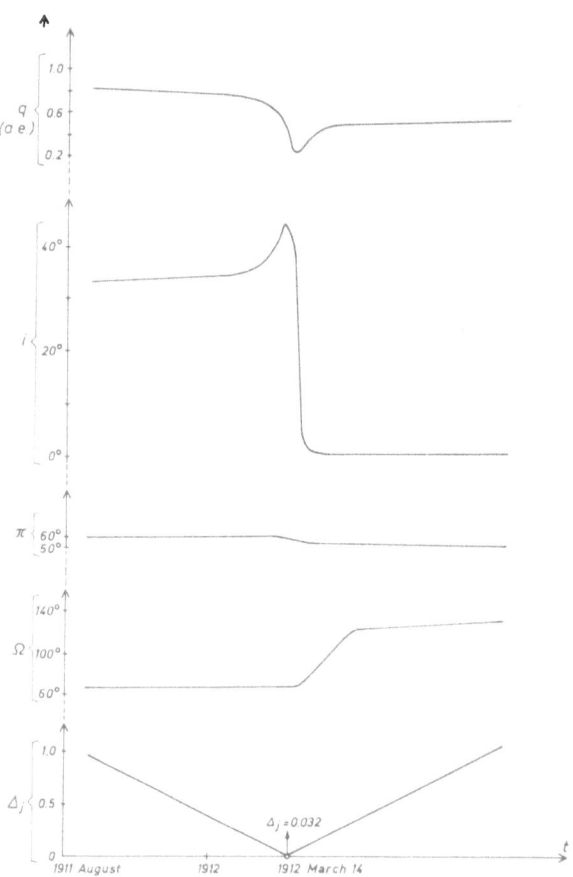

Fig. 6.   Evolution of the orbital elements of the main particle of the α Capricornid stream.

Integration of the equations of motion of four other particles in the stream gave
results and conclusions similar to those obtained in the case of the α Virginids.

4. Conclusions

On the strength of the foregoing results on the α Virginid and α Capricornid meteor streams the following conclusions may be drawn:

(1) The close approach of individual parts of meteor streams to Jupiter result in the production of systems of related meteor orbits, and in changes in the spatial density and structure of the streams and the conditions of the encounters of those streams with the Earth.

(2) The streams investigated were characterized by stability of the semimajor axes, periods of revolution and aphelion distances, but by considerable variations in the lines of nodes and apsides and in the inclinations.

(3) Consequently, the great perturbations in Jupiter's activity sphere are the main cause of the gradual dispersion of the basic streams and the origin of small meteor streams and sporadic meteors that may be observed at other times of the year and with different radiants.

It is interesting to study the changes in the radiants, for these changes reflect those in the orbital elements. For this purpose Terent'eva has developed algorithms for calculating by electronic computer the following quantities:

(1) the moment of closest approach of the Earth to the orbit of a meteor stream;

(2) the shortest distance between the orbits of the Earth and the meteor stream at that moment;

(3) the coordinates of the geocentric radiant and the geocentric velocity of a meteor shower; and

(4) ephemerides for the radiant of a shower.

Acknowledgment

The authors wish to express their deep appreciation to I. S. Astapovich for his valuable advice.

References

Belyaev, N. A.: 1972, this Symposium, p. 90.
Jacchia, L. G. and Whipple, F. L.: 1961, *Smithsonian Contr. Astrophys.* **4**, 97.
Kazimirchak-Polonskaya, E. I.: 1972, this Symposium, p. 95.
Terent'eva, A. K.: 1966, *Result. Issled. Meteor.* No. 1, 62.
Wright, F. W., Jacchia, L. G., and Whipple, F. L.: 1956, *Astron. J.* **61**, 61.

79. THEORETICAL COMETARY RADIANTS AND THE STRUCTURE OF METEOR STREAMS

E. N. KRAMER

Astronomical Observatory, Odessa State University, Odessa, U.S.S.R.

Abstract. Observational data on radiants of meteor showers and the structure of meteor streams may be used as an indirect source of information about the evolution of cometary and meteor orbits. Investigation of the structure of some meteor streams provides evidence for comparatively high velocities of ejection of meteoroids from cometary nuclei. Further evolution of the streams is determined by planetary perturbations and other factors, the influence of which depends on the size and composition of the meteoroids.

1. Introduction

The physical properties of comets and meteors seem to confirm their common origin. The relationship between young streams and individual comets is established in a straightforward way. In the long run, a meteor stream gradually dissipates, the comet and the meteoric particles are diverted into new orbits by planetary perturbations, and the link between the stream and the parent comet becomes less obvious. While the effect of dispersive forces leads to the complete destruction of the meteor stream, even before this ultimate result the orbital elements of individual meteoroids have changed so much that it is difficult to see anything in common with those of the comet. It is therefore reasonable to investigate, not the relationship between individual meteors and comets, but rather the link between meteor streams (or associations) and comets.

The cometary radiant method consists of the computation, using cometary orbital elements, of the direction and velocity of meteoric particles during the time the observer is passing by the hypothetical streams; the computational data are then compared with the observational results. When the observer meets the stream in the ecliptic plane the theoretical radiant is characterized by the following parameters:

(1) the ecliptic longitude of the cometary orbit's intersection with the ecliptic plane;

(2) the heliocentric distance of the intersection; and

(3) the orthogonal coordinates of the reversed heliocentric velocity vector.

For terrestrial observations the following theoretical radiant parameters are usually calculated:

(1) the time of closest approach of the Earth to the cometary orbit;

(2) the right ascension and declination of the hypothetical meteor radiant; and

(3) the heliocentric meteor velocity.

Detailed catalogues have been established for theoretical cometary radiants (Pokrovskij and Shajn, 1918; Porter, 1949; Kramer, 1953; Hasegawa, 1958; Zentsev, 1970). Many of these theoretical radiants have been identified with observed meteor

Chebotarev et al. (eds.), The Motion, Evolution of Orbits, and Origin of Comets, 472–481. All Rights Reserved.
Copyright © 1972 by the IAU.

data. Table I lists the theoretical radiants for comets observed between 1950 and 1961 (Zentsev, 1970). For some of them the corresponding observed radiants of meteor streams are also given.

TABLE I

Some cometary and meteor radiants

Comet/shower	Date	α	δ	v_g (km s^{-1})	Δ (AU)	Ref.
⌠1951 II	Aug. 3	22	−37	50	0.06	
⌡Sculptorids	Aug. 4	21	−39			Astapovich (1962)
1951 IV	Mar. 6	138	−27	12	0.22	
1953 II	Mar. 3	229	+20	51	0.22	
1953 VII	Nov. 8	248	−13	9	0.15	
⌠1954 VII	Dec. 10	198	+68	45	0.12	
⟨ κ Draconids	Dec. 18–28	194	+67			Astapovich (1962)
⌡195	Dec. 7	190	+75	34		Kashcheev et al. (1967)
⌠1954 X	Mar. 22	254	+57	34	0.02	
⌡ε Draconids	Mar. 12–31	257	+62			Astapovich (1962)
⌠1954 XII	June 6	299	+20	49	0.18	
⌡Sagittids	June 1–13	294	+13			Astapovich (1962)
1955 V	Sept. 2	78	−20	58	0.12	
1955 VII	Nov. 26	297	+32	12	0.16	
1957 IX	Apr. 20	332	−23	67	0.05	
1959 III	July 7	228	+34	11	0.25	
1959 VIII	Oct. 9	267	+52	21	0.06	
1960 VIII	Nov. 4	243	−32	11	0.18	
⌠1961 II	Sept. 19	101	+39	70	0.06	
⌡155	Sept. 24–27	100	+31	68		Kashcheev et al. (1967)

In addition to identifying the parental relationship between comets and meteor streams it is possible, on the basis of the relatively rich photographic and other observational data, to investigate the structure of individual meteor showers. The best known of the great showers are the Perseids and the Geminids. There are more than 320 photographs available for the former shower, and rather more than 130 photographs and over a thousand radio observations for the latter. The relationship between the Perseids and comet 1862 III was established by Schiaparelli soon after that comet was observed. The parent comet of the Geminids is unknown.

2. The Perseid Stream Structure

For investigating the structure of the Perseid stream data have been used relating to 320 Perseids, photographs of which had been taken in Odessa (Kramer et al., 1963; Babadzhanov and Kramer, 1963; Kramer and Markina, 1965), Dushanbe (Katasev, 1957; Babadzhanov, 1958; Babadzhanov and Kramer, 1965; Babadzhanov and Sosnova, 1960), Massachusetts and New Mexico (Wright and Whipple, 1953; Whipple, 1954; Jacchia and Whipple, 1961; McCrosky and Posen, 1961), and Ondřejov (Ceplecha et al., 1964; Ceplecha, 1958). The average elements of the stream are given in Table II.

TABLE II

Orbital elements of the Perseid stream

	a(AU)	e	q(AU)	P(yr)	i	ω	Ω
Comet 1862 III	24.28	0.960	0.963	119.6	113°.6	153°	139°
Perseids (Harvard)	21.56	0.956	0.946	100.1	113.2	150	138
Perseids (320 meteors)	33.56	0.974	0.948	194	113.0	150	138

Figure 1a shows the distribution of Perseids with solar longitude $\lambda_\odot = \Omega$ (around the ascending node). The shower activity slowly increases from the last days of July, reaches a sharp peak on August 10–11 and rapidly diminishes in only two days. If one ignores the displacement of the radiant caused by the orbital motion of the Earth, the remaining distribution of meteor velocities and the coordinates of the radiant may be explained, either by measurement errors or by the natural scatter of individual meteor orbits. In general, the instrumental measurement errors are two or three times

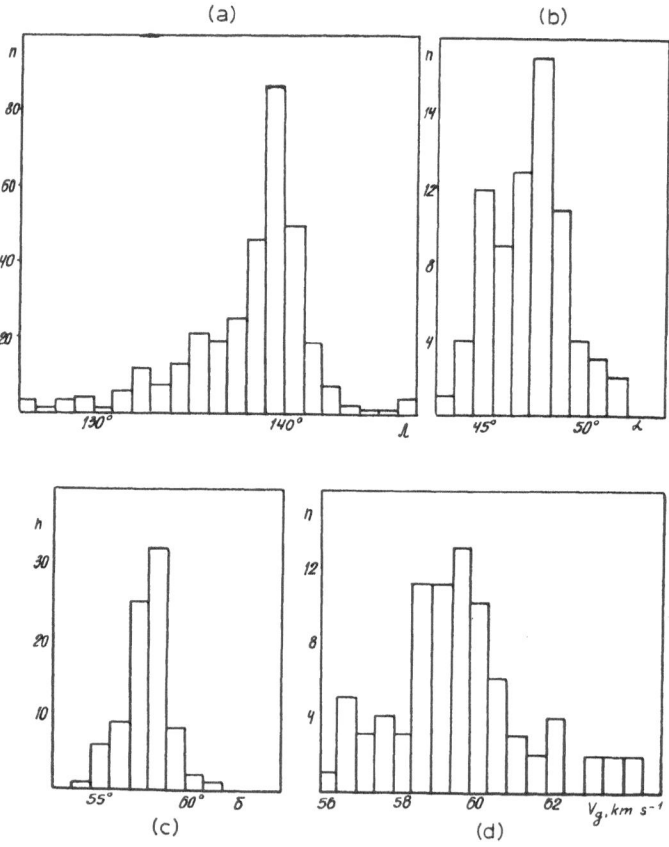

Fig. 1.   Distribution of Perseids in (a) Ω, (b) α, (c) δ, and (d) v_g.

less than the observed orbital scatter. We may therefore safely ascribe the observed distribution of stream parameters to orbit scatter inside the meteor stream itself. Figures 1b, 1c, and 1d are histograms of the distributions of the Perseids with respect to right ascension α, declination δ and geocentric velocity v_g. In order to exclude the effect of motion of the radiant these histograms were obtained using observations only during the 24 hours for which $139° < \lambda_\odot = \Omega < 140°$. The parameters of the distributions are given in Table III.

TABLE III

Distribution parameters for Perseid radiants

	α	δ	v_g (km s^{-1})
Average	47°8	57°9	59.7
Variance	4.3	1.6	3.3
No. of meteors	78	84	82

By grouping the observations according to Ω it is easy to compute average values of the orbital elements for separate parts of any cross-section of the meteor stream. We shall consider in some detail the dependence of the orbital inclination i on Ω, for the correlation coefficients between the other elements are very small. Instead of the elements Ω and i it is convenient to introduce the spherical coordinates $\lambda_k = \Omega_k - 90°$ and $\beta_k = 90° - i_k$ of the kth orbital pole. The average weighted values of $\bar{\beta}_k$ for average values $\bar{\lambda}_k$ are given in Figure 2. The nearly straight-line dependence of $\bar{\beta}_k$ on $\bar{\lambda}_k$ shows that the poles of the meteor orbits are approximately located along the arc of a great circle, the pole of which has coordinates λ_p, β_p and is the point of closest approach of all the meteor orbits.

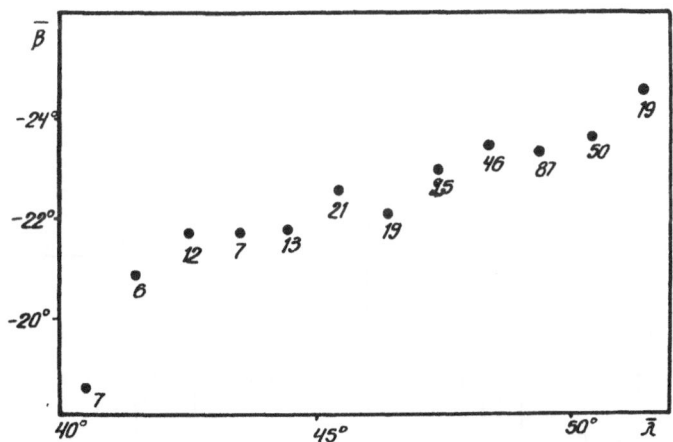

Fig. 2. The $\bar{\beta}$, $\bar{\lambda}$ dependence.

Solving the system of equations of the type

$$x \cos \lambda_k + y \sin \lambda_k = \tan \beta_k, \qquad k = 1, \ldots, n \tag{1}$$

for

$$x = \cos \lambda_p \cot \beta_p$$
$$y = \sin \lambda_p \cot \beta_p, \tag{2}$$

we obtained for the 320 meteors

$$\lambda_p = 5^h 50^m, \qquad \beta_p = + 61°03', \tag{3}$$

to which correspond an average true anomaly of 275°, an average heliocentric distance of 1.55 AU, and an average distance of 1.36 AU above the ecliptic plane.

Guigay (1947, 1948) obtained approximately the same result from visual observations. A detailed analysis of the observations shows that nearly 60% of the meteors pass within 2° of the point λ_p, β_p; more than 50% of the orbits pass within 0.15 AU of the point where the hypothetical decay of the parent comet took place.

3. The Geminid Stream Structure

In comparison with the Perseids the Geminid stream is very small, its semimajor axis not exceeding 1.5 to 1.8 AU and its revolution period being some 2.5 yr. The Geminid meteoroids are nearly uniformly dispersed along the orbit.

For our investigation we have used photographic observations of 134 meteors (Whipple, 1947; Ceplecha, 1957). Figure 3 illustrates the average dependence of the

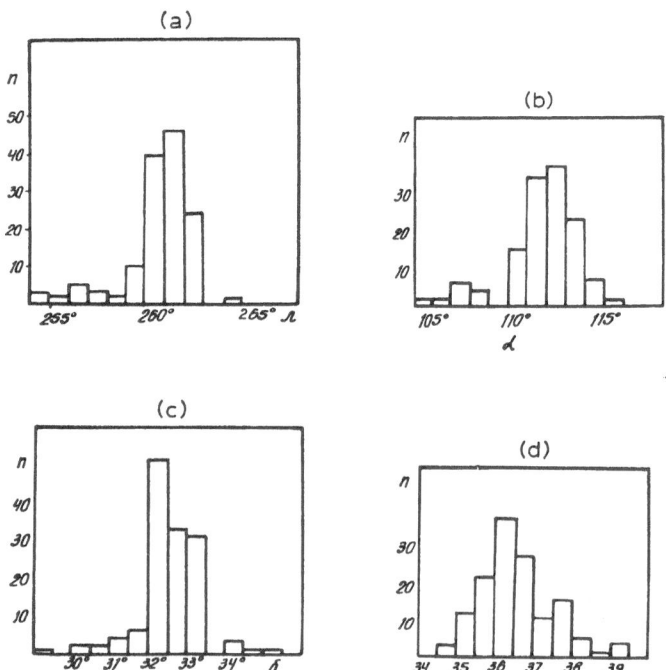

Fig. 3.   Distribution of Geminids in (a) Ω, (b) α, (c) δ, and (d) v_g.

number of photographed meteors on the solar longitude and gives histograms of the distributions in α, δ, and v_g. At solar longitude $\lambda_\odot = 260°$ the stream activity increases sharply, while at $\lambda_\odot = 263°$ the number of meteors is practically zero. The radiant is limited to an area of not more than 12 square degrees, the right ascension varying between 110° and 115° and the declination from 32°.0 to 33°.5. The relative range in geocentric velocity is somewhat greater, from 33 to 38 km s^{-1} with the peak near 36 km s^{-1}.

In the case of the Geminids the dependence of the orbital inclination on the longitude of the ascending node is weak, revealing itself as an increase in i with a decrease in Ω. The point of closest approach of the meteor orbits is located near the ascending node at a heliocentric distance of 0.16 AU.

4. Changes in the Structure of Meteor Streams Due to Gravitational Perturbations

A meteor stream gradually dissipates under the influence of various forces. Particles fine enough are affected by light pressure; while the orbits of all particles change in response to planetary perturbations. It cannot be excluded that the structure of a stream is also determined by other factors; their contribution, however, may be assessed only after one has taken into account all the gravitational forces and the effects related to the absorption and the re-emission of light.

The structure of a stream and its evolution obviously depend on the initial velocity of ejection of particles from the surface of the nucleus of the parent comet. The cometocentric velocity vector is a random quantity, so there is no need to compute the motions of individual particles by the methods applied for the calculation of the perturbations on planets, asteroids, and comets, where the theory of motion involves a large number of exact observational data. As for meteors, the only data available are their coordinates and the velocity at the moment of disappearance, and the most perfect classical theory for the motion of an individual meteoroid would be of no use, since it cannot be verified by observation. We thus conclude that in order to resolve the problems of meteor astronomy we should turn instead to the study of the distribution of random values characterizing the velocity vector and coordinates of a random meteoroid.

When studying relatively large meteoroids, which penetrate the Earth's atmosphere and give rise to relatively bright meteors, we must first find out how the planetary perturbations change the initial distribution of particles in the stream. To this effect, we may operate with some statistical model for the particle distribution.

The meteoroids are constantly under the attraction of the planets. The Perseid meteoroids, however, are generally located far from the planets, and they are subject to large perturbations only when near their nodes. Accordingly, it is convenient to divide the disturbances in the motions of Perseids into two classes: (1) perturbations far from the nodes; and (2) perturbations near the nodes.

The first class of perturbations is generally secular, and they do not reach high values. Moreover, their distribution in absolute value may be considered uniform. The secular perturbations on the Perseids were studied by Hamid (1950), although Southworth (1963) has pointed out their weak contribution to the orbital changes. Southworth, applying the Monte-Carlo method, computed the average variance increase in the observed Perseid radiant per revolution, starting from the Harvard reference orbit listed in Table II. These variance increase data are given in the first line of Table IV; the following two lines give the variances in α, δ, v_g and T (time duration) as computed from observations (I) by Harvard stations, and (II) from the 320 meteors of our list. Assuming the average period of revolution P to be 100 yr, it is easy to calculate the time needed for the variance to increase to the observational values. This time may be called the age of the stream. The last two lines of the table give the results of Southworth's calculations, and of our calculations based on the Observations II data, respectively. The age obtained by Southworth from the variances in α is an order of magnitude larger than that obtained from δ. Southworth apparently does not take into account the diurnal displacement of the radiant caused by the orbital motion of the Earth. The data in the final line correspond to the maximum of the shower

TABLE IV

Variance increase data

	σ_α^2	σ_δ^2	σv_g^2	σ_T^2
Calculations, per revolution	0.08	0.11	0.02	0.09
Observations I	21.9	2.8	1.0	11.6
Observations II	3.8	1.5	2.2	10.5
$u = 2.6$ km s^{-1}, $T_0 = -1000$ yr	23.1	2.0	1.1	10.5
Age I, $P = 100$ yr	27 000	2600	5 300	13 200
Age II, $P = 100$ yr	7 800	1360	11 000	11 700

$(139° < \Omega < 140°)$; here the diurnal displacement of the radiant can be ignored, and the age computed from α is three times that derived from δ. It is obvious that, in addition to the gravitational perturbations, other factors should be taken into account. For example, on the assumption that the ejection of the meteoroids took place at $T_0 = -1000$ yr with a velocity of 2.6 km s^{-1} at 1.5 AU from the Sun and 1.3 AU from the ecliptic plane, Southworth obtained at the time of the observations the data given in the fourth line of the table, showing that the assumed velocity of ejection is somewhat too high.

By the end of one revolution the gravitational perturbations cause the geocentric velocity of the meteoroids to vary on the average in the range $u = \sigma_v = \pm 0.14$ km s^{-1}. The heliocentric velocity varies approximately over the same range. The initial heliocentric velocity is $u_0 = 41.4$ km s^{-1}. From these data and applying the Monte-Carlo method, we may easily compute the time necessary for, say, 30 revolutions around the Sun. It is assumed that the period of revolution takes random values with random changes of velocity. We have calculated the motions of 100 random selected particles. The results are given as a histogram in Figure 4, where $\tilde{P}$ and $\tilde{P}_0$ are the times for 30 random perturbed and 30 mean unperturbed revolutions, respectively. It is evident that $\tilde{P}$ is usually smaller than $\tilde{P}_0$; moreover, the calculations show that about one-third of the initial particles reach parabolic velocity, i.e., after about 3000 yr almost 30% of the initial particles leave the solar system.

The fraction q of the initial meteoroids crossing the parabolic limit in unit time at the instant t may be determined from (Chandrasekhar, 1943)

$$q(x, t) = \frac{x}{t} \frac{1}{(4\pi Dt)^{0.5}} \exp\left(-\frac{x^2}{4Dt}\right), \tag{4}$$

where $x = u_p - u_0 = 0.6$ km s^{-1}, u_p being the parabolic velocity (42 km s^{-1}), and $D = \frac{1}{2}nu^2$, n being the number of 'steps' (revolutions) of the particle in unit time and $u = 0.14$ km s^{-1} is the average change in velocity. For the time unit we may choose the mean revolution period (about 100 yr), in which case $n = 1$, $D = 0.01$. By $t = 30$ we obtain $q(0.6, 30) \approx 0.007$; i.e., in the 30th century of the stream existence some 0.7% of the initial particles leave the solar system as a consequence of random perturbations.

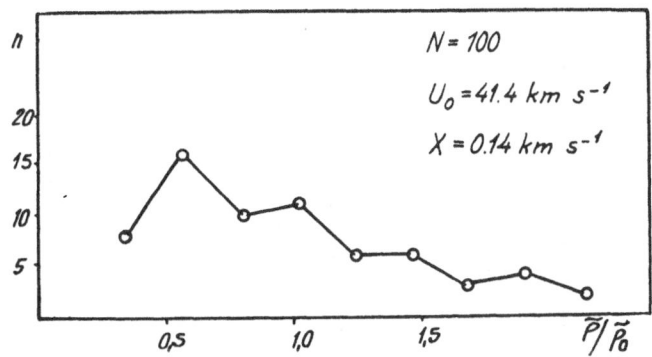

Fig. 4.   The theoretical age of the Perseids.

If a large enough statistical sample is available, then not only the variances, but also the theoretical and observational distribution laws can be compared.

To a first approximation the perturbations may be taken as proportional to the attractive force by the perturbing planet when the particle crosses the ecliptic plane. Let R be the mean heliocentric distance of the perturbing planet, r the heliocentric distance of the particle at the node, and ρ the distance of the planet from the node (assuming that the planet moves in a circular orbit in the plane of the ecliptic). We suppose that the heliocentric angle θ between the directions to the node and to the planet is a uniformly distributed random quantity in the range $(0, 2\pi)$. This means that at the instant the meteoroid crosses the ecliptic plane the planet can occupy any point in its orbit with equal probability. The probability density of random perturbations is then

$$p_1(W) = p[\theta(W)] \left| 2 \frac{d\theta}{dW} \right|, \tag{5}$$

where

$$p(\theta) = 1/2\pi \tag{6}$$

and

$$W = \frac{A}{\rho^2} = \frac{A}{R^2 + r^2 - 2Rr \cos \theta}, \tag{7}$$

where A is a constant. Hence

$$p_1(W) = \frac{1}{\pi W} \left\{ \left[2 \frac{W}{A} (R^2 + r^2) - \frac{W^2}{A^2} (R^2 - r^2) - 1 \right]^{1/2} \right\}^{-1}. \tag{8}$$

This is the first approximation to the theoretical distribution law of perturbations for a single meteoroid passing through the ascending node. For the Perseids we may assume $r = 11$ AU, $R = 10$ AU (in the case of Saturn), and if we suppose for simplicity that $A = 1$, then $W_{max} = (R - r)^{-2} = 1$; the corresponding distribution curve is given in Figure 5. From elementary calculations it follows that the probability

$$p_1(W < 0.01 \ W_{max}) > 0.65 \text{ and } p_1(W < 0.1 \ W_{max}) > 0.90;$$

i.e., insignificant perturbations prevail in the distribution.

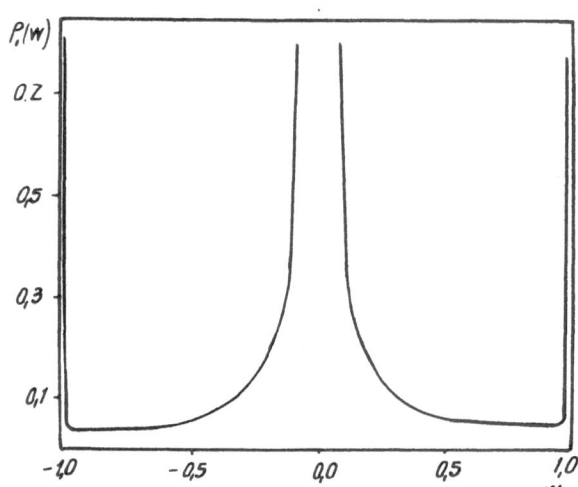

Fig. 5.   Theoretical distribution of perturbations on the Perseids.

The distribution of random perturbations after k passages through the ascending node is complicated and more easily calculated by the Monte-Carlo method. Preliminary results show the probability of random perturbations after 30 revolutions of the Perseid meteoroids to be $p_k(|W| > |2 \ W_{max}|) < 0.1$. Thus the gravitational perturbations after 3000 yr cannot disturb the stream to the observed extent, and either the ejection velocity from the comet is higher than commonly assumed or the stream has been perturbed by factors other than gravitational ones.

References

Astapovich, I. S.: 1962, *Astron. Kalendar*, part 5, p. 616.
Babadzhanov, P. B.: 1958, *Byull. Inst. Astrophys. Akad. Nauk Tadzhik. SSR* No. 26, 13.

Babadzhanov, P. B. and Kramer, E. N.: 1963, *Metody i Nekotorye Rezultaty Fotograficheskikh Issledovanij Meteorov*, Nauka, Moscow.
Babadzhanov, P. B. and Kramer, E. N.: 1965, *Smithsonian Contr. Astrophys.* **11**, 67.
Babadzhanov, P. B. and Sosnova, A. K.: 1960, *Byull. Inst. Astrophys. Akad. Nauk Tadzhik SSR* No. 2, 3.
Ceplecha, Z.: 1957, *Bull. Astron. Inst. Czech.* **8**, 51.
Ceplecha, Z.: 1958, *Bull. Astron. Inst. Czech.* **9**, 225.
Ceplecha, Z., Sežková, M., Novák, M., Rajchl, J., Sehnal, L., and Davies, J. G.: 1964, *Bull. Astron. Inst. Czech*, **15**, 144.
Chandrasekhar, S.: 1943, *Rev. Mod. Phys.* **15**, 1.
Guigay, G.: 1947, *J. Obs.* **30**, 33.
Guigay, G.: 1948, *J. Obs.* **31**, 1.
Hamid, S. E.: 1950, Ph.D. dissertation, Harvard University.
Hasegawa, I.: 1958, *Documentation Obs.* **11**, 1.
Jacchia, L. G. and Whipple, F. L.: 1961, *Smithsonian Contr. Astrophys.* **4**, 97.
Kashcheev, B. L., Lebedinets, V. N., and Lagutin, M. F.: 1967, *Meteornye Yavleniya v Atmosfere Zemli*, Nauka, Moscow.
Katasev, L. A.: 1957, *Fotograficheskie Metody Meteornoj Astronomii*, Gostekhizdat, Moscow.
Kramer, E. N.: 1953, *Izv. Astron. Obs. Odess. Gos. Univ.* **3**, 163.
Kramer, E. N. and Markina, A. K.: 1965, *Probl. Kosmich. Fiz.* No. 1.
Kramer, E. N., Vorob'eva, V. V., and Rudenko, O. A.: 1963, *Izv. Astron. Obs. Odess. Gos. Univ.* **5**, No. 2.
McCrosky, R. E. and Posen, A.: 1961, *Smithsonian Contr. Astrophys.* **4**, 15.
Pokrovskij, K. D. and Shajn, G. A.: 1918, *Trudy Astron. Obs. Perm. Univ.* No. 1.
Porter, J. G.: 1949, *J. Br. Astron. Assoc.* **60**, 1.
Southworth, R. B.: 1963, *Smithsonian Contr. Astrophys.* **7**, 299.
Whipple, F. L.: 1947, *Proc. Am. Phil. Soc.* **91**, 2.
Whipple, F. L.: 1954, *Astron. J.* **59**, 201.
Wright, F. W. and Whipple, F. L.: 1953, *Harvard Repr. Ser. 2* No. 47.
Zentsev, I. N.: 1970, *Astron. Tsirk.* No. 559.

B. POSSIBILITY OF COMMON ORIGIN

80. ON THE RELATION BETWEEN COMETS AND METEOROIDS

H. ALFVÉN
Royal Institute of Technology, Stockholm, Sweden

In order to understand the origin and evolution of the solar system it is especially important to study the small bodies – asteroids, comets and meteoroids – because in the planetesimal state (preceding the present state) the matter was dispersed.

An essential part of the cosmogonic problem is the state of the primeval plasma and how small bodies condensed from it. This directs the attention to plasma physics, which has developed very rapidly because of thermonuclear research. Plasma physics deals with the interaction of a large number of bodies. Many of its theorems are applicable also to the motion of neutral (solid) particles in a Coulomb field, and we can therefore make use of them when studying the behaviour of small particles in interplanetary space.

One of the results is that the interaction of particles will lead to the formation of jet streams, i.e., the particles will tend to move in similar orbits. This is contrary to the general belief that encounters produce a state of random motion. Some support for the jet-stream theory is provided by the results by Arnold and Danielsson that such streams exist in the asteroid belt.

In principle, meteor streams could be produced by the jet-stream mechanism. We thus have an alternative to the usual theory, and this alternative ought to be studied further. We can then proceed a step further and ask whether comets may be produced from jet streams. We can again take an analogy from plasma physics. An electron beam acted on by a variable electric field will 'bunch' so that its density in some regions will increase by many orders of magnitude. Hence a condensation could be formed in a meteor stream. It is desirable to explore whether the formation of comets may be explained in this way. The matter is especially interesting in the case of the short-period comets, for there seems to be little chance of explaining these comets on the basis of the capture theory.

Discussion

F. L. Whipple: The idea of meteor stream condensation is interesting, but I believe it is quite impossible in the present solar system – although it may have been important in the formation of the solar system. Meteoric particles are dispersed by gas when ejected by comets; further, they are dispersed by the differences in the orbital elements, especially the period, and because of the differential perturbations by the planets. They are embedded in a high-temperature medium, and this causes frequent collisions and dissipation. Solar radiation pressure quickly removes the fine particles (those with dimensions less than one micron). The solar wind quickly eliminates the atoms and molecules, so that all the dissipation products are removed. And finally, I know of no strong streams that were not associated with known comets and near their orbits.

H. Alfvén: I don't think these points are decisive. A jet stream has an intrinsic tendency to contract; all this requires is that the collision time be smaller than the differential perturbation time.

Chebotarev et al. (eds.), The Motion, Evolution of Orbits, and Origin of Comets, 485–486. *All Rights Reserved.*
Copyright © 1972 by the IAU.

The important thing is, as Trulsen will describe, that the planetary perturbations produce a set of density waves in the stream. These density waves could then produce large concentrations in density at different points, especially if there is a series of consecutive perturbations, because the waves will be damped rather slowly.

B. Yu. Levin: Different parts of the stream move with different velocities, and if there is an increase in density, it is only temporary, when one part passes through another part, and the point of intersection is not suitable for the formation of a comet. I agree with Whipple that this process was possible in the early stages of the solar system but not at present.

H. Alfvén: The important thing is if the density goes above the limit where inelastic collisions become frequent. Condensation will then take place.

L. Kresák: I should like to point out that the spatial density of most well-known meteor streams is lower than that of the sporadic background through which they move. Even if planetary perturbations could build up a significant cloud-like concentration within a stream, it would certainly be quickly destroyed during the next revolution due to the dispersion of semimajor axes by the tangential velocity component. Moreover, we have ample evidence that the high-dispersion streams are old and the low-dispersion streams are new. This conclusion is based not only on the orbits, but also on such features as the lack of small particles within old streams, produced by interplanetary erosion and drag effects, or peculiar physical properties of the meteoroids of most recent origin (the Draconids).

H. Alfvén: We really don't know the density of meteor streams, because we can never know whether the Earth has passed through the centre of a stream. Moreover, in the case of a comet observed soon after its formation there should be a decrease in the density of the stream.

A. Z. Dolginov: How could comets, consisting mainly of ice, be formed of meteor particles?

81. FORMATION OF COMETS IN METEOR STREAMS

J. TRULSEN

The Auroral Observatory, University of Tromsø, Tromsø, Norway

Abstract. The temporary focusing effect of planetary perturbations on the particles of a meteor stream is studied. It is indicated that allowance for collisions could result in more permanent focusing, leading to the possible formation of a comet.

The present work was started following a suggestion by Alfvén, who has recently stressed the possible importance of material streams in the processes leading to the formation of the solar system. It was felt that the traditional theories for the formation of short-period comets, both the capture theory and the eruption theory, were not completely satisfactory. The capture theory has been discussed by a number of authors, most recently in a series of papers by Everhart (1967a, 1967b, 1968, 1969), whereas Vsekhsvyatskij (1966) has worked out arguments in favour of eruption. By short-period comets we shall mean here comets with periods less than or about that of Jupiter.

It is well known that several comets are associated with meteor streams. The question was therefore posed as to whether the short-period comets could in fact be formed out of meteor streams. The existence of a large number of such streams out in space is readily inferred from observed meteor showers on the Earth (Lindblad, 1971). In order to substantiate the above picture it was necessary to find a mechanism that could focus a considerable fraction of the material dispersed in the stream into a smaller region and then prevent this material from subsequent spreading. The above high density region would then constitute the new-born comet, possibly existing in a kind of equilibrium with the remnants of the meteor stream. I should like to discuss here the possible initial phase of this creation process – the focusing of the material in the meteor stream.

Similar situations are well known in plasma physics. If, for example, an electron beam passes through a region where it is subject to a varying electric field, there is often observed bunching of the electrons after they have emerged from the perturbing region. Processes of this kind are indeed of great importance in the electronics industry.

In the present case the electron beam is, of course, replaced by the meteor stream, and it was thought that the necessary perturbations could be produced by Jupiter through a close approach to the stream. In order to study the problem quantitatively the following simple model was considered. A tenuous meteor stream is moving around a central body around which there is also a planet in a circular orbit. The mass ratio between the planet and the central body is chosen to correspond to that of the Jupiter-Sun system. In order to reduce the number of parameters involved the meteor stream and the planet are assumed to have the same orbital plane. Preliminary studies indicate that there is no qualitative difference between the results of this two-dimensional system and those of the more general case, except for the fact that the chance of a

Chebotarev et al. (eds.), The Motion, Evolution of Orbits, and Origin of Comets, 487–490. All Rights Reserved.
Copyright © 1972 by the IAU.

real close approach of part of the meteor stream to the planet is larger in the former case.

A group of particles in the stream is now selected, and the focusing or spreading of this group of particles relative to the unperturbed stream is studied numerically. The group selected consists of particles originally following the same elliptical orbit (but they are displaced slightly along that orbit on account of the perturbations by the planet). To specify the different groups of particles a rotating coordinate system was used, in which the planet is the origin and the central body is on the negative x-axis. The parameters chosen to identify each group are the position of the aphelion of the central particle in the group in this coordinate system and the eccentricity of the orbit at the time of aphelion passage.

Figure 1 illustrates some typical examples of the rms distance between particles for an eccentricity $e=0.4$, for three different aphelion distances and for six different rotations of the aphelion relative to the planet. As is easily seen, the focusing of the meteor stream by the planetary perturbations represents an appreciable effect. Reduction in intermeteoric distance by one order of magnitude relative to the unperturbed value is quite common. The maximum focusing achieved in the results presented here amounts to a factor of more than 20.

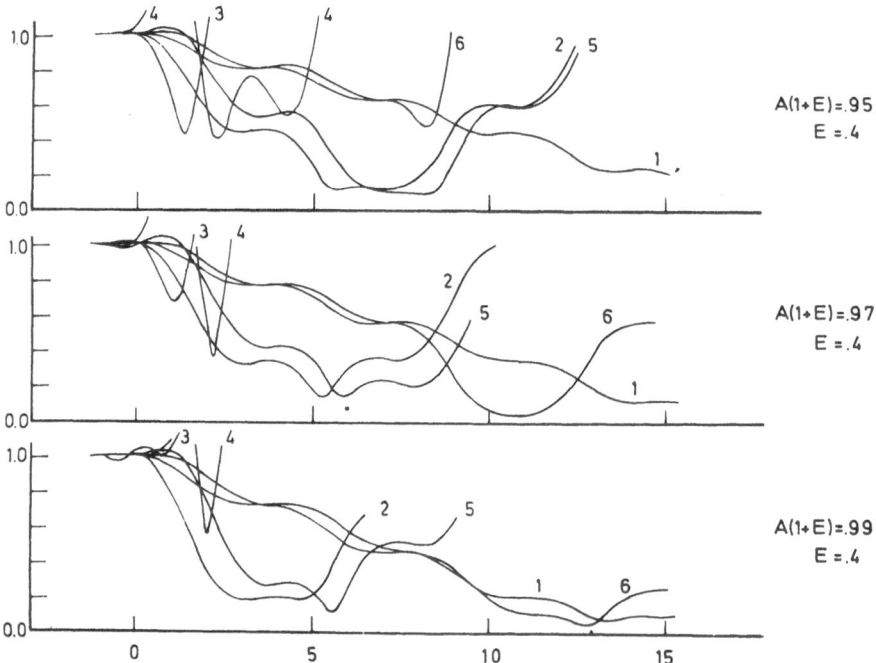

Fig. 1.   The rms distance between particles in each group, normalized so that in the absence of planetary perturbations the ordinate would be unity (i.e., the density variations in the stream due to the variable velocity in the elliptical orbit are taken out). The curves are for eccentricity $e=0.4$ and for aphelion distances $a(1+e)=0.95$, 0.97, and 0.99 (the distance between the Sun and the planet being unity). The curves, labelled 1–6, refer to the cases where the angle planet-Sun-aphelion is equal to $-30°$, $-20°$, $-10°$, $+10°$, $+20°$, and $+30°$, respectively.

The physical mechanism behind the focusing is very similar to the electron beam experiment referred to above. In order to achieve any focusing the particles in each group must be subject to a velocity modulation. In the case studied here this takes place around aphelion, when the distance meteor-planet is smallest. There is thus no effect before aphelion passage. After the velocity modulation has taken place the system needs some time before the maximum effect is achieved – the longer the time the smaller the modulation (see curves 1 and 6 in Figure 1). The modulation should not be too large, however (curves 3 and 4); in this case what is observed is rather the scattering of the group of particles. In this respect there is an important difference between the present theory for the origin of short-period comets and both the capture and the eruption theories. In our case the perturbations by Jupiter should not exceed a certain level. The mechanism works when the minimum distance between the particles and Jupiter is of the order of 1 AU. The calculations indicate that the maximum focusing achieved is relatively insensitive to the strength of the perturbations and that it is mainly the focusing time that varies.

It is possible to give a simple physical description of the resulting perturbations in the stream. To this end let us pick as a reference the particular meteor that in the rotating coordinate system has an orbit symmetrical about the planet-Sun line. The perturbations imposed on the stream may now be regarded as a density wave propagating along the stream in both directions from this reference meteor. For the parameters chosen here disruption of the stream takes place in the region adjacent to the reference meteor. Farther out, however, a density increment is gradually built up with a maximum amplitude about one order of magnitude above the unperturbed level. The wave crest then passes outwards and the group of particles spreads out again. This dispersion is discussed by Levin *et al.* (1972). For our purpose this eventual dispersion is of no interest. The important thing to note is that an appreciable focusing is produced, and that for the Jupiter-Sun system its duration is measured in years.

The work described here necessarily represents only a preliminary study of the model. Only the possible initial phase of this creation process has been discussed since the above mechanism can only produce a temporary focusing. To arrive at a more permanent focusing viscous effects have to be included – these being in the form of collisions between the meteors themselves and possibly also between meteors and released gas. These are highly nonlinear effects, and we should expect that there exists a limiting density above which such effects become of crucial importance, possibly in the way that the above density wave is arrested in a certain element of the stream. One possible way of achieving this critical density may be through interference between two such waves excited at consecutive close approaches to the planet. A study of viscous effects on the focusing is underway. It is hoped that this study will give insight into the conditions that must be satisfied for the present model to work. We might then have a better possibility to judge on the probability of this model even if very little is known of the conditions in a pre-comet meteor stream.

In trying to find connections between the present model and the first-apparition short-period comets two factors are of importance: the position of the pre-apparition orbits of the comets relative to Jupiter, and the time of first apparition. The latter factor is important because we should allow enough time for the focusing to develop. A preliminary study of this question has been undertaken, and it shows that for the majority of the first-apparition short-period comets these factors are indeed in favour of the present model.

References

Everhart, E.: 1967a, *Astron. J.* **72**, 716.
Everhart, E.: 1967b, *Astron. J.* **72**, 1002.
Everhart, E.: 1968, *Astron. J.* **73**, 1039.
Everhart, E.: 1969, *Astron. J.* **74**, 735.
Levin, B. Yu., Simonenko, A. N., and Sherbaum, L. M.: 1972, this Symposium, p. 454.
Lindblad, B. A.: 1971, *Smithsonian Contr. Astrophys.* **12**.
Vsekhsvyatskij, S. K.: 1966, *Mem. Soc. Roy. Sci. Liège Ser. 5* **12**, 495.

82. STATISTICS OF THE ORBITS OF METEOR STREAMS AND COMETS

V. N. LEBEDINETS, V. N. KORPUSOV, and A. K. SOSNOVA
Institute for Experimental Meteorology, Obninsk, U.S.S.R.

Abstract. From radar observations carried out during 1967–1968 a catalogue of 20 000 orbits was obtained. A method has been developed for identifying meteor streams and associations, and the orbits of the meteors recorded during January–April 1968 have been analysed. Among them 163 streams have been identified, and these contain 33.5% of the total number of orbits. Comparison with the orbits of the short-period comets of $q < 1.1$ AU shows considerable differences.

Radar measurements of the radiants and velocities of individual meteors were carried out at the Institute for Experimental Meteorology from September 1967 to August 1968 using the method described by Korpusov and Lebedinets (1970). The basic radar parameters were: wavelength 11.9 m, transmitter pulse power 75 kW, pulse repetition frequency 20 μs, transmitting and receiving antenna gain 16, maximum sensitivity of the basic receiver 0.5×10^{-14} W. The meteors recorded were mainly between magnitudes 5 and 8. As a result of these measurements the orbits of 20 000 meteoroids were calculated.

One of the main purposes of this work was the study of the orbits of meteor streams and associations. We developed a special method for the computer selection of meteor orbits in order to reveal the existence of meteor streams. In this method a meteor orbit is represented as a point in four-dimensional space $1/a$, e, i, ω. The orbits of sporadic meteoroids form a continuous background in the portion of this space satisfying the condition of collision with the Earth; inside a sufficiently small volume of the space the distribution of the sporadic background can be considered as accidental. Groups of orbits of meteoroids that are members of meteor streams are superimposed on the sporadic background.

The accidental $(O-C)$ deviations of the radiants of individual meteors from the mean radiants for a number of known streams were evaluated according to the results of photographic observations (Whipple and Wright, 1954); see Table I. The $(O-C)$ values are proportional to the stream-width and are several times less than the mean

TABLE I

$(O-C)$ deviations of radiants of individual meteors from mean radiants for known streams

Stream	O−C	Stream	O−C
Draconids	0°.12	Southern Taurids	0°.57
Leonids	0.22	Perseids	0.68
Geminids	0.22	Northern Taurids	1.05
Orionids	0.38	δ Aquarids	0.80

square errors of the radar measurements of the radiants of individual stream meteors
(Korpusov and Lebedinets, 1970). Consequently, we can consider as a first approximation that the spread of individual radar orbits is mainly due to measurement errors. It
will be necessary eventually to take into account that, at least for some streams, the
real spread of orbits of individual meteoroids can be compared with the spread due to
measurement errors.

Mean square errors of the measurements of the coordinates of radiants $(\sigma_\alpha, \sigma_\delta)$
and atmosphere-free velocities (σ_v) were determined by two methods: (1) by analysis of
the different sources of error, and (2) according to the spread of radiants and velocities
of individual meteors in well-known major meteor streams (Quadrantids, Geminids,
Orionids, etc.).

The σ_α, σ_δ, and σ_v values turned out to be dependent on the velocity and radiant
position of a meteor and were expressed analytically as functions of α, δ, and v. The
time of appearance of a meteor is determined with practically zero error. For each
meteor mean square errors $\sigma_{1/a}$, σ_e, σ_i, and σ_ω of the determination of the orbital
elements were calculated from the values of σ_α, σ_δ, and σ_v.

There were two stages to the selection of meteor-stream orbits for each period of
observation, the duration of which was 5 to 10 days. In the first stage meteor associations were outlined. A meteor was assumed to be related to a given association if
the following conditions were satisfied:

$$|(1/a) - \overline{(1/a)}| < 2\sigma_{1/a}, \qquad |i - \bar{i}| < 2\sigma_i$$
$$|e - \bar{e}| < 2\sigma_e, \qquad |\omega - \bar{\omega}| < 2\sigma_\omega.$$

Here, $1/a$, e, i, ω are the elements of the meteor orbit and $\overline{(1/a)}$, $\bar{e}$, $\bar{i}$, $\bar{\omega}$ are the mean
elements of the association orbit, calculated by successive approximations on the
electronic computer in the process of search for the association. At the same time a
region around the association was outlined, and meteors were assumed to be related
if their orbits did not deviate from the mean orbit by more than $4\,\sigma$. The main purpose
of the first stage was to decrease the volume of information necessary for the final
selection of meteor orbits.

In the second stage the membership of the associations and the mean orbits of the
associations were obtained, and the probability was calculated that an association is
merely an accidental fluctuation in the sporadic background. The mean orbit of N_K
association meteors confined by the sphere of radius r_K in the four-dimensional space
was calculated by successive approximations. This mean orbit is assumed to be the
centre of the sphere, and the radius of the point characterizing the orbit of each meteor
is calculated as

$$r = \left\{ \left[\frac{(1/a) - \overline{(1/a)}}{2\sigma_{1/a}} \right]^2 + \left[\frac{e - \bar{e}}{2\sigma_e} \right]^2 + \left[\frac{i - \bar{i}}{2\sigma_i} \right]^2 + \left[\frac{\omega - \bar{\omega}}{2\sigma_\omega} \right]^2 \right\}^{1/2}.$$

For each association K mean orbits were calculated, their radii from $r_{K_{\min}} = 0.2$ $(K=1)$
to $r_{K_{\max}} = 2.0$ $(K=10)$.

The mean number of background meteors contained inside the sphere $N_{fK} = n_f v_K$ is
calculated for each sphere, v_K being the volume of the sphere of radius r_K and n_f the

mean space density of background meteors in the vicinity of the association. This mean space density is taken to be the difference between the number of meteors corresponding to $K=7$ and $K=10$, divided by the difference between the volumes of these spheres, i.e.,

$$n_f = \frac{N(K=10) - N(K=7)}{v(K=10) - v(K=7)}.$$

If the sphere of radius r_K is partially outside the region satisfying the conditions of collision with the Earth (these conditions being $1+e \geqslant 1/a \geqslant 1-e$ for elliptical orbits and $1/a \leqslant 1-e$ for hyperbolic orbits), then v_K is equal to the volume of a truncated sphere.

The probability that the association is an accidental fluctuation in the background is found from the values of N_K and N_{fK} as

$$p(N_K) = \sum_{i=0}^{2} \frac{\exp(-N_{fK})(N_{fK})^{N_{K_i}}}{N_{K_i}!}.$$

The most probable radius of the sphere confining the association meteors is chosen according to $p(N_K)$. For the majority of the streams the radius was found to be in the range 0.8 to 1.2.

If the most probable radius of the sphere cannot be chosen according to the character of the variation of $p(N_K)$, it is assumed to be 0.8. The mean orbit of the meteors inside the sphere of most probable radius for the association is assumed to be the orbit of the association including N_K meteors, of which N_{fK} can be background meteors. Further, we consider only the associations with $N_K \geqslant 5$.

The mean values of the radiant coordinates, velocities, and a number of other quantities characterizing the association, as well as the diurnal variations and mean square errors of these values, were then calculated.

By means of this procedure, the catalogue of 3970 meteor orbits recorded during the period January–April 1968 was analysed, and 163 streams (comprising 1330 orbits, or 33.5% of the whole) were identified.

An analogous investigation of radar meteor streams was carried out by Kashcheev *et al.* (1967), using the results of observations at Kharkov in 1960; however, the Kharkov catalogue contains no data for the period January to March, and our results therefore make up for this deficiency.

Comparison of our catalogue of orbits of meteor streams with the catalogue of minor streams by Terent'eva (1963, 1967) and with a portion of the Kharkov catalogue enabled us to identify 36 streams, 20 of which are identified with some confidence.

Of the 163 streams selected 161 have revolution periods $P < 200$ yr. Thus, if we assume that meteoroids are the products of cometary disintegration, then for the majority of the streams we should look for a connection with short-period comets. Comparison of the stream orbits with cometary orbits enabled us to establish such a connection for only ten streams, the orbits of which are shown in Table II (which also includes two questionable cases). The coordinates of the cometary radiants, velocity, and orbital elements, the solar longitude $\lambda_\odot$ at the time of closest approach of the

TABLE II
Comet-meteor associations

Stream/comet	1/a (AU)	e	q (AU⁻¹)	i	ω	Ω	π	α	δ	v∞ (km s⁻¹)	λ☉	Δ (AU)
No. 9 P/Tuttle	0.37 / 0.18	0.64 / 0.82	0.97 / 1.03	53.4 / 55.0	193.5 / 207.0	282.7 / 269.7	116.2 / 116.7	232.5 / 223.0	+65.3 / +75.5	33.8 / 35.1	270.2	+0.096
No. 43 1943 II	0.29	0.76 / 1.00	0.84 / 0.76	165.8 / 161.3	48.8 / 36.4	103.4 / 82.6	152.2 / 119.0	178.4 / 164.0	−7.6 / −4.0	68.9 / 68.8	272.0	+0.10
No. 31 1870 IV	0.19	0.92 / 1.00	0.40 / 0.39	147.5 / 147.3	103.9 / 90.6	103.3 / 94.7	207.2 / 185.3	159.5 / 151.0	−5.5 / −2.0	63.5 / 63.6	275.8	−0.20
No. 35 1883 I	0.25	0.81 / 1.00	0.77 / 0.76	77.6 / 78.1	120.6 / 110.9	284.0 / 278.1	44.6 / 29.0	242.7 / 243.0	+30.8 / +28.0	46.5 / 45.0	279.0	+0.14
No. 39 1792 II	0.21	0.80 / 1.00	0.96 / 0.97	139.0 / 131.0	162.4 / 147.3	282.5 / 283.3	84.9 / 70.6	205.4 / 214.5	+14.6 / +16.5	66.8 / 64.8	285.5	+0.07
No. 11[a] 1939 I (?)	0.35	0.67 / 0.99	0.98 / 0.72	71.7 / 63.5	169.9 / 169.0	282.6 / 288.8	92.4 / 97.8	230.2 / 242.0	+49.1 / +46.0	42.6 / 35.4	285.5	−0.26
No. 31 1787 (?)	0.19	0.92 / 1.00	0.40 / 0.35	147.5 / 131.7	103.9 / 99.1	103.3 / 106.9	207.2 / 206.0	159.5 / 156.0	−5.5 / −10.0	63.5 / 61.8	289.0	−0.15
No. 65 868	0.44	0.80 / 1.00	0.46 / 0.42	69.7 / 65.0	281.6 / 277.0	330.7 / 305.0	252.3 / 222.0	204.8 / 185.7	+30.9 / +34.7	43.3 / 45.0	320.1	−0.03
No. 72 1947 III	0.23	0.77 / 1.00	0.99 / 0.96	123.2 / 129.1	178.9 / 182.1	330.6 / 322.3	149.5 / 144.4	244.9 / 236.0	+11.8 / +12.0	62.7 / 65.4	322.3	−0.03
No. 163[b] 1861 I	0.10 / 0.02	0.91 / 0.98	0.93 / 0.92	81.6 / 79.8	213.5 / 213.4	28.3 / 29.9	241.8 / 243.3	271.5 / 270.5	+32.3 / +33.5	48.5 / 47.7	29.9	0.00
No. 153 1844 II	0.51	0.51 / 1.00	0.97 / 0.86	134.3 / 131.4	201.0 / 211.2	27.7 / 31.7	228.7 / 242.9	290.4 / 287.0	+3.2 / +5.0	61.7 / 64.2	33.2	−0.08
No. 157 1748 II	0.21	0.86 / 1.00	0.66 / 0.63	68.8 / 67.1	254.6 / 245.6	27.3 / 33.1	281.9 / 278.7	251.9 / 255.0	+25.0 / +24.0	43.4 / 42.0	35.9	−0.12

[a] Quadrantids. [b] Lyrids.

Earth to the cometary orbit, and the distance $\varDelta$ between the Earth and the cometary orbit at that time are taken from Kramer's (1953) catalogue. For the majority of the streams the parent comets are not known.

The distributions of the orbital elements of the radar meteor streams and the short-period comets (Porter, 1961; Vsekhsvyatskij, 1967) are compared in Figures 1 to 3.

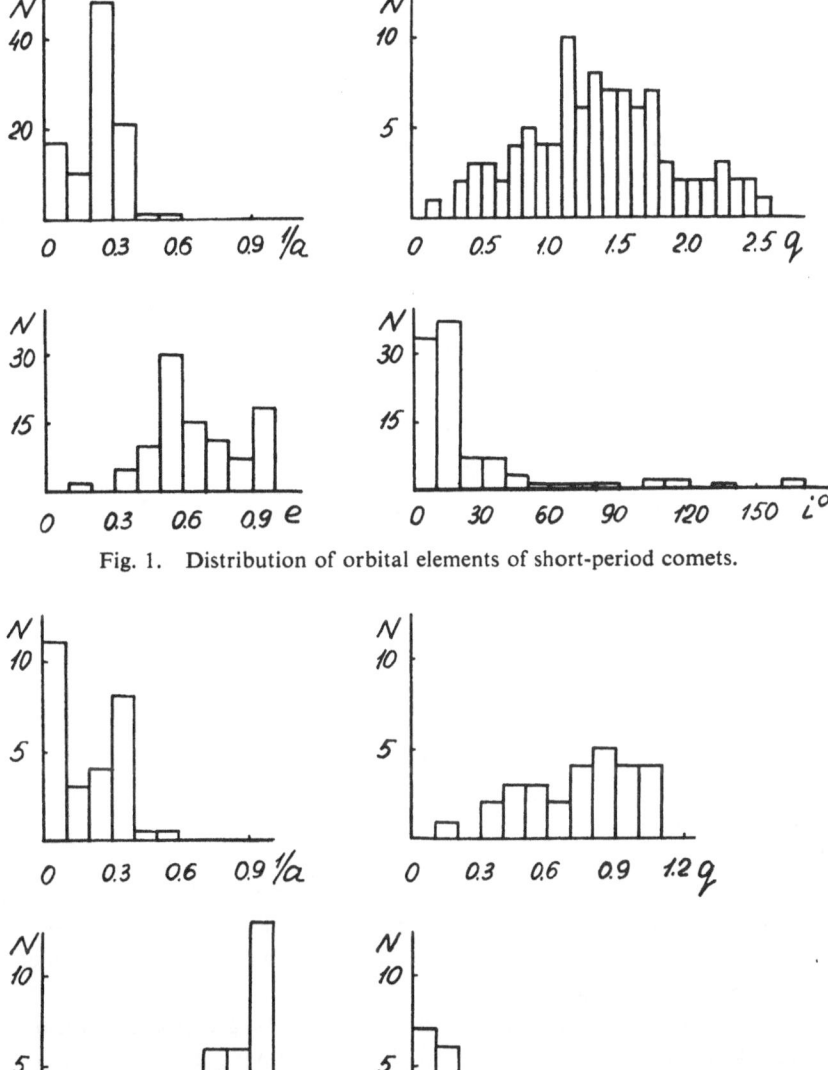

Fig. 1. Distribution of orbital elements of short-period comets.

Fig. 2. Distribution of orbital elements of short-period comets with $q < 1.1$ AU.

Figure 1 shows the distribution for all the short-period comets, while Figure 2 refers only to those with $q < 1.1$ AU. The orbital distribution for the meteors is in Figure 3. The agreement of the q-distributions is good, the essential difference being that there are many streams of very small perihelion distance ($q < 0.3$ AU), but only one comet (and it has an orbit with $a = 27.6$ AU). In the case of the distributions in $1/a$ and e the differences are essentially greater.

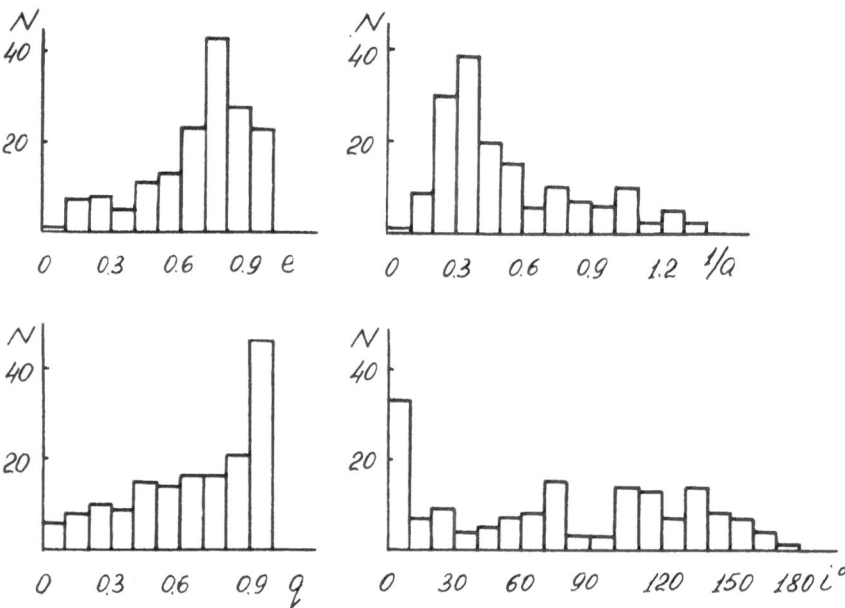

Fig. 3.   Distribution of orbital elements of radar meteor streams.

Comparing the distributions of orbital elements of the meteor streams and short-period comet orbits over the four elements $1/a$, e, q, and i, we deduce:

(1) Some of the observed meteor streams might have been generated as the result of distintegration of short-period comets;

(2) There is an essential difference between the orbits that cannot be explained solely by the evolution of the meteor orbits; this difference probably indicates that the character of the orbits of a considerable proportion of the meteor streams already differed considerably from that of the orbits of the known short-period comets at the time the streams first appeared.

The total observed complex of meteor streams with $P < 200$ yr could not have been generated as the result of the disintegration of comets moving in orbits similar to those of the known short-period comets.

References

Kashcheev, B. L., Lebedinets, V. N., and Lagutin, M. F.: 1967, *Meteornye Yavleniya v Atmosfere Zemli*, Nauka, Moscow.

Korpusov, V. N. and Lebedinets, V. N.: 1970, *Astron. Vestn.* **4**, 102.
Kramer, E. N.: 1953, *Izv. Astron. Obs. Odess. Gos. Univ.* **3**, 163.
Porter, J. G.: 1961, *Mem. Br. Astron. Assoc.* **39**, No. 3.
Terent'eva, A. K.: 1963, *Astron. Tsirk.* Nos. 249, 264.
Terent'eva, A. K.: 1967, *Astron. Tsirk.* Nos. 415, 423.
Vsekhsvyatskij, S. K.: 1967, *Komety 1961–1965*, Moscow.
Whipple, F. L. and Wright, F. W.: 1954, *Monthly Notices Roy. Astron. Soc.* **114**, 229.

83. ON THE PRODUCTION OF METEOR STREAMS BY COMETARY NUCLEI

L. A. KATASEV and N. V. KULIKOVA

Institute for Experimental Meteorology, Obninsk, U.S.S.R.

Abstract. An attempt is made to use Monte-Carlo techniques to model the process of formation of meteor streams. It is supposed that meteor streams are formed as the result of ejection of meteoroids from a cometary nucleus at perihelion. Possible ejection velocities are determined for the Draconids, Perseids, Leonids, and Taurids. In general the values do not exceed 100 m s^{-1}. The data obtained can also be used to estimate the ages of the streams.

The ejection of dust by cometary nuclei is thought to be the main source of the meteoric particles forming the meteor streams. This ejection of meteoric material usually takes place in the vicinity of perihelion (Orlov, 1960; Vsekhsvyatskij, 1966; Southworth, 1963). This present work is an attempt, by the Monte-Carlo technique, to model the process of formation of meteor streams by the isotropic ejection of meteoric material from the nuclei of the parent comets when the latter are at perihelion.

Let us introduce a right-handed rectangular coordinate system xyz having its origin at the cometary nucleus at perihelion, and with the x-axis along the prolonged radius vector of the comet, the y-axis in the direction of the comet's motion, and the z-axis perpendicular to the orbital plane. If C is the velocity of ejection of a particle, and C_x, C_y, C_z are its projections on the coordinate axes, we shall write

$$C_x = C \sin \theta \cos \varphi$$
$$C_y = C \sin \theta \sin \varphi \tag{1}$$
$$C_z = C \cos \theta,$$

where $\theta \in (0, \pi)$, $\varphi \in (0, 2\pi)$. We define C to be

$$C = A\xi_i, \tag{2}$$

where A is the upper limit of the ejection velocity, and ξ_i is a random number in the range $(0, 1)$.

The direction of the ejection of meteoroids depends on θ and φ, which can be found according to Neumann's (1951) formula:

$$\sin \varphi = \frac{2\xi_{i+1}\xi_{i+2}\zeta}{\xi_{i+1}^2 + \xi_{i+2}^2}$$

$$\cos \varphi = \frac{\xi_{i+1}^2 - \xi_{i+2}^2}{\xi_{i+1}^2 + \xi_{i+2}^2} \tag{3}$$

$$\cos \theta = 1 - 2\xi_{i+3},$$

where $\zeta = (\xi_{i+4} - \frac{1}{2})/|\xi_{i+4} - \frac{1}{2}| = \pm 1$ defines the sign of $\sin \varphi$ with probability 1/2 and $\xi_{i+1}, \xi_{i+2}, \xi_{i+3}, \xi_{i+4}$ are random numbers in $(0, 1)$.

Chebotarev et al. (eds.), The Motion, Evolution of Orbits, and Origin of Comets, 498–502. All Rights Reserved.
Copyright © 1972 by the IAU.

In order to obtain the deviations of the orbital elements of the ejected particles from those of the parent-comet we shall use the formulae (Plavec, 1955):

$$\delta(1/a) = -2v_0 C_y$$

$$\delta P = 3a_0 P_0 v_0 C_y$$

$$\delta p = 2q_0^2 v_0 C_y$$

$$\delta e = 2q_0 v_0 C_y$$

$$\delta u = C_x p_0^{1/2} e_0^{-1}$$

$$\tan i' = |C_z| v_y^{-1}$$

$$\delta i = i' \cos \omega_0 \tag{4}$$

$$\delta\omega = -\delta u - \delta\Omega \cos i_0$$

$$\delta\Omega = i' \sin \omega_0 \operatorname{cosec} i_0,$$

where a_0, P_0, q_0, e_0, i_0, ω_0 are the elements of the cometary orbit; i, Ω, ω are the ecliptic angular elements of the meteor orbit; u is the true anomaly, p the parameter of the meteor orbit, i' the inclination of the meteor orbit to the orbital plane of the comet, v_0 the velocity of the comet at perihelion, and $v_y = v_0 + C_y$ is the projection of the velocity of the ejected particle on the y-axis.

With the help of a high-speed computer we are able to obtain a great many possible orbits for meteoroids ejected with various velocities $C \leqslant A$ and at different angles. From the orbital elements of these particles it is possible to draw conclusions on the possibility of formation of specific meteor streams.

We have applied the procedure to the Perseid, Draconid, Leonid, and Taurid meteor streams (see Table I). For each variant no fewer than 200 000 orbits were calculated. In the computation process the total interval of ejection velocity was divided into K subintervals; in each subinterval the deviations of the orbital elements of the meteoroids were found according to Equation (4) and using Equations (1) to (3). The absolute values of these deviations were summed, and this sum was divided by N, where N is the number of particles within the specified velocity interval.

For the formation of the Draconid meteor stream the velocity of ejection of meteoroids from the nucleus of P/Giacobini-Zinner should probably not exceed 50 m s^{-1}. Deviations in the orbital elements of particles ejected at such velocities agree with the observational data for this stream in 1952 (Lovell, 1954). According to the observational data then the increments in eccentricity and semimajor axis were $\delta e = 0.0013$, $\delta a = 0.008$ AU; we obtained such values at $C \leqslant 50$ m s^{-1}. The duration of the maximum of the 1952 shower was a little more than 30 min, which corresponds to our result $\delta\Omega = 1'20''$ at $C = 50$ m s^{-1}.

The Leonid meteor shower, known from 889 (Astapovich, 1968), is still densely concentrated in an 18-AU arc of the orbit, the cross-section diameter there being 0.001 AU; about half the meteoroids are concentrated there (Murakami, 1961). From the observational data of the 1866 meteor shower Astapovich and Terent'eva

TABLE I

Results of Monte-Carlo calculations on meteor streams

C (m s^{-1})	δP (yr)	δa (AU)	δe	$\delta \Omega$
Draconids				
10	0.015	0.0050	0.00041	8″3
30	0.043	0.016	0.0013	25.0
50	0.074	0.026	0.0021	41.0
100	0.15	0.052	0.0042	83.0
200	0.30	0.10	0.0084	169.0
Leonids				
0.025	0.00057	0.00012	0.0000011	0.034
0.225	0.0053	0.0011	0.000010	0.31
0.525	0.012	0.0026	0.000024	0.71
1.025	0.24	0.0049	0.000045	1.42
5.000	1.15	0.024	0.000219	6.67
Perseids				
2.5	0.48	0.065	0.00011	0.35
12.5	2.49	0.33	0.00055	1.75
52.5	10.27	1.4	0.0023	7.37
102.5	20.42	2.8	0.0045	14.03
200.0	38.51	5.8	0.0086	27.21
Taurids				
50	0.041	0.018	0.0012	35.0
250	0.21	0.093	0.0064	180.0
1000	0.87	0.43	0.027	742.0
2000	1.68	1.1	0.052	1481.0
3000	2.44	3.8	0.075	2104.0

have calculated a new system of elements for the Leonids, which differ from the orbit of the parent comet P/Tempel-Tuttle in 1866 by $\delta a = 0.00383$ AU, $\delta e = 0.000316$, $\delta P = 0.07$ yr, $\delta \Omega = 3'$. Variations in the character of particle motion obtained by modelling the ejection of meteoroids from the nucleus of P/Tempel-Tuttle show that the maximum velocity of ejection of material must probably not exceed 2.5 m s^{-1}. The densest part of the stream could be formed by particles ejected with a velocity not greater than 25 cm s^{-1}.

The differences between the Perseid orbit calculated by Southworth (1963) and the orbit of comet 1862 III are $\delta a = 2.68$ AU, $\delta e = 0.004$, $\delta P = 19.5$ yr, $\delta \Omega = 48''$. We can obtain such deviations on the assumption of ejection velocities not greater than 100 m s^{-1}. At $C = 100$ m s^{-1} the changes in the revolution period are such that the stream could extend all along the orbit after a few revolutions, coinciding with South-worth's conclusion that the age of the Perseids is 1000 yr. As for Ω, it is small at all ejection velocities, and as Hamid (1950) pointed out, the stream is possibly influenced greatly by planetary perturbations.

It is supposed (Whipple and Hamid, 1952) that the Taurid meteor stream was formed as the result of ejection from the nucleus of P/Encke 4700 yr ago and also

1500 yr ago as subsequent ejection from a body having an orbit similar to that of the comet. The ejection velocity was 3 km s^{-1}. The results obtained by modelling the formation process of the Taurids show that with an ejection velocity $C=3$ km s^{-1} there would be substantial changes in semimajor axis ($\delta a=3.84$ AU) and revolution period ($\delta P=2.45$ yr). The investigations by Wright and Whipple (1950) give the following values for the orbital elements of the Northern and Southern Taurids, respectively:

$$a = 2.14 \text{ AU}, \quad e = 0.849, \quad P = 3.13 \text{ yr},$$

$$a = 2.3 \text{ AU}, \quad e = 0.7835, \quad P = 3.49 \text{ yr}.$$

For P/Encke these elements are $a=2.217$ AU, $e=0.847$, $P=3.3$ yr.

We obtain values for the orbital elements of ejected particles most appropriate to these data with velocities $C=250$ m s^{-1}, namely, $\delta a=0.09$ AU, $\delta e=0.0004$, $\delta P=0.21$ yr. However, the heliocentric velocities acquired by the ejected particles at this ejection velocity do not correspond to the heliocentric velocities of the particles in the Northern and Southern Taurids. The necessary heliocentric velocities require ejection at velocities greater than 1 km s^{-1}. Thus the question as to whether the Taurid meteors formed as the result of ejection from P/Encke is a complicated one, and to solve it we must consider the influence of planetary perturbations. Even at $C=250$ m s^{-1} the changes in revolution period are such that in a few revolutions the stream could extend along the whole orbit.

In conclusion, we point out that in modelling the formation process of a meteor stream it is the revolution period and semimajor axis that are subjected to the greatest changes with increasing ejection velocity.

To evaluate the age of a stream Plavec (1955) has given the following formula:

$$sP_0 = \frac{\Delta M}{3a_0 v_0 C_y}, \tag{5}$$

where sP_0 is the age of the stream in years, ΔM is the difference in mean anomaly, a_0 is the semimajor axis of the cometary orbit, v_0 is the heliocentric velocity of the comet at the point of ejection, and C_y is the projection of the ejection velocity on the y-axis.

In the course of our calculations we automatically obtain the value of C_y for every stream, but its uncertainty makes it difficult to use this formula directly. In this manner we found that the age of the Draconids is not greater than 60 yr. It is possible that this stream is still being formed at the present time. On the other hand, the age of the Leonids may be as much as 1580 yr.

References

Astapovich, I. S.: 1968, *Probl. Kosmich. Fiz.* No. 3, 103.
Hamid, S. E.: 1950, Ph.D. dissertation, Harvard University.
Lovell, B.: 1954, *Meteor Astronomy*, Clarendon Press, Oxford.
Murakami, T.: 1961, *Publ. Astron. Soc. Japan* 13, 51, 212.
Neumann, J.: 1951, *Natl. Bur. Sci. Appl. Math. Sci.* 12, 36.
Orlov, S. V.: 1960, *O Prirode Komet*, Akad. Nauk SSSR.
Plavec, M.: 1955, *Bull. Astron. Inst. Czech.* 6, 20.

Southworth, R. B.: 1963, *Smithsonian Contr. Astrophys.* 7, 299.
Vsekhsvyatskij, S. K.: 1966, *Fizika Komet i Meteorov*, Kiev, p. 32.
Whipple, F. L. and Hamid, S. E.: 1952, *Bull. Roy. Obs. Helwan* No. 41.
Wright, F. W. and Whipple, F. L.: 1950, *Harvard Repr. Ser. 2* No. 35.

Discussion

F. L. Whipple: In my 1951 investigation I gave a formula for the distribution of ejection velocities, although without knowledge of the sizes of the particles it is difficult to use this formula. But it seems that the velocities amount to some tens of metres per second.

B. Yu. Levin: The velocities obtained from Whipple's formula and by other methods are less than those obtained in this investigation. This is probably because perturbations have not been taken into account here.

V. V. Fedynskij: What value did you adopt for the upper limit of the ejection velocities?

N. V. Kulikova: It was different for each stream and determined according to the best representation of the observations.

L. M. Shul'man: The creation of a meteor stream is not an instantaneous act, but it takes place over a definite period of time. Further, the ejection cannot be considered isotropic.

N. V. Kulikova: I agree. We intend in future to make a more sophisticated formulation of the problem that takes these points into account.

84. ON THE DIVIDING LINE BETWEEN COMETARY AND ASTEROIDAL ORBITS

L. KRESÁK

Astronomical Institute, Slovak Academy of Sciences, Bratislava, Czechoslovakia

Abstract. A simplified form of the Jacobi integral in the three-body system Sun-Jupiter-comet or asteroid provides an excellent method for discriminating between cometary and asteroidal orbits. Omitting the librating bodies, unambiguous separation is obtained for all known objects with reliable orbital data, i.e., about 600 comets and 1800 asteroids. The only exception is the peculiar asteroid 944 Hidalgo – which is presumably a comet. The intermediate region is occupied exclusively by bodies revolving in resonance with Jupiter, and the value of the libration argument yields a sharp secondary criterion in these cases. Besides the direct perturbational capture of long-period comets from high-eccentricity orbits into Jupiter's family, a ring of nearly circular orbits between Jupiter and Saturn is suggested as another significant source of short-period comets. For these comets the subsequent operation of nongravitational effects gives a better chance of injection into small orbits of the Apollo type and for the formation of short-period meteor streams. Some phenomena (the outbursts of P/Schwassmann-Wachmann 1, the probable recent splitting of one parent body into P/Whipple and P/Shajn-Schaldach) give reasons for speculation about the population of this region, too distant for the discovery of typical asteroids or comets, by interplanetary particles up to sizeable solid bodies.

The clear physical distinction between comets and minor planets is connected in a regular way with characteristic differences in their orbits. However, there are also some bodies moving in cometary orbits, the appearance of which is entirely asteroidal (944 Hidalgo) or intermediate (P/Arend-Rigaux, P/Neujmin 1, P/Väisälä). Some orbits of very short period are at variance both with those of normal asteroids and short-period comets (P/Encke; limiting Apollo asteroids like 1566 Icarus and Adonis; meteor streams like the Geminids and Arietids, presumably generated by unknown comets), while in some problematical cases the orbits are essentially indeterminate (P/Wilson-Harrington, P/Kulin). The evolutionary significance of the limiting objects has been stressed by Öpik's (1963) suggestion that most Apollo asteroids are extinct comet nuclei, and by Marsden's (1970) suggestion of a transitional phase between the two types of objects, represented by comets in librating motion avoiding approaches to Jupiter. A correct discrimination is of particular importance in the domain of small interplanetary particles, cometary or asteroidal fragments observed as meteors, where the orbital data available are strongly biased by the Earth-crossing condition, and the physical differences appear in a more subtle and complicated form (Cook *et al.*, 1963; Jacchia *et al.*, 1967; Ceplecha, 1967; Kresák, 1968, 1969; Verniani, 1969).

For the discrimination between cometary and asteroidal orbits, the set of six conventional elements can be divided into three groups:

(1) Semimajor axis a and eccentricity e, determining the size and shape of the orbit, are undoubtedly of primary significance.

(2) The regularities impressed on the angular elements i, Ω, ω by the origin and evolution – the ecliptical concentration of orbital planes, direct motion, and alignment

Chebotarev et al. (eds.), The Motion, Evolution of Orbits, and Origin of Comets, 503–514. All Rights Reserved.
Copyright © 1972 by the IAU.

of the lines of apsides with that of Jupiter – are practically irrelevant. As far as only short-period orbits are concerned, the effects are very much alike in both systems and the distributions in i and $\pi = \Omega + \omega$ overlap so widely that individual values are useless.

(3) The time of perihelion passage T is often erroneously disregarded as insignificant. However, in conjunction with a it implicitly involves the position relative to Jupiter, which is of fundamental importance for the character of resonant orbits.

Two-dimensional distributions of minor planets and short-period comets in a and e are intercompared in Figure 1. All numbered minor planets and all Apollo and Albert asteroids ($q < 1.25$) are shown as black dots, except for the librating bodies which are denoted by triangles. The latter include, from above, 14 Trojans around the triangular libration points with Jupiter (resonance 1:1), 279 Thule (4:3), 19 minor planets of the Hilda group (3:2), 1101 Clematis and 1362 Griqua (2:1), 887 Alinda and 1953 EA (3:1). A few additional asteroids are either very near the libration limit or the accuracy of their elements does not allow a check on the stability. These are 334 Chicago and 1256 Normannia (3:2), 978 Aidamina and 1125 China (2:1), 1381 Danubia and 1722 = 1938 EG (3:1). Further data on the librating asteroids can be found in the recent papers by Schubart (1968), Schweizer (1969), Sinclair (1969), and Marsden (1970). The heavy circles denote the present state of the system of short-period comets under observation, i.e., the osculating elements of the comets determined during the last or last but one revolution. The last observed returns of the other comets are indicated by light circles; these include disrupted comets (P/Biela, P/Taylor), those ejected by perturbations into unobservable orbits (P/Lexell, P/Oterma), lost by fading or, more frequently, due to the insufficient accuracy of predictions for recovery. Six comets were omitted because their orbits were considered too inaccurate (P/La Hire, P/Grischow, P/Perrine, P/Kulin, P/Wilson-Harrington, and P/Anderson).

The distribution of the objects in the a/e diagram is interesting in many respects. As regards the dividing line between the comets and the asteroids, we see that this can be drawn quite easily in the range of medium eccentricities, say $0.3 < e < 0.6$. For $e < 0.3$ data on comets are lacking because greater perihelion distances of Jupiter comets make only exceptionally bright objects observable. For $e > 0.6$ the separation is good, but the asteroid side is occupied by the Apollo objects, the original nature of which is uncertain. Formal criteria for meteor orbits, fitting arbitrary functions of the elements to the statistics of known comets and asteroids, the K-criterion ($K = \log a(1 + e)(1 - e)^{-1} - 1 \lesssim 0$; Whipple, 1954) and Pe-criterion ($Pe = a^{3/2}e \lesssim 2.5$; Kresák, 1967), deviate markedly from one another just in these two regions.

While it appears that there are no more live comets in the latter area, except for the marginal case of P/Encke and the doubtful case of P/Wilson-Harrington, some information on the population of the former area is available from the integrations of the prediscovery and future comet orbits. The computations by Kazimirchak-Polonskaya (1967) and Belyaev (1967), extending from 1660 to 2060, reveal besides P/Schwassmann-Wachmann 1 two other comets that revolved, not long ago, in nearly circular orbits between Jupiter and Saturn: P/Whipple and P/Shajn-Schaldach. A fourth comet, P/Oterma, left this region in 1937 and returned back in 1963 (see also Marsden, 1961). The orbits of these comets for 1660 and 2060 are plotted as open squares. The elements of P/Whipple and P/Shajn-Schaldach strongly suggest that these are two parts of a single body that split apart about 250–300 yr ago. This evidence, together with the

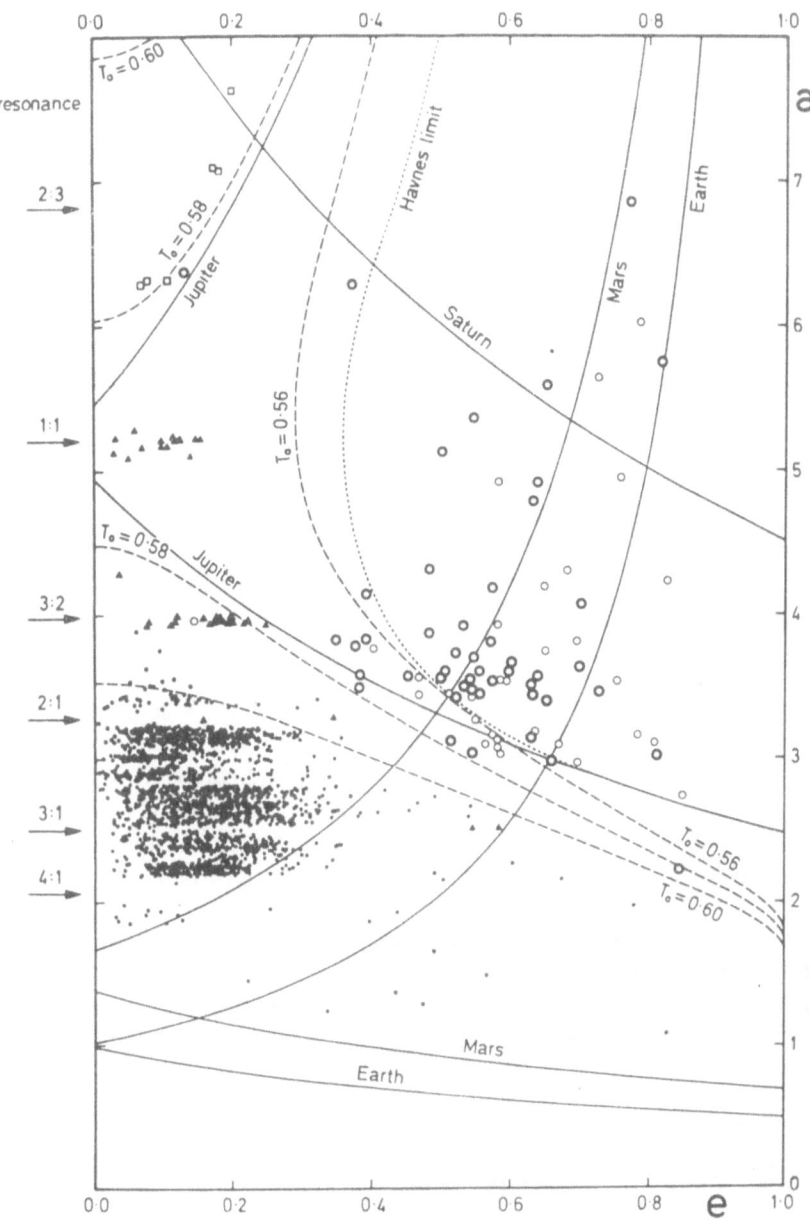

Fig. 1. The a/e diagram.

repeated outbursts of P/Schwassmann-Wachmann 1, shows that even in this region, far remote from the Sun, comets exhibit significant activity.

The full lines in Figure 1, labelled with the names of planets on the inner side, delimit the areas where close approaches to the respective planets are possible. Of the two principal areas of avoidance of any encounter, the lower one is occupied by the asteroid belt between Mars and Jupiter, the upper one by the hypothetical comet belt between Jupiter and Saturn. Its largest member is P/Schwassmann-Wachmann 1; its only other known present member, P/Oterma, owes its discovery to a temporary change of orbit in 1937–1963. Comets of normal size would remain there undiscovered as long as strong perturbations would not force them into orbits of much smaller perihelion distance. The comets captured in this way apparently evolved tending to avoid the 2:3 resonance with Jupiter. This fact, however, does not rule out the presence of librating objects near exact commensurability, which would be prevented from stronger perturbations as in the Neptune-Pluto case, and hence could not move into the range of visibility. It may be noted that P/Neujmin 1 was found to librate in the resonance 2:3 for at least 3400 yr (Marsden, 1970) in spite of having perihelion distance $q = 1.5$ and almost intersecting the orbit of Jupiter. Possible librating orbits between Jupiter and Saturn pose intriguing problems (e.g., the effect of the Jupiter-Saturn resonance 5:2) and would repay a closer study. Anyway, it can be inferred that this region contains a number of invisible comets, and one can speculate whether it does not also contain some asteroids and an abnormal concentration of meteor dust.

The Jupiter-Saturn belt is important also as a potential alternative source of the Jupiter family of comets. The results on perturbational capture by Jupiter obtained by Havnes (1969), with a refinement accounting for the solar perturbations on the jovicentric hyperbolic arc, delimit the elements which can result from captures from nearly parabolic orbits. This limit in a and e is indicated in Figure 1 by the dotted line. Evidently, neither the Jupiter-Saturn belt nor a considerable part of the Jupiter family satisfies this condition. 30% of known comets with $a < 8$ lie outside, but if we take into account the selection effect of perihelion distance on discovery, we can estimate that the real contribution is as high as 60 to 70%. This figure is in surprising agreement with the number of three comets known to have been captured from the Jupiter-Saturn belt (P/Whipple in 1852, P/Shajn-Schaldach in 1875, P/Oterma in 1937), against one comet captured from a nearly parabolic orbit (P/Kearns-Kwee in 1855 and 1961).

It has been shown earlier (Kresák, 1967, 1969) that the value of the Jacobi constant in the restricted three-body problem Sun-Jupiter-comet/asteroid, or the Tisserand invariant, provides a good dividing line between cometary and asteroidal orbits. Neglecting the mass ratio Jupiter:Sun, Jupiter's orbital eccentricity and inclination, we have

$$T = a^{-1} + 2a_J^{-3/2}a^{1/2}(1 - e^2)^{1/2}\cos i, \tag{1}$$

and neglecting also the inclination of the third body,

$$T_0 = a^{-1} + 2a_J^{-3/2}a^{1/2}(1 - e^2)^{1/2}. \tag{2}$$

The latter quantity, although involving an additional approximation, appears preferable for two reasons:

(1) It can be directly plotted on the a/e diagram, as is done in Figure 1 by the dashed lines for $T_0 = 0.56$, 0.58, and 0.60.

(2) The families of librating bodies are more compact in T_0 than in T. For direct short-period orbits the difference $T_0 - T$ is generally insignificant, exceeding 0.02 for 10% of the short-period comets ($a < 8$) and 5% of the minor planets; $T_0 - T > 0.05$ for no more than two short-period comets (P/Pigott, P/Tuttle) and four asteroids (944 Hidalgo, 1208 Troilus, 1373 Cincinnati, 1580 Betulia). The osculating elements obviously have to be determined at sufficient distances from Jupiter to make the additional terms in the Jacobi integral, with Jupiter's mass in the numerator, negligible.

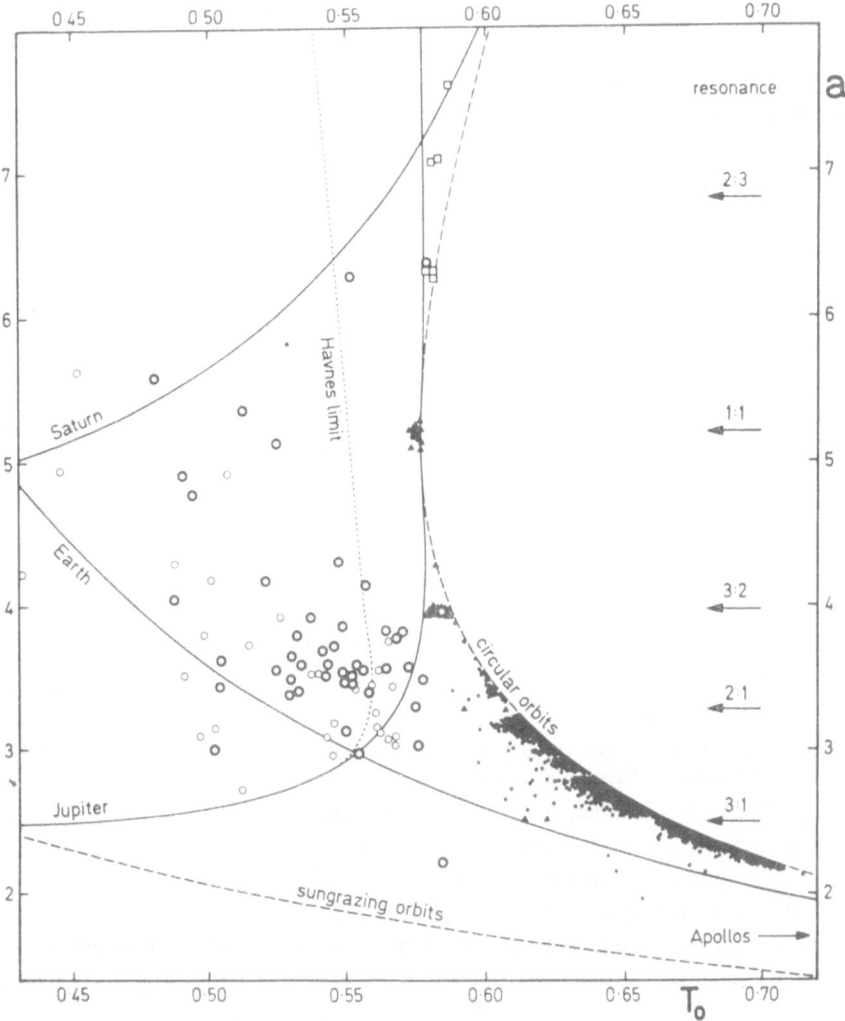

Fig. 2.   Plot of T_0 against a.

The values of T_0 are plotted against a in Figure 2 for the objects shown in Figure 1, using the same notation. The area of real orbits is limited by the two dashed lines (circular and sungrazing orbits); the regions where encounters with the Earth, Jupiter, and Saturn are possible are limited by the full lines, Havnes' condition by the dotted line. The displacement of individual objects by the perturbing action of Jupiter follows the vertical direction. The separation of the comets from the asteroids by $T_0 \simeq 0.58$ and the restriction of Jupiter's capture from elongated orbits by $T_0 < 0.56$ is clearly recognized, as well as the position of the Trojans and the Hilda group near the critical limit. Of special interest is the concentration of the extrapolated orbits in the Jupiter-Saturn belt (the segment running in the centre from the upper edge), with $0.58 < T_0 < 0.59$. These orbits would be kept by Jupiter at the critical limit, as demonstrated by the transitional orbit of P/Oterma (1958) embedded in the Hilda group. If this comet had passed at a slightly smaller distance from Jupiter in 1937, so as to come at the next encounter in 1963 in front of instead of behind it, there might have occurred the interesting case of injection of a comet into an orbit of extremely short period, required for the subsequent evolution into an object of Encke or Apollo type. Such an evolution of comets captured from the Jupiter-Saturn belt may present a clue to the origin of the Apollo objects and of the parent bodies of short-period meteor streams. Perturbations in inclination affecting the difference $T_0 - T$ may become important in this process, in particular at low eccentricities.

Since the parameter T_0 implicitly involves the perihelion distance, the observed distribution in T_0 is subject to selection effects of discovery. Their operation is illustrated by Figure 3, showing T_0 plotted against the absolute magnitudes. The comet values, $H_{10} = M - 5 \log \Delta - 10 \log r$, are taken from Vsekhsvyatskij's catalogue and its supplements (Vsekhsvyatskij, 1958, 1962, 1966), plus recent data from other sources. For the comets of more than one apparition the median value is used, irrespective of the secular variation. For the asteroids, the absolute magnitudes g are taken from the list by Gehrels (1967; Chebotarev, 1969). The notation is analogous to Figures 1 and 2; the limits of apparent magnitude M for a given absolute magnitude H_{10} or g are indicated by the curves. For the comets, $r = q$, $\Delta = q - 1$ is assumed, so that a comet of absolute magnitude H_{10} situated below the curve cannot appear brighter than apparent magnitude 15, unless irregular flares occur. Three cases are indicated, with separate scales of perihelion distance q at the bottom: (1) parabolic orbit (dashed line); (2) elliptical orbit with $n/n_J = 1.5$ (resonance 2:3, centre of the Jupiter-Saturn belt); (3) elliptical orbit with $n/n_J = 0.5$ (resonance 2:1, main commensurability gap near the inner fringe of Jupiter's family). For the asteroids the dotted line refers to the mean opposition magnitude at $r_c = a$, $\Delta = a - 1$ (circular orbit, $i = 0$), so that an asteroid of higher eccentricity situated below the line may on favourable occasions appear considerably brighter even than apparent magnitude 18.

The figure demonstrates that quasi-parabolic orbits with $T_0 > 0.58$ are virtually unobservable because their perihelia must lie beyond the orbit of Jupiter. For the comets of the Jupiter-Saturn belt the situation is similar, and only the absolutely brightest of all short-period comets, P/Schwassmann-Wachmann 1, is observable at present in this region. For Jupiter comets around the resonance 2:1 the situation is much more favou rable. While the absence of objects near the lower left edge ($T_0 < 0.53$, $q_{III} < 1.1$) indicates that the statistics are essentially complete in this range (as a matter of fact, only one of 21 comets with $q < 1$, $P < 100$ yr, P/Honda-Mrkos-Pajdušáková, has been discovered since 1902), a number of faint comets with $0.56 < T_0 < 0.58$ ($1.5 < q_{III} < 2$) evidently remain undetected. On the other hand, the absolute absence of short-period comets near $T_0 = 0.60$ ($2 < q_{III} < 2.5$) can hardly be attributed to observational selection. Similarly, the discovery conditions obviously tend to reduce the proportion of known asteroids with lower values of T_0, but the sharp cut-off of nonlibrating objects near $T_0 = 0.60$ ($r_c \simeq 3.5$) is evidently real.

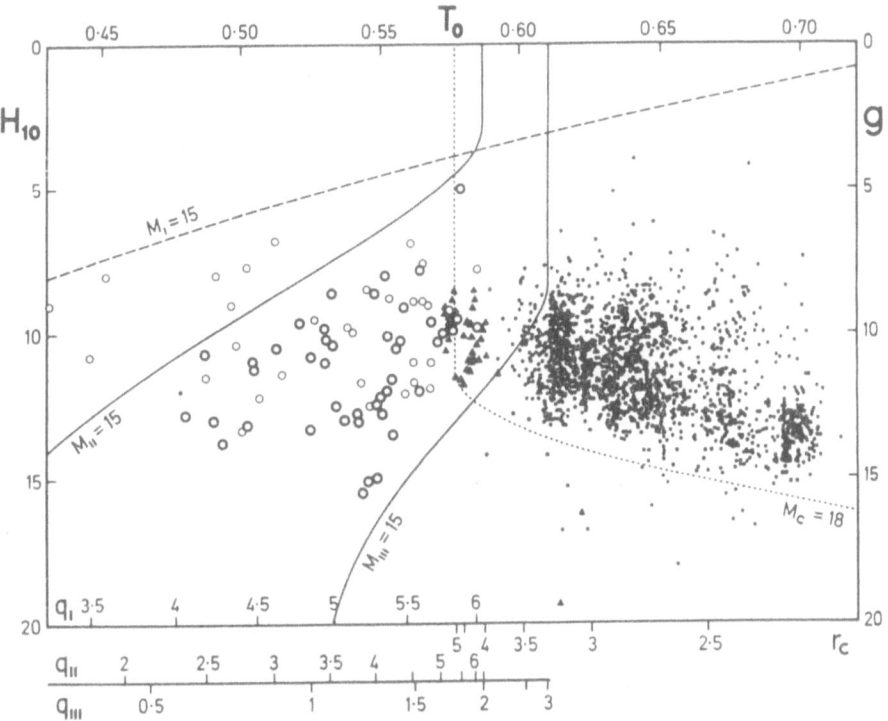

Fig. 3.   Plot of T_0 against absolute magnitude.

The comets with the highest values of T_0 and the asteroids with the lowest values of T_0 are listed in Table I. We see that there is practically no overlapping, the only object for which the classification according to T_0 evidently disagrees with the physical appearance being 944 Hidalgo. This is almost certainly an extinct comet nucleus. There exist undoubtedly also other objects of this type, but their detection is very difficult in photographic work, including the minor planet searches, and utterly impossible in visual comet searches. It may also be noted that the extreme cases of T_0 include all comets detached recently from the Jupiter-Saturn belt, and that the remaining cases, with aphelia distinctly inside the orbit of Jupiter, may represent results of similar evolution in the past. On the other hand, librating bodies strongly prevail among the extreme asteroidal cases.

TABLE I

Comets with highest and asteroids with lowest values of T_0

	T_0	q(AU)	Q(AU)	n/n_J	Note
P/Oterma	1958	0.584	3.39	4.53	1.51	resonance 3:2
	1660	0.587	6.01	9.14	0.57	J-S belt
	2060	0.583	5.88	8.31	0.63	J-S belt
P/Encke	1967	0.584	0.34	4.09	3.60	Apollo region
P/Schwassmann-Wachmann 1	1957	0.579	5.54	7.21	0.74	J-S belt
	1660	0.580	5.79	8.35	0.63	J-S belt
	2060	0.581	5.85	6.74	0.75	J-S belt
P/Schwassmann-Wachmann 2	1968	0.577	2.15	4.83	1.82	Jupiter family
P/Tempel 2	1967	0.576	1.37	4.69	2.25	Jupiter family
P/Tempel 1	1967	0.575	1.50	4.73	2.15	Jupiter family
P/Johnson	1970	0.574	2.20	4.96	1.75	Jupiter family
P/Whipple	1970	0.570	2.47	5.16	1.59	Jupiter family
	1660	0.580	5.64	7.00	0.75	J-S belt
P/Neujmin 2	1927	0.568	1.34	4.84	2.18	Jupiter family
P/Shajn-Schaldach	1949	0.565	2.23	5.28	1.63	Jupiter family
	1660	0.581	5.85	6.78	0.75	J-S belt
Other comets (~600)		0.018	0.01	4.81	0.00	
		0.567	4.71	∞	2.64	
944 Hidalgo		0.479	2.00	9.64	0.84	cometary orbit
Trojans (14)		0.572	4.40	5.32	0.98	libration 1:1
		0.577	5.07	5.98	1.04	
Hilda group (19)		0.579	2.97	4.23	1.48	libration 3:2
		0.587	3.65	4.88	1.52	
279 Thule		0.582	4.14	4.42	1.34	libration 4:3
1256 Normannia		0.588	3.65	4.22	1.52	near 3:2
1373 Cincinnati		0.588	2.31	4.51	1.88	
334 Chicago		0.589	3.66	4.11	1.55	near 3:2
1362 Griqua		0.592	2.16	4.40	2.00	libration 2:1
1144 Oda		0.592	3.41	4.09	1.63	
225 Henrietta		0.594	2.37	4.32	1.94	near 2:1
Other belt asteroids (~1700)		0.596	1.39	1.92	1.72	
		0.770	3.37	4.26	4.72	
Apollo + Albert groups (23)		0.607	0.19	1.66	2.74	
		1.026	1.22	4.10	10.60	

The discrimination according to T_0 becomes ambiguous only for the librating objects like the Trojans or the Hilda group. The oversimplified definition that a comet is a body which can closely approach Jupiter (say, to within 1.0 AU) whereas an asteroid cannot, is also reflected in this anomaly. Moreover, as pointed out by Marsden (1968, 1970), the two comets which for centuries avoided encounters with Jupiter (P/Arend-Rigaux and P/Neujmin 1) are conspicuous by nearly asteroidal appearance and the absence of nongravitational effects in their motions. In the limiting cases of T_0 the value of the libration argument with respect to Jupiter,

$$\sigma = (A - B)\pi - A\lambda_J + B\lambda, \tag{3}$$

where A, B are relatively prime integers, $A/B = n/n_J$, yields a sharp secondary comet-asteroid criterion.

Figure 4 shows the absolute value of the libration argument σ plotted against the mean diurnal angular motion n (in degrees) around the exact resonances 1:1, 3:2, 2:1, and 3:1 (full horizontal lines). The dashed lines indicate the combinations of $|\sigma|$, n for which the orbits of the respective perihelion distance cross the orbit of Jupiter, assumed circular. For the comets only the elements from observed returns are plotted, and the connecting lines show which positions refer to the same body. The last observed returns are denoted by heavy circles. The positions of the asteroids refer to the epochs for which the elements are listed in the latest issue of the Minor Planet Ephemerides (Chebotarev, 1969).

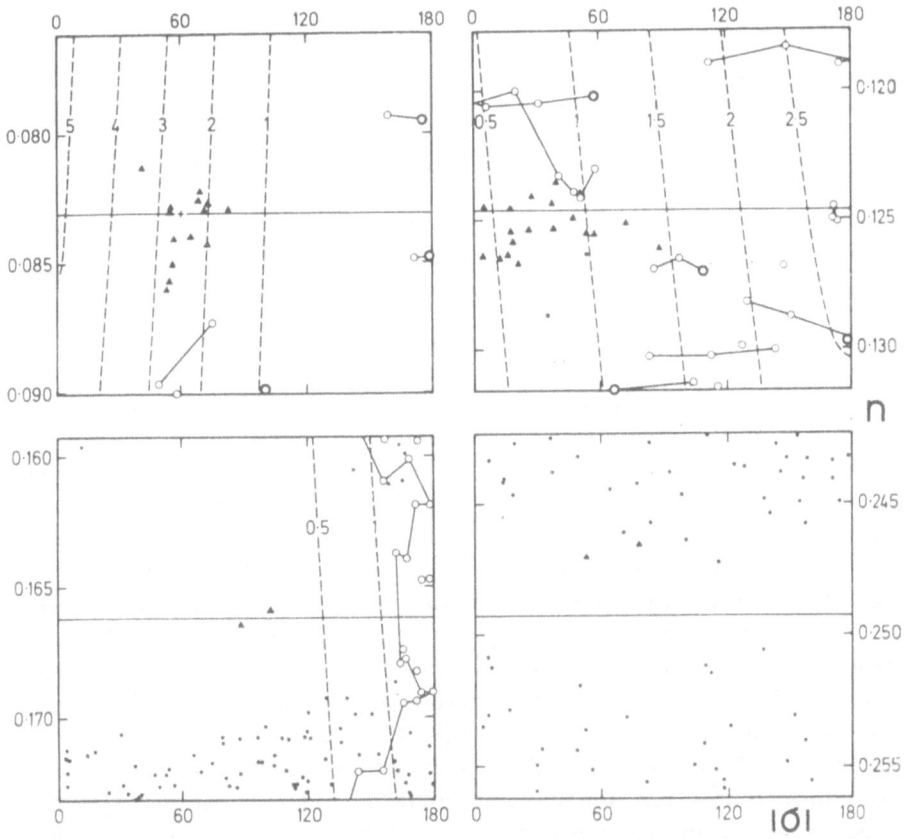

Fig. 4. Plot of libration argument σ against mean daily motion n around the resonances 1:1, 3:2, 2:1, and 3:1 with Jupiter.

At the resonance 1:1, the cluster of Trojans surrounds the libration centres L_4, L_5 ($|\sigma| = 60°$, a cross). Two comets in temporary resonance, P/Slaughter-Burnham and P/Van Biesbroeck, are situated near $|\sigma| = 180°$, i.e., in the region of no encounters with Jupiter, and three other comets are slightly beyond the zone of significant resonance. At 3:2 the Hilda group extends up to $|\sigma| = 90°$, and a number of comets are irregularly distributed over the whole area. The fact that any value of σ permits an approach to Jupiter for a value of $q < 3$ is the reason why no distinct gap is formed. The temporary resonance of P/Oterma appears at the extreme right; the next comets closest to the exact resonance with Jupiter are P/Schaumasse and P/Arend. At 2:1 and 3:1 the distribution of asteroids reverses and resonance gaps are observed, much wider and asymmetric in the former case. Comets appear only around the 2:1 resonance. All six of them are pronouncedly concentrated towards $|\sigma| = 180°$ where the lengthy passage of P/Pons-Winnecke from 1869 to 1951 appears. The two librating asteroids, 1101 Clematis and 1362 Griqua, are prevented from entering this zone; for Griqua $|\sigma|$ is always less than 100° (Marsden, 1970). A similar restriction is put on the motion of the two librating asteroids of the last diagram, 887 Alinda and 1953 EA, where $|\sigma|$ steadily remains greater than 40° (Marsden, 1970).

Figure 4 includes only those comets whose osculating mean motions were near the resonance at the time of perihelion passages. There were many other cases where a comet crossed the resonance zone or even several zones during a single encounter with Jupiter, while moving along the outer arc of the orbit. The change of n of P/Lexell was as large as $+0°.069$ in 1767 and $-0°.173$ in 1779, that of P/Oterma $+0°.070$ in 1937 and $-0°.073$ in 1963, P/Brooks 2 $+0°.103$ in 1886, P/Kearns-Kwee $+0°.092$ in 1961, etc., much greater than the total range of $\pm 0°.007$ represented in each diagram of Figure 4. These rapid passages obviously concentrate to the region of lowest stability, e.g., to $|\sigma| = 180°$ for the 2:1 case with $q \geq 1.5$. The long-term integrations of 10 comets published by Kazimirchak-Polonskaya (1967) and Belyaev (1967) give an average number of five crossings of the principal resonance zones (2:3, 1:1, 3:2, and 2:1) per millennium, one of them over more than one zone during a single revolution. The data are not quite representative, as there was a tendency to select the most complex, and hence most interesting, orbital evolutions, and also because comets are often discovered after drastic changes in their orbits, during the return immediately following a reduction of the perihelion distance. Nonetheless, a characteristic time interval between two crossings is several centuries, or less than 100 revolutions for a typical short-period comet. Most resistive to such changes are those comets which are temporarily librating around higher-order resonances, like P/Arend-Rigaux (7:4) P/Tempel 2 (9:4); the interval is shortest when a comet happens to be injected into a resonant orbit and ejected again at the next occasion, like P/Oterma (3:2) or P/Lexell (2:1). The population of the resonance gaps ($\Delta n = \pm 0°.002$ to $0°.003$) can be estimated at about 30 to 40% of the surroundings. This contrasts with the asteroid system where no crossings occur, except for the low-amplitude periodic variations of the librating bodies, and the resonance gaps within the main belt are much more pronounced. In the 2:1 gap the relative population is 5% of the outer adjacent zone (around 1.9:1), and 1% of the inner adjacent zone (around 2.1:1).

Thus the libration parameters can be applied as a secondary comet-asteroid criterion in the very rare cases where the Jacobi constant (T_0-test) does not give a clear answer. An excellent example is the orbit of P/Oterma, 1937–1963. Its orbital elements were in fact entirely unrecognizable from the Hilda asteroids (see Figures 1 and 2, with P/Oterma embedded right in the middle of the Hilda group). But its libration argument $|\sigma|$ was greater than 170°, whereas it is less than 90° for all Hilda asteroids. Consequently, the resonance of P/Oterma persisted only for 25 yr, or 1/10 of one libration period of the Hilda asteroids.

The T_0-criterion locates the Apollo objects at the opposite extreme of the asteroid system from the comet limit, thus qualifying them as indubitable minor planets. At the same time, if we consider whether they may have detached from the main asteroid belt or from Jupiter's family of comets, a preceding capture from the Jupiter-Saturn belt seems to represent a possible mechanism whereby their aphelia could cross Jupiter's perturbational barrier and come to the outskirts of the Apollo region. Subsequent reduction of semimajor axes might be due to nongravitational forces, which affect the motions of live comets and meteoroids, but hardly of extinct comets and minor planets.

The Apollo objects obviously need not all be of the same nature and origin. Marsden (1970) suggests that periodic variations in brightness, betraying rotating fragments of irregular shape (like those observed in 433 Eros or 1620 Geographos) can be used for discriminating the true asteroids from the extinct comet nuclei. This approach appears promising, but unfortunately 14 of 23 asteroids with $q < 1.25$ have been lost and some others are too faint for high-precision photometry. Even so, the collection of photometric data for as many Apollo objects as possible would be very desirable. Physical observations of 944 Hidalgo and the few comets of almost asteroidal appearance would be of great interest, as well as the search for possible physical differences between the short-period comets captured from nearly parabolic orbits and those transferred from the Jupiter-Saturn belt. The discovery of additional members of this belt would be very important, as their existence and orbital history may remove some difficulties of the capture theory. Model computations of the evolution of orbits similar to P/Oterma may assist in explaining the injection of comets into the region of the terrestrial planets and the origin of meteor streams of very short period.

References

Belyaev, N. A.: 1967, *Astron. Zh.* **44**, 461.
Ceplecha, Z.: 1967, *Smithsonian Contr. Astrophys.* **11**, 35.
Chebotarev, G. A.: 1969, *Efemeridy Malykh Planet na 1970 God*, Leningrad.
Cook, A. F., Jacchia, L. G., and McCrosky, R. E.: 1963, *Smithsonian Contr. Astrophys.* **7**, 209.
Gehrels, T.: 1967, *Trans. IAU* **13B**, 121.
Havnes, O.: 1969, *Astrophys. Space Sci.* **5**, 272.
Jacchia, L. G., Verniani, F., and Briggs, R. E.: 1967, *Smithsonian Contr. Astrophys.* **10**, 1.
Kazimirchak-Polonskaya, E. I.: 1967, *Astron. Zh.* **44**, 439.
Kresák, L.: 1967, *Smithsonian Contr. Astrophys.* **11**, 9.
Kresák, L.: 1968, in L. Kresák and P. M. Millman (eds.) 'Physics and Dynamics of Meteors', *IAU Symp.* **33**, p. 217.

Kresák, L.: 1969, *Bull. Astron. Inst. Czech.* **20**, 177, 231.
Marsden, B. G.: 1961, *Astron. J.* **66**, 246.
Marsden, B. G.: 1968, *Astron. J.* **73**, 367.
Marsden, B. G.: 1970, *Astron. J.* **75**, 206.
Öpik, E. J.: 1963, *Adv. Astron. Astrophys.* **2**, 219.
Schubart, J.: 1968, *Astron. J.* **73**, 99.
Schweizer, F.: 1969, *Astron. J.* **74**, 779.
Sinclair, A. T.: 1969, *Monthly Notices Roy. Astron. Soc.* **142**, 289.
Verniani, F.: 1969, *Space Sci. Rev.* **10**, 230.
Vsekhsvyatskij, S. K.: 1958, *Fizicheskie Kharakteristiki Komet*, Moscow.
Vsekhsvyatskij, S. K.: 1962, *Astron. Zh.* **39**, 1094.
Vsekhsvyatskij, S. K.: 1966, *Astron. Zh.* **43**, 1292.
Whipple, F. L.: 1954, *Astron. J.* **59**, 201.

Discussion

G. A. Chebotarev: How closely do you think the Jupiter-Saturn comet belt resembles the Jupiter-Mars asteroid belt?

L. Kresák: Unfortunately, all our information on the Jupiter-Saturn belt is based on the orbits of four comets, and only one of the comets is actually observable there. Nevertheless, the low probability of the strong perturbations experienced by the others suggests that the number of invisible comets within the belt is very great indeed. The belt is evidently confined to the region between Jupiter and Saturn and to the vicinity of the ecliptic plane. It possibly includes a slight decrease in the distribution of semimajor axes near the 2:3 resonance with Jupiter ($a = 6.8$, $P = 17.8$).

S. K. Vsekhsvyatskij: A number of asteroids have been observed to exhibit diffuse envelopes. What are the orbital characteristics of these asteroids? Do you agree with Marsden about the transformation of a number of the Jupiter-family comets into asteroids?

L. Kresák: Slight indications of diffuse envelopes have been observed on a few occasions, but without recurrence, around normal belt asteroids. To my knowledge, there exists no photographic record of one, and although some of these phenomena have been reported by very experienced observers (e.g., Hind), I do not think that we can accept them as observational evidence before we have an unambiguous confirmation. These envelopes have never been seen around any Apollo asteroid or Hidalgo. The fact that the comets of quasi-asteroidal appearance have librating orbits is very interesting and can hardly be explained by chance. The two examples, P/Neujmin 1 and P/Arend-Rigaux, are quite normal comets as to their values of T_0. On the other hand, as I have shown in my paper, the T_0-criterion may fail if the body librates.

B. G. Marsden: How certain are you of a physical connection between P/Whipple and P/Shajn-Schaldach? It would be very exciting to find that two distinct periodic comets were formerly one single comet, but this is of course extremely difficult to establish unequivocally. Other pairs of comets with orbits that were formerly very similar are P/Borrelly and P/Daniel, P/Perrine-Mrkos and P/Honda-Mrkos-Pajdušáková.

L. Kresák: The suggestion of a common origin is based on the striking similarity of all six orbital elements for several decades around 1700. P/Shajn-Schaldach is a one-apparition comet, and both it and P/Whipple underwent very strong perturbations by Jupiter in the nineteenth century. This makes the backward computations less reliable, but all things considered, the agreement is very satisfactory indeed.

85. ON THE POSSIBLE COMMON ORIGIN OF MINOR PLANETS, COMETS, AND METEORS

S. GĄSKA

University of Toruń Observatory, Toruń, Poland

For the minor planets, comets, and meteors whose orbital elements are known values of the dispersions $\sigma_i(a)$ of the inclination i for intervals of the semimajor axis a have been calculated. Calculations on the minor planets show that, from the point of view of secular perturbations, $\sigma_i(a)$ is more stable than $\bar{i}(a)$, and we have therefore taken it as a fundamental relation for further considerations. It has been found that the relations $\sigma_i(a)$ are linear for minor planets, comets, and meteors with $a > 2$ AU. The lines are parallel, and for $a > 2$ AU the $\sigma_i(a)$ are proportional to a. Meteors have a second straight line for $a < 2$ AU, and in this case $\sigma_i(a)$ is proportional to $1/a$. It is concluded that minor planets, comets, and meteors could have a common origin.

The results are shown in Figure 1, where full dots denote minor planets, 𝄾 – comets, crosses – fireballs, open circles – photographic meteors, and triangles – radio meteors.

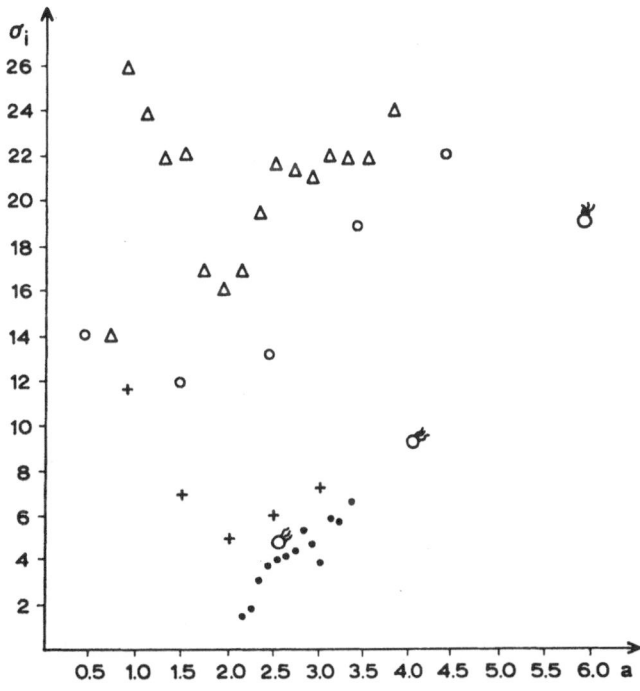

Fig. 1. The dispersions σ_i (a) for various objects.

Chebotarev et al. (eds.), The Motion, Evolution of Orbits, and Origin of Comets, 515. All Rights Reserved.
Copyright © 1972 by the IAU.

CONCLUDING DISCUSSION

G. A. Chebotarev: IAU Symposium No. 45, on 'The Motion, Evolution of Orbits, and Origin of Comets', is almost over. It is, of course, quite impossible for me to make on the spot a proper summary of the scientific results of the Symposium. I can only state that the reports delivered have contributed greatly to the further advance of cometary astronomy.

Of extreme importance also are the personal scientific contacts among the astronomers working in this field, and we sincerely hope that these contacts will be maintained in the future.

A number of the papers presented have raised and formulated new problems for investigation. We shall be able to discuss plans for further international cooperation on cometary matters a fortnight hence in Brighton, at the session of IAU Commission 20 devoted to comets, and also by direct correspondence between the individuals involved.

E. I. Kazimirchak-Polonskaya: I should first like to say how sorry I am that Dr E. Roemer, that indefatigable cometary observer, has not been able to participate in this Symposium.

In my introductory comments I spoke of the necessity for closer contacts between observers and theoreticians. The same idea was expressed later by Candy, Chernykh, and others. Theories of cometary motion cannot be completed unless positional observations are available. Observations are often published a very long time after they have been made. It is most desirable that observers make their observations available to the theoreticians as quickly as possible.

B. G. Marsden: In defence of some of the observers I must point out that the production of positions from plates taken with long-focus reflectors can be a very time-consuming business. In the easiest case the reduction can be made using the *Astrographic Catalogue*, but this is not at all the same straightforward matter as using the AGK$_2$, the Yale zones, or the *S.A.O. Catalogue*, and further, one frequently runs into problems involving reference stars with unknown, but evidently quite large, proper motions. Field plates could be taken, but this requires use of another telescope, and it is only proper that the observer should put his greatest effort into making the maximum use of the long-focus instrument, with which he is able to record comets when they are extremely faint.

S. K. Vsekhsvyatskij: At the Prague meetings in 1967 a committee was set up to consider the compilation of a new cometography. On behalf of the Soviet investigators on both the physics and the motions of comets I have proposed to Marsden, the Chairman of that committee, that the cometography should be supplemented with two additional volumes, one containing a completely revised catalogue of cometary orbits, with all the appropriate information from the primary sources, and the other consisting

Chebotarev et al. (eds.), The Motion, Evolution of Orbits, and Origin of Comets, 519–521. All Rights Reserved.
Copyright © 1972 by the IAU.

of ephemeris data for all cometary apparitions. I ascertained that the Institute for Theoretical Astronomy would be willing and able to participate in this monumental work. I should like to emphasize the great need for the volume of ephemeris data, for it will enable us to study again all the ancient comets and the results hitherto obtained; and this would be most valuable, not only as far as the individual comets are concerned, but as it involves all the important problems of cometary cosmogony.

B. G. Marsden: It was indeed the consensus of the cometography committee that the principal effort should be put toward the preparation of reliable orbital and ephemeris data. We have even considered the possibility of recalculating the orbits of all the comets anew. Bielicki and Sitarski have proposed a method for the uniform treatment of observations, and this could certainly be valuable in this connection. As I see it, the main drawback at present is that we do not know enough about the nongravitational effects on comets. This lack of knowledge influences, not only the preparation of a new orbit catalogue, but also the investigations on the orbital evolution of comets. I wish to urge all those concerned with studying the motions of comets to pay particular attention to the matter, making such numerical experiments as are necessary, and collaborating with astronomers involved in the physical study of comets, in the hope that we may eventually come up with a standard, physically meaningful procedure for handling the nongravitational effects.

Concerning the cometography proper, the complete observational account of all the comets ever observed, my feeling is that the main task is the collection and correction of errors in the existing compilations, such as those by Holetschek, Vsekhsvyatskij, and the annual reports of the Royal Astronomical Society. This work could sometimes be facilitated by calculating accurate ephemerides, for there are several instances where somebody has erroneously reported the recovery of a comet or a positional observation made long after everyone else has ceased observing it, and such a report can obviously give a very misleading impression of the comet's absolute brightness.

E. I. Kazimirchak-Polonskaya: I certainly don't insist that the new cometography should be compiled immediately. I merely wish to state our willingness at the Institute for Theoretical Astronomy to take an active part in the venture.

S. K. Vsekhsvyatskij: For more than a week now we have often heard in this room the name of the celebrated Polish astronomer Michael Kamieński; and a large number of his pupils are among those present here. I should like to propose to IAU Commission 20 that periodic comet Wolf, also known as Wolf 1, on which Kamieński has worked so extensively since the very earliest years of this century, should in future be given the name Wolf-Kamieński.

G. A. Chebotarev: We are not authorized to approve such matters here, but the proposal will be put to Commission 20 in Brighton.*

M. Bielicki: I wish to thank Professor Chebotarev and all those who have contributed to the success and superb organization of this Symposium. My very special

---

* At the meeting in Brighton the proposal was tabled. Although the Commission highly appreciated Kamieński's splendid work on the orbit of this comet, it was felt that the renaming of the comet would be at variance with the practice generally adopted and might represent an undesirable precedent for the future.

thanks are due to Dr Kazimirchak-Polonskaya for her selfless labour to the benefit of the Symposium and to us all. I leave for home, and I hope that others do too, newly inspired to continue cometary research. Comets continue to hold many secrets, but ultimately we shall know whence they originated and how they survive.

G. A. Chebotarev: Thank you. I declare our final session closed. Dear guests, I wish a happy journey to you all!